Biotechnology
for Sustainable Utilization of Bioresources

The Editors

Dr Hrudayanath Thatoi is serving as a Professor in Department of Biotechnology of North Orissa University. He has obtained his M.Sc. M. Phil. and Ph.D in Botany from Utkal University. He is also working as the Director, Centre for Similipal Studies, North Orissa University. Dr. Thatoi has implemented several research projects funded by UGC-DAE Govt. of India, DST Govt. of Odisha and Forest Department, Govt. of Odisha. Till date, Dr. Thatoi is credited with publication of 10 books including two text books and more than 230 research papers in National and International journals, proceedings of conferences and book chapters. He serves as member of several scientific societies and editorial boards of the National and International journals. He is a recipient of Fellow of Society of Applied Biotechnology. So far, he has guided 15 Ph.D. scholars and several M. Tech and M.Phil students. He has visited Bangkok and Nepal on International Conferences for Biodiversity and Biotechnology. Dr. Thatoi is also a Functional Area Expert in Ecology and Biodiversity being accredited by QCI, Govt of India

Dr Sukanta Kumar Nayak is currently working as Reader in the Department of Biotechnology, North Orissa University, Takatpur, Baripada. Dr. Nayak has more than 20 years of research experience in the field of microbiology immunology and biotechnology. Dr. Nayak has passed M. Sc. In Microbiology from Orissa University of Agriculture and Technology, Odisha while obtained his Ph.D. degree in Life Science from Utkal University, Bhubaneswar, Odisha. To his credit, he has published several scientific papers in reputed International and National Journals. He is also a member of several Scientific Societies, Reviewers for Nature groups, Springer, Elsevier's, *etc.* Journals.

Biotechnology for Sustainable Utilization of Bioresources

– *Editors* –

Prof. Dr Hrudayanath Thatoi

Dr Sukanta Kumar Nayak

Department of Biotechnology
North Orissa University
Odisha

2019

Daya Publishing House®

A Division of

Astral International Pvt. Ltd.

New Delhi – 110 002

Reviewers & Subject Experts:

Prof. (Dr.) H. N. Thatoi, Professor, Dept. of Biotechnology

Dr. B. Rath, Reader & HOD, Dept. of Biotechnology

Dr. S. K. Nayak, Reader, Dept. of Biotechnology

Dr. S. K. Sahu, Assistant Professor, Dept. of Biotechnology

Dr. G. Dhangdamajhi, Assistant Professor, Dept. of Biotechnology

Mr. R. N. Munda, Assistant Professor, Dept. of Biotechnology

Editorial Assistance:

Mr. Tapan Panigrahi, Technical Assistant, Dept. of Biotechnology

Mr. H. Gourabh, Inspire Fellow, Dept. of Biotechnology

Miss. S. Ray, Research Scholar, Dept. of Biotechnology

ISBN: 978-93-88982-00-9 (HB)

Published by : **Daya Publishing House®**
A Division of
Astral International Pvt. Ltd.
– ISO 9001:2015 Certified Company –
4736/23, Ansari Road, Darya Ganj
New Delhi-110 002
Ph. 011-43549197, 23278134
E-mail: info@astralint.com
Website: www.astralint.com

Digitally Printed at : **Replika Press Pvt. Ltd.**

Prof. Pradeep K. Chand
Vice Chancellor
North Orissa University
Takatpur, Baripada, Odisha 757003

Foreword

Bioresources imply any resource of biological origin. According to the Biological Diversity Act 2002, bioresources encompass plants, animals and microorganisms, or parts thereof, their genetic material and by-products with actual or potential value. As such, biological resources can fulfill all the essential needs of human beings. They play a key role in present and future socio-economic evolutions. A country's survival and growth is hugely dependent on its biological resources and also on its capacity to sustainably utilize these resources for the benefit of its citizens. Unfortunately, the loss of bioresources is occurring at an alarming pace as a consequence of ruthless over-exploitation, agricultural land degradation, urbanization and, above all, neglect and ignorance. Thus, conservation and sustainable management of bioresources has become a growing realization worldwide.

Bioresources – living organisms, cells, genes, and the related information – are the raw materials, fundamental to the advancement of biotechnology. This fast proliferating technology based on biological principles and processes can be applied to study genetic diversity in crop plants, livestock, wild animals, industrially important microbes, bio-waste recycling, ex-situ conservation and mass multiplication of rare, endangered and threatened (RET) plant species, large-scale production of therapeutically important substances from medicinal plants and microbes, genetic transformation, cryopreservation, DNA banking, etc. Modern biotechnology being armed with advanced tools and techniques is poised to ensure sustainable utilization of bioresources to meet the increasing demand owing to population pressure. The biotechnological revolution has endowed us with an enhanced ability to obtain and to willfully modify these biological resources for the benefit of mankind.

In this backdrop, I am delighted to introduce the book "Biotechnology for Sustainable Utilization of Bioresources", edited by Prof (Dr.) H.N. Thatoi and Dr. S.K. Nayak of the Department of Biotechnology, North Orissa University. The book deals with 38 important and interesting articles in the field of animal, agricultural, environmental, and microbial biotechnology. The entire work has been divided into four valuable sections. The contents of the book make it an appropriate source of information for conservation scientists, academicians, social activists and all those who are zealously engaged in product oriented research on bioresources. The present volume will be of considerable benefit to researchers in the fields of plant-based alternative medicines, enzymes, anti-microbials, anti-cancer and anti-inflammatory molecules, medicinal significance of polyphenols containing fermented products, etc. Undoubtedly, this volume will be of great use to those pursuing biotechnological research towards combating maladies, which presently threaten human life, as well as towards sustainably utilizing the bioresources without disturbing the environmental equilibrium of nature.

(Prof. Pradeep K. Chand)

Vice Chancellor

North Orissa University

Preface

Bioresources are the resources derived from biological origin or the total biological variation manifested as individuals such as animals, plants, microorganisms and their gene pools which have been exploited by human beings since time immemorial. Bioresources are used in food, fuel, cosmetics, pharmaceuticals, traditional medicines, nutraceutical industries, paper manufacturing industries to name a few. A country's survival and growth is hugely dependent on its biological resources and also on its capability to sustainably utilize these resources for the benefit of its people. Therefore bio-resources which forms a major component of the natural resource, occupy a central position in economy. Bio-resources are the base for several industries contributing to the national economy and human well being. Human securities depend on bio-resources which play a key role in present and future socio-economic evolutions.

Now days, bioresources are getting depleted fast and are under threat. The rate of loss of natural habitants including forest and bioresources are not only a serious but a complex issue due to several reasons viz. deforestation, agricultural land degradation, ignorance of people, escalating population density, immigration of people toward urban areas etc. Thus, bioresources management and utilization for human welfare is very important for the optimum utilization of the bioresources. In this context, conservation and sustainable use of bioresources is the need of the hour which is the centre of attention around the world. The preferens of conservation methods and technologies depends upon the prevailing objective, conservation efforts, breeding methods, adoption and behaviour of the species in question as well as their availability.

Biotechnology is broadly defined as the application of knowledge in the life sciences to create products or services that are beneficial to humans. Biotechnology is used in a variety of industries, from enhancing the quality and efficiency of food production to producing new medicines to and vaccines. Biotechnological tools can be used to improve and conserve bioresources. The latest advancement in biotechnology plays an important role to create awareness, conservation and sustainable utilization of bioresources. Biotechnological tools perform an significant role in creating effective *ex-situ* and *in-situ* conservation strategies, groupings of bioresources through molecular lineages, identify useful genes through gene maps and develop a genetically modify bioresources. Biotechnology plays gigantic role for genetic improvement of bioresources.

Biotechnology can be applied to study genetic diversity in crop plants, domesticated animals, industrially important microbes, biowaste recycling, propagation and mass multiplication of threatened genetic resources, large scale production of therapeutically important substances, genetic transformation, cryopreservation, DNA banking, etc. It involves manipulation of genetic information of a particular bioresource and the application, production, maintenance, conservation, security and various other aspects of components of biodiversity. There is a growing realization world over that the biotechnology along with information technology is going to be the major thrust area in the new millennium. With the announcement of the 'Human Genome Sequence' and the progress made in genetic engineering in the past few decades, biotechnology, as an industrial activity, is set to increase exponentially in the future.

During the past decades, development of biotechnological tools in manipulating genetic framework empowered the cell factories as a reliable bioresource of recombinant protein production. The number of organisms that can be used as expression system is expanding. *Escherichia coli* was the most studied model organism, then followed by the implementation of yeast as a eukaryotic organism. Recently, algal cells, insect cells, and mammalian cells are catching up with their abilities of post-translational modifications. In addition to the living cells, *in vitro* protein production is also emerging as one important tool for recombinant protein production, especially in membrane proteins, toxins, and aggregation prone proteins. Using cell lysate or purified proteins as the source of transcription and translation machinery, corresponding recombinant proteins of certain genes can be produced within several hours. The boosts of recombinant protein production develop along with the biophysics methods. Nuclear magnetic resonance and liquid chromatography-mass spectrometry are now routinely used as identification and characterization of recombinant proteins.

Biotechnology is the path breaking technological revolution of this millennium. It has been said that the 21st century will belong to biotechnology. Research in this area in recent times has revolutionized the way in which scientists view living matter. Scientists and researchers are working in different aspects of bioresources using biotechnological tools. These information needs to be compiled and disseminated for the benefit of the scientific communities. In this endeavour, effort was made to invite articles from potential authors on broad range of topics and compiled in this volume for benefit of students, researchers, and scientists.

We extend our sincere thanks to all the contributors. We also thank authorities of North Orissa University, Baaripada, Mayurbhanj, and Department of science and Technology, Odisha for financial support. Last but not the least, we are thankful M/S Astral Publication, New Delhi for publication of the book in due time.

Dr. Hrudayanath Thatoi

Dr. Sukanta Kumar Nayak

Contents

Foreword *v*

Preface *vii*

AGRICULTURAL BIOTECHNOLOGY

1. **Climate Change and Agricultural Vulnerability** 3
   M. Dash, K.C. Samal, L. Behera and G.R. Rout
2. **Molecular Markers: A New Approach for Pulse Improvement** 11
   B.S. Pavithra, L. Behera and K.C. Samal
3. **Seed Germination and Survival of *Careya arborea* under Varying Micro-habitat Conditions of Similipal Biosphere Reserve, Odisha: Conservation Implications** 23
   R.K. Mishra, P.K. Nayak and R.C. Mohanty
4. **Phytochemical Screening and Evaluation of Antioxidant and Carbohydrate Metabolism Enzyme Inhibition Potential of a Mangrove Plant, *Avicennia officinalis*** 31
   D. Samantaray, S.K. Das and H.N. Thatoi
5. **Antimicrobial and Phytochemical Activity of Leaf Extracts of *Nyctanthes arbor-tristis*** 41
   S.B. Ojha, S. Roy and G. Dhangadamajhi
6. **Ethnomedicinal Value and Phytochemical Studies of Traditionally Used Plant Species in Udala, Mayurbhanj, Odisha** 51
   S. Jena and H.K. Sahu

ALLIED BIOTECHNOLOGY

7. **Ethnomedicinal Uses and Bioactivities of Mangrove Plant Resources with Special Reference to Antidiabetic Studies of *Avicennia officinalis* from Mangroves of Mohanadi Delta, Odisha** **65**

H.N. Thatoi and S.K. Das

8. **Application of Nanotechnology in Biology: Current Trends, Future Scopes, Benefits and Ethical Issues** **77**

Priyabrat Swain

9. ***In silico* Identification and Characterization of Abiotic Stress Genes and Proteins of *Oryza sativa*** *85*

B. Patro and S.K. Pradhan

10. **Application of Different Pretreatment Strategies on Grass Biomass for Enhanced Delignification and Cellulose Exposure for Bioethanol Production** **105**

S. Mohapatra, S.S. Mishra and H.N. Thatoi

11. **Review of Machine Learning Applications in Bioinformatics** **121**

S. Dash

12. **Rural Livelihood Through Aquaculture Development** **133**

P. Mishra

13. **Medicinal Plants Diversity with Special Reference to Phytosociological Aspect of Kuldiha Wildlife Sanctuary, Balasore, Odisha** **141**

S.D. Rout, S.K. Panda and T. Panda

14. **Ethnomedicinal Uses of Plants from Lulung, Similipal Biosphere Reserve, Odisha** **151**

D. Nayak and S.C. Sahu

15. **Status of Wild Tasar Silkworm Populations in Mayurbhanj Revealed Through GIS and Remote Sensing** **167**

S.N. Achary, Debabrata Nandi and Prasanta Kumar Kar

16. **Cryptogamic Flora: A Prospect in Environmental Toxicity Remediation with Physiological Interventions** **181**

M.K. Adak, C. Mandal, K. Das and A.K. De

ANIMAL BIOTECHNOLOGY

17. **Anuran Tail Resorption as a Model to Study Lysosomal Enzyme Mediated Cell Death** 191

C. Mahapatra

18. **Characteristics Features of Mesenchymal Stem Cells** 203

Seema Tripathy and Shikha Singh

19. **Bioprospecting of Horseshoe Crabs and Biotechnological Approaches Towards Sustainable Utilization** 213

B.P. Dash

20. **Biotechnology Intervention in Tropical Tasar Culture: The Need of the Hour** 219

P.K. Kar

21. **Study on the Anticancer Properties of Four Locally available Medicinal Plants in Human Glioma Cells (T98G)** 225

S. Hansdah, S. Mohapatra and P.R. Debata

22. **Anti-Proliferative and Cytotoxic Effect of Chloroform Extract of *Shorea robusta* Resin on Cancer Cell Lines** 239

G.Pradhan, N.S. Pal and R.N.Munda

23. **Effect of Protein, Lipid, Reduced Glutathione, Lipid Peroxidation, Superoxide Dismutase and Catalase Level in Earthworm Exposed to Nickel Chloride** 249

P. Parida and S.P. Dhupal

24. **Colour Enhancement and Growth of Goldfish, *Carassius auratus* (Linnaeus) through Natural Carotenoids Feed Supplement** 257

T. Raj, K. Bisai, S. Bhattacharjee and M.K. Pati

25. **Preliminary Studies on Determination of Efficacy of some Plant Extracts against Red Spider Mites (*Tetranychus urticae*) in Brinjal Plant (*Solanum melongena*)** 267

S. Nayak, A. Behera, M.K. Chand and G. Mohanty

MICROBIAL BIOTECHNOLOGY

26. **Therapeutic Potentials of Wild Mushrooms with a Note on Mushrooms of Simlipal Biosphere Reserve, Odisha** 279
M.K. Panda, K. Tayung and H.N. Thatoi

27. ***Termitomyces* Mushrooms: A Natural Resource of Biomolecules with Enormous Bioactive Potentials for Biotechnological Exploitation** 295
K. Singha, D.K. Sahoo, A. Banerjee and P.K. Das Mohapatra

28. **Preparation of Giant Unilamellar Vesicles (GUVs) from Bacterial Polar Lipid Extract: Developing a Prokaryotic Model Membrane System** 309
H.G. Behuria, N.S. Pal, R.N. Munda and S.K. Sahu

29. **Structural Diversity of Heterotrophic and Aerobic Denitrifying Bacteria in Domestic Wastewater** 321
S.K. Padhi, S. Tripathy, S. Mohanty and N.K. Maiti

30. **Impact of Heavy Metal on Growth and Amelioration of Antioxidant Defense Mechanism of *Terribacillus goriensis* Strain MNSMP5 with Improved Production of Acid Protease** 331
M.M. Panda, S.K. Sahoo and S. Sahoo

31. ***In vitro* and *In vivo* Characterization of Bacteria Isolated from different Sources for Possible Probiotic Use in Aquaculture Practices** 343
M. Behera, L. Nayak, Lopamudra and S. K. Nayak

32. **Bioremediaton of Selenite using Salt Tolerant Bacteria from Mangrove Soils of Bhitarkanika, Odisha** 355
R.R. Mishra and H.N. Thatoi

33. **Enumeration of Soil Microbiota from Paddy Field and its Relation to Soil Microbial Biomass Carbon** 367
M. Das, P. Bhattacharyya and S.K. Nayak

34. **Effect of Natural Plant Oils on *Candida albicans* Clinical Isolates from Endophthalmitis Cases** 373
G.C. Patra, P.S. Sahu, P. Panda and S. Sahoo

35. **Phytochemistry and Antibacterial Activity of *Spirogyra bichromatophora* (Randhawa) Transeau, Sourced from Barehipani Water Fall** 385
B.S. Sipra, Emamee Sahu, Sukumar Bhakta and Akshaya Kumar Bastia

36. Eco-Friendly Bioenergy Options in India: Role of Micro-Algal Biofuels 393

D.C. Bhattacharya

37. Antimicrobial Activity of Certain Marine Cyanobacteria from Talasari Coast, Odisha 399

S. Dash, S. Parida and B. Rath

38. Climate Change, Methane Emission from Rice Fields and Possible Mitigation Options 409

A.K. Rath and S.R. Mishra

Index 419

Color Plates

Agricultural Biotechnology

1

Climate Change and Agricultural Vulnerability

M. Dash, K.C. Samal, L. Behera and G.R. Rout*

Department of Agricultural Biotechnology, College of Agriculture,
Orissa University of Agriculture and Technology,
Bhubaneswar – 751 003, Odisha
****E-mail: grrout@rediffmail.com***

ABSTRACT

Agriculture and allied sector across the globe is exposed to the potential hazards of climate change. Climate change factors viz. surging temperature, erratic rainfall patterns and outbreak of pests and diseases negatively affect agricultural productivity and food security. Indian agriculture has been the gamble of monsoons because more than 50 per cent of the country's farm land is rain-fed and most farmers are resource poor. There is demand to conduct crop and region specific extensive studies involving small and marginal farmers to assess the real challenges so that the measures can be planned effectively. A climate resistant agriculture is the need of hour to achieve the objectives of maximizing farm productivity while keeping the effects of climate change at bay. The mitigation and adaptation strategies intend reversal of the ill impacts of climate change and promotion of sustainable development. The mitigation and adaptation strategies can be categorized under two plans i.e. novel crop improvement techniques and management technologies. The management technologies are broadly categorized under crop based technologies and resource conservation technologies. The key resource conservation-based technologies are in situ moisture conservation, rainwater harvesting and recycling, efficient use of irrigation water, conservation agriculture, energy efficiency in crop production and irrigation. Crop based technologies emphasized on promoting cultivation of such crops and varieties which thrive well into the new cropping systems and seasons under the climate-stress condition.

Keywords: *Climate change, Climate resistant agriculture, Mitigation and adaptation strategies, Food security.*

INTRODUCTION

Today the most debated topic in major international forums across the globe is undoubtedly climate change and its profound impact on the very survival of man and other life-forms on earth. The agenda has gathered top priority because of its far-reaching consequences on food security of world particularly on the developing countries of 3rd world. The Inter Governmental Panel on Climate Change (IPCC) defines climate change as "change in climate over time, either due to natural variability or as a result of human activities". In other words, it can be defined as the statistical record of observed variances in climate and weather characteristics over a long temporal expansion. The main indication of such change is alarming increase in temperature of the globe, popularly termed as "Global Warming". The World Meteorological Organization viewed that the change in climate has progressively risen in past decade (2001-10) and it was declared "The warmest decade on record till now". The impact was marked by extreme variation in pattern of precipitation that led to devastating floods as well as severe droughts in almost all continents particularly in the African and Latin-American countries. Due to warmer winters and hotter summer months the glacial runoff is getting reduced threatening the existence of the permanent flow rivers like the Ganges.

Sea-levels are predicted to rise up by one meter by 2100 according to a study which means increasing risk of losing land through inundation. The low-lying areas of South–Asia including Bangladesh, India and Vietnam are going to face major problems due to inundation of coastal areas, submergence of shore lines and salinity of water table. A statement by the Indian Government in the parliament says, there has been a 1.29 mm rise in the sea-level along the Indian coastline. The chief reason behind this altered climate phenomena is the continuous hike in the level of greenhouse gas (GHG) emissions. A recent study by Stockholm Environment Institute concluded that, the damage to oceans by 2100 will amount to 2 trillion USD per annum, given the ongoing upward spiraling of Carbon-emissions continues. There is no surprise in it as the IPCC confirms, "Rising GHGs have already started affecting production of food grains and effect is more profound in the poor tropical countries". Major possible impact of climate change are warmer winters with shortened duration; rise in temperature of land and oceans; increased frequency of extreme climatic events like floods, cloud bursting, hurricanes, tornadoes and droughts; rising sea-levels; disappearing glaciers; impact on agriculture and allied sectors; food insecurity and hike in food prices.

Indian Scenario in the Backdrop of Changing Climate

Estimation of the bright side as well as the down sides of an altering climate in the Indian context is rather difficult on account of diversity prevalent in cropping systems adopted and technology followed. There has been no such studies till now that has been able to put forth concrete data on performance of the agri-allied sector under changing environment of climatic stress, though some observations indicated adverse impacts on production of crops. Climate change studies pertaining to agriculture holds importance for India because:

- ☆ India is one of the largest agrarian economies in the world and it provides food for more than one billion people.
- ☆ Two third of the country's work force depends on agriculture
- ☆ Indian agriculture has been the gamble of monsoons traditionally as more than 50 per cent of the country's farm land is rain-fed.
- ☆ Most farmers of the country are resource poor and transfer of technology (TOT) is poor.
- ☆ Shock-absorbing capacity of farmers and farm land is less.
- ☆ Diversity of growing season, crops grown and farming systems in use.
- ☆ Close link between climate and water availability.
- ☆ As the main crop is rice hence methane gas emission from flooding rice fields due to methanogenesis becomes a problem.

Climate Change Trends for India

The data released by World Development Report, 2010 attributed reduction in the productivity of rice in India due to adverse climate, global warming and local pollution. The extrapolated data warned that the output of major crops might fall with figures accounting to 4.5-9 per cent in the coming 30 years even after considering the mitigation measures adopted. A World Bank Study from 2006-2009 found that the rapid hike in temperature and uneven and erratic precipitation might interfere in production of crops like rice, groundnut and jowar in the arid regions of Deccan. Another Report by Washington based international Food Research Institute (IFRI) threatened worsening of the current condition by 2050 owing to higher temperature and lesser precipitation leading to increased chances of severe droughts in India.

ICAR's Network project on climate change (NPCC) report emphasized that rice and wheat yield may be reduced by 6 per cent and 18 per cent respectively by the year 2020. According to a parliamentary report severe droughts and erratic rains lowered total food grain production in particular rice production and sugar production by 14 per cent and 13 per cent respectively in the year 2010-11 consequently leading to 11 year record rise in food prices. Indian Institute of Tropical Meteorology, Pune came out with facts regarding stress on agriculture sector in Post 2050 scenario owing to Solar in temperature by 3–4 °C over present coupled with variations in intensity and amount of rainfall received across time and space limits.

Climate Change Impact on Insect Pests

In order to cope with the variable climate, changes may occur in local to regional insect population dynamics and species composition. Hence there may be migration of new types of insect pests into a locality not previously seen in that area under conditions of climate change. Global warming is said to responsible for recent increase in populations of *Helicoverpa armigera*. The growth rate of insects like *Nephotettix cincticeps, Nezara viridula, Leptocorisa acuta* **etc.** in rice eco system is likely to sprawl up due to reductions in winter mortality. The implications of such changes mean huge for the farmers. With increase in temperature and rainfall, the microclimate will be much more favourable for the insect-pest invasion. But at the same time the most damaging insect pests of rice-ecosystems like BPH, *Nilaparvata lugens* and rice leaf folder *Cnaphalocrosis medinalis* is likely to decline giving some relief.

Climate Impact on Farm Animals

Direct impact on livestock population include physical ailments like increased respiration, lack of appetites *etc.* while indirect impact may come in several ways *viz.*, reduction in grazing land and availability of good quality water. Reduced N_2 content in grazing plants due to climate change has shown to affect digestion process in sheep. All these factors tend to decrease the productivity of farm animals.

Adaptation and Mitigation Strategies

The mitigation and adaptation strategies can be categorized under two plans *i.e.* novel crop improvement techniques and management technologies. The mitigation and adaptation strategies intend reversal of the ill impacts of climate change and promotion of sustainable development. The mitigation and adaptation practices are broadly categorized under crop based technologies and resource conservation technologies.

Crop-based Technologies

- Crop based technologies emphasized on promoting cultivation of such crops and varieties which thrive well into the new cropping systems and seasons under the climate-stress condition.
- Improved and novel agronomic and crop production practices like adjustment of planting dates to minimize the effect of high temperature induced spikelet sterility, to reduce yield instability by avoiding flowering to coincide with the hottest period.
- Changing the crop calendar to take advantage of the wet period and to avoid extreme weather events during the growing season is another option.
- Adoption of varieties tolerant to lodging, high temperature, drought, salinity and submergence; high CO_2 concentration adoption of varieties with high fertilizer and radiation use efficiency
- Crop rotations and intercropping, integrated pest management, supplemented with agro forestry and afforestation schemes will be an important component in strategic adaptation to climate change in India.
- A thorough re-evaluation of all the wild relatives, land races, extant varieties, modern varieties and breeding stocks to find out earlier ignored traits that can be useful in the present scenario.
- Establishment of seed banks are of crucial importance in highly variable and unpredictable environments. This facility will provide a practical means to re-establish crops obliterated by major disasters and extreme climate events. This will also help in plant community dynamics, as

differential plant germination strategies to buffer against inter-annual variability in growing conditions.

- Commendable progress has been made in finding out the genetic background of flowering time, inflorescence architecture, and temperature and drought tolerance by Indian scientists in certain model plant systems using comparative genomics.
- Identification and validation of putative marker genes for high temperature tolerance in addition to higher yield potential can prove effective for adapting to climate change needs particularly in case of pulses.
- In grazing lands, pasture improvement is essential to combat impeding changes through planned grazing processes, enclosures for recovery, or enrichment planting.
- The promotion of scientific agro forestry forms a key component in the war against climate change. Agro forestry systems buffer farmers against climate variability, and reduce atmospheric loads of greenhouse gases. Agro forestry can both sequester carbon and produce a range of economic, environmental, and socio-economic benefits.

Resource Conservation Technologies

- The key resource conservation-based technologies are *in situ* moisture conservation, rainwater harvesting and recycling, efficient use of irrigation water, conservation agriculture, energy efficiency in crop production and irrigation.
- Contingency crop planning to minimize loss of production during drought/flood years is also a stone to be turned. Zero tillage (ZT) has effectively reduced the demand for water in rice-wheat cropping systems in more than 1 million ha of area in the Indo-Gangetic plains.
- With ZT technology, farmers can realize higher yields and reduce production costs. In addition, ZT has a direct mitigation effect as it converts the green house gases like CO_2 into O_2 in the atmosphere and carbon, and enriches soil organic matter.
- Bed-planting is widely adopted in the Indo-Gangetic plains, proved to be a successful conservation technology. The main advantages are: increased water use efficiency, reduced water logging, better access for inter-row cultivation, weed control and banding of fertilizers, better stand establishment, less crop lodging and reduced seed rates.

Integrated Nutrient Management (INM) to Mitigate Effects of Climate Change

Demonstrated benefits of these technologies are:

- Increased rice yields and thereby increased net CO_2 assimilation, 30-40 per cent increase in nitrogen use efficiency.

- One of the key emerging technologies to reduce GHG emissions from paddy fields is the use of zymogenic bacteria, acetic acid and hydrogen-producers, methanogens, CH_4 oxidizers, and nitrifiers and denitrifiers in rice paddies which help in maintaining the soil redox potential in a range where both N_2O and CH_4 emissions are low (Hou *et al.*, 2000).
- The application of urease inhibitor, hydroquinone (HQ), and a nitrification inhibitor, dicyandiamide (DCD) together with urea also is an effective technology for reducing N_2O and CH_4 from paddy fields.
- Use of neem-coated urea is another simple and cost effective technology which can be practised in the entire South Asia by small farmers. Promotion of integrated farming systems for marginal and small farmers will also be a viable and effective alternative in combating climate change.
- Multiple-Enterprise agriculture wherein crop, livestock, poultry, fish farming and trees in a single unit of land will minimize risk.
- Soil conservation by means of carbon sequestration is a recent technique that helps mitigating adverse effect of altering climate by removing CO_2 from atmosphere.

Finding the Way Through: A Hope for Future

Agriculture and allied sector across the globe is exposed to the potential hazards of climate change. In developing countries, stress on agriculture is even more. Hence, it is of utmost necessity to understand clearly and thoroughly the possible impacts of climate change on agriculture in the years to come. There is demand to conduct crop and region specific extensive studies involving small and marginal farmers to assess the real challenges so that the measures can be planned effectively. A climate resistant agriculture is the need of hour to achieve the objectives of maximizing farm productivity while keeping the effects of climate change at bay.

References

Adger, W.N. (2006). Vulnerability. Glob Environ Chang, 16: 268–281.

Ahuja, L.R, Andales, A.A., Ma, L. and Saseendran, S.A. (2007). Whole-system integration and modeling essential to agricultural science and technology for the 21st century. Journal of Crop Improvement, 19(1–2): 73–103

Anandhi, A., Steiner, J.L. and Bailey, N. (2016). A system's approach to assess the exposure of agricultural production to climate change and variability. Climate Change, 136: 647–659.

Baker, J.T., Mahan, J.R. and Gitz, D.C. (2013). Comparison of deficit irrigation scheduling methods that use canopy temperature measurements. Plant Biosystem - An International Journal Deal with all Aspects of Plant Biology, 147: 40–49.

Climate Change Science Program, CCSP(2008). Preliminary review of adaptation options for climate-sensitive ecosystems and resources (SAP 4.4). Washington, D.C.

Günther, F. (2005). Socio-economic and climate change impacts on agriculture: an integrated assessment, 1990–2080. Biological Sciences, 360: 1463.

Hatfield, J.L, Boote, K.J., Kimball, B.A. (2011). Climate impacts on agriculture: implications for crop production. Agron Journal, 103: 351.

Hoffmann, I. (2010). Climate change and the characterization, breeding and conservation of animal genetic resources. Animal Genetics, 41(Suppl 1): 32–46.

Howden, S.M., Soussana, J.F. and Tubiello, F.N. (2007). Adapting agriculture to climate change. Proc Natl Acad Sci USA, 104: 19691–19696.

IPCC. (2007). Technical summary. In: Parry, M.L., Canziani, O.F., Palutikof, J.P. (Eds.), Climate Change 2007: Impacts, Adaptation, and Vulnerability. Contribution of Working Group II to the Fourth Assessment Report of the Intergovernmental Panel on Climate Change. Cambridge University Press, Cambridge.

Marshall, N.A. (2010). Understanding social resilience to climate variability in primary enterprises and industries. Global Environmental Change, 20: 36–43.

Mbuthia L.W., Acosta-Martínez, V. and DeBruyn, J. (2015). Long term tillage, cover crop, and fertilization, effects on microbial community structure, activity: implications for soil quality. Soil Biological Biochemistry, 89: 24–34.

Mora, C., Wei, C-L. and Rollo, A. (2013). Biotic and human vulnerability to projected changes in ocean biogeochemistry over the 21st century. PLoS One, 11: e1001682.

National Agricultural Statistics Service (2014). 2012 census of agriculture. https: // www.agcensus. usda.gov/Publications/2012/Accessed 6/9/16

Xue, Q., Rudd, J.C. and Liu, S. (2014). Yield determination and water-use efficiency of wheat under water-limited conditions in the U.S. Southern High Plains. Crop Science, 54: 34–47.

Zheng, (2012). Climate Change: new dimensions in disaster risk, exposure, vulnerability, and resilience Prog. Inqui. Mut. Clim., **8**: 79-83.

2

Molecular Markers: A New Approach for Pulse Improvement

B.S. Pavithra, L. Behera and K.C. Samal*

Department of Agricultural Biotechnology,
Orissa University of Agriculture and Technology, Bhubaneswar – 751 003, Odisha
****E-mail: samalkcouat@gmail.com***

ABSTRACT

Pulses are important sources of proteins in vegetarian diet. Genetic improvement of the crop plants has been done through development of new generation of molecular markers. Molecular markers offer several advantages over traditional phenotypic markers and having high discrimination power. These markers are phenotypically neutral, independent of developmental stage and environment. These methods are rapid, relatively cheap and eliminate the need to grow plants to maturity. Molecular markers can also be used to confirm the parents of hand-pollinated crosses and to determine the paternity of seedlings derived from open-pollinated crosses. This approach should enable us to produce much greater number of crosses and that will accelerate the breeding programme. Utilization of molecular markers and their applications place important role in development of novel traits and helps in identification of resistance varieties in pulses. The present view of researchers is that the effectiveness and efficiency of conventional breeding can be significantly improved by using molecular markers. Now a days, molecular markers are routinely utilized worldwide in all major crops as a component of breeding, the pace of development of molecular markers and other Gld enomic sources has been accelerated in pulses such as mungbean, chickpea, pigeon pea etc. In this article the progress made and application of marker assisted selection in genetic improvement of pulse crops was presented.

***Keywords:** Pulses, Molecular markers, Marker assisted breeding.*

Introduction

Pulses have an imperative contribution in the world Agriculture and also considered as indispensable part of Indian Agriculture as India is the largest producer and consumer of pulses in the world. India ranks first on pulse production.

According to 2016-2017 data total pulse production in India is 17.82 million tones. India's contribution towards total global acreage and production of pulses is at 35 per cent and 25 per cent. Pulse crops play crucial role in the agricultural economy of India by virtue of their competency to fix atmospheric nitrogen in symbiotic association with *Rhizobium*, also consider as mini fertilizer factory by committing substantially to the enrichment of soil. The second captivating character of pulse crops is their deep penetrating root system which contributes to their efficiency in utilizing the limited available moisture and also to loosen up the soil. Thirdly their role in human diet, as a source of protein is globally well established. Pulses are considered to be an excellent source of dietary protein important for the human diet. Pulses are not only important sources of proteins but also offer vitamins and minerals. For an active normal body pulses requirement is about 40 g per day or 14.6 kg per person per year. Pulses are the basic ingredient in the diets of a vast majority of the Indian population, as they provide a perfect mix of vegetarian protein component of high biological value when supplemented with cereals. It is generally grown for its edible seeds, sprouts, noodles and consumed as "Dal" in Asian subcontinents. It is also popularly admitted as "Poor man's meat" and "rich man's vegetable "and also reasonable source to conquer protein malnutrition as they having 20 to 25 percent protein on dry seed basis which is almost 2.5 - 3.0 times higher than cereals. For an active normal body pulses requirement is about is 60 g and 55 g for adult male and female per day respectively, but the per capita availability of pulses is 42 g per day. Pulses availability decrease with the period of time, which was 60.5 g/day during 1950-51 and 42 g per day during 2015-16. Hence to address future global food security, nutrition and environmental sustainability needs UN declared 2016 as International Year of Pulses. It gains its importance mainly because of the reason that it requires a low input, minimal care, short duration (65-75 days) and photo insensitive crop which fits well in different cropping situations. Odisha has a rich diversity of Urdbean landraces possibly due to the agro-climatic situation desirable for Urdbean cultivation.

Morphological traits can be used to asses phenotypic variation in growing environment and are used as tools for the indirect analysis of genetic variability and diversity (Tabasum *et al.*, 2010) where molecular markers place significance role in identification and development of novel traits in pulses and various molecular markers have been utilized in breeding programmes. The power of discrimination of DNA-based markers is so high that very closely related varieties can be differentiated between the parents and must be accurately reproduced in the progeny.

Molecular Markers

Molecular markers are genetic loci that can be simply traced and quantified in a population, correlated with a particular gene of interest (Alice *et al.*, 2015). Molecular marker is a gene or DNA sequence with a known location on a chromosome that can be used to identify individuals or species. Assessment of genetic variation in mungbean has been carried out using different types of molecular markers including random amplified polymorphic DNA (RAPD) (Santalla *et al.*, 1998; Lakshanpaul *et al.*, 2000), amplified fragment length polymorphism (AFLP) (Bhat *et al.*, 2005), inter-simple sequence repeat (ISSR) (Reddy *et al.*, 2008), and simple sequence repeat (SSR) (Gwag *et al.*, 2010) *etc.* DNA-based molecular markers have acted as versatile tools and have found their own position in various fields like plant breeding, genetic engineering, taxonomy (Joshi *et al.*, 2011). Molecular markers include biochemical constituents (*e.g.* secondary metabolites in plants) and macro-molecules vizproteins and deoxyribonucleic acid (DNA). Over the last few decades plant genomics has been studied extensively bring about a revolution in this area, making molecular markers useful for plant genomic analysis, therefore becoming an important tool in this revolution (Joshi *et al.*, 2011). Molecular markers are indispensable for genomic study, among various marker systems such as RFLP, RAPD, STS, AFLP, SSRs have pivotal importance because of their reproducibility, co-dominant inheritance, good genetic coverage, multiallelic nature, relative abundance. SSRs are the clusters of short tandem repeated nucleotide bases distributed throughout the genome. Major features that make SSRs very popular are their abundant distribution in the genomes examined to date and their hypervariable nature (Singh *et al.*, 2013). According to Joshi *et al.* (2011), ideal DNA markers should however possess the following properties such as abundant occurrence in genome; co-dominant inheritance; easy and fast assay; highly polymorphic; selective natural behavior (the DNA sequences of any organism are neutral to environment conditions or management practices) and having high reproducibility.

Types of Molecular Markers

a. DNA-DNA Hybridization: Restriction Fragment Length Polymorphism

Genetic marker systems based on DNA-DNA hybridization were developed in early 1980 (Farooq and Azam, 2002). This marker was first reported by Botstein *et al.* (1980); in the detection of DNA polymorphism (Agarwal *et al.*, 2008). Plants are able to replicate their DNA with high accuracy and rapidity, but many mechanisms causing changes (mutation) in the DNA are operative (Joshi *et al.*, 2011). To begin, total cellular DNA is digested with a restriction endonuclease, which reduces the genome to a large pool of restriction fragments of different sizes. An RFLP probe is a labeled DNA sequence that hybridizes with one or more complementary sequence in the DNA bound to the membrane. The DNA probing relies on complementary base pairing; both the DNA fragments on the nylon membrane and the probe.

Restriction fragment length polymorphism analysis has been applied to study: Phylogenetic relationships (Wagner *et al.*, 1991, 1992; Tsumura *et al.*, 1995); Genetic variation within species (Ali *et al.*, 1991; Strauss *et al.*, 1993; Ponoy *et al.*, 1994); Modes of organelle DNA inheritance; status of genetic diseases; paternity tests or criminal investigations and genetic mapping *etc.* The limitation/disadvantages of RFLP markers are requirement of relatively large quantity of pure DNA; time consuming process, low level of polymorphism and use of hazardous radioactive probes.

b. PCR (Polymerase Chain Reaction) Based Markers

PCR is an *in vitro* method of nucleic acid synthesis by which a particular segment of DNA can be specifically replicated (Mullis and Faloona, 1987). The process involves two oligonucleotide primers that flank the DNA fragment of interest and amplification is achieved by a series of repeated cycles of heat denaturation of the DNA, annealing of the primer to their complementary sequences, and extension products themselves are also complementary to primers, successive cyclesof amplification essentially double the amount of the target DNA synthesized in the previous cycle and the result is an exponential accumulation of the specific target fragment. A decade after the emergence of AFLP, there was another breakthrough which involves the use of PCR in 1990 (Farooq and Azam, 2002).

c. AFLP (Amplified Fragment Length Polymorphism)

AFLP is a PCR-based tool used in genetics research, DNA fingerprinting, and in the practice of genetic engineering. It Amplified Fragment Length Polymorphism (AFLP) was developed in the early 1990s by Keygene and a highly sensitive method for detecting polymorphisms in DNA. In this approach the sample of DNA is enzymatically cut up into small fragments (as with RFLP analysis), but only a fraction of fragments are studied following selective PCR amplification (Liu *et al.*, 1994). It is a combination of RFLP and RAPD methods. AFLP technique shares some characteristic with both RFLP and RAPD (Farooq and Azam, 2002) and combines the specifically of restriction analysis with PCR amplification. AFLP is extremely sensitive technique and added use of fluorescent primers for automated fragment analysis system and software packages to analyze the biallelic data makes it well suitable for high thorough put analysis. The major advantages of AFLP techniques (Farooq and Azam, 2002) are: high polymorphism rate; highly specific, co-dominant nature and high reproducibility. The application of AFLP markers includes; monitoring inheritance of agronomic traits, diagnostic in genetically inherited disease, pedigree analysis, forensic typing (parentage analysis), identifying hybrids, and species level relationship.

d. Randomly-Amplified Polymorphic DNA Marker (RAPD)

The randomly-amplified polymorphic DNA markers (RAPD) were the PCR-based markers and were developed independently by Welsh and McClelland (1990) and Williams *et al.* (1990). The RAPD marker system is easy to apply as no prior DNA sequence information is needed for designing PCR primers as is required for other PCR-based genetic marker systems. It detects nucleotide sequence polymorphism in DNA by using a single primer of arbitrary nucleotide sequence (Oligonucleotide primer, mostly ten bases long) (William *et al.*, 1991). In this marker system, a PCR reaction is carried out using a very small amount of template DNA (usually less than 10 nanograms) and a single decamer RAPD primer. RAPD marker system has the potential to randomly survey a large portion of the genome for the presence of polymorphisms. The small amount of DNA needed is a big advantage of the RAPD. Advantages of RAPD analysis include: no sequence information required; non-involvement in blotting or hybridization; requirement of low quantity of pure DNA, multi locus profiling, fast and efficient. The limitations associated with RAPD markers: inability to differentiate heterozygote or homozygote; allelic relationship not defined and low reproducibility. Applications of RAPD markers are gene mapping, diversity study and identification of genotypes. Variants of the RAPD technique include Arbitrarily Primed Polymerase Chain Reaction (AP-PCR) which uses longer arbitrary primers than RAPDs, and DNA Amplification Fingerprinting (DAF) that uses shorter, 5-8 bp primers to generate a larger number of fragments. Multiple Arbitrary Amplicon Profiling (MAAP) is the collective term for techniques using single arbitrary primers.

e. Simple Sequence Repeat Repeats (SSRs) or Microsatellites

SSR also known as microsatellite were developed during 1990s. These are mono-, di-, tri-, tetra-, and penta-nucleotides (A,T, AT, GA, AGG, AAAG *etc.*) with different lengths of repeat motifs. SSR markers are taken from non-coding genomic regions. This method is inexpensive, easy and simple detection through gel electrophoresis for large number of samples. Polymorphism is depend on the differences in the number of repeats in different genotypes. Compared with RFPL and RAPD, the high degree of polymorphism of SSR markers their locus specific nature and co-dominant nature make them unique for different purposes as well as in practical plant breeding. Therefore due to the vast genome coverage and higher variable nature SSR or microsatellite markers become markers of choice for any array of application in plant. PCR (polymerase chain reaction)-based SSR and SNP markers have been preferred by breeders because these markers can easily be employed in genotyping of large segregating populations in a cost-effective manner and with minimum infrastructure facilities. Also, high reproducibility and user-friendliness make them preferential to traditional plant breeders for MAS (Gupta *et al.*, 2010).

f. Inter Simple Sequence Repeats (ISSR)

ISSR marker include DNA segments amplification available in between two identical microsatellite repeat regions oriented in an opposite direction. This uses SSR as primers in a single primer PCR reaction targeting multiple genomic loci. For ISSRs, these microsatellite repeats used as primers. The used primers can be unanchored or mostly anchored at 3′ or 5′ end. As compared to RAPD primers, ISSRs use longer primers (15–30 mers), which allow the further use of high Tm resulting to higher stringency. Tm depends on the GC content of the used primer and that vary from 45 to 65°C. These amplicons are 200–2000 bp long and can be detected by both polyacrylamide gel electrophoresis and agarose gel electrophoresis. ISSRs need no prior primer sequence information [. This technique is quick, simple, and radioactivity use is not mandatory. Usually, ISSR markers show high polymorphism. ISSRs segregate mostly as dominant markers. It was reported that the reproducibility level of this technique ranged 86 per cent to 99 per cent.

g. Sequence Characterized Amplified Region (SCAR)

The SCARS are PCR-based markers that represent genomic DNA fragments at genetically defined loci that are identified by PCR amplification using sequence specific oligonucleotide primer. SCARs are DNA fragments amplified by the PCR using specific 15-30 bp primers, designed from nucleotide sequences established from cloned RAPD fragments linked to a trait of interest. By using longer PCR primers, SCARs do not face the problem of low reproducibility generally encountered with RAPDs. Obtaining a codominant marker may be an additional advantage of converting RAPDs into SCARs, although SCARs may exhibit dominance when one or both primers partially overlap the site of sequence variation.

h. Simple Primer Amplification Reaction (SPAR)

SPAR uses the single SSR oligonucleotide principles. PCR based single primer amplification reaction (SPAR) methods that are useful for genetic diversity studies in plants and collectively provide a comprehensive description of the nature and extent of the diversity. These SPAR methods have been used for diversity analysis in several plants such as mulberry, neem and papaya (Bhattacharya and Ranade, 2001; Ranade and Farooqui, 2002; Saxena *et al.*, 2005), blackgram (Ranade and Gopalkrishna, 2004).

i. Randomly Amplified Microsatellite Polymorphism (RAMP)

This is a micro satellite based marker which shows a high degree of allellic polymorphism, but they are labor-intensive. The technique involves a radio labeled primer consisting of a 51 anchor and 31 repeats which is used to amplify genomic DNA in the presence or absence of RAPD primers (Agarwal and Shrivastava, 2008).

j. Sequence-Related Amplified Polymorphism (SRAP)

The aim of SRAP technique (Li and Quiros, 2001) is the amplification of open reading frames (ORFs). It is base on two-primer amplification using the AT- or GC-rich cores to amplify intragenic fragment for polymorphism detection (Agarwal and Shirvastava, 2008).

k. Target Region Amplification Polymorphism (TRAP)

The TRAP technique (Hu and Vick, 2003) is a rapid and efficient PCR-based technique, which utilizes bioinformatics tools and expressed sequence tag (EST) database information to generate polymorphic markers, around targeted candidate gene sequences.

Application of Molecular Markers in Pulses

Molecular markers have been look upon as a tools for a large number of applications ranging from localization of a gene to improvement of plant varieties by marker-assisted selection called genome analysis which has generated a vast amount of information and a number of databases are being generated to preserve and popularize it (Joshi *et al.*, 2011). In the modern era of research on molecular biology, molecular marker-assisted section is a major achievement for scientists and breeders. Molecular markers have been used in legumes for crop improvement programmes;

a. Soybean

Tremendous advances in all aspects are being made in advanced laboratories particularly in the USA. Therefore, a high priority for example, could be the use of marker assisted breeding for selecting lines with the ability to cause suicidal germination of *Strigaher monthica*, a parasitic weed affecting maize but not soybean (trap crop). In the longer term, increased nodulation and resistance to pod shattering would be highly important candidates for MAS systems (Ortiz, 2004).

b. Mungbean

Studies reveal that molecular markers widely used in development of resistance to various disease of mungbean such as powdery mildew, cercospora leaf spot, YVMV. SSR markers are useful in assessment of mungbean diversity and selection of a core collection to enhance the efficiency of genotypes management for use in mungbean breeding and conservation (Muthusamy *et al.*, 2008; Singh *et al.*, 2013).

The number of these SSRs is still very limited in mungbean (Swag *et al.*, 2006). However, SSRs from azuki bean [*V. angularis* (Willd.) Ohwi and Ohashi] (Wang *et al.*, 2004), common bean (Blair *et al.*, 1997) and cowpea (Li *et al.*, 2001) can be used in both mungbean and blackgram. As high as 72.70 per cent and 78.20 per cent of the azuki bean SSRs amplify the genomic DNA of mungbean and blackgram respectively (Chaitieng*et al.*, 2006) while, 60.6 per cent of common bean SSRs amplify the genomic DNA of mungbean. Methods for the isolation of SSR loci in *Vigna* have been developed (Kumar *et al.*, 2002; Miyagi *et al.*, 2004 and Wang *et al.*, 2004).

Wang *et al.* (2004) used an SSR enrichment method based on oligo-primed second-strand synthesis to develop SSR markers in azuki bean (*V. angularis*). Using this methodology 49 primer pairs were made to detect dinucleotide (AG) SSR loci. The average number of alleles in complex, wild and town populations of azuki bean was 3.0 to 3.4, 1.1 to 1.4 and 4.0 respectively. The genome size of azuki bean is 539 Mb; therefore, the number of (AG) n and (AC) n motif loci per haploid genome were estimated to be 3500 and 2100 respectively.

Kajonphol*et al.* (2012) used the SSR markers to construct a linkage map and identify chromosome regions controlling some agronomic traits in mungbean. Twenty QTLs controlling major agronomic characters including days to first flower (FLD), days to first pod maturity (PDDM), days to harvest (PDDH), 100 seed weight (SD100WT), number of seeds per pod (SDNPPD) and pod length (PDL) were located on to the linkage map. Most of the QTLs were located on linkage groups 7 and 5.

Gupta *et al.* (2013) studied molecular tagging of MYMIV resistance gene in blackgram by using 61 SSR markers, 31 were found polymorphic between the parents. Marker CEDG 180 was found to be linked with resistance gene following the bulked segregant analysis. This marker was mapped in the F2 mapping population of 168 individuals at a map distance of 12.9 cm.

c. Urdbean

Molecular markers have been used to study genetic diversity (Souframanien and Gopalakrishna, 2004) and to tag the MYMV resistance gene in blackgram (Basak *et al.*, 2004; Souframanien and Gopalakrishna, 2006; Gupta *et al.*, 2013). Random amplified polymorphic DNA (RAPD) markers and Inter Simple Sequence Repeat (ISSR) markers have been successfully used for the construction of linkage maps in many crop plants including grain legumes (Kalo *et al.*, 2000; Lambrides *et al.*, 2000 and Cobos *et al.*, 2005). Only recently, microsatellite or simple sequence repeat (SSR) markers, a marker system of choice, have been developed from mungbean

(Kumar *et al.*, 2002 and Miyagi *et al.*, 2004). Simple Sequence Repeat (SSR) markers, because of their ubiquitous presence in the genome, highly polymorphic nature and co-dominant inheritance, are another marker of choice for constructing genetic linkage maps in plants (Flandez *et al.*, 2003; Han *et al.*, 2005 and Chaitieng *et al.*, 2006).

d. Chickpea

Ascochyta rabiei (pass) Labri is the most severe fungal disease limiting chickpea production and studies in Syria revealed the occurrence of three pathotypes for *A. rabiei*. A set of micro satellite and RAPD markers were also used which lead to identification of suitable RAPD markers, allowing a more precise determination of the pathotypes. Furthermore, the availability of markers for pathotype I and II allow the monitoring of the pathotype distribution, which gives the recommendation for the planting of suitable chickpea cultivars (Baum, 2003)

e. Bambara Groundnut

The results showed that Bambara groundnut landraces from Tanzania form a genetically diverse population. Therefore, AFLP markers can be effectively employed to assess genetic diversity and to measure relationship among accessions (Ntundu *et al.*, 2004). Amplified fragment length polymorphism was used to assess genetic diversity among 100 selected Bambara groundnut (*Vigna subterranean* L Verdc).

Molecular Markers and Pulse Breeding

Molecular markers can play an important role in accelerating the introgression of genes conferring to target insects and pathogens into high-yielding cultivars, understanding the nature of gene action, and reducing the deleterious effects introgressing unwanted genes from wild species through linkage drag. Molecular breeding also offers the opportunity to pyramid different sources of resistance that could not be effectively selected through conventional breeding due to identical phenotypes and thereby accumulate levels of resistance and/or create potentially more durable resistant cultivars. Considerable has been made in developing genetic linkage maps of chickpea, cowpea, and soybean, while much remains to be done in pigeonpea, beans, lentil, and fieldpea. Preliminary identification of molecular markers resistance to insects in soybean, chickpea, mungbean, fieldpea, and cowpea has been reported. However, no distinct advantage has been observed by using assisted selection resistance to insect pests over the conventional approach, and in most cases, the epistatic are also quite high. Thus, a new paradigm approach may be required to combine conventional approaches and marker-assisted selection in such way as to create systems better than either approach.

Breeding of Polygenic Traits

By using dense genetic marker maps, the contributions of separate regions of the genome on the trait values can be estimated once the mapping population is sufficiently large. Due to utilization of markers can obviously prevent loss of quantitative trait loci (QTL) common with some crops DNA markers and this allow us to unravel the genetic basis of traits expressing continuous phenotypic variations as they are abundant and scattered throughout the genome. In addition agronomic important novel traits like yield, flower time, nutritional quality, durable resistance which appear to follow complex, polygenic inheritance patterns with multiple genes having small effects on the trait value can be easily by analyzed using markers. Evidence obtained from the various crops indicates that even such complex traits appear to be determined by only a few major factors or genes. (Frary *et al.*, 2000 and Thornsberry *et al.*, 2001)

Conclusion

Molecular markers and their application are pushing its way into all areas of modern biology, from genomics to breeding, from transgenic to developmental biology, from systematics to ecology. These are helpful in effectively used in early selection parent, disease resistance, genetic variation in germplasm and gene bank management especially during the last few decades. New promising approach such as allele mining associated genetic and comparative genomics to obtain insight the organization and variation of genes that affect relevant phenotypic traits. Some traits and markers are conserved across related species and so comparative genomics of traits and markers between crops also helps in breeding process. Development of new technology and approach sucked up by the combining expertise of several disciplines, including statistics, molecular genetics, and bioinformatics *etc.* The desirable properties of molecular markers are high polymorphism, co-dominant inheritance, frequent occurrence and even distribution throughout the genome, selectively neutral behavior, open access, easy and fast assay, low cost, high reproducibility and transferability between laboratories, populations, and/or species. No molecular markers are available yet that fulfill all these requirements, so it needs careful selection of molecular marker, which combines at least some of these desirable properties. Therefore, RFLP, SSR, RAPD, AFLP, and ISSR are the only markers that could be used for a wide range of applications in plants. Allele mining, associated genetics and comparative genomics are promising new approaches to obtain insight the organization and variation of genes that affect relevant phenotypic traits. These developments exploited by combining expertise from several disciplines, including molecular genetics, statistics and bioinformatics.

References

Agarwal, M., Shrivastava, N. and Padh, H. (2008). Advances in Molecular marker techniques and their applications in plant science. Plant Cell Report, 27: 617- 613.

Alice, C.H., Reece, T., Jessica, D-M. and Jacqueline, B. (2015). Molecular Marker Applications in Plants, Chapter 2, Jacqueline Batley (ed) *Plant Genotyping Methods and Protocols.* Methods in Molecular Biology, pp. 1245.

Anderson, J.A., Sorrells, M.E. and Tanksley, S.D (1993). RFLP analysis of genomic regions associated with resistance to pre-harvest sprouting in wheat (*Triticum aestivum*). Crop Science, 33: 453-459.

Berloo, R.V. and Stam, P. (1999). Comparison between marker - assisted selection and phonotypical selection in a set of Arabidopsis thaliana recombinant inbred lines. Theoretical and Applied Genetics, 98: 113-118.

Bhat, K.V., Lakhanpaul, S. and Chadha, S. (2005). Amplified fragment length polymorphism (AFLP) analysis of genetic diversity in Indian green gram [*Vigna radiata* (L.) Wilczek] cultivars. Indian Journal of Biotechnology, 4: 55–64.

Blair, M. and Mccouch, S.R. (1997). Microsatellite and sequence-tagged site markers diagnostic for the bacterial blight resistance gene xa-5. Theoritical Applied. Genetics, 95: 174–184.

Bostein, D., White R.L, Skolnick, M, Davis R.W (1980). Construction of a genetic linkage map in man using restriction fragment length polymorphisms. American Journal of Human Genetics, 32: 314-333.

Datta, S., Kaashyap, M. and Kumar, S. (2010). Amplification of chickpea-specific SSR primers in *Cajanus* species and their validity in diversity analysis. Plant Breed, 129, 334–340.

FAO (2004). Scientific facts on genetically modified crops.

Farooq, S. and Azam, F. (2002). Molecular markers in plant breeding–1: concepts and characterization. Pakistan Journal of Biological Sciences, 5 (10): 1135- 1140.

IBPGR (International Board for Plant genetic resources). Report on International workshop on Okra genetic resources held at the National Bureau for Pant Genetic Resources New Delhi, India, 8-12 October, 1990.

Joshi, S.P., Prabhakar, K., Ranjekar, P.K and Gupta, V.S. (2011). Molecular markers in plant genome analysis. *http/www.ias.ac.in/currsci/jul25/articles 15. htm*. Pp. 1-19.

Kiss, G.B., Osanandi, G., Kalman, K., Kalo, P/and Okresz, L. (1993). Construction of a basic linkage map of alfalfa using RFLP. RAPD, isozyme and morphological markers. Molecular Genetics and Genet, 238: 129137.

McDermoth, J.M., Brandle, U., Dutly, F., Haemmerli, U.A., Keller, S., Muller, K.E. and Wolf, M.S. (1994). Genetic variation in powdery mildew of barley: development of RAPD, SCAR and VNTR markers. Phytopathology, 84: 1316-1321.

Ortiz, R. (2004). Biotechnology with horticultural and agronomic Crops in Africa. Acta Horticulture 642: 4356.

Powell, W., Morgante, M. and Andre, R. (1996). The comparison of RFLP, RAPD, AFLP and SSR (microsallite) markers for germplams analysis. Molecular Breeding, **2**: 225-238.

Prasanna, B.M., and Pixley, K. (2010). Molecular marker-assisted breeding options for maize improvement in Asia. Molecular Breeding, 26: 339-356.

Reddy, K.S., Souframanien, J., Reddy, K.S. and Dhanasekar, P. (2008). Genetic diversity in green gram as revealed by SSR and ISSR markers. Journal of Food Legumes, 21: 15-21.

Saxena, R.K., Prathima, K.B., Saxena, D.A., Hoisington, N.K., Singh, and Varshney, R.K. (2010). Novel SSR Markers for polymorphism detection in pigeon pea (*Cajanus* spp.). Plant Breeding, 129, 142-148.

Shanmugasundaram, S. (2001). New breakthrough with green gram. Centerpoint, 19: 1–2.

Singh, A., Dikshit, H.K., Jain, N., Singh, D. and Yadav, R.N. (2014) Efficiency of SSR, ISSR and RAPD markers in molecular in molecular characterisation of green gram and other *Vigna* species. Indian Journal of Biotechnology, 13: 81–84.

Singh, N., Singh, H. and Nagarajan, P. (2013). Development of SSR markers in mungbean, *Vigna radiata* (L.) Wilczek using *in silico* methods. Jornal of Crop Weed, 9: 69–74.

Souframanien, J., and Gopalkrishna, T. (2004). A comparative analysis of genetic diversity in blackgram genotypes using RAPD and ISSR markers. Theoritical Applied Genetics, 109: 1687-1693.

Thottappilly, G., Magonouna, H.D. and Omitogun, O.G. (2000). The useof DNA markers for rapid improvement of crops in Africa. African Crop Science Journal, 8(1): 99- 108.

3

Seed Germination and Survival of *Careya arborea* under Varying Micro-habitat Conditions of Similipal Biosphere Reserve, Odisha: Conservation Implications

R.K. Mishra[1], P.K. Nayak[2] and R.C. Mohanty[2]*

[1]Department of Wildlife and Biodiversity Conservation, North Orissa University, Baripada – 757 003, Odisha
[2]Department of Botany, Utkal University, Bhubaneswar – 751 004, Odisha
**E-mail: rabikumishra@gmail.com*

ABSTRACT

Seed germination and survival of an important species, Careya arborea Roxb. (family-Lecythidaceae) was studied under various micro-habitat conditions of Similipal Biosphere Reserve, Odisha. Mature seeds in two states i.e. with and without pulp were placed in forest floor under four micro-habitat conditions viz. undisturbed close canopy, undisturbed open canopy, disturbed close canopy and disturbed open canopy (4 micro-habitats × 2 seed states × 3 replicates of 50 seeds each). The heterogeneity in micro-environment due to combination of factors caused variation in average seed germination (31.25 ± 4.21 to 88.46 ± 8.59 per cent) and survival (18.56 ± 3.23 to 70.23 ± 6.43 per cent). The de-pulped seeds had significantly higher germination as well as seedling survival across all the micro-habitats as compared to those with intact pulp. Open canopy cover in undisturbed forest emerged as the best micro-habitat for germination, survival and extension growth (shoot length in cm) of C. arborea provided the seeds are liberated from the fruit pulp. Such a condition seems to confirm with dispersal by wild animals in nature. Conservation implications are discussed.

Keywords: *Careya arborea, Extension growth, Germination, Mean Germination Time, Similipal, Survival*

Introuction

Seed germination, survival and establishment of any tree species on forest floor is confined to a particular range of habitat conditions (Pokhriyal *et al.*, 2010) and is influenced by environmental factors (Khan and Shankar, 2001: Sharma *et al.*, 2009). Both germinating seeds and seedlings are most vulnerable to predation and damage as they constitute important resource for the pathogens. Seedling predation, disturbance in the form of trampling and micro-habitat alternations have direct implication for recruitment and multiplication of species. Consistency in micro-habitat condition enhances seedlings survival. One of the important micro-habitat factors in the forest floor is light which is regulated by tree canopy cover. Tree species vary considerably in terms of light requirements at seedling stage (Davies, 2001: Pacala *et al.*, 1994). *Careya arborea* Roxb. (Lecythidaceae) commonly called as Kumbhi is a deciduous dioecious species up to 6-8m high and distributed in moist deciduous forest covers of Odisha (Mishra, 2006). It flowers during March to April and fruiting during May. The fruits are large, green, fleshy, berry and globose. It requires approximately three to four months for attaining maturity. Seeds numerous, embedded in fleshy pulp. Medicinally the plant is valuable as the bark of the plant is traditionally used in the treatment of tumors, epileptic fits, antidote to snake-venom and skin disease (Kumar, 2016). Root of this plant has been traditionally used as oral contraceptive agent after delivery and the calices of flowers are well-known Indian remedies, and are valued on account of their astringent and mucilaginous properties, being administered internally in coughs and colds and applied externally as an embrocation (Behera *et al.*, 2008: Satish *et al.*, 2010). From wildlife point of view fruit of *C. arborea* is used as food by hog deer, chital, sambhar, barking deer and monkeys (Bharadwaj and Seth, 2017). The bark of the plant is utilized as food by the Asiatic elephants (Santra *et al.*, 2008). Considering the wildlife and medicinal potential, and dwindling natural population of *C. arborea* in sanctuaries and national parks of Odisha, this species has been recommended for in-situ as well as ex-situ conservation by several authors (Biswal *et al.*, 2005: Behera *et al.*, 2008). In view of this, a comprehensive study on the patterns of seed germination and seedling establishment of *C. arborea* under varied micro-habitat conditions inside Similipal Biosphere Reserve (SBR) was undertaken towards its conservation potential.

Material and Methods

Site Description

The study was conducted in Similipal Biosphere reserve (SBR) located in Mayurbhanj district of Odisha (21°28′- 22° 08′ N latitude and 86°04′ - 86°37′ E longitude). The climate of the reserve is influenced by a monsoon pattern of rainfall. Maximum rainfall occurs from mid June to October accounting for 75-80 per cent of annual rainfall. In spite of high annual rainfall summer and winter are relatively dry generally with <10mm monthly rainfall (Mishra *et al.*, 2008). The amount of average annual rainfall generally ranges between 28.11cm to 344.96 cm. Summer is not unbearable, as the maximum temperature rarely goes above 40 ^{0}C. Winter is severe and the temperature comes down to 4^{0}C in parts with frosts in valleys (Mishra *et al.*, 2008). Spring is very pleasant. Because of luxuriant vegetation cover and a network of perennial streams Similipal is relatively moist throughout the year. Humidity of Similipal at 0600 hrs is around 40 per cent and at 1800 hrs is around 81 to 93 per cent (Mishra *et al.,* 2008).The whole forest cover is divided into two parts. One part of the forest was accessible to anthropogenic activities,and cattle grazing and hence categorized as disturbed whereas the remaining part was relatively undisturbed (Mishra *et al.,* 2006). The two type of forests differed considerably in soil characters, tree density, canopy cover and light intensity on forest floor (Table 3.1).

Table 3.1: Comparison of Biophysical Characters at Disturbed and Undisturbed Sites of Similipal Biosphere Reserve (SBR), Odisha

Character	*Disturbed*		*Undisturbed*	
	Open Canopy	*Close Canopy*	*Open Canopy*	*Close Canopy*
Soil moisture (per cent)	11.74 ± 1.62	14.53 ± 1.81	32.42 ± 3.31	34.16 ± 4.56
pH	6.66 ± 0.14	6.53 ± 0.33	6.23 ± 0.27	6.19 ± 0.18
Organic Carbon (per cent)	0.96 ± 0.04	1.09 ± 0.07	1.18 ± 0.023	1.25 ± 0.06
NO_3^- -N (Kg ha^{-1})	14.26 ± 3.27	17.26 ± 3.39	21.51 ± 5.19	22.68 ± 5.56
Total Nitrogen (per cent)	0.16 ± 0.014	0.25 ± 0.02	0.31 ± 0.03	0.38 ± 0.037
Available -P (Kg ha^{-1})	19.03 ± 4.72	18.86 ± 3.29	18.73 ± 3.45	18.51 ± 3.76
Available -K (Kg ha^{-1})	205.33 ± 31.55	197.61 ± 31.27	182.21 ± 24.31	171.65 ± 19.84
Tree density (Plants ha^{-1})	655 ± 23.53	750 ± 21.56	840 ± 29.43	920 ± 31.41
Shrub density (individuals ha^{-1})	1665 ± 96.77	1560.9 ± 77.31	1120 ± 67,21	1005 ± 40.66
Herb density (individuals m^{-2})	27.96 ± 4.71	15.5 ± 3.12	6.5 ± 1.21	4.5 ± 1.08
Tree crown area (m^2 ha^{-1})	21423.45 ± 2031.47	26043.21 ± 2045.77	38551.53 ± 4047.46	41988.01 ± 40 68.08
Light intensity (μmol/m^2/Sec.)	344.91 ± 33.74	273.64 ± 29.68	158.16 ± 22.84	59.36 ± 9.13

The four micro-habitats *i.e.* open and close canopy in disturbed and undisturbed forest sites of Similipal were selected. At each micro-habitat two types of mature seeds (with and without pulp) were planted making eight possible micro-environments for seed germination and seedling establishment. Each micro-environment, therefore, represented a combination of three variables: seed with or without pulp, close and open canopy and disturbed or undisturbed forest stands. One hundred numbers of ripe fruits were collected from a few trees within the study area in the month of August. Of these 600 seeds were de-pulped while the other 600 seeds were intact with pulp. To study the germination, survival and growth of the seedlings of *C. arborea* under each condition, 3 replicates of 50 seeds each were placed on the forest floor after removing all naturally fallen seeds from the ground in an area measuring 1m × 1m (4 micro-habitats × 2 seed states × 3 replicates of 50 seeds each). The seed plots were roofed with fine nylon net to avoid natural seed deposition until the seeds germinated which were sown on the forest floor. After emergence of seedling nylon net was replaced with PVC pipes just to demarcate the seed plots for subsequent observations.

Observation on seed germination was made at a interval of 10 days for a total period of two months. Seedling survival and shoot growth were measured at one month interval for a period of one year. Seed germination was recorded at each site when epicotyls were visible above ground. Mean Germination Time (MGT) was calculated using the formula given by Czabat (1962), where MGT =Total number of emerged seeds/Total number of days. For extension growth five seedlings at each type of seed plot were randomly selected and shoot length was measured using centimeter scale. Means with standard deviation were calculated for each data set. Statistical test such as Analysis of Variance (ANOVA) was calculated at $p \leq 0.05$ level of significance to determine the overall influence of the micro-environments on germination, seedling survival, mean daily germination and extension growth (shoot length) of *Careya arborea*.

Results and Discussion

Seeds shown in the experimental area started germination within a time interval of 5 to 6 days with a significant variation in average seed germination (31.25 ± 4.21 to 88.46 ± 8.59 per cent) across the seed plots *i.e.* micro-environments (Table 3.2).

Table 3.2: Germination, Survival, Mean Germination Time and Shoot Length Extension (Mean ± SD) of Seeds of *Careya arborea* in different Seed States and Site Conditions on the Forest Floor of SBR

Parameters	*Seed State*	*Site Condition*			
		Undisturbed		*Disturbed*	
		Open Canopy	*Close Canopy*	*Open Canopy*	*Close Canopy*
Germination (per cent)	De-pulped	88.46 ± 8.59	65.41 ± 6.23	70.23 ± 6.81	82.45 ± 7.46
	Pulped	41.52 ± 5.67	31.25 ± 4.21	35.45 ± 4.65	38.44 ± 4.37
Survival (per cent)	De-pulped	70.23 ± 6.43	55.25 ± 6.75	25.61 ± 4.11	45.36 ± 4.27
	Pulped	36.75 ± 4.81	26.38 ± 4.15	18.56 ± 3.23	21.36 ± 3.27
MGT (in days)	De-pulped	5.4 ± 1.64	8.3 ± 2.4	7.4 ± 2.2	6.3 ± 2.1
	Pulped	6.7 ± 1.86	9.6 ± 2.31	8.5 ± 2.23	7.6 ± 1.73
Extension growth (Shoot length increase in cm per month)	De-pulped	18.55 ± 3.52	9.87 ± 2.29	13.65 ± 3.46	16.61 ± 2.85
	Pulped	15.72 ± 3.11	8.32 ± 1.85	11.26 ± 2.24	12.68 ± 2.39

MGT: Mean Germination Time.

Seed germination was constantly high (88.46 ± 8.59 per cent) in case of de-pulped seeds in undisturbed open canopy forest. Second highest germination (82.45 ± 7.46 per cent) was observed in case of de-pulped seeds in close canopy disturbed forest. This was followed by de-pulped seeds under open canopy disturbed forest (70.23 ± 6.81 per cent) and de-pulped seeds under close canopy undisturbed forest (65.41 ± 6.23 per cent). In the remaining seed state and site conditions seed germination was nearly equal to 31-42 per cent and all these seed plots had seeds with intact pulp. Thus, the seed state *i.e.* seeds without or with intact pulp, acted as most important determinant of seed germination. Along with the seed state the micro habitats (disturbed vs. undisturbed forest and close canopy vs. open canopy) also regulate the germination percentage of seeds. Analysis of variance of seed germination between seed state (pulped vs. de-pulped), canopy cover (open vs. close) and site condition (disturbed vs. undisturbed) was found significant (Table 3.3).

Table 3.3: Analysis of Variance (ANOVA) of Seed Germination, Survival, MGT and Extension Growth between Seed State, Canopy Cover, Site Condition and their Interactive Effect of *Careya arborea* on the Forest Floor of SBR

Parameter	*Seed State (Pulped and de-pulped)*	*Canopy Cover (Open and close)*	*Site Condition (Disturbed and undisturbed)*	*Seed State x Canopy Cover x Site Condition*
Germination (per cent)	91.36***	4.84*	5.36*	4.66*
MGT	63.55***	4.43*	5.11*	12.67**
Seedling survival (per cent)	60.18***	5.12*	21.6***	43.88***
Extension growth (Shoot length in cm)	73.75***	12.68**	9.06**	11.15**

***: Significant at P ≤ 0.001: **: Significant at P≤ 0.01: *: Significant at P ≤ 0.05.

The features like site condition, canopy cover and the seed state also have interactive significant effect on seed germination of *C. arborea* (Table 3.3). In contrast to total seed germination, the rate of germination, as evidenced by mean germination time (MGT), was less in de-pulped seeds (Table 3.2) than pulped ones indicating significant differences in the germination behaviour of two states of seeds. Fruit pulp has been reported as inhibitor of seed germination in a large number of species (Yagihasi *et al.*, 2000: Shrama *et al.*, 2009). Mahmud *et al.* (2005) have made similar observation on eight multipurpose tree species, which are companion species of *C. arborea* grown in tropical forest covers.

Impact of canopy cover on seed germination in disturbed and undisturbed forest was assessed. Results of the study indicated that impact of canopy cover on seed germination was different in the disturbed and undisturbed forest. For example, canopy cover regulate seed germination in the undisturbed forest as the percentage of seed germination was significantly higher under open canopy than in close canopy whereas the seed germination in the disturbed forest was in the reverse order under open canopy and close canopy. Thus the effect of canopy cover on seed germination of *C. arborea* was an important aspect (Sharma *et al.*, 2009). In the present investigation such trend may arise due to significant difference in the densities of upper storey vegetation between the forest stands (Table 3.1). The variation in ground flora and litter thickness on the forest floor between disturbed and undisturbed forest stands may also have significant impact on seed germination (Table 3.1). In addition to these facts availability of light on the forest floor is regarded as one of the important determinants of seed germination (Khan and Shankar, 2001).

As compared to average seed germination, average seedling survival on the forest floor was considerably low (18.56 ± 3.23 per cent to 70.23 ± 6.43 per cent) and varied with micro-environment conditions (Table 3.2). Similar to seed germination, the seedling survival was also more in case of de-pulped seeds than pulpy seeds.

Maximum percentage of seedling survival (70.23 ± 6.43 per cent) was found in de-pulped seeds than in pulped seeds under open canopy undisturbed forest and the differences were statistically significant for all other 7 conditions. Overall seedling survival was higher in undisturbed forests than in disturbed forests (Table 3.2). Seedling survival on forest floor is governed by the availability of light, water and nutrients (Kitajima, 2007). In addition, plant pathogens are also known to affect the micro-environment and survival of seedlings (Fenner and Kitajima, 1999). Increase in light and nutrients in open canopy of disturbed forest stimulate the growth of the competing shrubs and fast growing herbaceous plants (Pages *et al.,* 2003). In general varying micro-environment had a significant effect on survival of seedlings (Table 3.3).

The extension growth of seedlings on the forest floor also varied with the micro-environments. Mean height of seedlings from pooled data after 12 months across the micro-environments were different. Maximum mean height of the seedlings was recorded under open canopy undisturbed forest in case of de-pulped seeds after 12 months (Table 3.2). The result also indicated that the mean height of the seedlings in open canopy disturbed forest and close canopy undisturbed stands of the two seed states was comparatively very low as compared to the undisturbed open canopy and disturbed close canopy micro-environments. Low extension growth of seedlings in open canopy disturbed micro-habitat could be due to biotic interferences in terms of grazing activity of domestic animals though the sites included under this micro-habitat are located close proximity to human habitations. However, in undisturbed close canopy sites, low extension growth of seedlings may be the result of trampling of seedlings by small herbivores. Kitajima (2007) also concluded that complex interactions between herbivores and seedlings mediate or regulate the growth and survival of the plants on the forest floor.

From the foregoing account it is evident that heterogeneity in micro-habitat on the forest floor caused significant variation in germination and survival of *C. arborea* seedlings. Presence or absence of fruit pulp decisively determined seed germination and survival. Though presence of pulp on the seed hampered the seed germination and survival in most of the seed plots, as high as 50 per cent seeds with intact seed pulp could germinate in undisturbed open canopy forest. This could be due to presence of small consumers of the fruit pulp including micro-fauna on the forest floor that may facilitate removal of fruit pulp and emergence of seedlings.

Conclusion

The study reveals that best regeneration of *C. arborea* was in the undisturbed forest with open canopy cover provided the seeds are de-pulped. Such forests had highest seed germination, highest degree of seedling survival, shorter germination time and greater extension growth of the seedlings as compared to rest of the micro-habitats. *C. arborea* is one of the important species in the forest covers of Similipal as well as different tropical deciduous forest covers of Odisha required some extent of canopy openness without any anthropogenic disturbance and light availability on the forest floor for its better germination, survival and extension growth of the seedlings. In addition to its high medicinal importance, this species provided excellent habitat for large number of small and large herbivores and mammals, which facilitate its regeneration in turn. Multiplication of this species in degraded habitats of Similipal Biosphere Reserve would not only provided additional resources for the local people in terms of ethno-medicinal activities but also improve the habitat of several faunal species, thereby maintaining the local biodiversity.

Acknowledgements

We are thankful to CSIR, Govt. of India for providing financial assistance to carry out the research work and also thankful to the Field Director and other Officials of Similipal Biosphere Reserve for providing logistic support in the field for data collection and maintenance of seed plots.

References

Behera, K.K., Sahoo, S. and Patra, S. (2008). Floristic and medicinal uses of some plants of Chandaka denuded forest patches of Bhubaneswar, Orissa, India. Ethnobotanical Leaflets, 12: 1043 -1053.

Bhardwaj, J. and Seth, M. K. (2017). Edible wild plant resources of Bilaspur, Hamirpur and Una districts of Himachal Pradesh, India. International Journal of Botany Studies, 2: 9-17.

Biswal, A.K., Thatoi, H.N. and Sahu, D. (2005). Floral diversity in Chandaka wildlife sanctuary. Journal Economic taxonomic Botany, 29 (2): 385-402.

Czabat, F.J. (1962). Germination values an index combining speed and Completeness of pine seed germination. Forest Science, 8: 386-396.

Davies, S.J. (2001). Tree mortality and growth in 11 sympatric *Macaranga* species in Borneo. Ecology, 82: 920-932.

Fenner, M. and Kitajima, K. (1999). Seed and seedling ecology. In: Pugnaire, F.I. and Valladares, F. (Eds.) Functional Plant Ecology. Marcel Dekker Inc., USA. pp. 589-619.

Khan, M.L. and Shankar, U. (2001). Effect of weight, light regime and substratum microsite on germination and seedling growth *of Quercus semiserrata* Roxb. Tropical Ecology, 42(1): 117-125.

Kitajaima, K. (2007). Seed and seedling ecology. In: Pugnaire, F.I. and Valladares, F. (Eds.) Functional Plant Ecology. 2nd Edn., CRC Press, Tylor and Francis Group. pp. 549-580.

Kumar, V. (2016). Effect of pre-sowing seed treatment on germination and seedling growth of *Terminalia bellirica* (Gaertn.) Roxb. Indian Journal of Ecology, 43: 233-238.

Mahmud, S., Rafiqul, A.T.M and Mohiuddin, M. (2005). Germination behavior and initial growth performance of eight multipurpose tree species. International Journal of Agriculture and Biology, 7(4): 539-542.

Mishra, R.K., Upadhyay, V.P. Mohapatra, P.K., Bal, S. and Mohanty, R.C. (2006). Phenology of species of moist deciduous forest sites of Similipal biosphere reserve. Lyonia, 11: 5-17

Mishra, R.K., Upadhyay, V. P. and Mohanty, R. C. (2008). Vegetation ecology of the Similipal biosphere reserve, Odisha, India. Applied Ecology and Environmental Research, 6(2): 89-99.

Pacala, S.W., Canham, C.D., Silander, J.A. and Kobe, R.K. (1994). Sapling growth as a function of resources in a north temperate forest. Canadian Journal of Forest Research, 24: 2172-2183.

Pages, J.P., Pache, G., Joud, D., Magnan, N. and Michalet, R. (2003). Direct and indirect effects of shade on four forest tree seedlings in the French Alps. Ecology, 84: 2741-2750.

Pokhriyal, P., Uniyal, P., Chauhan, D.S. and Todaria, N. P. (2010). Regeneration status of tree species in forest of Phakot and Pathri Rao watersheds in Gharwal Himalaya. Current Science, 98: 171-175.

Santra, A.K., Pan, S., Samanta, A.K., Das, S. and Halder, S. (2008). Nutritional status of forage plants and their use by wild elephants in South West Bengal, India. Tropical Ecology, 49: 251-257.

Satish, K.B. N., Vrusabendra, S.B.M., Kamal, K.G. and Behera, G.M. (2010). Review on *Careya arborea* Roxb. International Journal of Research in Ayurbeda and Pharmacy, 1: 306-315.

Sharma, G., Nautiyal, B.P. and Nautiyal, A.R. (2009). Seedling emergence and survival in *Cinnamomum tamala* under varying micro-habitat conditions: conservations implications. Tropical Ecology, 50: 201-209.

Yagihasi, T., Hayashida, M. and Miyamoto, T. (2000). Inhibition by pulp juice and enhancement by ingestion on germination of bird dispersal of *Prunus* seed. Journal of Forest Research, 5: 213-215.

4

Phytochemical Screening and Evaluation of Antioxidant and Carbohydrate Metabolism Enzyme Inhibition Potential of a Mangrove Plant, *Avicennia officinalis*

***D. Samantaray*[1]*, S.K. Das*[1*] *and H.N. Thatoi*[2]**

[1]Department of Biotechnology,
College of Engineering and Technology, Bhubaneswar – 751 003, Odisha
[2]Department of Biotechnology, North Orissa University,
Baripada – 757 003, Odisha
****E-mail: das.swagat@gmail.com***

ABSTRACT

The present work is aimed at evaluating the antioxidant and carbohydrate metabolizing enzyme inhibition potential of Avicennia officinalis, a medicinal mangrove plant abundantly found in Odisha coast. Petroleum ether and aqueous bark extracts of the plant were screened for antioxidant activity using 2, 2-diphenyl-1-picrylhydrazil (DPPH): 2,2-Azino-bis-(3-ethylbenzothiazoline-6-sulfonic acid (ABTS) superoxide and hydrogen peroxide (H_2O_2) scavenging assay. The carbohydrate metabolism enzyme inhibition potential was studied by using α-amylase and α-glucosidase inhibition studies. Screening of phytochemicals revealed the presence of tannins, steroids, terpenoids, saponins, flavonoids and phenols in petroleum ether and aqueous bark extracts of A.officinalis which could be attributed to the antioxidant and carbohydrate metabolizing enzyme inhibition activities of this plant. Further, the petroleum ether and aqueous bark extracts of A.officinalis also demonstrated in vitro antioxidant and *carbohydrate metabolizing enzyme inhibition activities. However, a detailed investigation is needed to identify and characterize the active principle from this plant source that can contribute to discovery of novel molecule(s) to combat the diseases related to oxidative stress including diabetes mellitus.*

Keywords: *Avicennia officinalis, Antioxidant, Glycolytic enzymes, Phytochemical.*

Introduction

Mangrove forests are one amongst the world's most productive tropical ecosystems. Mangrove plants are highly potential because they always thrive under stress conditions which enhance the production of different compounds for their survival. Although medicinal uses of a number of mangroves species have been reported, yet very few studies have been undertaken so far to evaluate the antioxidant and carbohydrate metabolizing enzyme inhibition potentials of Indian mangroves. *Avicennia offiicnalis,* is a tropical mangrove plant belonging to family *Acanthaceaé* and is found far away from salt water, unlike other mangrove species. The bark of this plant is traditionally used as contraceptive, astringent, diuretic, antiulcer, treatment for snake bites, rheumatism, small pox, skin diseases, hepatitis, leprosy, antitumor, bronchial asthma, antibacterial, gastroprotective, aphrodisiac, boils and abscesses (Sumithra *et al.,* 2011: Thirunavukkarasu *et al.,* 2011: Das *et al.,* 2018). The pharmacological studies have also reported the antimicrobial, antiulcer, anti-nociceptive, anti-inammatory, anticancer and antioxidant activities of different leaf and bark extracts of *A. offcinalis* (Bakshi and Chaudhuri, 2014; Das *et al.,* 2018). The present study reports the antioxidant and carbohydrate metabolism enzyme inhibition activities of bark extracts of *A. offiicnalis* along with the screening of its phytochemical constituents which might be responsible for such bioactivity.

Materials and Methods

Collection of Plant Sample

Fresh barks of *A. officinalis* were collected from the Mahanadi delta region of Odisha coast. The specimen was identified by Dr. Prasanna Kumar Nayak, ICZMP, Forest Department, Govt. of Odisha. The barks were cleaned and air dried in shade. The dried plant materials were pulverized.

Extraction Process using Maceration

Extraction was carried out with maceration technique. 10 g of dried sample was taken and mixed separately with petroleum ether and water and kept under maceration for 72 h. The filtered liquid obtained from maceration was transferred to a petriplate and allowed to evaporate completely. The obtained crude extracts *i.e.,* petroleum ether (PEB) and aqueous (AQB) were stored in airtight container in refrigerator below 4 °C for further studies. The final test samples were prepared by taking 20 mg of crude sample and dissolving it in 1 ml of dimethyl sulfoxide (DMSO) and stored at 4 °C for further use.

Qualitative Phytochemical Analysis

The crude extracts were subjected to preliminary qualitative phytochemical analysis by following standard protocols for identification of different phytochemical constituents (Harborne, 1998).

Test for Steroids

In a test tube, 20 mg of ethanolic plant extract was taken and to it 2 ml of chloroform was added. Then H_2SO_4 was added drop wise. Formation of a reddish brown color gives positive result to the test.

Test for Saponins

About 20.0 mg of ethanolic plant extract was vigorously shaken with 2 ml of water in a test tube. The persistent foam indicates the presence of saponins.

Test for Phenols

1.0 mg ethanolic plant extract was dissolved in 1 ml of methanol in a test tube and to it 2 ml of distilled water was added followed by few drops of 10 per cent aqueous $FeCl_3$. Formation of a blue/green color sample will give a positive result to the test.

Test for Terpenoids

To 1.0 mg of ethanolic plant extract, 2 ml of chloroform was added followed by 1ml of conc. H_2SO_4. Reddish brown color sample will give a positive result to the test.

Test for Tannins

About 0.5 gm of ethanolic plant extract was stirred with about 10 ml of distilled water and then filtered. Few drops of 1 per cent ferric chloride solution were added to 2 ml of the filtrate. Occurrence of a blue-black, green or blue-green precipitate indicates the presence of tannins.

Test for Flavonoids

0.1 ml of alcoholic extract was taken in a test tube and 5 to 10 drops of diluted HCl was added to it. Traces of zinc/magnesium was added to the solution, mixed and boiled for few minutes. A reddish pink/dirty brown color gives a positive result to the test.

Total Phenol Content

0. 25 ml of plant extract (1mg/ml) concentration was prepared in methanol. 1.25 ml of a 10 fold Folin phenol reagent followed by 7.5 per cent Na_2CO_3 was introduced to the test tube containing sample. The test tubes were covered with parafilm and allowed to stand for 30 mins at room temperature. Then absorbance was taken at 760 nm. Standard curve was done by taking gallic acid at different concentrations (10, 25, 50, 80,100 μg/ml).

Total Flavonoid Content

1mg/ml of plant sample (dissolved in DMSO) was taken and 1.5ml ethanol (95 per cent), 0.1ml aluminium chloride (10 per cent) and 0.1ml potassium acetate (1M) were added. The sample was then incubated for 30 minutes at room temperature followed by absorbance at 415nm using UV-Vis spectrophotometer. Similarly, standard curve for Querctin was determined for 6.5, 12.5, 25, 50 µg/ml.

Total Tannin Content

The tannins were determined by follins –ciocalteu method.About 0.1 ml of sample extract was taken in volumetic flask. Then 7.5ml of distilled water and 0.5 ml of folin phenol reagent, 1ml of 35 per cent Na_2CO_3 solution and dilute to 10 ml with distilled water. The total mixture was shaken well and kept at room temperature for 30 min. The standard solutions of Gallic acid (20, 40, 60, 80 and 100 µg/ml) were prepared in the same manner of follin-ciocalteu method as a set of reference. Absorbance for test and standard solutions were measured against the blank at 725 nm.The tannin content was determined in terms of mg of GAE/g of extract.

Antioxidant Studies

DPPH (2, 2-diphenyl-1-picrylhydrazil) Radical Scavenging Activity

Free radical scavenging activity was determined by using DPPH, as described by Brand-Williams *et al.* (1995) with slight modification. 2 ml of DPPH solution (0.1 mM DPPH dissolved in methanol), 50 µl of the plant extract of different concentrations (0.05, 0.1 and 0.15 mg/ml) were added and 150 µl DMSO were added. The reaction mixture was shaken and incubated in the dark at 37 °C for 30 min. The absorbance was read at 517 nm against methanol as a blank by using UV-Vis spectrophotometer.

ABTS Radical Scavenging Study

2,2-Azino-bis-(3-ethylbenzothiazoline-6-sulfonic acid (ABTS) free radical-scavenging activity was done following the method of Rath *et al.* (2011) with some modifications. Briefly, ABTS (75 mM) and potassium persulfate (1.225 mM) were mixed overnight and the mixture was diluted with ethanol to 10-fold before use. The samples of different concentrations (0.05, 0.1 and 0.15 mg/ml) and diluted ABTS reaction mixture were incubated at room temperature for 60 min. The scavenging ability was evaluated by measuring the absorbance at 734 nm.

Super-oxide Dismutase Scavenging Activity

The superoxide scavenging potential was evaluated following the method of Kono (1978) with minor modifications. 50 µl of plant extract of different concentrations (0.05, 0.1 and 0.15 mg/ml), 1 ml Na_2CO_3 (50 mM) 0.4 ml NBT and 0.2 ml of EDTA were added. The reaction was initiated by adding 0.4 ml of hydroxyl amine hydro chloride (3 mM). Zero time absorbance was taken at 560 nm using spectrophotometer followed by recording the absorbance after 15 min at 25 °C. The control was simultaneous run without plant extract.

Hydrogen Peroxide Activity

The ability of the extracts to scavenge hydrogen peroxide (H_2O_2) was determined by the reported method (Rath *et al.*, 2011). Hydrogen peroxide (43 mM) was prepared in phosphate buffered saline (pH 7.4). Ascorbic acid as positive control and extract solutions were prepared at concentrations of 0.05, 0.1 and 0.15 mg/ml. Aliquots of standard or extract solutions (3.4 ml) were added to 0.6 ml of H_2O_2 solution. The reaction mixture was incubated at room temperature for 10 min, and the absorbance was determined at 230 nm.

Carbohydrate Metabolism Enzyme Inhibition Study

α-Amylase Inhibition Assay

α-amylase inhibitory assay was carried out with different solvent extracts of bark as per Ali *et al.* (2006). To each of the test tube, 100 µl crude plant extract of 0.1, 0.5 and 1 mg/ml concentrations were added. Then 100 µl distilled water followed by 100µl enzyme (4 units/ml α-amylase) was added and mixed. This reaction was carried out at 25°C for 5 min followed by addition of 200 µl starch (0.5 per cent). It was again re-incubated at 25°C for 5 min. For terminating the reaction 200 µl DNS (dinitro salicylic acid) was added and incubated in water bath at 85°C for 15 min. After incubation the mixture was diluted with 1800 µl distilled water. The α-amylase inhibitory activity was determined by measuring absorbance at 540 nm using UV- Vis spectrophotometer.

α-Glucosidase Inhibition Assay

α–Glucosidase inhibitory assay was carried out using the method of Apostolidis *et al.* (2007) with slight modification. To each of the test tube, 50 µl crude plant extract of 0.1, 0.5 and 1 mg/ml concentrations were added. Then 125 µl of enzyme solution (1units/ml α-glucosidase) and 150 µl of phosphate buffer (pH-6.9) was added and mixed. This was carried out at 37 °C for 10 min incubation. 100 µl of PNPG (3mM) as a substrate was added to the reaction mixture to start the reaction and the reaction was re-incubated at 37 °C for 20 min and stopped by adding 2 ml of Na_2CO_3. α glucosidase activity was determined by measuring the p-nitro phenol release from PNPG at 405 nm using UV-Vis spectrophotometer.

Statistical Analysis

All measurements were performed in triplicate and the results of the data were expressed as mean (± SD).

Results and Discussion

Phytochemical Screening

The phytochemicals act as remedy for various ailments and have been known for centuries which form the basis for their use in ethnomedicinal practices around the world. Preliminary qualitative phytochemical screening for presence of secondary metabolites in *A. officinalis* bark extracts was carried out and the results are shown in Table 4.1. The result depicted that both the extracts contains all the secondary metabolites such as steroids, terpenoids, saponins, phenols, flavonoids and tannins which were being studied in the present study. The quantitative phytochemical study demonstrated that amongst the two extracts AQB exhibited higher phenol, flavonoid and tannin content than PEB extract. The phenol, flavonoid and tannin content in AQB extract were recorded as 22.57 GAE in mg/g bark, 33.7 QAE in µg/g bark and 32.2 GAE in mg/g bark, respectively (Table 4.2). The present study is in agreement with the earlier study that demonstrated the presence of different secondary metabolites in the *A. officinalis* bark extracts (Das *et al.*, 2018; Thatoi *et al.*, 2016). Nevertheless, the results of the present study supported the ethnomedicinal uses of *A. officinalis*.

Table 4.1: Preliminary Phytochemical Screening of Extracts of *A. officinalis*

Sample	*Steroids*	*Saponins*	*Phenols*	*Terpenoids*	*Flavonoids*	*Tannins*
PEB	+	+	+	+	+	-
AQB	+	+	+	+	+	+

Table 4.2: Quantitative Phytochemical Assay of all the Extracts of *A. officinalis*

Sample	*Total Phenol (GAE in mg/g DW)*	*Total Flavonoid (QAE in µg/g DW)*	*Total tannin (GAE in mg/g DW)*
PEB	2.31 ±0.45	18.0 ±1.48	27.26 ± 2.5
AQB	22.57 ±1.02	33.7 ±1.47	32.2 ± 2.69

Antioxidant Activities

In the present study DPPH, ABTS, superoxide and H_2O_2 scavenging assays were used for the determination of antioxidant activity of different extracts. The concentrations of 0.05, 0.1 and 0.15 mg/ml of the two extracts were analyzed. Antioxidant capacities of the extracts were expressed in terms of IC_{50} value of the extracts and low IC_{50} value corresponds to a high antioxidant capacity.

Table 4.3: Bioactivities of Bark Extracts of *A. officinalis* (Expressed as IC_{50} values in mg/ml)

Bioactivity Assay	*PEB*	*AQB*
α-Amylase inhibition	3.03	2.71
α-Glucosidase inhibition	4.04	1.32
DPPH scavenging	0.22	0.17
ABTS scavenging	0.52	0.25
Superoxide scavenging	0.61	0.38
H_2O_2 scavenging	0.96	0.59

DPPH is a stable free radical, which has been widely used for determining free radical scavenging activity. When DPPH radical accepts an electron from an antioxidant compound, the DPPH becomes colourless from deep purple colour, which can be quantitatively measured from the changes in absorbance at 517 nm.

The free radical scavenging ability of the two solvent extracts (PEB and AQB) of *A. officinalis* were measured by stable free radical DPPH (Table 4.3; Figure 4.1a). The present study showed that both the extracts of *A. officinalis* could scavenge DPPH radical in a dose dependent manner. However, the AQB demonstrated better DPPH scavenging potential with IC_{50} value of 0.17 mg/ml. The DPPH scavenging potential of AQB varies between 30 to 47 per cent.

$ABTS^+$ scavenging assay is a widely accepted model to determine the total antioxidant activity of plant extracts. This assay is based on the inhibition of the absorbance of radical cation, $ABTS^+$ by antioxidants. The principle behind the assay involves the reaction between ABTS and potassium per sulphate to produce the ABTS radical cation ($ABTS^+$) which is a blue green chromogen. In the presence of antioxidants the coloured radical is converted back to colourless ABTS (Sreejayan and Rao, 1996). The present study showed that both the extracts of *A. officinalis* could scavenge ABTS radical. However, AQB showed better ABTS scavenging potential than PEB extract. The $ABTS^+$ scavenging activity of AQB varied between 11 to 30 per cent with IC_{50} value of 0.25 mg/ml (Figure 4.1b).

Superoxide radical is known to be very harmful to cellular components and acts as a precursor of the more reactive oxygen species. It contributes to tissue damage and various diseases (Hazra *et al.*, 2008). In superoxide radical scavenging assay, superoxide derived from hydroxylamine hydrochloride reduces NBT to a chromogenic complex formazan. Colour reduction indicated the consumption of

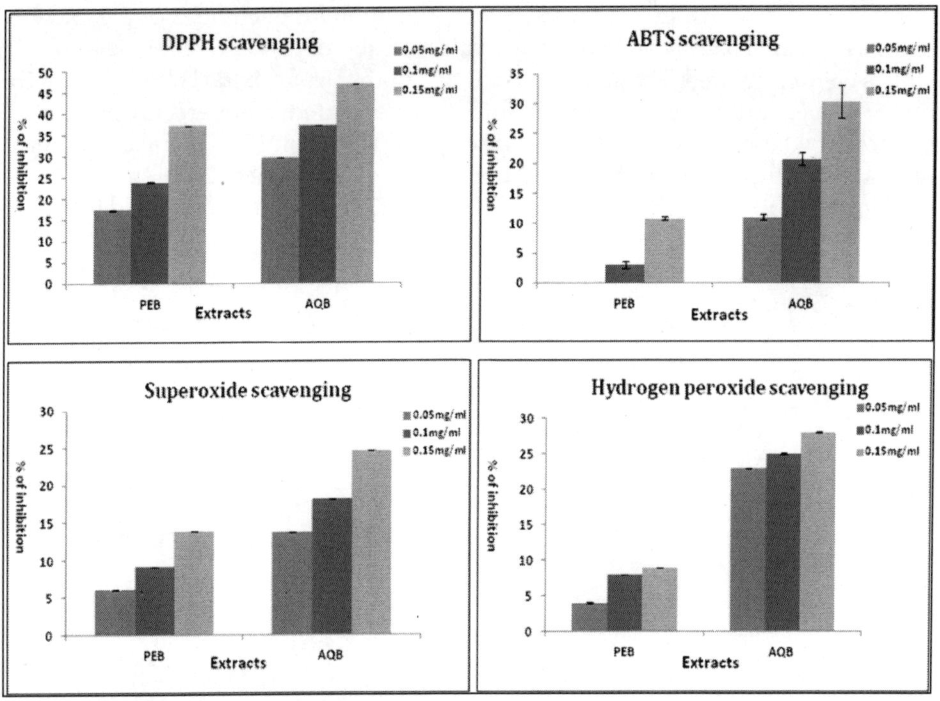

Figure 4.1: DPPH (a), ABTS (b), Superoxide (c) and Hydrogen Peroxide (d) Scavenging Activities of Bark Extracts of *A. officinalis*.

superoxide anion by antioxidants. The present study showed that both PEB and AQB extracts of *A. officinalis* showed superoxide scavenging potential in a dose dependent manner (Table 4.2; Figure 4.1c). The AQB extract possessed the highest scavenging activity of 30 per cent at highest concentration 0.15 mg/ml of the extract.

Hydrogen peroxide (H_2O_2) is a by-product of respiration and is made in all living cells. It is harmful and must be removed as soon as it is produced in the cell. The generation of even low levels of H_2O_2 in biological systems may be important. Hydrogen peroxide scavenging activity depends upon the phenolic content of the extract, which can donate electrons to H_2O_2 and thereby neutralizing it in to water. The PEB and AQB extract were capable of scavenging H_2O_2 in a dose dependent manner (Figure 4.1d). The AQB showed better scavenging potential than PEB. At a concentration of 0.15 mg/ml of the AQB showed 28 per cent scavenging activity with IC_{50} value of 0.59 mg/ml.

Carbohydrate Metabolism Enzyme Inhibition Activity

Management of the blood glucose level is an essential approach in the control of diabetes complications. Inhibitors of carbohydrates hydrolysing enzymes (α-amylase and α-glucosidase) have been helpful as oral hypoglycemic medicines for the control of hyperglycemia exclusively in patients with type-2 diabetes mellitus (Das *et al.*, 2016). Inhibition of these enzymes holds of carbohydrate digestion and extends the total carbohydrate digestion time, leading to a decrease in the rate of glucose absorption and therefore reducing the postprandial plasma glucose rise (Wild *et al.*, 2004). In the present work, the inhibition activities of the extracts obtained from *A.officinalis* were investigated on the α -amylase enzyme and IC_{50} values were calculated. The study showed that both the two extracts (PEB and AQB) of *A.officinalis* could inhibit α-amylase in a dose dependent manner. However, AQB extract showed better α-amylase inhibition potential than PEB. The AQB showed highest activities of 20 per cent at 1.0 mg/ml with IC_{50} value of 2.71 mg/ml (Table 4.3; Figure 4.2a). The *in vitro* α-glucosidase inhibitory study demonstrated that both

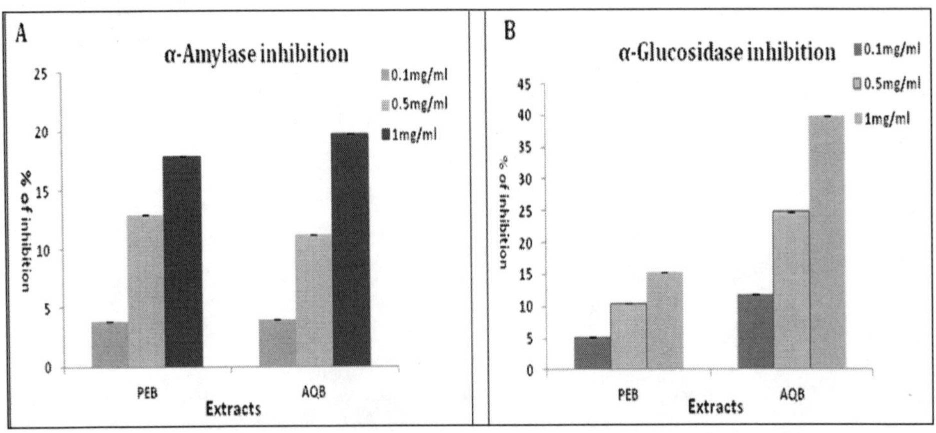

Figure 4.2: **α-Amylase (A) and α-Glucosidase (B) Inhibitory Activities of Bark Extracts of *A.officinalis*.**

PEB and AQB extracts of *A. officinalis* have potential of α-glucosidase inhibitory properties. The inhibitory activity was compared on the basis of their IC_{50} values. It was observed that all the extracts displayed inhibitory activity in a dose-dependent manner (Table 4.3; Figure 4.2b). However the highest activity was observed in AQB extract that showed 40 per cent with IC_{50} value of 1.32 mg/ml.

Conclusion

The findings of the present study supported the traditional use of *A. officinalis* but the present work on *A. officinalis* needs further extension to isolate and characterize the active principle(s) for its pharmaceutical applications in treatment of oxidative stress and its associated complications including diabetes mellitus.

Acknowledgements

The authors are thankful to PCCF (Wildlife), Govt. of Odisha, for giving the necessary permission for the research work. The authors are also thankful to the DFO, Rajnagar, Odisha and their field staff for their kind help and cooperation during the field study.

References

Ali, H., Houghton, P.J. and Soumyanath, A. (2006). α-amylase inhibitory activity of some Malaysian plants used to treat diabetes: with particular reference to *Phyllanthus amarus*. Journal of Ethnopharmacology, 107: 449-455.

Apostolidis, E., Kwon, Y. and Shetty, K. (2007). Inhibitory potential of herb, fruit, and fungal-enriched cheese against key enzymes linked to type 2 diabetes and hypertension. Innovative Food Science and Emerging Technologies, 8: 46-54.

Bakshi, M. and Chaudhuri, P. (2014). Antimicrobial potential of leaf extracts of ten mangrove species from Indian Sundarban. International Journal of Pharmacy and Biological Science, 5(1): 294-304.

Brand-Williams, W., Cuvelier, M.E. and Berset, C. (1995). Use of free radical method to evaluate antioxidant activity. LWT-Food Science and Technology, 28: 25-30.

Das, S.K., Samantaray, D., Mahapatra, A., Pal, N., Munda, R. and Thatoi, H.N. (2018). Pharmacological activities of leaf and bark extracts of a medicinal mangrove plant *Avicennia officinalis* L. Clinical Phytoscience, 4: 1-10.

Das, S.K., Samantaray, D. and Thatoi, H.N. (2016). Antidiabetic potential of mangrove plants: a review. Frontier in Life Sciences, 9: 75-88.

Harborne, J.B. (1998). Phytochemical methods: A guide to modern technique of plant analysis, Champman and Hall, London. Hazra, B., Biswas, S. and Mandal N. (2008) Antioxidant and free radical scavenging activity of *Spondias pinnata*. BMC Complementary and Alternative Medicine, 8: 1-10.

Kono, Y. (1978). Generation of superoxide radical during autoxidation of hydroxylamine and an assay for superoxide dismutase. Archives of Biochemistry and Biophysics, 186: 189-195.

Rath, S., Patra, J.K., Mohapatra, N., Mohanty, G. and Dutta, S. (2011). *In vitro* antibacterial and antioxidant studies of *Croton roxburghii* L., from Similipal Biosphere Reserve. Indian Journal of Microbiology, 51: 363-368.

Sreejayan, N. and Rao, M.N. (1996). Free radical scavenging activity of curcuminoids. Arzneimittelforschung, 46: 169-171.

Sumithra, M., Anbu, J., Nithya, S. and Ravichandiran, V. (2011). Anticancer activity of methanolic leaves extract of *Avicennia offcinalis* on ehrlich ascitis carcinoma cell lines in rodents. *International Journal of PharmTech Research,* 3: 1290-1292.

Thatoi, H.N., Samantaray, D. and Das, S.K. (2016). The genus *Avicennia,* a pioneer group of dominant mangrove plant species with potential medicinal values: a review. Frontier in Life Sciences, 9: 267-291.

Thirunavukkarasu, P., Ramanathan, T., Ramkumar, L., Shanmugapriya, R. and Renugadevi, G. (2011). The antioxidant and free radical scavenging effect of *Avicennia offcinalis.* Journal of Medicinal Plants Research, 5: 4754-4758.

Wild, S., Roglic, G., Green, A., Sicree, R. and King, H. (2004). Global prevalence of diabetes Estimates for the year 2000 and projections for 2030. Diabetes Care, 27: 1047-1053.

5

Antimicrobial and Phytochemical Activity of Leaf Extracts of *Nyctanthes arbor-tristis*

S.B. Ojha, S. Roy and G. Dhangadamajhi*

Department of Biotechnology, North Orissa University, Baripada – 757 003, Odisha

****E-mail: gunarmrc@gmail.com***

ABSTRACT

Nyctanthes arbor-tristis L. commonly known as Harsinger or Night jasmine, is a well known medicinal plant belonging to the family Oleaceae. Different parts of N. arbor-tristis are known to possess various ailments by rural mainly tribal people of India especially in Odisha and Bihar states. In the present study different extracts of the leaf of N. arbor-tristis plant were evaluated for their in vitro antimicrobial activity against certain clinically significant human pathogens like Escherichia coli, Staphylococcus aureus, Bacillus subtilis and Pseudomonas aeruginosa. The result indicated that all of the extracts showed antimicrobial activity inhibiting at least one of the test pathogens. The methanolic, ethyl acetate and chloroform extract displayed considerable zone of inhibitions against all the four pathogens. Among all the extracts chloroform extract showed highest antimicrobial activity (20.3±0.5 mm) against S. aureus and petroleum benzene extract showed least activity (12±1.0 mm) against E. coli. Also phytochemical analysis of leaf extracts of N. arbor-tristis was carried out applying various phytochemical tests. The findings revealed that metabolites extracted from N. arbor-tristis could be an excellent source of antimicrobial agents and can be exploited for further therapeutic application.

***Keywords**: Antimicrobial, Nyctanthes arbor-tristis, Phytochemical.*

Introduction

As several health hazards or side effects are associated with random use of synthetic drugs, now a day most of the world's populations choose herbal drugs for treatment of several diseases. Plants are always a great source of chemical compounds which serve as drugs, and supply newer leads for useful drug synthesis. More than 5,000 phytochemicals have been identified, and many are still unknown. Many plants have been investigated for their beneficial properties like antimicrobial, antioxidant and phytochemical assays. Natural microbial can be derived from different parts of plant (Gordan *et al.*, 2001). Phytochemicals are the secondary metabolites highly reported in plants that have several sub groups possessing various bioactivities such as antioxidants, antimicrobial, antivirus, anticancer *etc.* (Duffy *et al.*, 2001). Undoubtedly the plant kingdom still holds many species of plants containing substance of medicinal value which have yet to be discovered. So, it is necessary to evaluate, in a scientific base, the potential bioactivity of different plants for the treatment of several diseases.

In India many plants are used for the traditional system of medicine from ancient age. *Nyctanthes arbor-tristis* L. commonly known as Harsinger or Night jasmine (Oleaceae) is one of them. It is a native of India, distributed wild in sub-Himalayan region and also found in Indian garden as ornamental plant. Juice of the leaves is used as digestives, antidote to reptile venoms, mild bitter tonic, laxative, enlargement of spleen, diaphoretic diuretic (Nadkarni, 1982; Kirtikar and Basu, 2000) and antimicrobial activities (Khandelwal *et al.*, 1999). The whole plant is used for treatment of cancer, root for fever, sciatica, anorexia; bark as expectorant, leaf for control fever, diabetes and as cholagogue, diaphoretic and anthelmintic. Various extracts of the plant is used to treat arthritis (Saxena *et al.*, 1987; Mathuram and Kundu, 1991, Chetty *et al.*, 2008), malaria, intestinal worms tonic, laxative, antitrypanosomal, anti-inflammatory and antioxidant activity (Omkara *et al.*, 2006; Amarite *et al.*, 2007; Rathee *et al.*, 2007).

However, the contribution of *N. arbor-tristis* has not been validated scientifically and to date, there are only limited studies assessing the antimicrobial activities and phytochemicals screening of *N. arbor-tristis* leaf. The use of leaf against microbes can further be substantiated if more studies are done on their potential. Therefore, it is the purpose of this study to determine the various phytochemicals present in leaf and access the antimicrobial activity against the bacterial pathogens.

Materials and Methods

Collection of Sample

Healthy leaves of *N. arbour-tristis* were collected from different locality of Baripada, Mayurbhanj, Odisha. Those leaves were collected in sterile plastic bags and then processed in laboratory by washing these plant parts under tap water and then with distilled water within 24 hr of collection. These were then allowed to dry under shed for 10-15 days and the dried plant part was pulverized into coarse powder through grinder.

Extraction of Metabolites from Leaves of *N. arbor-tristis*

The pulverized plant parts (leaf) were then used for crude extract preparation using soxhlet apparatus with different solvents with increasing polarity (petroleum benzene, chloroform, ethyl acetate, methanol and aqueous). The crude extracts thus obtained were then allowed to evaporate in room temperature. After complete evaporation the dried plant extract were collected in a sterile vials and kept for further assay.

Antimicrobial Activity of the Plant Extracts

Antimicrobial activity of various solvent extract of leaf was carried out by agar cup method (Perez *et al.*, 1990) against four bacterial pathogens (*Escherichia coli, Bacillus subtilis, Staphylococcus aureus and Pseudomonas aeruginosa,* obtained from Institute of Microbial Technology (IMTECH) Chandigarh, India). Nutrient agar (NA) plates were swabbed with broth culture of respective bacteria. The bacterial suspensions were adjusted with sterile saline to an absorbance of 0.08-0.12. Wells were made in each of these plates using sterile cork borer. Stock solution of each plant extract (petroleum benzene, methanol, chloroform, ethyl acetate and aqueous), prepared at a concentration of 10 mg/ml in DMSO except for the aqueous extract which was prepared with sterile water, were added into the wells (Agar cup method). The plates were then incubated at 37 °C for 18-24 h and after incubation, the diameter of the inhibition zone (mm) was measured. The experiment was repeated thrice and the inhibitory zone size as mean (±SD) was calculated.

Phytochemical Analysis of Extract Prepared from different Plant Parts

Test Performed for the Presence of Phytoconstituents

A. *Test for Alkaloids*

i. Dragendorff's Test

To 1 ml of each of the sample solution taken in a test tube few drops of Dragendorff's reagent (potassium bismuth iodide solution) was added. A reddish brown precipitate was observed indicating the presence of alkaloids.

ii. Mayers Test

To 1ml of each of the sample solution few drops of Meyer's reagent (potassium mercuric chloride solution) was added. A creamish white precipitate was formed indicating the presence of alkaloids.

iii. Hagers Test

To 1 ml of each of the sample solution few drops of Hager's reagent (Picric acid) was added. Yellow precipitate was formed reacting positively for alkaloids.

iv. Tannic Acid Test

When few ml of 10 per cent Tannic acid was added to 1 ml of each sample, a buff colour precipitate was formed giving positive result for alkaloids.

v. Ferric Chloride Test

One drop of **ferric chloride** solution was added to each of the test sample, formation of yellow precipitate was resulted reacting positively for alkaloids.

B. Test for Glycosides

i. Legals Test

When the test samples were treated with pyridine and sodium nitroprusside solution blood red colour appears.

ii. Kellar Kiliani Test

1ml of concentrated sulphuric acid was taken in a test tube then 5ml of extract and 2ml of glacial acetic acid with one drop of ferric chloride were added, reaction shows formation of a blue colour.

iii. Concentrated Sulphuric Acid test

Conc.H_2SO_4 was added to test sample which resulted in appearance of reddish colour.

C. Test for Tannins and Phenolic compounds

i. Ferric Chloride Test

When few drops of ferric chloride were added to sample solution a blackish precipitate appears.

ii. Gelatin Test

When gelatin and water were added to test samples formation of white precipitate was resulted.

iii. Alkaline Reagent Test

When sodium hydroxide solution was added to the sample solution results in the formation of yellow to red precipitate.

iv. Ellagic Acid Test

When 5 per cent glacial acetic acid and 5 per cent sodium nitrite were added to test samples a muddy niger brown colour appears, which is a positive result for phenols.

D. Test for Flavonoids

i. Ferric Chloride Test

To few ml of test samples taken separately, few drops of ferric chloride were added which resulted in the formation of blackish red precipitate.

ii. Shinoda Test

To the test solution few fragements of magnesium ribbon and concentrated hydrochloric acid were added drop wise and reddish to pink colour was resulted.

iii. Alkaline Reagent Test

To few ml of test samples taken separately, few drops of ferric chloride were added which resulted in the formation of blackish red precipitate.

E. Test for Sterol

i. Libermann-Buchard Test

When samples were treated with few drops of acetic anhydride, boiled and few drops of concentrated sulphuric acid from the sides of the test tube were added, shows a brown ring at the junction of two layers and the upper layer turns green which shows the presence of steroids.

ii. Salkowski Test

Few drops of concentrated sulphuric acid were added to the test samples in chloroform, a red colour appears at the lower layer indicates the presence of sterols.

F. Test for Lignins

Furfural Dehyde Test

When furfuraldehyde is added to the test sample a red colour appears indicating the presence of lignin.

G. Test for Saponins

Foam Test

5ml of extract was shaken vigorously to obtain a stable persistent froth. The froth was then mixed with three drops of olive oil and observed for the formation of an emulsion, which indicated the presence of saponins.

Results

Antimicrobial Activity

Different extracts (petroleum benzene, chloroform, ethyl acetate, methanol and aqueous) of *N. arbor-tristis* leaves were subjected for their antimicrobial activity. The result indicated that all of the extracts except aqueous extract showed antimicrobial activity inhibiting at least one of the test pathogens (Table 5.1). The methanolic, ethyl acetate and chloroform extracts displayed considerable zone of inhibitions against all the four pathogens. Among all the extracts chloroform extract showed highest antimicrobial activity (20.3 ±0.5 mm) against *S. aureus* and petroleum benzene extract showed least activity (12.0 ±1.0 mm) against *E. coli*.

Phytochemical Assay

The Phytochemical screening was carried out by different methods for the qualitative identification of different bioactive compounds present in the extracts. Among all the extracts chloroformic extracts had the maximum phytochemicals followed by ethyl acetate extract. The petroleum benzene extract had the least number of phytochemicals documented. While alkaloids were present in all solvent fractions except petroleum benzene and methanol, glycosides were present in all fractions except ethyl acetate and methanol. Phenolic compounds (tannin and phenol) were present mostly in chloroform and ethyl acetate fractions.

Table 5.1: Antimicrobial Activity of Various Leaves Extracts of *Nyctanthes arbor-tristis* against different Bacteria

Name of Extract	*Bacillus subtilis*	*Staphylococcus aureus*	*Escherichia coli*	*Pseudomonas aeruginosa*
Petroleum benzene	13.3±1.5	17.3±0.5	12±1.0	No Zone
Ethyl acetate	13.3±2.0	16.6±1.5	13.6±0.5	14.0±2.0
Methanol	17.3±1.5	16.0±1.0	14.0±1.0	15.0±1.0
Chloroform	19.0±1.0	20.3±0.5	16.6±1.1	14.3±1.1
Aqueous	No zone	No zone	No zone	No zone

*Zone of inhibition in mm.

Table 5.2: Analysis of Phytochemicals from Leaf Extracts of *N. arbor-tristis*

Name of the Compounds with Test Name		*Petroleum Benzene Extract*	*Chloro-form Extract*	*Ethyl Acetate Extract*	*Methanol Extract*	*Aqueous Extract*
Alkaloids	Dragendorff	–	+	–	–	–
	Mayer	–	+	–	–	–
	Hager	–	+	+	–	+
	Tannic acid	–	+	+	–	–
	Ferric Chloride	–	–	+	–	–
Glycosides	Kellar Kiliani	+	+	–	–	–
	Conc. Hydrogen Sulphite	+	+	–	–	–
Tannins	Gelatin	–	+	+	–	–
	Alkaline Reagent	–	+	+	–	–
	Potassium dichromate	–	+	--	--	--
Phenol	Ferric Chloride	–	+	--	–	–
	Ellagic acid	–	+	+	–	–
Flavonoids	Ferric Chloride	–	+	–	–	–
	Alkaline Reagent	–	+	–	+	+
Sterols	Salkowski	+	–	+	+	–
Lignin	Furfuraldehyde	+	–	–	–	–

Discussion

Different parts of *N. arbor-tristis* are widely used for the treatment of several human ailments. Of which, leaf is commonly used for the treatment of fever including malaria, as antidote to reptile venom (Nadkarni, 1982), used in diabetes and other medical conditions by the tribal people. All these bioactive potentials are due to the presence of different types of phytochemicals. Previous studies on phytochemicals screening of *N. arbor-tristis* leaf using three different solvents (i.e methanol, chloroform and water) has revealed for the presence of flavonoids, tannin, saponin, glycosides and sterol (Ladumor *et al.*, 2017). It's a successful prediction of phytochemicals/compounds from plant material is largely dependent on the type of solvent used in the extraction procedure. In the present study we used four different organic solvents and water for the extraction of phytochemicals in a sequential soxhlet extraction methods based on polarity *i.e.*, from less polar solvent to high polar solvent (petroleum benzene, chlororform, ethyl acetate, methanol and finally with highly polar water). Our qualitative screening of phytochemicals results showed the presence of alkaloids, glycosides, phenol, flavonoids, tannin, sterol and lignin in different solvent fractions as presented in Table 5.1. We report for the presence of alkaloid mostly in chloroform and ethyl acetate fractions followed by aqueous fraction which is in contrast to Ladumor *et al.* (2017) who did not find any alkaloid. Using, Chouhan *et al.* (2014) also reported the presence of flavonoid, saponin and terpenoids from the acetone, methanol and water leaf of *N. arbor-tristis* but no alkaloid. Similar to our observation, Agarwal and Goyal (2013), also found alkaloids in chloroform fraction, whereas Geetha *et al.* (2014) documented the presence of alkaloids in both ethanol and aqueous fractions. The above differences in documenting phytochemicals from *N. arbor-tristis* leaf could be due to variations in methods of extractions, solvent used and geographical differences in plant contents of phytochemicals.

Antibacterial activity screening of different solvent extracts of *N. arbor-tristis* against four different bacterial pathogens (*B. subtilis, S. aureus, E. coli* and *P. aeroginosa*) revealed that all of the extracts except aqueous extract showed antimicrobial activity inhibiting the growth of at least one of the test pathogens. The methanolic, ethyl acetate and chloroform extract displayed considerable zone of inhibitions against all the four pathogens. Previous studies on antimicrobial activity of different solvent extracts of *N. arbor-tristis* leaf against different microbial pathogens including some of the pathogens of our study have also shown considerable zone of inhibition of methanolic, ethyl acetate fractions (Chouhan *et al.*, 2014; Paikara *et al.*, 2015; Agarwal and Goyal., 2013;). Among all the extracts in the present study, chloroform extract showed highest antimicrobial activity (20.3 mm) against *S. aureus* and petroleum benzene extract showed least activity (12.0 mm) against E. coli. Similar to our observation, Paikara *et al.* (2015) also have found the maximum zone of inhibition of chloroform extract against *S. aureus*. In contrast, the activity of chloroform extract has been shown to be least against various microbial pathogens in other studies Chouhan *et al.* (2014); Agarwal and Goyal (2013). We obtained highest activity of methanolic extract against *P. aeroginosa*, whereas the methanolic extract has been reported to exhibit the least activity against *P. aeroginosa* in another study (Prasad *et al.*, 2014). The results obtained by Agarwal and Goyal (2013) clearly shows that most of the antimicrobial activity are present in polar fractions than non-polar fractions of *N. arbor-tristis* leaf indicating that the active ingredients could be polar by nature. These discrepancy in antimicrobial activity of different solvent extracts of *N. arbor-tristis* could be attributed to the types of bioactive secondary metabolites they possess and thus responded differently against various test pathogens.

Conclusion

Our study reveals that the leaf of *N. arbor-tristis* is a rich source of phytochemicals with antimicrobial activity and the extraction of potent bioactive compound is largely dependent on the type of solvents and methods adopted in the extraction procedure. Further studies need to be conducted to explore the other bioactivities of these solvent extracts.

References

Agarwal, S. G. and Goyal, S. (2013). *Nyctanthes arbor-tristis* against pathogenic bacteria. Journal of Pharmacognosy and Phytochemistry, 2(3): 124-127.

Amarite, O., Bhuskat, P., Patel, N. and Gadgoli, C. (2007). Evaluation of antioxidant activity of carotenoid from *Nyctanthes arbor-tristis*. International Journal of Pharmacy and Biological Sciences, 2: 57-59.

Chetty, K. M. (2008). Flowering plants of Chittoor. Student Offset Printers, Tirupati India, Edn 1, Andhra Pradesh. 132.

Chouhan, A., Gokhale, M. and Bansal, S. (2014). Antimicrobial potential of Nyctanthes arbor-tristis and isolation of *Colletotrichum gleosporioides* –an endophyte. Journal of Pharmacy Research, 8(8): 1082-1090.

Duffy, C.F. and Power, R.F. (2001). Antioxidant and antimicrobial properties of some Chinese plant extracts. International Journal of Antimicrobial Agents, 17: 527-529.

Geetha, D.H., Jayashree, I. and Rajeswari, M. (2014). Anti- bacterial activity of leaf of *Nyctanthes arbor-tristis* linn. International Research Journal of Pharmaceutical and Applied Sciences, 4(4): 4-6.

Gordan, M.C. and David, J.N. (2001). Natural product drug discovery in the next millennium. Pharmaceutical Biology, 39: 8-17.

Khandelwal, K.R., Kadam, S.S. and Singhama. (1999). Antibacterial activity of the leaves of *Nyctanthes arbor-tristis* Linn. Indian Journal of Natural Products and Resources, 15: 18-20.

Kirtikar, K.R. and Basu, B.D. (2000). Indian Medicinal Plants. Vol.VII, Sri Satguru Publications. New Delhi, 2110-2113.

Ladumor, V.C., Shrisha, R.S., Makwana, C.N., Modi, C.M., Patel, H.B., Patel, U.D. and Bhatt, P.R. (2017). *In vitro* free radical scavenging activity of *Nyctanthes arbor-tristis* L. leaf extracts. The Pharma Innovation Journal, 6(10): 391-393.

Mathuram, V. and Kundu, A.B. (1991). Occurrence of two new ester of 6-Hydroxyloganin in *Nyctanthes arbor-tristis*. Journal of the Indian Chemical Society, 68: 581-584.

Nadkarni, A.K. (1982). Indian Materia Medica. Vol.I, 3rd ed. Popular Prakashan Pvt. Ltd, 857-858.

Omkar, A., Jeeja, T. and Chhaya, G. (2006). Evaluation of anti-inflammatory activity of *Nyctanthes arbor-tristis* and *Onosma echiodes*. Pharmacognosy Magazine, 8: 258-260.

Paikara, D., Singh, S. and Pandey, B. (2015). Antimicrobial activity of *Nyctanthes arbor-tristis*. Indian Journal of Life Science, 5(1): 029-032.

Perez, C., Paul, M. and Bazerque, P. (1990). An antibiotic assay by the agar well diffusion method. Acta Biology and Medicine Experiments, 15: 113-115.

Prasad, M.P. and Shekhar, S. (2014). *In-Vitro* Phytochemical and Antimicrobial activity of *Nyctanthes arbor-tristis* Linn. against human pathogens. International Journal of Pure and Applied Bioscience, 2 (1): 1-5.

Rathee, J.S., Hassarajani, S. and Subrata, C. (2007). Antioxidant activity of *Nyctanthes arbour-tristis* leaf extract. Food Chemistry, 103: 1350-1357.

Saxena, R.S., Gupta, B., Saxena, K.K., Srivastava, V.K. and Prasad, D.N. (1987). Analgesic, antipyretic and ulcerogenic activities of *Nyctanthes arbour-tristis* leaf extract. Journal of Ethnopharmacology, 19: 193- 200.

6

Ethnomedicinal Value and Phytochemical Studies of Traditionally Used Plant Species in Udala, Mayurbhanj, Odisha

*S. Jena and H.K. Sahu**

Department of Zoology, North Orissa University, Takatpur, Baripada – 757 003, Odisha

****E-mail: hks_nou@yahoo.com***

ABSTRACT

Ethnobiology is one of the flourished branches since the Vedic culture to the present time, which includes a variety of information regarding both the medicinal plants and animal species. These traditional species, which were being used in the ancient time have high medicinal potency and periodically used by the people belonging to different groups and communities. Although several studies has been carried out to document the ethnomedicinal values of various medicinal plants in many parts of India, still there are many areas especially in Odisha, where documentations of traditionally used plants and animal species are yet to be explored. Therefore, the present investigation is carried out to document such plant species in different sites of Udala Block, Mayurbhanj district of Odisha along with their bioprospecting evaluation with special reference to phytochemical analysis. In the present investigation a thorough survey in the four regions viz. Kundabai, Radho, Sridamchandrapur and Adapal in Udala block, Mayurbhanj, Odisha was done. Based on survey, 11 medicinal plants species were processed for phytochemical studies. The analysis showed the presence six different phytochemicals in the selected plant species. As the discovery of new drugs and antibiotics is a current approach in the field of Bio-medical science the plant sources could be a substitute to design efficacious medicines fight against the emerge of resistant microbes and their side effects.

Keywords: *Ethnobiology, Medicinal plants, Phytochemicals.*

Introduction

Ever since the birth of mankind there has been a relation between life, disease and plants. One of the earliest human activities has been the study of plants and animals, which were the most common things they could find in their vicinity and used them as a source of food and cure of various human ailments. The majority of plants were suitable as food, and few are poisonous or medicinally important. The knowledge of herbal remedies was transferred through generation as folk medicines. The plant-derived drugs have serving through age as the mainstay in the treatment of human ailments and preservation of health. Since time immortal, plants and herbs have been used in virtually all cultures as a source of medicine. The widespread of herbal remedies and healthcare preparations, as those described in ancient texts such as the Vedas and the Bible, and obtained from traditional or folklore practices, has been traced to the occurrence of natural products with medicinal properties. Many of the homeopathic drugs are crude extracts of plant or plant parts. Some of plant part and their extracts are directly used as medicine for curing some human diseases. Instinctive urges, intuition and the accumulated knowledge have guided man to discover remedies for common ailments from plants as natural sources. The use of traditional medicine and medicinal plants in most developing countries, as a normative basis for the maintenance of good health, has been widely observed and about 80 per cent of the world population relies on herbal medicines (UNESCO, 1996). Furthermore, an increasing reliance on the use of medicinal plants in the industrialized society has been traced for extraction and development of several drugs and chemotherapeutics from these plants as well as from traditionally used rural herbal remedies (UNESCO, 1998). Moreover, in other societies, herbal remedies have become more popular in the treatment of minor ailments, and an account of the increasing costs of personal health maintenance. Indeed the market and public demand has been so great that there is a great risk that many medicinal plants today may face either extinction or less of genetic diversity.

The Indian system of medicines itself is of great importance and is believed to be one of the most ancient, India has been identified as one of the top 12 mega diversity centers of the world with an immensely rich medicinal and aromatic plant population occurring in diverse ecosystems. These medicinal plants are used both for primary health care and for treating chronic diseases such as AIDS, cancer, hepatitic disorders, heart disease, and age-related diseases such as memory loss, osteoporosis, and diabetic wounds *etc.* It is estimated that out of 3.6 lakh plant species available worldwide 40 per cent available in India. Recently, the WHO compiled a list of 10,000 plants are used for phytotherapy in Indian system of medicine (Manavalan and Manian, 2001). In the Indian coded system (Ayurveda, Unani, Siddha, Amchi), Ayurveda currently utilizes as many as 1000 single drugs and over 8000 compound formulations of recognized merit. Similarly 600-700 plants are utilized by other systems such as Unani, Siddha, and Amchi. About 70 per cent of Indian medicinal plants are found in tropical and sub-tropical forest and less than 30 per cent are found in the temperate and high altitude forest (Kumar *et al.*, 1997). These medicinal plant species belong to a wide range of plant types, including trees, herbs, lianes, woody climbers, and twiners (Singh *et al.*, 2003). Several crude drugs are manufactured from different parts of these plants *viz.* roots, rhizomes, stem, bark, seed, wood, leaves, flowers, fruits and whole plants *etc.* (Manavalan and Manian, 2001). In India more than 90 per cent of plant species used by industry are collected from the wild and over 70 per cent of the collection involves destructive harvesting using different parts of the plants (root, stem, bark, wood, whole plant) (Ved and Tandem, 1998). The domestic Indian System of Medicine (ISM) market, comprising Ayurveda, Unani, Siddha, and homeopathy, has been estimated to exceed Rs. 42 billion (US$950 million) and India at present exports herbal medicines and materials to the tune of Rs.5.5 billion (US$124 million) (Padhi and Panda, 2017). The world trade in medicinal plants is estimated to be about US$ 62 billion, with the major players being the European Union at 45 per cent, Asia 17 per cent, and Japan 16 per cent (Ahmad *et al.*, 2006). Conservative estimate suggests that there are more than 2,50,000 species of higher plants existing in this plant, and only a very small percentage of plants have been exhaustively studied of its potential value as a source of drugs. Therefore, natural products are believed to be extremely important source of medicinal agents. However, there are many new approaches are also available such as combinational chemistry and computer based molecular modeling design, but none of them can replace the important role of natural products in drug discovery and development. Due to absent of sufficient modern health care system, particularly in the rural areas people prefer to visit traditional healer (used medicinal plants). Most of these plants, although they are not investigated scientifically, can cure certain infection; it can also be assumed that the major part of traditional therapy involves the use of plant extracts or their active principles. With proper investigations, these may serve as a source of modern drugs. Based on above sighted facts the present study was undertaken to survey different plant species and their phytochemical analysis used the traditional medicine in Udala of Mayurbhanj district, Odisha.

Materials and Methods

Survey

Mayurbhanj district in Odisha is well known for its deep forest, plant diversity and traditional contents. A great percentage of tribal people live here who use plants as medicines for their daily ailments. In the investigation fourregions of Udala block in Mayurbhanj district were selected *viz.* Kundabai, Radho, Sridamchandrapur and Adapal (Figure 6.1). Questionnaire survey was conducted with field study by individual interaction and group discussion at different localities of Udala block, a northeastern part of Mayurbhanj. In total 188 informants were interviewed and the selection of plants was based on their recognition as having sound knowledge related to health care practice. The informants were asked about the ailments cured by plant based medicines and their mode of preparation and administration including right length of medication. According to them, their traditional ethno-botanical knowledge was mainly acquired through parental heritage and experience about medicinal value of plants to heal themselves.

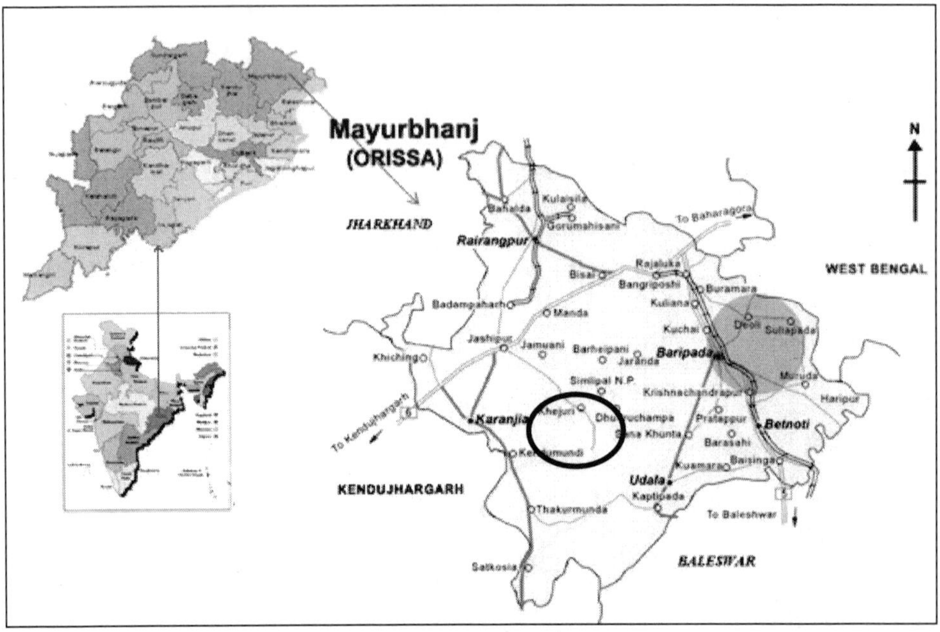

Figure 6.1: Udala Region (Encircled) in the Map of Mayurbhanj District, Odisha.

Phytochemical Analysis

The plant samples like leaves, bark, rhizoids and roots were freshly collected in zip pockets from four selected localities of Udala based on the traditional data. All the collected samples from different plant species were allowed to dry in the oven at 80°C for 2-3 days and then crushed to turn them into powder form and dissolve in the distilled water and then heat to get the plant extract.

The well-prepared plant extracts were subjected to detect the presence of various phytochemicals like proteins, tannins, alkaloids, phenols, tannic acid and quinone by using standard methods (Sabandar *et al.*, 2013).

Results

In the present investigation out of 188 informants interacted in four different sites of Udala 38 were positive respondent. As per the interaction, the informants were found to use11 plant species for different medicinal purpose in their day-to-day life (Figures 6.2a-k). The details of 11plants species along with their phytotherapeutic information are shown in the Table 6.1. The *Vitex negundo* (Begunia) has the maximum use among the eleven plant species followed by *Phyllanthus emblica* (Pani Anla), followed by *Butea monosperma* (Palas) and *Saraca asoca* (Ashoka) and *Azadirachta indica* (Neema) placed at fourth position in the percentage of use. The percentage of use of *Kalanchoe pinnata* (Amar Poi) was found least in Udala.

Table 6.1: Traditional Use of Various Plant Species for different Aliments Used by the People in Udala Areas

Plant Name	*Tridax procumbens*	*Phyllanthus emblica*	*Kalanchoe pinnata*	*Rouvolfia serpentine*	*Terminalia arjuna*	*Saraca asoca*
Parts Used	Leaf	Leaf	Leaf and seed	Root	Bark	Bark
Ailment (Use for disease)	Wound healing, blood clotting, cough and cold, piles	Stop bleeding	Bruises, wound, boils and burns, indigestion	High blood pressure, chest and abdominal pain	Cardiac tonic, hypertension, blood dysentery, chronic fever, fracture, insect sting	Uterine disorders, gynecological disorders, fever, fracture,worms
Used as	Medicine	Medicine	Medicine	Medicine	Medicine	Medicine
Mode of Preparation	Leaves are to grinded well and the paste is applied to the affected areas, the juice extracted from the leaves is to be consumed for other remedies	Paste of the leaves to be applied over the cut to stop bleeding	Crushed leaves with water is used for Dropsy and the oil extracted from seeds is applied over skin to cure diseases	Roots has to be grinded well with palm sugar to be consumed for the remedies	Powder of dried barks are soaked in hot water and after that the extracted water is used and also the remain bark can be used to cure other diseases	The bark has to be dried or blended well with some boiled water and consumed for the remedies
Results	It will take two days to heal the injury completely.	It will take 10 to 20 minutes to stop bleeding	Shows the effectiveness within two days in both the cases	Effectiveness will find within 1 to 2 months if continues periodically at an interval of one day	Effectiveness varies with the aliments to which it is applied *i.e.* 10 to 15 days for cardiac tonic, 2 to 3 months for hypertension, 5 to 7 days for blood dysentery, 2 to 3 days for chronic fever, 1 to 2 week for facture andpain relief within 3 to 4 hrs in insect sting.	It will show the effectiveness within one month for uterine and gynecological disorder and will take one week for fever, fracture and worm infection.

Plant Name	*Azadirachta indica*	*Vitex negundo*	*Pterocarpus marsupium*	*Curcuma longa*	*Butea monosperma*
Parts Used	Leaf, fruit and bark	Leaf and stem	Wood	Rhizome	Bark and seed
Ailment (Use for disease)	Antimicrobial, antifungal, antiseptic, skin and mouth infection, wound healing, menstrual problems	Body pain, cough, skin diseases, fever, pain, ear-ache	Diabetes, skin diseases, arthritis	Jaundice, antiseptic, blood purifier, cough and worm	Dropsy, skin diseases, diabetes
Used as	Medicine and food	Medicine and tooth cleaner	Medicine	Medicine	Medicine
Mode of Preparation	Leaves are consumed as food by cooking, paste made out of its leaves and water is used to cure wounds and fruit is crushed with water and consume for other remedies, when its bark extraction is mixed with garlic it cure some menstrual problems	Stem is used to clean tooth and the paste made by crushed leaves is applied and also boiled to get the extract to be consumed for its prescribed diseases	The heartwood powder is soaked in fresh water overnight and the filtered water is then consumed	Rhizome is soaked with warm water and then dried to make it in powder form and consumed	Crushed bark with water is used for Dropsy and the oil extracted from seeds is applied over skin to cure diseases
Results	It will show effectiveness within a week in all the cases	It will give the result with in one day. The stem directly washed and used, which makes the oral cavity healthy and germs free	Found to be effect within one to two year for diabetic and arthritis and for skin infection itwill take one to two days to cure	For jaundice it will take two to three month, for cough and worm infection it will take two to three and the regular use can make the blood pure and also it a good antiseptic	Regular use can bring the result

Figures 6.2a-h: List of Plants Available in and Around Udala Area Used for Traditional Medicines [a. *Kalanchoe pinnata* (Amarpoi); b. *Butea monosperma* (Palas); c. *Pterocarpus marsupium* (Piasal); d. *Azadirachta indica* (Neem); e. *Phyllanthus emblica* (Anla); f. *Vitex negundo* (Begunia); g. *Terminalia arjuna* (Arjuna); h. *Tridax procumbens* (Bisalyakarani)].

Figures 6.2i-k: List of Plants Available in and Around Udala Area Used for Traditional Medicines [i. *Saraca asoca* (Ashoka); j. *Rouvolfia serpentine* (Patalagaruda); k. *Curcuma longa* (Haladi)].

The phytochemical analysis of plant extract results show the presence of tannic acid and tannins in all the plant extract, the both test *i.e.* Wagner test and Mayer test for alkaloids show the positive (presence) in ten plant extract except *Saraca asoca* (Ashoka). The presence of protein found in five plant extract *i.e. Curcuma longa, Pterocarpus marsupium, Phyllanthus emblica, Vitex negundo* and *Butea monosperma*, which shows the nutritive value of these plants. The phenolic content was found in ten plant extract except the *Vitex negundo* and the Quinone is found to be present three plant extract *i.e. Saraca asoca, Kalanchoe pinnata* and *Vitex negundo* (Table 6.2).

Discussion

All most all the 11 plant species are used as medicine for different aliments but only the leaves of the Neema plant is used in the diet and the Begunia is used for the tooth. The plant products like the leaves and the barks are more preferable for medicinal purpose only the seeds of Amar Poi are used as the medicine. Out the eleven plants only 5 plants (Bisalyakarani, Amarpoi, Arjun, Ashoka and Haladi) are available in all the seasons and rest of them are commonly available in the dry season. In most of the case the plant extract is used for enteric problem, blood related disorder, skin infection and also for the injury healing, which signifies that all plants which are traditionally collected have ethnobotanical value. The presence of the phytochemicals like tannic acid and tannin in these plants indicates their possible uses in treating blood pressure and obesity (Huang *et al.*, 2013). The presence of

Table 6.2: Different Plants Used in Traditional Medicine along with their Availability in and Around Udala Area, Mayurbhanj

Sl.No.	*Scientific name*	*Order*	*Family*	*English Name*	*Local Name*	*Habitat*	*Seasonal Availability*
1	*Tridax procumbens*	Asterales	Asteraceae	Mexican Dai sy	Bisalyakarani	Tropical, subtropical and mild temperate regions	All season
2	*Phyllanthus emblica*	Malpighiales	Phyllanthaceae	Indian gooseberry	PaniAnla	Rainforests	Rainy season
3	*Kalanchoe pinnata*	Saxifragales	Crassulaceae	Good luck leaf	Amar Poi	Banks, hummocks, waste grounds and thickets in dry and wet regions	All season
4	*Rouvolfia serpentine*	Gentianales	Apocynaceae	Rouvolfia	Patalagaruda	Tropical areas and foot of hills(plains)	Summer
5	*Terminalia arjuna*	Myrtales	Combretaceae	Arjuna tree	Arjun	Deciduous tropical areas	All season
6	*Saraca asoca*	Fabales	Fabaceae	Ashok tree	Ashoka	Tropical areas	All season
7	*Azadirachta indica*	Sapindales	Meliaceae	Neem	Neema	Statuesque woodland or tropical regions	Flowering season
8	*Vitex negundo*	Lamiales	Lamiaceae	Indian Privet	Begunia	Near bodies of water, recently disturbed land, grassland, and mixed open forests	Dry damp season
9	*Pterocarpus marsupium*	Fabales	Fabaceae	Gum-kino tree	Piasal	Deciduous forests at low elevations	Dry season
10	*Curcuma longa*	Zingiberales	Zingiberaceae	Turmeric	Haladi	Dry areas	All season: perennials
11	*Butea monosperma*	Fabales	Fabaceae	Flame of forest	Palas	Deciduous areas	Dry season

Table 6.3: Various Types of Phytochemicals Present in different Plant Species, Available in and Around Udala Area, Mayurbhanj

Name of Plant	Tannic Acids	Tannins	Alkaloids		Proteins	Phenols	Quin-ones
			Wagner's Test	Mayer's Test			
Curcuma longa	+	+	+	+	+	+	-
Rouvolfia serpentine	+	+	+	+	-	+	-
Saraca asoca	+	+	-	-	-	+	+
Kalanchoe pinnata	+	+	+	+	-	+	+
Pterocarpus marsupium	+	+	+	+	+	+	-
Terminalia arjuna	+	+	+	+	-	+	-
Tridax procumbens	+	+	+	+	-	+	-
Phyllanthus emblica	+	+	+	+	+	+	-
Vitex negundo	+	+	+	+	+	-	+
Azadirachta indica	+	+	+	+	-	+	-
Butea monosperma	+	+	+	+	+	+	-

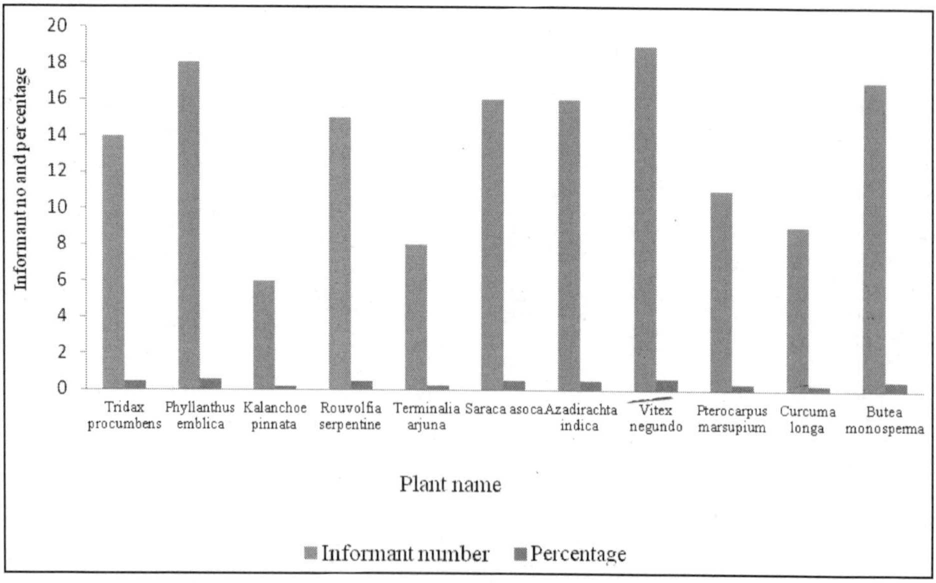

Figure 6.3: Total Number of Informant and Percentage of Use of Plants for Ailments in Udala, Mayurbhanj.

alkaloid in ten plant extracts besides the Ashoka plant indicates that these ten plants are the source of nitrogen in the nature (Liu *et al.*, 2010). Presence of proteins in five plants *i.e.* Haladi, Piasala, Panianla, Begunia and Palas states that these can be used to extract out different proteins, which could be a supplementary product from the nutritional point of view. As we found all most all plant species except the Begunia are having the phenolic content, so that these plants can be used to prepare the drugs and medicines for physiological disorders. The three plant species are dominated by the presence of the quinone *i.e.* Ashoka, Begunia and Amarpoi these three plants can be used to design the physiological medicines in the field of drug design in the Bioscience.

Conclusion

The medicinal use of plant extract showed by tribes of Udala, Mayurbhanj, Odisha for different ailment conclude that the efficiency of cure and medicinal value of these eleven plant extract. The tribal people use this plant extract for numbers of ailment. In the current investigation, the different plant extracts were prepared from leaf, bark, flower and rhizoids. This study justifies the claimed uses of plants in the traditional system of medicine to treat various diseases like Jaundice, common cold, cough, skin infection, enteric problem *etc.* However, further studies are needed to better evaluate the effective potential of the extracts as the antimicrobial agents. The present results will form the basis for selection of plants parts for further investigation in the potential discovery of new natural bioactive compounds. As from the social data we found the use of these plants in the tribal people is an integral part of their life, which have been a source of food, medicine, a pest control agent and also a source to earn money. The traditional knowledge on ethnologically important plants regarding its use for various diseases can help us to put a step ahead on the path of creating new era in pharmacological science and drug designing.

References

Ahmad, I., Aqil, F., Ahmad, F. and Owais, M. (2006). Herbal medicines: prospects and constraints. In: I. Ahmad, F. Aqil, M. Owais eds. Modern Phytomedicine: Turning Medicinal Plants into Drugs. WileyVCH Verlag GmbH and Co. pp.59-77

Culture, U. N. E. S. C. O. (1996).health Orientation texts-World decade for cultural development 1988-1997.*DocumentCLT/DEC/PRO-1996*, 129.

Huang, W.Y., Davidge, S.T. and Wu, J. (2013). Bioactive natural constituents from food sources potential use in hypertension prevention and treatment. Critical Reviews in Food Science and Nutrition, 53(6): 615-630.

Kumar, S., Singh, J., Saha, N.C. and Ranjan, V. (1997). Indian medicinal and aromatic plants facing genetic erosion. Central Institute of Medicinal and Aromatic Plants (CIMAP), Lucknow, India.

Liu, Y., Wei, J. and Che, C.M. (2010). [Fe (F 20 TPP) Cl] catalyzed intramolecular C–N bond formation for alkaloid synthesis using aryl azides as nitrogen source. Chemical Communications, 46(37): 6926-6928.

Manavalan, A.R.S. and Manian, K. (2001). Medicinal and aromatic plants diversity and utility: Conservation and utilization of medicinal and aromatic plants. Allied Publisher Ltd., New Delhi.

Padhi, L. and Panda, S.K. (2017). Medicinal plants: Potential to fight superiority diseases. Annals of Pharma Research, 1(1): 8-17.

Sabandar, C.W., Ahmat, N., Jaafar, F.M. and Sahidin, I. (2013). Medicinal property, phytochemistry and pharmacology of several Jatropha species (Euphorbiaceae): a review. Phytochemistry, 85: 7-29.

Singh, J., Singh, A.K. and Pravesh, R. (2003). Proc. First NIM on Medicinal Aromatic Plants, 50-58.

UNESCO, F. (1998). 504-RAF-48 Terminal Report: Promotion of Ethnobotany and the Sustainable Use of Plant Resources in Africa.

Ved, D. K. and Tandon, V. (1998). CAMP report for high altitude medicinal plants of Jammu-Kashmir and Himachal Pradesh. Foundation for Revitalization of Local Health Traditions, Bangalore, India.

Allied Biotechnology

7

Ethnomedicinal Uses and Bioactivities of Mangrove Plant Resources with Special Reference to Antidiabetic Studies of *Avicennia officinalis* from Mangroves of Mohanadi Delta, Odisha

H.N. Thatoi[1]* *and S.K. Das*[2]

[1]*Department of Biotechnology, North Orissa University,*
Baripada – 757 003, Odisha
[2]*Department of Biotechnology, College of Engineering and Technology,*
Ghatikia, Bhubaneswar – 751 003, Odisha
**E-mail: hn_thatoi @ rediffmail.com*

ABSTRACT

Mangroves are salt tolerant plants capable of growing in conditions of high salinity, extreme tides, strong winds, high temperatures, muddy and anaerobic soil at the land and sea interface in tropical and sub-tropical regions of the world. Mangroves growing under inhospitable condition are rich in phytochemicals and exhibit diverse medicinal properties. A number of studies conducted in the past have revealed the medicinal potential of mangroves. Systematic studies are being undertaken to isolate and characterise the active principles from medicinal mangrove plants responsible for curing different diseases. Extracts from mangroves and mangrove-dependent species have proven activity against human, animal and plant pathogens. The present paper discusses the ethnomedicinal uses and bioactivities of important medicinal mangrove plants. Besides, this paper also highlights the evaluation of antidiabetic activity of A. officinalis, a medicinal mangrove plant from Mahanadi delta of Odisha.

Keywords: *Mangrove, Ethnomedicinal, Phytochemicals, Bioactivities, Antidiabetic, Pharmaceuticals.*

Introduction

Mangroves are salt tolerant plant communities that grow at the interface between land and sea in tropical and sub-tropical latitudes of the world. The plants, and animals along with microbes constitute the mangrove forest community or mangal. MacNae (1969) proposed that "mangal" should refer to the forest community while "mangroves" should refer to the individual plant species. Mangroves are highly dynamic and fragile ecosystem. Mangrove ecosystems support genetically diverse groups of aquatic and terrestrial organisms. The mangrove ecosystems support genetically diverse groups of aquatic and terrestrial organisms. They include diversified habitats such as core forests, litter-forest floors, mudflats, water bodies (rivers, bays, intertidal creeks, channels and backwaters), and adjacent coral reefs and sea grass ecosystems. The mangal flora comprises of macro and microphytic communities which exhibit heterogeneity in its distribution in different formative zones (Bandaranayake, 1998). They vary from grasses, herbs, twines, climbers, shrubs to tall trees and phytoplankton (Kaladharan *et al.*, 2005). The ecosystem is evergreen throughout the year; self-sustaining and renewable. The networks of root system of plants help in binding the nutrient laden soil. The fauna include resident and migrant arboreal, terrestrial, aquatic and semi aquatic species. Mangroves serve as breeding, nursery and feeding grounds for a large number of terrestrial and aquatic organisms. It also provides refuge for juveniles of many species. The mangrove wetland ecosystems are the most productive and diverse habitats in the world. It supports varieties of life forms which form a good genetic treasure house. Never the less they are less explored.

Global Scenario of Mangroves

Mangroves are distributed over 123 countries and territories in the tropical and sub-tropical region of the world and cover a total of 152,000 sq. kms (Spalding *et al.*, 2010). The total mangrove area accounts for 0.7 per cent of total tropical forests of the world. The largest extent of mangroves is found in Asia (42 per cent) followed by Africa (20 per cent), North and Central America (15 per cent), Oceania (12 per cent) and South America (11 per cent) (Giri *et al.*, 2011). Asia has large mangrove areas because the environmental conditions are especially favorable for mangrove growth. According to the 2010 World Atlas of Mangroves, there are a total of 73 species and hybrids, which are considered to be true mangroves – those that have adapted to mangrove environments and are rarely found elsewhere. Thirty-eight of these species might be considered '*core species*' that typify mangroves and dominate in most locations. The remainder are either not so abundant or are found mostly on the fringes of mangrove habitats.

Indian Scenario

Mangroves in India account for about 3 per cent of the global mangroves and 8 per cent of Asian mangroves (FAO, 2007). It is mainly distributed along all the maritime states, except the union territory of Lakshadweep, covering an area of about 4461 km^2 along the 7,500 km long Indian coastline (Naskar and Mandal, 1999).

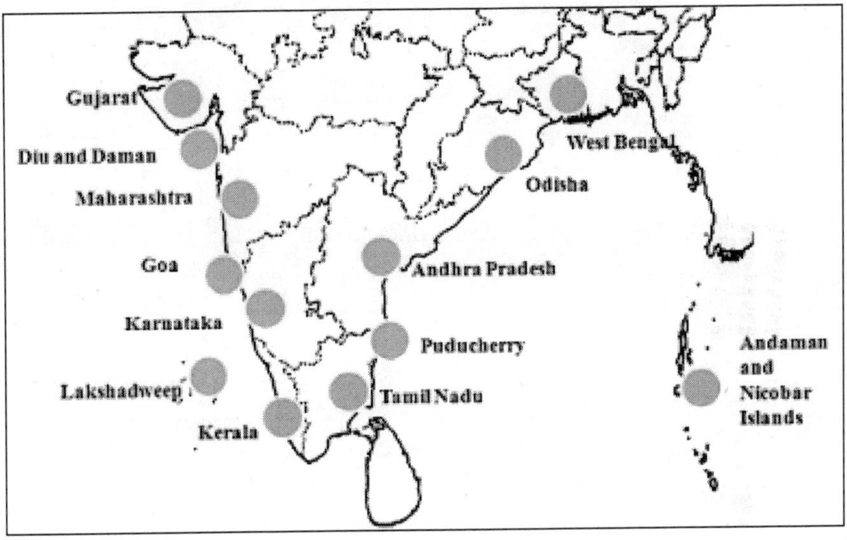

Figure 7.1: Map Showing Mangrove Habitats in India.

The current assessment shows that mangrove cover in the country is 4921 sq km which is 0.15 per cent of the India's total geographical area. Mangrove forests are categorised on the basis of their canopy cover. Mangrove cover which canopy density of more than 70 per cent has been categorised into very dense, canopy density between 40-70 per cent are considered as moderately dense and canopy density between 10-40 per cent are considered as open mangrove. In India, the very dense mangrove comprises 1481 sq km (30.10 per cent) of mangrove cover, moderately dense mangrove comprises1481 sq km (30.07 per cent) while open mangrove constitute an area of 1960 sq km (39.83 per cent) of mangrove cover (FSI, 2017). Based on density, mangrove forests are categorised into different categories. In India, about 60 per cent of the mangroves occur on the east coast along the Bay of Bengal, 27 per cent on the west coast bordering the Arabian Sea, and 13 per cent on Andaman and Nicobar Islands (Ingole, 2005). Distribution of mangroves along the Indian coast is presented in Figure 7.1. Gradual topography along the east coast is said to have an extensive intertidal expanse which favours major formation of mangroves in the deltaic regions (Jagtap *et al.*, 2003). On the other hand, less mangroves are found in narrow and steep in slop along the west coast. Mangroves are relatively well known for their floral diversity which is comprised of only 65-69 species of vascular plants which have several specific adaptations to the dynamic coastal environment (Kathiresan and Bingham, 2001). Nevertheless, because mangroves grow where nothing else grows, they are always useful, even where they cannot be managed as productive forests

Table 7.1: Assessment of Mangrove Cover in India during 2017 (Area in sq km) (FSI, 2017)

Sl.No.	*States/UTs*	*Very Dense Mangrove*	*Moderately Dense Mangrove*	*Open Mangrove*	*Total Mangrove*
1	Andhara Pradesh	0	213	191	404
2	Goa	0	20	6	26
3	Gujarat	0	172	968	1140
4	Karnataka	0	2	8	10
5	Kerala	0	5	4	9
6	Maharashtra	0	88	216	304
7	Odisha	82	94	67	243
8	Tamil Nadu	1	25	23	49
9	West Bengal	999	692	423	2114
10	A and N Islands	399	169	49	617
11	Daman and Diu	0	0	3	3
12	Puducherry	0	0	2	2
Total		**1481**	**1480**	**1960**	**4921**

Ethnomedicinal Uses and Bioactivity Studies of Mangroves

Mangroves are biochemically unique plants producing a wide array of novel natural products and are therefore used popularly in traditional medicinal practices (Bandaranyake, 1998; Kathiresan, 2000). Different parts of the mangrove plants have been used for treatment of various human ailments. Several mangrove plants such as *Acanthus illicifolius, Acanthus marina, B. cylindrica, E. agallocha, R. mucronata, R. lamarkii. Salicornia brachiata, Sesuvium portulacastum, Sueda maritima, Pongamia pinnata* and *S. monica* have been reported for antiviral activity against TMV (Premnathan *et al.*, 1992; Elanchezhiyan *et al.*, 1993; Padamakumar *et al.*, 1993). Fruits of the large glabrous shrub *Lumnitzera racemosa* are curative in skin disorders. This plant also exhibited antihypertensive activity (Lin *et al.*, 1993). The root of *Balanites aegyptiaca* is used as anthelmintic, and the edible leaf has been regarded as an effective medicine for sleeping sickness. *Clerodendron inerme* is reported to exhibit uterine stimulant activity and febrifugal and pesticidal properties (Bandaranyake, 1998). Several antifungal compounds have been characterized from *H. littoralis and X. granatum* (Chittawong, 1987). According to the Indian 'materia medica' a soft reddish substance obtained from the lower part of the trunk of another mangrove plant *Excoecaria agallocha* was reputed as an 'aphrodisiac tonic'. Being a member of the Euphorbiaceae, it exudes an acrid milk sap, or latex rich in alkaloids (Kawashima *et al.*, 1971). In traditional medicine this sap and decoctions from different parts of the plant (mainly the leaves) are used for different purposes, such as purgatives, and against epilepsy. *Excoecaria* paste is used as a cure for leprosy. A paste made from the wood is applied externally on ulcers and leprous sores or, alternatively, the ulcers and sores are exposed to the smoke from the wood. Extracts from the bark of *Rhizophora apiculata, R. mucronata, Ceriops tagal, C. decandra, Xylocarpus granatum, X. moluccensis* and *Laguncularia racemosa* are considered to have astringent, antidiarrhoea, antiemetic and haemostatic properties. The bark pressings of *Xylocarpus granatum* and *X. moluccensis,* referred to as the 'puzzle-nut trees' by Fijians are a cure for fevers including those caused by malaria (Alvi *et al.,* 1991).

Table 7.2: Ethnomedicinal Uses of some of the Important Mangrove Plants

Mangrove Species	*Ethnomedicinal Uses*
Acanthus illicifolius	Diarrhoea, febrifuge
Carapa moluccensis	Diarrhoea, febrifuge
Ceriops decandra	Hepatitis, ulcers
Ceriops tagal	Stops hemorrhages
Concocarpus erecta	Catarrh, febrifuge, gonorrhoea, malaria, stops bleeding
Excoecaria agallocha	Epilepsy, conjunctivitis, dermatitis, haematuria, leprosy, purgative, toothache
Heritiera littoralis	Diarrhoea
Hibiscus tiliaceus	Ear infections
Ipomoea pes-caprae	Jelly fish sting dermatitis
Kandelia rheedii	Diabetes
Lumnitzera racemosa	Antifertility, asthma, diabetes, snake bite
Nypa fruiticans	Asthma, diabetes, leprosy, rheumatism, snake bite
Pluchea indica	Fever, gangrenous ulcers, rheumatism, scabies, sinusitis
Rhizophora apiculata	Antiemetic, antiseptic, diarrhoea, haemostatic, hepatitis, stops bleeding, typhoid
Rhizophora lamarckii	Hepatitis
Rhizophora mucronata	Elephantiasis, febrifuge, haematoma, hepatitis, ulcers
Rhizophora racemosa	Stops bleeding
Sonneratia alba	Poultice in swellings and sprains
Sonneratia apetella	Hepatitis
Sonneratia caseolaris	Bleeding, hemorrhages, piles, sprain poultices
Xylocarpus granatum	Cholera, fever, malaria
Xylocarpus moluccensis	Aphrodisiac, fever, malaria

A triterpenoidal saponin isolated from *Acanthus illicifolius* has revealed antileukemic activity (Pezzuto *et al.,* 1983). Extracts of *Brugueira sexangula* bark were active against two tumors, Sarcoma 180 and Lewis Lung carcinoma (Loder and Russell, 1969). *Acanthus illicifolius, Avicennia marina and E. agallocha* showed significant analgesic activities (Padamakumar *et al.,* 1993). A triterpenoidal saponin isolated from *A. illicifolius* is useful in the treatment of paralysis, asthma, rheumatic pains and has revealed analgesic and anti-inflammatory activities (Minocha and Tiwari, 1981). The *Clerodendron inerme* and *A. illicifolius* extracts were reported for their anti-inflammatory activity (Bandaranayake 1999). The young unripe fruits of *X. moluccensis* strongly inhibit the respiratory reactions of mitochondria from rat liver (Kubo *et al.*, 1976). Several mangrove plants are well documented for their diverse ethnomedicinal properties as reported in Table 7.2 (Walsh, 1977; Rollet, 1981; FAO, 1985; Tomlinson, 1986; Barr *et al.*, 1988; Premanathan *et al.*, 1992; Field, 1995).

Evaluation of Antidiabetic Activities of *Avicennia officinalis* from Mangroves of Mahanadi Delta, Odisha

Odisha is a maritime state located in the East coast of Indian peninsula and having a long coast line of 480 km. The mangroves of Odisha are distributed in six major river deltas such as the Budhabalanga, Subernarekha, Brahamani-Baitarani (Bhitarkanika), Mahanadi and Devi river mouths. The Mangroves of Bhitarkanika (Brahmani-Baitarani river deltas) are quite extensive and in the luxuriant state where as mangroves of the Mahanadi delta are in degraded state. As per an estimate, the state has a 243 sq. km total mangrove area is distributed over four coastal districts (State of Forest Report, 2017) (Table 7.3).

Table 7.3: Mangroves of Odisha (FSI, 2017)

Sl.No.	*District*	*Area in Km^2*
1	Balasore	5
2	Bhadrak	32
3	Jagatsinghpur	8
4	Kendrapara	197
5	Puri	1
	Total	**243**

In Mahanadi delta the mangroves occupy about 38.56 sq. km and are part of Bhitarkanika Wildlife Sanctuary (672 sq. km). It is under the administrative control of Rajnagar Mangrove Forest Division, Kendrapara. From ecological point of view, Mahanadi mangroves constitute one of the significant mangrove formations in Odisha next to Bhitarkanika. Mahanadi delta, Odisha is the second largest mangrove ecosystem in Odisha next to Bhitarkanika. Some parts of this forest exhibit very rich mangrove vegetation as seen in Bhitarkanika. About 34 true mangroves, 42 mangrove associates and 20 obligate mangroves are occurring in Mahanadi delta (Thatoi and Biswal, 2008). The commonly occurring floral mangrove species are *Rhizophora apiculata, R. mucronata, Avicennia alba, A. officinalis, Excoecaria agallocha, Sonneratia opetala, Aegiceras corniculatum etc.*

Out of the eight different species of *Avicennia* plants, the antidiabetic potential is only reported in *A. marina* species. The leaves (Babuselvam *et al.*, 2013) and pneumatophores (Mahera *et al.*, 2011 and 2013) were reported for their anti-hyperglycaemic activities. However, no studies have been done to assess the antidiabetic activities of this plant. Hence an attempt was made to evaluate the antidiabetic activity of leaves of *A. officinalis* employing *in vitro* assay methods from Mahanadi delta. For this purpose, leaves of *A. officinalis* were collected from the mangrove forest of Mahanadi delta region of Odisha coast. After collection, the plant samples were air dried under shade at room temperature and grounded. The samples were extracted by maceration using petroleum ether, 90 per cent ethanol and distilled water at room temperature for 72 h with constant shaking. After the extraction, the crude extracts were collected and evaporated to dryness. Then the extracts were dissolved in DMSO for further studies.

Phytochemical Analysis

Phenolic compounds, such as flavonoids, phenolic acid and tannins possess diverse biological activities, such as anti-inflammatory, anticarcinogenic, antidiabetic and anti-atherosclerotic activities which might be related to their antioxidant activity (Chung *et al.*, 1998). Therefore in the present study, different phytochemicals from PEL, ETL and AQL in *A. officinalis* were analyzed by qualitative phytochemical assays (Harborne, 1998). The qualitative phytochemical analysis showed the presence of steroids, saponins, phenols, terpenoids, tannins and flavonoids in PEL, ETL and AQL extracts of *A. officinalis* with the exception of saponins and terpenoids in ETL extracts (Table 7.4).

Table 7.4: Preliminary Phytochemical Screening of Leaf Extracts of *A. officinalis*

Sample	*Steroids*	*Saponins*	*Phenols*	*Terpenoids*	*Flavonoids*	*Tannins*
PEL	+	+	+	+	+	+
ETL	+	-	+	-	+	+
AQL	+	+	+	+	+	+

In vitro Carbohydrate Metabolizing Enzyme Inhibition Assay

The study has endeavoured to investigate the possible inhibitory potential of *A. offcinalis* leaf extracts to modulate key enzymes involved in diabetes *i.e.*, α-amylase and α- glucosidase. The *in vitro* antidiabetic activity of the leaf extracts of *A. officinalis* were evaluated employing the α-amylase enzyme and α-glucosidase enzyme inhibition assay (Gella *et al.*, 1997; Apostolidis *et al.*, 2007). The experiments were performed by taking plant extracts at 0.1, 0.5 and 1 mg/ml concentrations. Acarbose was used as positive control. All the experiments were performed in triplicate. The IC_{50} values were determined.

The inhibitory potential of crude petroleum ether, ethanol and aqueous leaf extracts of *A. officinalis* against both the enzymes were carried out carried out at 0.1, 0.5 and 1 mg/ml. The results showed that PEL, ETL and AQL extracts could inhibit the α-amylase enzyme in a dose dependent manner (Figure 7.2). However, both AQL and PEL didn't show any α-amylase potential at 0.1 and 0.5 mg/ml concentrations and could only inhibit α-amylase at 1.0 mg/ml. The ETL could inhibit α-amylase at all the concentrations studied in the present condition. The IC_{50} value of AQL and PEL was >1 mg/ml and for ETL it was found to be 0.29 mg/ml (Table 7.5).

The α-glucosidase enzyme inhibition study demonstrated that all the three extracts of *A. officinalis* could inhibit α-glucosidase in a dose dependent manner (Figure 7.3). However, the IC_{50} value for α-glucosidase enzyme inhibition activity of all the three solvent extracts *i.e.* PEL, ETL and AQL were >1 mg/ml.

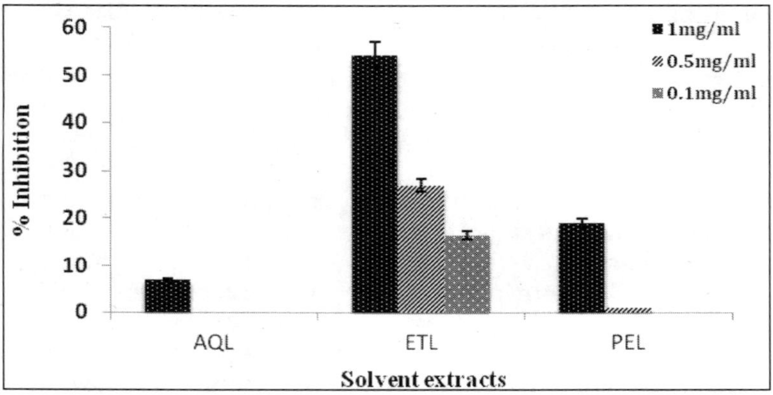

Figure 7.2: α-Amylase Inhibition Activities of Leaf Extracts of *A. officinalis*.

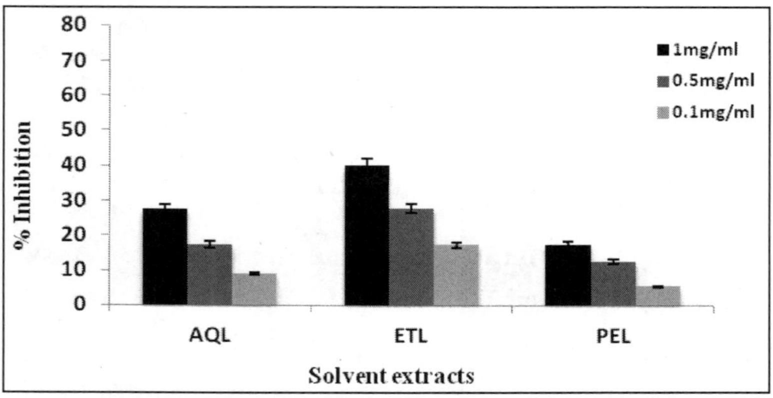

Figure 7.3: α-Glucosidase Inhibition Activities of Leaf Extracts of *A. officinalis*.

Table 7.5: Antidiabetic activities (as IC_{50} value in mg/ml) of leaf extracts of *A. officinalis*

Sample	*α-amylase inhibition*	*α-glucosidase Inhibition*
PEL	>1.0 mg/ml	>1.0 mg/ml
ETL	0.29 mg/ml	>1.0 mg/ml
AQL	>1.0 mg/ml	>1.0 mg/ml
Acarbose	0.15 mg/ml	0.11 mg/ml

Amongst various therapeutic approaches for treating diabetes, decrease in the postprandial hyperglycemia is the major one. This condition can be achieved by retarding the absorption of glucose through the inhibition of carbohydrate hydrolysing enzymes (α-amylase and α-glucosidase) in the digestive tract. Presences of inhibitors of these enzymes delays and prolong carbohydrate digestion time. The effect causes a reduction in the rate of glucose absorption and consequently blunts the postprandial plasma glucose rise (Ali *et al.*, 2006). The carbohydrate metabolism

8

Application of Nanotechnology in Biology: Current Trends, Future Scopes, Benefits and Ethical Issues

Priyabrat Swain

Principal Scientist and National Fellow
Fish Health Management Division,
ICAR- Central Institute of Freshwater Aquaculture,
Kausalyaganga, Bhubaneswar – 751 002, Odisha
E-mail: pswainy2k@yahoo.co.in

ABSTRACT

Nanotechnology involves the application of materials at the nanoscale to produce new products or processes, and with many potential benefits to society. Recent research has highlighted the potential for nanotechnologies' use in wide ranging food applications, including improving supplements, novel food packaging, increasing the range of food textures, colours and tastes, increasing the efficiency of liquid filters, cooking oil catalysation and targeted crop pesticides. Because of these new developments it is likely that radical changes in the way food is perceived, stored, packaged, transported, monitored, consumed and processed. Available literature suggests that many uncertainties remain about nanomaterials, including the potential for bioaccumulation and potential human health risks. While proposed applications of nanotechnologies are wide and varied, developments are met with some caution, while progress may be stifled by lack of governance and potential risks. Herein in this review detailed applications, risks and benefits of nanotechnology, especially in life sciences have been discussed.

***Keywords**: Antimicrobial, Nanotechnology, Nanovaccine, Immunomodulation.*

Introduction

Nanotechnology involves the application of materials at the nanoscale to produce new products or processes, and with many potential benefits to society (Roco and Bainbridge, 2005). The term nanotechnology is an umbrella term that includes an immense scope of innovations over various controls and accordingly can be a debilitate to any talk about social and ecological ramifications that might be explicit to specific applications. The "nano" prefix derives from the Greek noun "nanos", meaning dwarf. A nanometer(nm) is one billionth (1×10^{-9}) of ameter: the length of about ten atoms placed side-by-side, or 1/80,000th of the thickness of a human hair. Nanotechnology is commonly understood as a technology involving the manipulation and application of matter, based on its properties at the atomic scale. The term covers a family of technologies, including nanosciences and nanotechnologies. There are several interesting facts of nanomaterials such as they behave very differently in the world of atoms and molecules and have different physical properties. For example, the metal copper is transparent on the nanoscale while gold, which is normally unreactive, becomes chemically very active. Carbon, which is quite soft in its normally occurring form (graphite), becomes incredibly hard.

The nanotechnology industry has grown rapidly. Initial estimates (NSF, 2001) predicted a global economic value of nanotechnology at around one trillion U.S. dollars by about 2015, with a significant proportion on the theme of sustainability, including food production and agriculture. These early predictions may have been over-optimistic, but nonetheless, the global investment in nanotechnology was around 4·6 billion U.S. dollars by governments and 4·5 billion U.S. dollars fromcommercial companies in 2005. The growth of the industry is exemplified by over a thousand different products containing NMs in 2009 and estimates of sales from nanotechnology worth around 1500 billion U.S. dollars in 2010. Clearly, the nanotechnology industry is an expanding market. The applications, risks and benefits have been discussed for agriculture and food production, most notably in an initial report from the Woodrow Wilson International Centre for Scholars (Kuzma and VerHage, 2006). Recent research has highlighted the potential for nanotechnologies' use in wide ranging food applications, including improving supplements, novel food packaging, increasing the range of food textures, colours and tastes, increasing the efficiency of liquid filters, cooking oil catalysation and targeted crop pesticides. Because of these new developments, it is likely that radical changes in the way food is perceived, stored, packaged, transported, monitored, consumed and processed. Available literature suggests that many uncertainties remain about nanomaterials, including the potential for bioaccumulation and potential human health risks. While proposed applications of nanotechnologies are wide and varied, developments are met with more or less forethought, while progress may be strangled by lack of governance and potential dangers.

Biological Applications

A. Antimicrobial Activity of different Metal Oxide Nanoparticles

The emergence of the antibiotic resistance in microbial pathogens now becomes a serious health issue and numerous studies have been reported to improve the current antimicrobial therapies. It is known that over 70 per cent of bacterial infections are resistant to one or more of the antibiotics that are generally used to eradicate the infection. Development of new and effective antimicrobial agents seems to be of paramount importance. Recently, nanotechnology offered great possibilities in Pharmaceutical nanotechnology with numerous advantages have growingly attracted the attention of many researchers. Several types of metal nanoparticles have received great attention for their potential antimicrobial effects such as Ag, silver oxide (Ag_2O), titanium dioxide (TiO_2), silicon (Si), copper oxide (CuO), zinc oxide (ZnO), Au, calcium oxide (CaO) and magnesium oxide (MgO) were identified to exhibit antimicrobial activity (Dizaj, 2014; Swain *et al.*, 2014 and 2016). The exact mechanisms for antibacterial effect of nanometals are still being investigated, but there are two more popular proposed possibilities in this regard: (a), free metal ion toxicity arising from the dissolution of the metals from the surface of nanoparticles and (b), Inducing oxidative stress via the generation of reactive oxygen species (ROS) which disrupts cell integrity.

Furthermore, morphological and physicochemical characteristics of the nanometals have been proven to exert an effect on their antimicrobial activities. It is known that the smallest nanoparticles have the strongest bactericidal effect. The positive surface charge of the metal nanoparticles facilitates their binding to the negatively charged surface of the bacteria which may result in an enhancement of the bactericidal effect. The shape of the nanoparticles also influences their antimicrobial effects. The antimicrobial nanoparticles could be benefitted in the pharmaceutical and biomedical industries for sterilization of the medical devices. These nanostructures could also be directed for the preparation of chemical disinfectants, coating-based applications and food preparation processes. Finally, it may be concluded that, the metal oxide nanoparticles with the minimized toxicity possibly will be extensively used in the near future for eradicating several infectious conditions. We believe that development of the simple and low cost inorganic antimicrobial agents such as metal and metal oxide nanoparticles as an alternative to traditional antibiotics might be promising for future of pharmaceutics and medicine.

B. Immunomodulation by Metal Nanoparticles

As an important protective system to defend organisms from foreign matters and danger signals inside the body, the immune system plays a critical role in keeping homeostasis in the human body. The immune system exerts its function through innate immunity and adaptive immunity. Innate immunity is the first line of defense against microbial invasion, which interacts with the foreign materials. The function of innate immunity was realized by the phagocytic cells (macrophages, dendritic cells (DCs), neutrophils, and mast cells, *etc.*), which phagocytose antigen and release cytokine to clear pathogen. If the pathogen cannot be effectively cleared by innate immunity, the adaptive immunity, as the second line of defense in the human body, will be activated. During these processes, some phagocytic cells act as antigen-presenting cells (APCs) and present specific antigens to specialized cells, which are responsible for adaptive immunity, such as T cells and B cells. By this antigen NPs interact with the immune system and modulate its function, leading to immunosuppression or immunostimulation. Immunosuppression can be caused by impairment of any component of the immune system, which results in a decreased immune response and that why antigens cannot be effectively cleared from the body. Immunostimulation could enhance the ability to resist pathogen, but it may ensue in a strong adverse response such as autoimmune disease if it was sensitized When nanomedicines are applied *in vivo*, they function as foreign materials and cause the immune response, immunosuppression, or immunostimulation. However, the modulations of the immune system caused by NPs are undesirable in most cases when nanomedicine is applied. Furthermore, the immune modulations by NPs could be causeing an adverse effects like disease prevention and treatment such as a vaccine adjuvant and anti allergy therapeutic agents. Therefore, NPs play as a Janus' double-face in nanomedicine applications.

Many factors contribute to immunomodulation of NPs such as composition, surface chemistry, size, shape, and protein-binding ability dominates these interactions. Besides, individual difference and exposure route also contribute to immunomodulation of NPs. NPs interact with both innate and adaptive immune cells and affect their functions. Inflammation is an important response of immune system, which can be induced by NPs, evidenced by the production of cytokines or chemokines. So NPs with immunosuppression effects might be used as anti-inflammatory or anti-autoimmune therapeutic agents. On the contrast, NPs which activate immune system might be used as vaccines, or vaccine adjuvants. Different NP like CNT, silver, gold Quantum dot, Ceo2and magnetic NP may either stimulate or suppress immune system depending on their immune cell target. In a recent study revealed that when MWCNTs injected intravenously activated Th2 immune response by elevating Th2 cytokines and increasing number of $CD4^+$ and $CD8^+$ T cell in the spleen. In other studies, CNTs showed an allergy adjuvant effect in the OVA-sensitized mice, which aggravated asthma and induced fibrosis in OVA-sensitized lungs but showed no response with healthy pulmonary. The results indicated that these NPs could bring harm to asthma patients, but not healthy ones. So, the immune response of NPs is like a double-edged sword in nanomedicine applications by bringing both benefits and harms. We should take advantage of the benefits from the immunomodulating properties of NPs and, on the other hand, avoid the undesirable immune responses in order to minimize the systemic side effects.

C. Nano-Vaccine and Drug Delivery System

Nanotechnology recently emerged as one of the most promising fields in vaccine delivery. Vaccines are mainly composed of live attenuated organisms, killed whole organisms, or inactivated toxins. Sometimes these conventional vaccines are poorly immunogenic and ineffective due to lack of appropriate adjuvants and/or suitable vaccine carrier. Therefore, there is an urgent need for the development of potent and safe adjuvants and delivery systems for modern vaccinology. Recently, introduction of nanotechnology has been the most exciting and challenging in vaccine delivery that have been providing various delivery systems These delivery systems improve the stability and adsorption, and permit a long-term release of the vaccine within the body to increase the immunostimulatory properties of the vaccine. Nanotechnology provides innumerable nanocarrirer for delivery of vaccine along with other therapeutics. These delivery systems play a major role to create a novel nanovaccine, due to their confirmed capacity to interact with biological barriers and transport the bioactive molecule without altering its antigenicity agents. Various types of vaccine delivery systems such as nanoparticles, microparticles, micelles liposomes, dendrimers are the reward of nanotechnology. Polymeric micro- and nanoparticles have been used to develop vaccines against many infectious diseases. The immune response produced by polymeric particles, mostly influenced by a number of factors, such as the inner structure of particulate carriers, use of adjuvant, surface charge, particle size, and surface hydrophobicity. These parameters can also affect differential uptake by macrophages and dendritic cells, which in turn play an important role in generating the right kind of immune response. The biodegradable, biocompatible polymers PLGA (poly-(D,L-lactide-co-glycolide) and PLA (poly lactide) used in the nanocarrier formulation of vaccine antigens are safe and Food and Drug Administration (FDA) approved for animal use.

D. Nanotechnology in Disease Diagnosis and Therapy

Correct and rapid identification and quantification of disease is important to effective treatment, and is a major emphasis of biotechnological and medical research currently. Nanotechnology may enhance disease screening by improving sensitivity, selectivity, time to diagnosis, and the availability of testing equipment. Nano-biosensors are the measurement system for the detection of an analyte that combines a biological component with a physicochemical detector. Several nanomaterials that are widely employed include nanotubes, nanowires, nanorods, nanoparticles, and thin films made up of nanocrystalline matter. For example most popular sensing device is the designs of glucose biosensors that involve the use of CNT nanotubes as immobilizing surfaces for enzyme glucose oxidize; this enzyme is used for estimation of glucose from the several body fluids even from the scarce body fluids such as tears and saliva. This will help in the diagnosis of diabetes in patients.

Protein Detection

Gold nanoparticles are widely used in immunohistochemistry to identify protein-protein interaction. Nevertheless, the multiple simultaneous detection capabilities of this technique are pretty limited. Surface enhanced Raman scattering spectroscopy is a well-established technique for detection and identification of individual dye molecules. By mixing both methods in a single nanoparticle probe one can drastically improve the multiplexing capabilities of protein probes.

Cancer Therapy

Photodynamic cancer therapy is based on the destruction of the cancer cells by laser generated atomic oxygen, which is cytotoxic. A greater quantity of a special dye that is used to generate the atomic oxygen is taken in by the cancer cells when compared with a healthy tissue. Hence, only the cancer cells are destroyed, then exposed to a laser radiation. Unfortunately, the remaining dye molecules migrate to the skin and the eyes and make the patient very sensitive to the daylight exposure. This effect can last for up to six weeks. To avoid this side effect, the hydrophobic version of the dye molecule was enclosed inside a porous nanoparticle. The dye stayed trapped inside the Ormosil nanoparticles and did not spread to the other parts of the body. At the same time, its oxygen generating ability has not been affected and the pore size of about 1 nm freely allowed for the oxygen to diffuse out.

E. Nanotechnology in Nutraceuticals

One significant application of nanotechnology in food and nutrition is to design and develop of novel functional food ingredients (that is, Nutraceuticals) with improved water solubility, thermal stability, oral bioavailability, sensory attributes, and physiological functioning. Nutraceuticals, which are food supplements with health benefits, are commonly used as part of the everyday diet. Because of their low solubility, many Nutraceuticals are poorly assimilated by the human body, thus one of the most significant and interesting applications for encapsulation of Nutraceuticals is to enhance the bioavailability of Nutraceuticals by changing the pharmacokinetics and biodistribution. In recent years, extensive research has been extended away to analyze the health promotion properties of different Nutraceuticals and to devise novel encapsulation materials and methods, trying to contain functional ingredients in foods. One of the suitable example is that of Curcumin which is the major curcuminoid compound found in the rhizome of the turmeric plant (*Curcuma longa*). There are four major properties of curcumin such as a nutraceutical: antioxidant, anticancer, anti-inflammatory, and antimicrobial activity. Moreover, clinical studies have shown that curcumin is able to improve the expression of biomarkers in the digestive tract, especially the colon, of cancer patients. On the other hand, curcumin is well known as a nutraceutical with low bioavailability. It is water - insoluble and undergoes rapid metabolism in the body. Considering the potential health benefits of curcumin and its low bioavailability, nanotechnology - based formulations are being used to achieve better bioavailability and thus health benefits. To enhance nutritional quality and stability of the nutraceuticals, one option is to encapsulate the functional ingredients using food-grade or "generally recognized as safe" (GRAS) materials that can exhibit controlled-release behavior. Encapsulation and controlled-release of active food ingredients is important applications in food and nutrition that can be attained with nanotechnological approaches. To overcome instability, poor water solubility, and to enhance the bioavailability of active components, one option to entrap the compound of interest into a nanoemulsion. Nanoemulsions are a class of extremely small droplet emulsions that appears to be transparent or translucent and usually in the range 50 to 200 nm, much smaller than the range (from 1 to 100μm) for conventional emulsions. Compared with conventional methods, such as co-solvent addition, micronizing/milling, spray drying, and salt formation, the use of lipid based delivery systems, offers many advantages: (i) high kinetic or thermodynamic stability, which provides significantly better stability over unstable dispersions, such as conventional emulsions and suspensions; (ii) because of the small droplet sizes, active component can be transported through the cell membranes much more easily, resulting in an increased concentration in plasma and bioavailability. Nanotechnology is being used to devise novel formulations that may overcome the problem of oral bioavailability, especially the solubilization problem. Nanotechnology - based formulations have already shown that the oral bioavailability of nutraceuticals can indeed be improved, demonstrating a promising outlook for the application of nanotechnology in food science.

F. Tagging, Monitoring and Tracking using Nanotechnology

Nanobarcodes

A "nanobarcode" is an alternative tagging or monitoring device that works more like the UPC code, merely on the nano-scale. One type of nanobarcode - developed by Nanoplex Technologies - is a nanoparticle consisting of metallic stripes, where mutations in the striping provide the method of encoding information. Nanoplex changes the length and width of the particles and the number, width and composition of each stripe to make billions and billions of variations. Recent advances in nanotechnology have allowed scientists to track millimeter sized organisms. Small organisms can be labelled with quantum dots, a microscopic fluorescent probe, the core of which is a semiconductor material with high photostability. In one experiment, amine proteins on the exoskeleton of Daphnia magna were biotinylated and streptavidin was attached to the quantum dots. This allowed for a simple bioconjugation to tag the organisms with the quantum dots, by taking advantage of the high affinity interaction between streptavidin and biotin. *D. magna* individuals were successfully tracked with each having a unique quantum dot that fluoresced and emitted light of a specific wavelength that could be detected using cameras.

G. Environmental Application of Nanotechnology

One of the main environmental applications of nanotechnology is in the water sector. As fresh water sources become more and more scarce due to overconsumption and contamination, scientists have started to consider seawater as another source for drinking water. The majority of the world's water supply has too much salt for human consumption and desalination is an option but expensive method for removing the salt to create new sources of drinking water. Carbon nanotube membranes have the potential to reduce desalination costs (White *et al.*, 2007). Similarly, nanofilters could be applied to repair or clean up dry land water or surface water contaminated with chemicals and hazardous substances. Also remediation of contaminated groundwater using nanoparticles containing zero-valent iron is one of the most emerging technologies with considerable potential benefits. Air pollution is another possible area where nanotechnology has great promise. Nanofilters could be applied to automobile tailpipes and factory smokestacks to separate out contaminants and prevent them from entering the atmosphere. Finally, nanosensors could be trained to detect toxic gas leaks at extremely low concentrations. Overall, in that respect there is a large number of promising environmental applications for nanotechnology.

Social, Ethical and Economical Issues of Nanotechnology

There are some hefty social concerns about nanotechnology too. Bioethical researchers believe that nanotechnology could be manipulated to harm the human body structure rather than curing it.How would the use of a technology that can't be seen be regulated? What if the nanomedicine technology is used for terrorism purposes? Will the materials used for the nano-medicinal technologies be non-toxic and eco-friendly? It is widely recognized that, because the applications of nanotechnology are not yet clear, neither are the ethical issues clear. And still, many argue, the nano community must begin to come up to these matters at once, before they overwhelm nanotechnology and derail potential benefits. The technology is being regulated by the government with the primary concern being safety of the people. We can surely state that application of nanotechnology, if practiced carefully and meticulously, will guarantee great benefits. Similarly, several economical issues are in that location; Is there a certain patent for the usage of nanotechnology? Does nanotechnology favor the rich or the poor? Will the poor get equal access to nano-medicinal technologies?

Future Directions

In recent years there is increasing optimism that nanotechnology applied to biomedical science will bring significant advances in the diagnosis, treatment, and prevention of diseases. Growing interest in the future biological applications of nanotechnology is leading to the emergence of a new field called Nanobiotechnology which can overcome the challenges for its application and bring more sophisticated, preventive properties.

REFERENCES

Behera, T. and Swain, P. (2011). Antigen adsorbed calcium phosphate nanoparticles stimulate both innate and adaptive immune response in fish, *Labeo rohita* H. Cellular Immunology, 271(2): 350-359.

Dizaj, S.M. (2014). Antimicrobial activity of the metals and metal oxide nanoparticles. Materials Science and Engineering, 44: 278–284.

Roszek, W.H. and Geertsma, R.E. (2005). Nanotechnology in medical applications: state-of-the-art in materials and devices RIVM report, 265001001/2005.

Swain P., Nayak S. K., Sasmal A., Behera T., Barik S. K., Das J. K. and Jayasankar P. (2014) Antimicrobial activity of metal based nanoparticles against microbes associated with diseases in aquaculture. World Journal of Microbiology and Biotechnology, 30: 2491-2502.

Swain S., Barik S. K., Behera T., Nayak S. K. Sahoo S. K. Mishra S. S. and Swain P. (2016) Green synthesis of gold nanoparticles using root and leaf extracts of *Vetiveria zizanioides* and *Cannabis sativa* and its antifungal activities. BioNanoScience, 1-9.

White, A.A., Best, S.M. and Kinloch, I.A. (2007) Hydroxyapatite-carbon nanotube composites for biomedical applications: A review. *International Journal of Applied Ceramic Technology*, 4(1), 1-13.

Zhang, X. (2003). Nanoscale iron particles for environmental remediation: An overview, Journal of Nanoparticles Research, 5: 323-332.

9

In silico Identification and Characterization of Abiotic Stress Genes and Proteins of *Oryza sativa*

B. Patro and S.K. Pradhan*

Department of Bioinformatics, Centre for Post Graduate Studies, Orissa University of Agriculture and Technology, Bhubaneswar – 751 003, Odisha
****E-mail: ksukantapradhan@gmail.com***

ABSTRACT

Rice belongs to the genus Oryza and has two cultivated and 22 wild species. The cultivated species are Oryza sativa and Oryza glaberrima. O. sativa (rice) is the staple food for 2.5 billion people, and, after maize, the grain with the second highest worldwide production. In addition to its agronomic importance, O. sativa is also an important biological model species for monocot plants, and major cereal crops such as maize, wheat, barley and sorghum. O. sativa has a compact genome of approx. 500 Mbase (n=12) compared with the multi-gigabase genomes of maize, wheat and barley. Two sub-species of O. sativa have been sequenced; the Japonica Group, and the Indica Group. Drought remains as one of the major constraints limiting rice productivity. Conventional breeding approaches are slow in progress towards developing drought tolerant rice varieties due to complex nature of drought tolerance mechanism. The understanding of plant responses to abiotic stresses at the genomic level provides a foundation for gene discovery and genetic engineering of rice for drought tolerance. Identifying quantitative trait loci (QTLs) controlling drought tolerance related traits will help in precise introgression of desired genomic regions through marker assisted selection and genetic engineering. This study elucidates different candidate genes inducing drought resistant in O. sativa (rice), its related proteins and the QTLs associated with the drought resistant using bioinformatics tools and techniques.

***Keywords**: Abiotic stress, Oryza sativa, Quantitative trait loci, In silico.*

Introduction

Rice is central to the lives of billions of people around the world. Possibly the oldest domesticated grain (~10,000 years), rice is the staple food for 2.5 billion people (International Year of Rice, 2004, Gender and rice factsheet) and growing rice is the largest single use of land for producing food, covering 9 per cent of the earth's arable land. Rice provides 21 per cent of global human per capita energy and 15 per cent of per capital protein(International Rice Research Institute,Rice Almanac, 3rd Edition, 2002). As expected, Asia accounts for over 90 per cent of the world's production of rice, with China, India and Indonesia producing the most. Rice belongs to the genus Oryza and has two cultivated and 22 wild species. The cultivated species are *Oryza sativa* and *Oryza glaberrima*. *O. sativa* is grown all over the world while *Oryza glaberrima* has been cultivated in West Africa for the last ~3500 years (International Rice Research Institute, Rice Research and Production in the 21st Century, 2001). Rice is grown under many different conditions and production systems, but submerged in water is the most common method used worldwide. Rice is the only cereal crop that can grow for long periods of time in standing water (International Year of Rice, 2004, Rice and water: a long and diversified story). 57 per cent of rice is grown on irrigated land, 25 per cent on rainfed lowland, 10 per cent on the uplands, 6 per cent in deepwater, and 2 per cent in tidal wetlands (Chopra and Prakash, 2002; International Year of Rice, Aquatic biodiversity in rice fields, 2004). Rice can be a short, medium or long grain size. It can also be waxy (sticky) or non-waxy. Some rice varieties are considered aromatic (Alford and Duguid, 2003; Chaudhary, 2001). Rice also comes in many different colors including brown, red, purple and black (International Year of Rice, Rice and human nutrition, 2004). *O. sativa* (rice) is the staple food for 2.5 billion people, and, after maize, the grain with the second highest worldwide production. In addition to it's agronomic importance, *O. sativa* is also an important biological model species for monocot plants, and major cereal crops such as maize, wheat, barley and sorghum. *O. sativa* has a compact genome of approx. 500 Mbase (n=12) compared with the multi-gigabase genomes of maize, wheat and barley. Two sub-species of *O. sativa* have been sequenced; the Japonica Group, and the Indica Group. The Japonica sequence, which has become the most widely used reference genome for rice. Rice (*O. sativa*) is the important cereal food crop and model crop for plant genome analysis. Drought remains as one of the major constraints limiting rice productivity. Conventional breeding approaches are slow in progress towards developing drought tolerant rice varieties due to complex nature of drought tolerance mechanism. The understanding of plant responses to abiotic stresses at the genomic level provides a foundation for gene discovery and genetic engineering of rice for drought tolerance. Identifying quantitative trait loci (QTLs) controlling drought tolerance related traits will help in precise introgression of desired genomic regions (QTLs) through marker assisted selection and genetic engineering. The genomic region on chromosome 9 (73 – 95cm) controlling drought tolerant related traits were selected and attempts were made through *in silico* methods to understand the putative candidate genes controlling drought tolerance. Through literature survey and database search, QTL located on chromosome 9 (73 to 95cM) linked to drought tolerance related traits was selected. About 145 drought responsive genes were found to be located in this region as reported in a publicly available microarray data. The drought responsive genes were classified according to their function and presence of cis elements. Detailed analyses revealed that majority of these drought responsive genes are involved in important cellular functions and enriched with DRE-cis elements. Aquaporin (OsPIP2; 7) and cytochrome P450 (OsCYP78A3) genes were selected for functional validation. Semi-quantitative RT-PCR analysis revealed the differential expression of these genes between contrasting rice genotypes IR64 and Nootripathu for drought tolerance at vegetative stage (http://web.expasy.org/docs/relnotes/relstat.html).

Materials and Methods

Identification of Drought Resistant Genes and Proteins

The NCBI in houses genome sequencing data in GenBank and an index of research articles in PubMed Central and PubMed, as well as other information relevant to biotechnology. The NCBI's GenBank database was queried to find out the drought resistant gene sequences of *O. sativa* (Figure 9.1). The gene sequences of *O. sativa* were retrieved in fasta format and analyzed. Similarly the annotated protein sequences involved in drought resistance were download from protein knowledgebase (UniProtKB) database (https://www.uniprot.org/) (Figure 9.2).

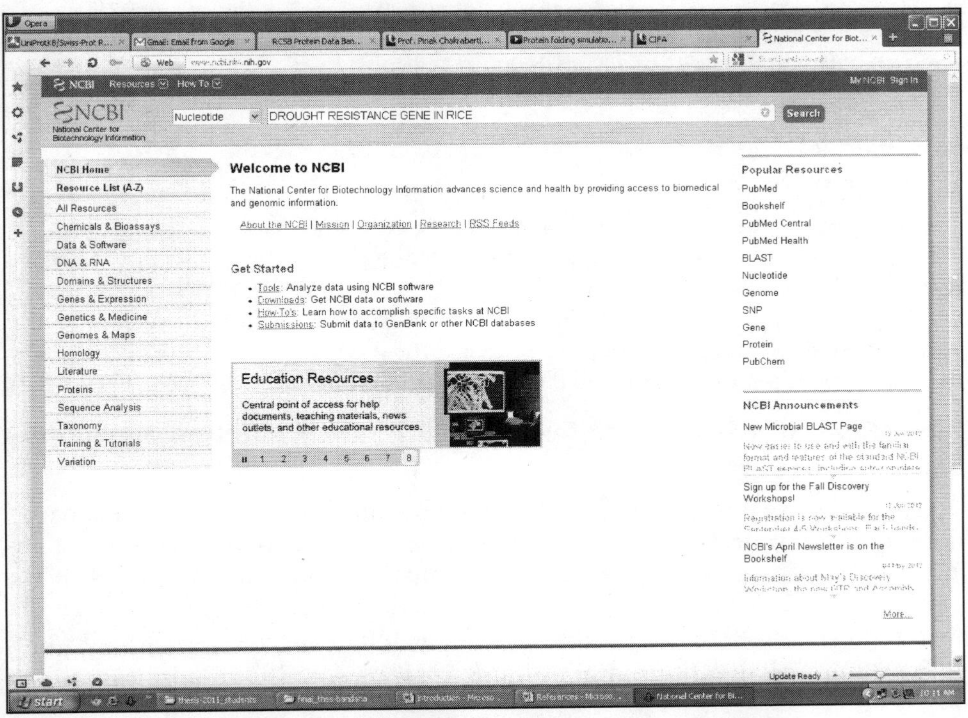

Figure 9.1: Home Page of National Centre for Biotechnology Information.

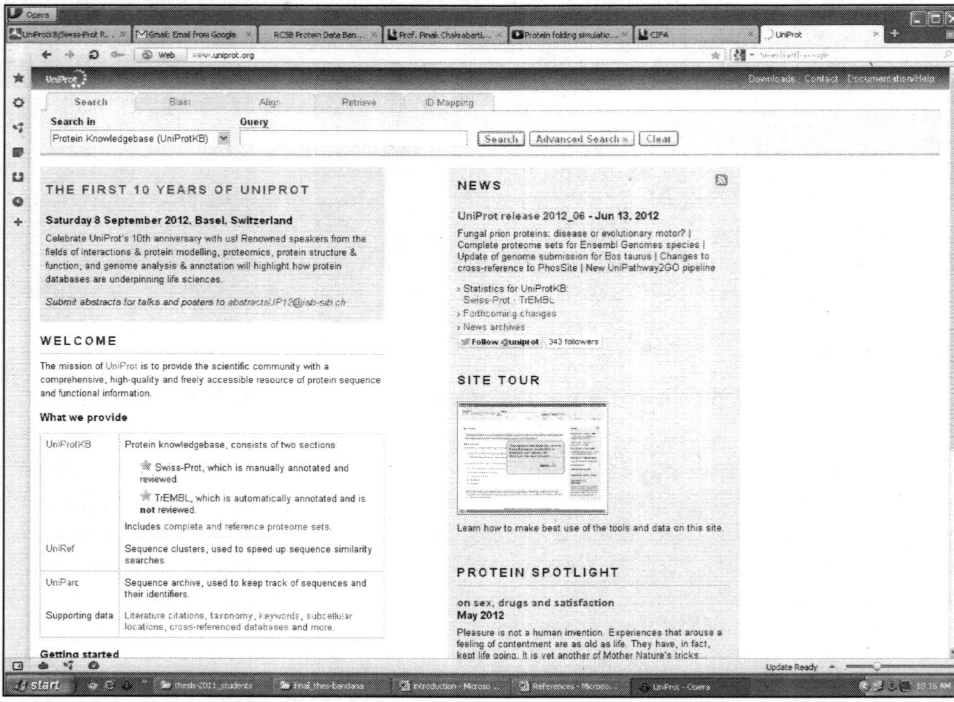

Figure 9.2: Home Page Protein Knowledgebase (UniProtKB).

Analysis of Physiochemical Properties and Identification of the Functional Domains

The physiochemical properties such as no. of amino acids, Molecular Weight, PI, Instability Index and Hydrophobicity of drought reststant proteins of *O, sativa* were calculated in order to understand its nature using EBI-Prot Param tool(http://web.expasy.org/protparam/) (Figure 9.3). The presence of different protein domains were scanned by using EBI-Inter Pro Scan tool(http://www.ebi.ac.uk/Tools/pfa/iprscan/) to correlate the annotated functions and family classification (Figure 9.4).

Identification of Sub-cellular Location and Prediction Secondary Structure of Protein

RSLPred is a SVM based prediction method (http://crdd.osdd.net/raghava/rslpred/home_middle.html) was used to predict four major target proteins (Chloroplast, Cytoplasm, Mitochondria and Nucleus) of *O. sativa* (Figure 9.5). The secondary structure of these proteins were predicted using GORIV tool in order understand the arrangements of secondary structure elements such as alpha helices, beta sheets, random coils *etc.* and their role in stress resistant.

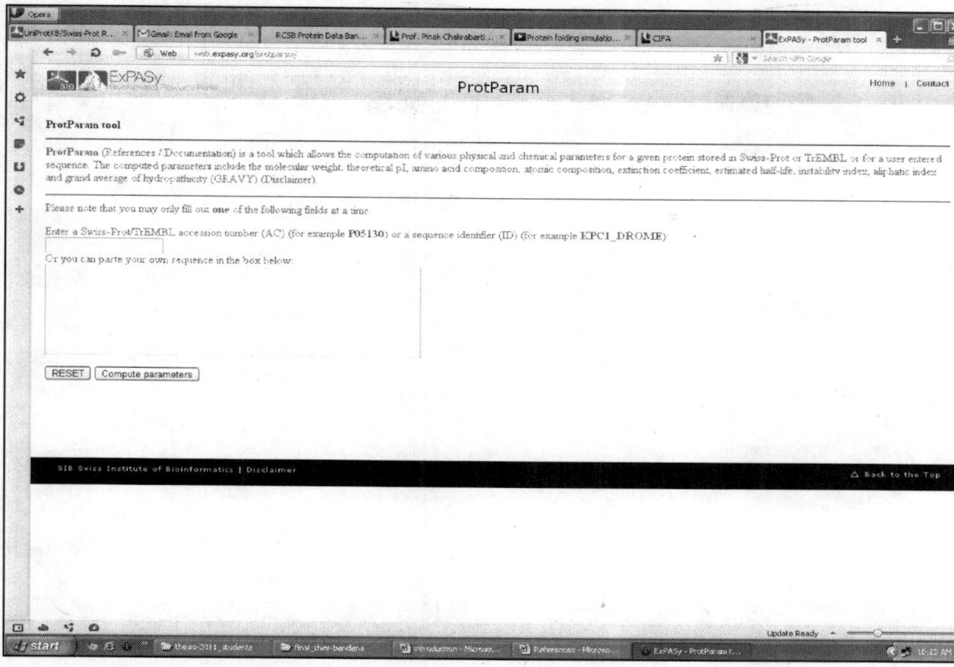

Figure 9.3: Home Page of EBI-ProtParam Tool.

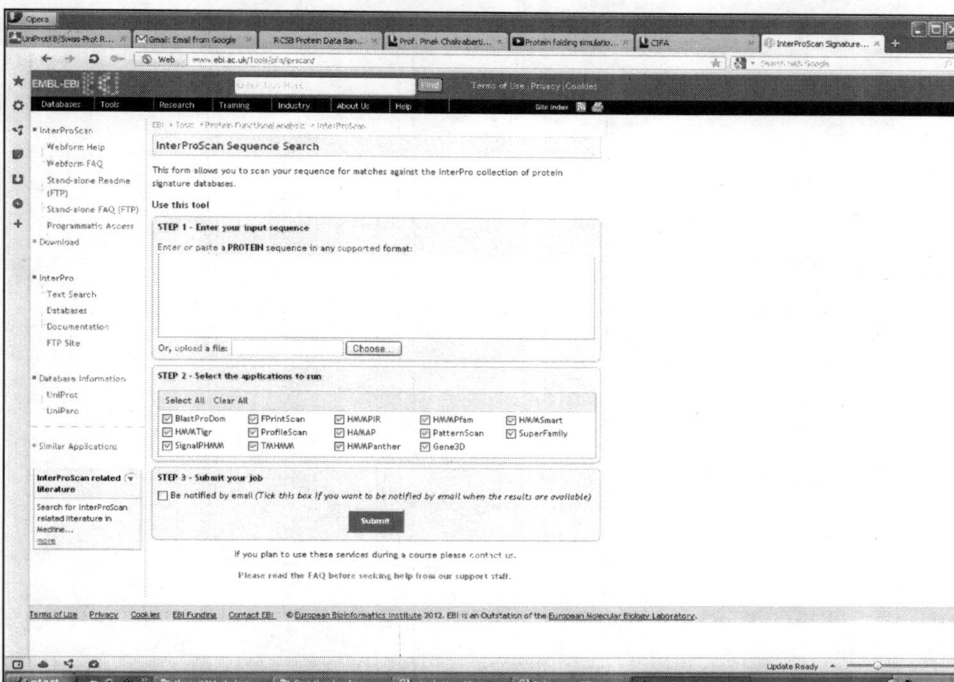

Figure 9.4: Home Page of EBI-InterProScan Tool.

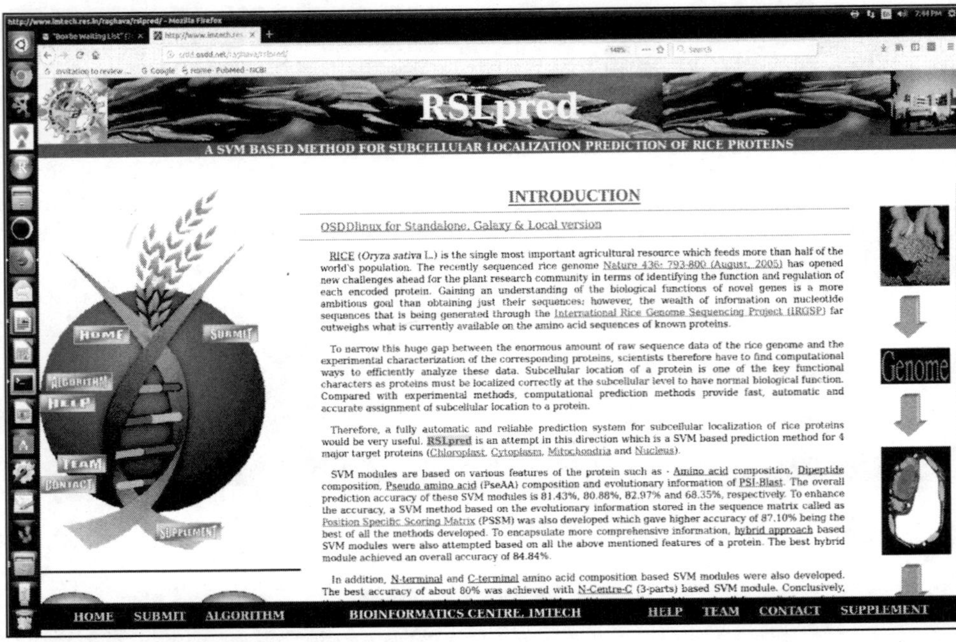

Figure 9.5: Home Page of RSLPred Tool.

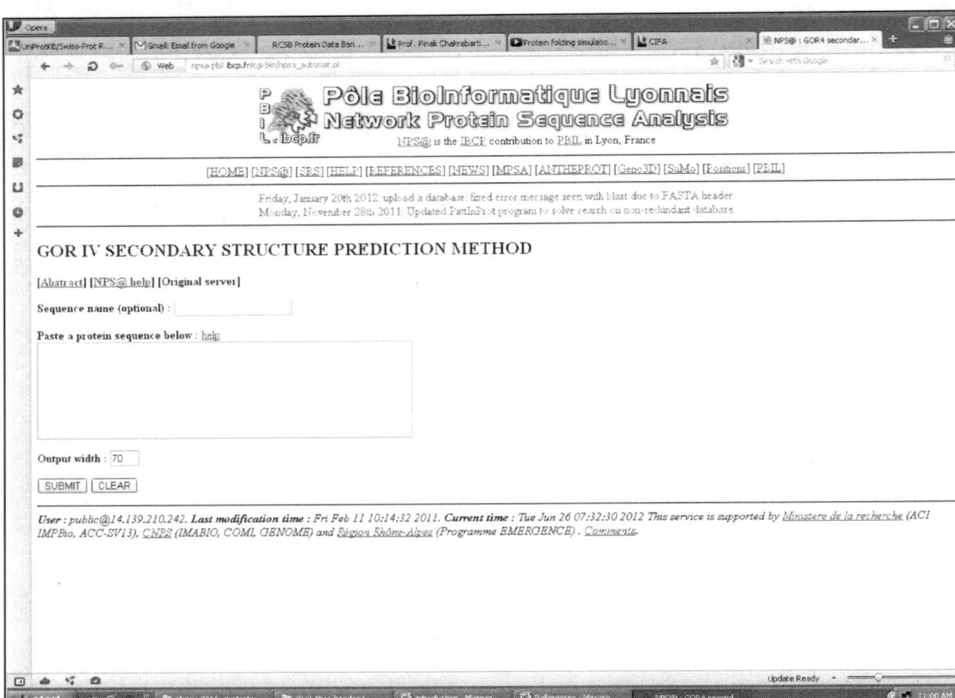

Figure 9.6: Home Page of GOR IV Tool.

Mining of Drought Resistant QTLs

Quantitative Trait Loci (QTLs) identifies a particular region of the genome containing one or more genes and that region is associated with the particular trait being measured. It is shown as an interval across a chromosome where the probability of association is plotted for each marker used in the mapping experiment. The Gramene QTLs database includes QTLs identified for numerous agronomic traits in rice, maize, barley, oat, sorghum, pearl millet, fox tail millet and wild rice. The rice genome was scanned for identifications of QTLs associated with the drought resistant (http://www.gramene.org/qtl).

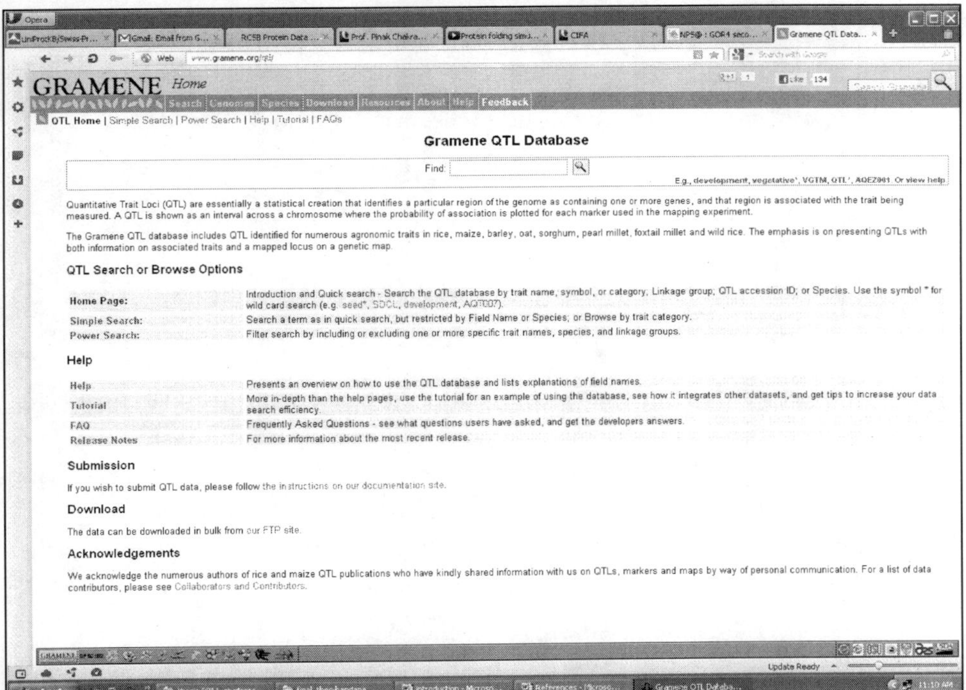

Figure 9.7: Home Page of Gramene QTL Database.

Results and Discussion

Drought Resistant Genes and Proteins

The twenty two numbers of candidate genes associated with drought resistant in *O. sativa* were retrieved with the gene length, its location in the chromosome and putative functions. There are 17 numbers of *O. sativa* alpha expansin and one number of Rab21, Rab16, Cis, P5cs and Rwc3 drought resistant genes have been retrieved from the Gene Bank (Table 9.1). Most of the genes are located in Chromosome-2 and Chromosome-3 and are larger in size. There are twenty four numbers of proteins involved in the drought resistant mechanism in *O. sativa,* out of which 23 are conceptual translation (reviewed) except *O. sativa* clone (14b) osmotin protein

Table 9.1: List of Drought Resistant Genes of *Oryza sativa*

Sl.No.	*Candidate Gene*	*Name*	*Accession No.*	*Length*	*Chromosome Location*	*Function*
1	OsExpa-1	*O. sativa* alpha expansin	AF 394543	2637 DNA Linear	Chromosome-4	Expression of alpha – expansin like genes in deep water rice.
2	OsExpa-2	*O. sativa* alpha expansin	AF394544	5000 DNA Linear	Chromosome-1	They are cell elongation proteins. Their root specific expression facilitates loosening of cell wall hence root length.
3	OsExpa-4	*O. sativa* alpha expansin	AF394545	5851 DNA Linear	Chromosome-5	Expansins were first identified as cell-wall- loosening proteins that, at least in part, mediate pH-dependent extension of the plant cell wall and growth of the cell.
4	OsExpa-5	*O. sativa* alpha expansin	AF394546	6774 DNA Linear	Chromosome-2	Cell wall biogenesis/degradation
5	OsExpa-7	*O. sativa* alpha expansin *Oryza sativa* alpha expansin	AF394547	4985 DNA Linear	Chromosome-3	Expansins were first identified as cell-wall- loosening proteins that, at least in part, mediate pH-dependent extension of the plant cell wall and growth of the cell.
6	OsExpa-12	*O.sativa* alpha expansin	AF394548	2998 DNA Linear	Chromosome-3	Expansins were first identified as cell-wall- loosening proteins that, at least in part, mediate pH-dependent extension of the plant cell wall and growth of the cell.
7	OsExpa-13	*O. sativa* alpha expansin	AF394549	4996 DNA Linear	Chromosome-2	Expansins were first identified as cell-wall- loosening proteins that, at least in part, mediate pH-dependent extension of the plant cell wall and growth of the cell.
8	OsExpa-14	*O. sativa* alpha expansin	AF394550	2000 DNA Linear	Chromosome-2	Expansins were first identified as cell-wall- loosening proteins that, at least in part, mediate pH- dependent extension of the plant cell wall and growth of the cell.
9	OsExpa-15	*O. sativa* alpha expansin	AF394551	4105 DNA Linear	Chromosome-3	Expansins were first identified as cell-wall- loosening proteins that, at least in part, mediate pH-dependent extension of the plant cell wall and growth of the cell.
10	OsExpa-18	*O. sativa* alpha expansin	AF394553	2908 DNA Linear	Chromosome- 3	Expansins were first identified as cell-wall- loosening proteins that, at least in part, mediate pH-dependent extension of the plant cell wall and growth of the cell.

Sl.No.	Candidate Gene	Name	Accession No.	Length	Chromosome Location	Function
11	Os Expa-19	*O. sativa* alpha expansin	AF394554	4913 DNA Linear	Chromosome-3	Expansins were first identified as cell-wall- loosening proteins that, at least in part, mediate pH- dependent extension of the plant cell wall and growth of the cell.
12	Os Expa-20	*O. sativa* alpha expansin	AF394555	4996 DNA Linear	Chromosome-3	Expansins were first identified as cell-wall- loosening proteins that, at least in part, mediate pH- dependent extension of the plant cell wall and growth of the cell.
13	OsExpa-21	*O. sativa* alpha expansin	AF394556	3929 DNA Linear	Chromosome-3	Expansins were first identified as cell-wall- loosening proteins that, at least in part, mediate pH-dependent extension of the plant cell wall and growth of the cell.
14	OsExpa-23	*O.sativa* alpha expansin	AF394558	4759 DNA Linear	Chromosome-2	Expansins were first identified as cell-wall- loosening proteins that, at least in part, mediate pH-dependent extension of the plant cell wall and growth of the cell.
15	OsExpa-24	*O. sativa* alpha expansin	AF394559	4189 DNA Linear	Chromosome-2	Expansins were first identified as cell-wall- loosening proteins that, at least in part, mediate pH-dependent extension of the plant cell wall and growth of the cell.
16	OsExpa-25	*O. sativa* alpha expansin	AF394560	3848 DNA Linear	Chromosome-3	Expansins were first identified as cell wall- loosening proteins that, at least in part, mediate pH-dependent extension of the plant cell wall and growth of the cell. Expansins were first identified as cell-wall-loosening proteins that, at least in part, mediate pH- dependent extension of the plant cell wall and growth of the cell.
17	Os Expa-26	*O. sativa* alpha expansin	AF394561	6312 DNA Linear	Chromosome-12	Expansins were first identified as cell-wall- loosening proteins that, at least in part, mediate pH-dependent extension of the plant cell wall and growth of the cell.
18	Rab-16	*O. sativa* dna, cis acting	AF333275	2828 DNA Linear	Chromosome-1	Stresses in various tissues of vegetative and floral organs.
19	Rab-21	Rice rab for water stress inducible protein	Y00842	2537 DNA Linear	Chromosome-11	These genes are responsive to absiscic acid of and rab16a promoter was induced by aba and osmotic adjustment.

Sl.No.	*Candidate Gene*	*Name*	*Accession No.*	*Length*	*Chromosome Location*	*Function*
20	Cis	*O. sativa* sub species japonicarab16 b gene.	AB001682. 1	1207 DNA Linear	Chromosome-5	Cis acting element of gene pwsi 18 promoter showed transient Expression in response to water stress which makes wsi gene express during water defict.
21	P5cs	*O.sativa* mrna for deltal pyrroline 5- carboxylate synthetase	D49714	2549 DNA Linear	Chromosome-12	Induced by dehydration ABA and high salt environment .responsible for the production of proline which acts as osmoprotectant, during water deficit.
22	Rwc3	*O. sativa* gene for water channel protein rwc3 promoter region.	AB029325	5325 DNA Linear	Chromosome-6	They are water channel proteins which are permeable to water and are root –specific.

which is conceptual translation (unreviewed). Most of the proteins are extracellular in origin however very few are membrane bound and nuclear in origin. All the proteins are low molecular weight and stable in the cellular conditions (Table 9.2).

Table 9.2: List of Drought Resistant Proteins

Sl. No.	*Protein Name*	*Swiss Prot ID*	*Molecular Weight*	*PI*	*Aliphatic Index*	*Size*	*Sub Cellular Location*
1	*O.sativa* alpha expansin (OS Expa-1)	Q7XWU8	27.74	7.44	65.48	261	EXTRA CELLULAR
2	*O. sativa* alpha expansin (OS Expa-2)	Q40636	26.67	6.78	64.14	251	EXTRA CELLULAR
3	*O. sativa* alpha expansin (OS Expa-4)	Q0DHB7	25.88	8.13	61.14	246	PLASMA MEMBRANE
4	*O. sativa* alpha expansin (OS Expa-5)	Q6ZGU9	30.17	9.43	68.76	291	EXTRA CELLULAR
5	*O. sativa* alpha expansin (OS Expa-7)	Q852A1	28.22	9.45	66.89	264	EXTRA CELLULAR
6	*O. sativa* alpha expansin (OS Expa-12)	Q7G6Z2	26.42	8.51	59.40	250	EXTRA CELLULAR
7	*O. sativa* alpha expansin (OS Expa- 13)	Q4PR52	27.97	9.19	70.84	262	EXTRA CELLULAR
8	*O. sativa* alpha expansin (OS Expa-14)	Q4PR51	28.38	8.96	64.85	262	EXTRA CELLULAR
9	*O. sativa* alpha expansin (OS Expa-15)	Q4PR50	27.79	8.92	71.31	260	EXTRA CELLULAR
10	*O. sativa* alpha expansin (OS Expa- 18)	Q4PR48	26.43	6.50	74.10	249	EXTRA CELLULAR
11	*O. sativa* alpha expansin (OS Expa-19)	Q7G6Z5	26.79	5.11	72.17	249	MEMBRANE BOUND CHLOROPLAST
12	*O. sativa* alpha expansin (OS Expa- 20)	Q10RK1	25.81	5.65	76.46	240	MEMBRANE BOUND CHLOROPLAST
13	*O. sativa* alpha expansin (OS Expa- 21)	Q10KN4	27.18	9.53	70.23	264	EXTRA CELLULAR
14	*O. sativa* alpha expansin (OS Expa- 23)	Q4PR43	28.57	9.24	66.25	267	EXTRA CELLULAR
15	*O. sativa* alpha expansin (OS Expa- 24)	Q4PR42	30.04	8.90	67.45	278	EXTRA CELLULAR
16	*O. sativa* alpha expansin (OS Expa-25)	Q4PR41	27.08	8.37	72.00	255	MEMBRANE BOUND VACUOLE
17	*O. sativa* alpha expansin (OS Expa-26)	Q2QP13	29.39	8.69	61.13	284 GB 290SP	EXTRA CELLULAR
18	Rice rab21 gene for water stress inducible protein.	A2ZDX9	16.54	9.40	28.16	163 GB 172SP	CYTOPLASM

Sl. No.	*Protein Name*	*Swiss Prot ID*	*Molecular Weight*	*PI*	*Aliphatic Index*	*Size*	*Sub Cellular Location*
19	Rab-16 *O. sativa* sub species japonica	Q2R4Z5	16.53	9.27	32.26	164	CYTOPLASM
20	Deltal –pyrroline 5 carboxylate synthetase	O04226	77.69	6.34	103.49	716	CYTOPLASM
21	Water channel protein rwc3	Q9SXF8	30.87	9.44	93.55	287 GB 288SP	PLASMA MEMBRANE
22	Lipid transfer protein precursor partial *O. sativa*	A2ZAS9	10.02	8.92	64,14	99 GB 121SP	EXTRA CELLULAR
23	*O. sativa* water stress inducible protein (kcdl 917) mrna.	O24227	16.17	11.19	66.30	154	NUCLEAR
24	*O. sativa* clone (14b) osmotin protein	Q40630	22.43	7.97	71.24	218	EXTRA CELLULAR

Analysis of Functional Domains and Secondary Structures of the Drought Resistant Proteins

The OSEXPA (*O. sativa* alpha expansin) group of proteins consist of IPR007112, IPR007118, IPR002963, IPR007117, IPR014733 and IPR009009 domains, whereas RAB proteins consist of IPR000167(Dehydrin) domain. P5CS consists of different types of functional domains such as IPR005715, IPR000965, IPR016163, IPR016162, IPRO16161, IPR015590, IPR019797, IPR005766, IPR020593, IIPR012134, IPR001057 and IPR001048 involved in wide range of biochemical reactions. RWC3 contains IPR023276, IPR012269, IPR022357, IPR015686 and IPR023271 domains. Domains named IPR013770, IPR000528 and IPR016140 were found in the LTP protein and KCDL contains IPR011991, IPR005818 and IPR005819 domains. It was observed that expansin/poolen allergen, DPBB domain, Expansin/lol pl and Brawin-related endoglucanase were commonly present in all the drought resistant proteins of *O. sativa*. These domains are conferring drought resistant in rice by adopting various biochemical reactions (Table 9.3). The secondary structural analysis revealed that most of the proteins are rich in extended strands and random coils, while proteins *viz.* Delta P5CS, RWC3 and KCDL possess more numbers of alpha helices than that of extended strands and random coils.

Analysis of Drought Resistant QTLs

QTL analysis revealed whether phenotypic differences are primarily due to a few loci with fairly large effects, or to many loci, each with minute effects. It appears that a substantial proportion of the phenotypic variation in many quantitative traits can be explained with few loci of large effect, with the remainder due to numerous loci of small effect. The genome wide analysis of *O. sativa* suggested only 38 Nos of QTLs associated with the drought resistant. These DTLs positions, co- localized markers and neighbouring markers also been identified. Once QTL have been identified, molecular techniques can be employed to narrow the QTL

Table 9.3: List of Functional Domains Present in the Proteins

Protein Name	*Functional Domain*
OSEXPA1	**IPR007112**-EXPANSIN/POOLEN ALLERGEN,DPBB DOMAIN, **IPR007118**-EXPANSIN/LOL PL **IPR002963**-EXPANSIN, **IPR007117**- EXPANSIN/POOLEN ALLERGEN,C-TERMINAL, **IPR014733**-BARWIN-LIKE ENDOGLUCANASE, **IPR009009**-BRAWIN –RELATED ENDOGLUCANASE
OSEXPA2	**IPR007112**-EXPANSIN/POOLEN ALLERGEN,DPBB DOMAIN, **IPR007118**-EXPANSIN/LOL PL **IPR002963**-EXPANSIN, **IPR007117**- EXPANSIN/POOLEN ALLERGEN,C-TERMINAL, **IPR014733**-BARWIN-LIKE ENDOGLUCANASE,**IPR009009**-BRAWIN –RELATED ENDOGLUCANASE
OS EXPA4	**IPR007112**-EXPANSIN/POOLEN ALLERGEN,DPBB DOMAIN, **IPR007118**-EXPANSIN/LOL PL **IPR002963**-EXPANSIN, **IPR007117**- EXPANSIN/POOLEN ALLERGEN,C-TERMINAL, **IPR014733**-BARWIN-LIKE ENDOGLUCANASE, **IPR009009**-BRAWIN –RELATED ENDOGLUCANASE
OSEXPA5	**IPR007112**-EXPANSIN/POOLEN ALLERGEN,DPBB DOMAIN, **IPR007118**-EXPANSIN/LOL PL **IPR002963**-EXPANSIN, **IPR007117**- EXPANSIN/POOLEN ALLERGEN,C-TERMINAL, **IPR014733**-BARWIN-LIKE ENDOGLUCANASE, **IPR009009**-BRAWIN –RELATED ENDOGLUCANASE
OSEXPA7	**IPR007112**-EXPANSIN/POOLEN ALLERGEN,DPBB DOMAIN, **IPR007118**-EXPANSIN/LOL PL **IPR002963**-EXPANSIN, **IPR007117**- EXPANSIN/POOLEN ALLERGEN,C-TERMINAL, **IPR014733**-BARWIN-LIKE ENDOGLUCANASE, **IPR009009**-BRAWIN –RELATED ENDOGLUCANASE
OSEXPA12	**IPR007112**-EXPANSIN/POOLEN ALLERGEN,DPBB DOMAIN, **IPR007118**-EXPANSIN/LOL PL **IPR002963**-EXPANSIN, **IPR007117**- EXPANSIN/POOLEN ALLERGEN,C-TERMINAL, **IPR014733**-BARWIN-LIKE ENDOGLUCANASE, **IPR009009**-BRAWIN –RELATED ENDOGLUCANASE
OSEXPA13	**IPR007112**-EXPANSIN/POOLEN ALLERGEN,DPBB DOMAIN, **IPR007118**-EXPANSIN/LOL PL **IPR002963**-EXPANSIN, **IPR007117**- EXPANSIN/POOLEN ALLERGEN,C-TERMINAL, **IPR014733**-BARWIN-LIKE ENDOGLUCANASE, **IPR009009**-BRAWIN –RELATED ENDOGLUCANASE
OSEXPA14	**IPR007112**-EXPANSIN/POOLEN ALLERGEN,DPBB DOMAIN, **IPR007118**-EXPANSIN/LOL PL **IPR002963**-EXPANSIN, **IPR007117**- EXPANSIN/POOLEN ALLERGEN,C-TERMINAL, **IPR014733**-BARWIN-LIKE ENDOGLUCANASE, **IPR009009**-BRAWIN –RELATED ENDOGLUCANASE
OSEXPA15	**IPR007112**-EXPANSIN/POOLEN ALLERGEN,DPBB DOMAIN, **IPR007118**-EXPANSIN/LOL PL **IPR002963**-EXPANSIN, **IPR007117**- EXPANSIN/POOLEN ALLERGEN,C-TERMINAL, **IPR014733**-BARWIN-LIKE ENDOGLUCANASE, **IPR009009**-BRAWIN –RELATED ENDOGLUCANASE

OSEXPA18	**IPR007112**-EXPANSIN/POOLEN ALLERGEN,DPBB DOMAIN, **IPR007118**-EXPANSIN/LOL PL IPR002963-EXPANSIN, **IPR007117**- EXPANSIN/POOLEN ALLERGEN,C-TERMINAL, **IPR014733**-BARWIN-LIKE ENDOGLUCANASE, **IPR009009**-BRAWIN –RELATED ENDOGLUCANASE
OSEXPA19	**IPR007112**-EXPANSIN/POOLEN ALLERGEN,DPBB DOMAIN, **IPR007118**-EXPANSIN/LOL PL **IPR002963**-EXPANSIN, **IPR007117**- EXPANSIN/POOLEN ALLERGEN,C-TERMINAL, **IPR014733**-BARWIN-LIKE ENDOGLUCANASE, **IPR009009**-BRAWIN –RELATED ENDOGLUCANASE
OSEXPA20	**IPR007112**-EXPANSIN/POOLEN ALLERGEN,DPBB DOMAIN, **IPR007118**-EXPANSIN/LOL PL **IPR002963**-EXPANSIN, **IPR007117**- EXPANSIN/POOLEN ALLERGEN,C-TERMINAL, **IPR014733**-BARWIN-LIKE ENDOGLUCANASE, **IPR009009**-BRAWIN –RELATED ENDOGLUCANASE
OSEXPA21	**IPR007112**-EXPANSIN/POOLEN ALLERGEN,DPBB DOMAIN, **IPR007118**-EXPANSIN/LOL PL **IPR002963**-EXPANSIN, **IPR007117**- EXPANSIN/POOLEN ALLERGEN,C-TERMINAL, **IPR014733**-BARWIN-LIKE ENDOGLUCANASE,**IPR009009**-BRAWIN –RELATED ENDOGLUCANASE
OSEXPA23	**IPR007112**-EXPANSIN/POOLEN ALLERGEN,DPBB DOMAIN, **IPR007118**-EXPANSIN/LOL PL **IPR002963**-EXPANSIN, **IPR007117**- EXPANSIN/POOLEN ALLERGEN,C-TERMINAL, **IPR014733**-BARWIN-LIKE ENDOGLUCANASE, **IPR009009**-BRAWIN –RELATED ENDOGLUCANASE
OSEXPA24	**IPR007112**-EXPANSIN/POOLEN ALLERGEN,DPBB DOMAIN, **IPR007118**-EXPANSIN/LOL PL **IPR002963**-EXPANSIN, **IPR007117**- EXPANSIN/POOLEN ALLERGEN,C-TERMINAL, **IPR014733**-BARWIN-LIKE ENDOGLUCANASE, **IPR009009**-BRAWIN –RELATED ENDOGLUCANASE
OSEXPA25	**IPR007112**-EXPANSIN/POOLEN ALLERGEN,DPBB DOMAIN, **IPR007118**-EXPANSIN/LOL PL **IPR002963**-EXPANSIN, **IPR007117**- EXPANSIN/POOLEN ALLERGEN,C-TERMINAL, **IPR014733**-BARWIN-LIKE ENDOGLUCANASE, **IPR00900**9-BRAWIN –RELATED ENDOGLUCANASE
OSEXPA26	**IPR007112**-EXPANSIN/POOLEN ALLERGEN,DPBB DOMAIN, **IPR007118**-EXPANSIN/LOL PL **IPR002963**-EXPANSIN, **IPR007117**- EXPANSIN/POOLEN ALLERGEN,C-TERMINAL, **IPR014733**-BARWIN-LIKE ENDOGLUCANASE, **IPR009009**-BRAWIN –RELATED ENDOGLUCANASE
RAB21	**IPR000167**-DEHYDRIN
RAB 16	**IPR000167**-DEHYDRIN
P5CS	**IPR005715**- Glutamate 5-kinase/delta-1-pyrroline-5-carboxylate synthase, **IPR000965**-Gamma-glutamyl phosphate reductase GPR, **IPR016163**- Aldehyde dehydrogenase, C-terminal, **IPR016162**- Aldehyde dehydrogenase, N-terminal, **IPRO16161**- Aldehyde/histidinol dehydrogenase, **IPR015590**- Aldehyde dehydrogenase domain, **IPR019797**-Glutamate 5-kinase, conserved site, **IPR005766**- Delta I-pyrroline-5-carboxylate synthetase, **IPR020593**- Gamma-glutamyl phosphate reductase GPR, conserved site, **IIPR012134-** Glutamate-5-semialdehyde dehydrogenase, **IPR001057**- Glutamate/acetylglutamate kinase,**IPR001048**- Aspartate/glutamate/uridylate kinase

RWC3	**IPR023276-** Aquaporin 5, **IPR012269-** Aquaporin, **IPR022357-** Major intrinsic protein, conserved site, **IPR015686-** Aquaporin 7, **IPR000425-**Major intrinsic protein, **IPR023271-** Aquaporin-like
LTP	**IPR013770**- Plant lipid transfer protein/hydrophobic protein helical domain, **IPR000528**- Plant lipid transfer protein/Par allergen, **IPR016140** -Bifunctional inhibitor/plant lipid transfer protein/seed storage helical domain
KCDL	**IPR011991**-Winged helix-turn-helix transcription repressor DNA- binding, **IPR005818**- Histone H1/H5 **IPR005819**-Histone H5

Table 9.4: List of Secondary Structural Elements

Sl.No.	*Name*	*Alpha-Helix*	*Extended Strand*	*Random Coil*
1	OS EXPA-1	6.13 per cent	41.00 per cent	52.87 per cent
2	OS EXPA-2	9.96 per cent	25.50 per cent	64.54 per cent
3	OS EXPA-4	11.79 per cent	25.20 per cent	63.01 per cent
4	OS EXPA-5	19.24 per cent	23.37 per cent	57.39 per cent
5	OS EXPA-7	9.85 per cent	25.38 per cent	64.77 per cent
6	OS EXPA-12	10.00 per cent	31.20 per cent	58.80 per cent
7	OS EXPA-13	16.03 per cent	28.24 per cent	55.73 per cent
8	OS EXPA-14	10.69 per cent	29.01 per cent	60.31 per cent
9	OS EXPA-15	9.62 per cent	28.85 per cent	61.54 per cent
10	OS EXPA-18	4.82 per cent	34.54 per cent	60.64 per cent
11	OS EXPA-19	7.23 per cent	31.73 per cent	61.04 per cent
12	OS EXPA-20	4.58 per cent	34.58 per cent	60.83 per cent
13	OS EXPA-21	15.91 per cent	26.14 per cent	57.95 per cent
14	OS EXPA-23	17.60 per cent	26.59 per cent	55.81 per cent
15	OS EXPA-24	14.03 per cent	28.06 per cent	57.91 per cent
16	OS EXPA-25	5.88 per cent	35.69 per cent	58.43 per cent
17	OS EXPA-26	12.32 per cent	30.28 per cent	57.39 per cent
18	RAB 21	14.72 per cent	28.22 per cent	57.06 per cent
19	RAB16	16.46 per cent	20.73 per cent	62.80 per cent
20	Delta P5CS	41.62 per cent	14.80 per cent	43.58 per cent
21	RWC3	30.13 per cent	23.00 per cent	46.69 per cent
22	LTP	9.09 per cent	20.20 per cent	70.71 per cent
23	KCDL	62.99 per cent	4.55 per cent	32.47 per cent
24	OSMOTIN	5.05 per cent	31.65 per cent	63.30 per cent

down to candidate genes. One important emerging trend in these analyses is the prominent role of regulatory genes, or genes that code for transcription factors and other signalling proteins (Table 9.5).

Table 9.5: List of Drought Resistant QTLs

QTL accession id	AQHP062	AQHP058	AQHP059	AQHP082	AQHP083	AQHP070	AQHP065
linkage group	8	2	4	6	7	4	1
Position	66.9-66.9 CM	104.3-104.3 CM	11.2-11.2 CM	24.9-24.9 CM	11.4-11.4 CM	11.2-11.2 CM	60.1-60.1 CM
Co- localized markers	RG1	RG544	RG21F	RZ192	CD0533	RG214	RG462
Neighbouring markers	RG978,RZ6 6	RG437,RZ6 6	RG620,RG4 76	RZ144,RZ6 4	RG351,CD0 405	RG620,RG4 76	RG101,CD0 920
QTL accession id	AQHP084	AQHP069	AQHP068	AQAN00 1	AQHP078	AQHP067	AQHP085
linkage group	8	3	2	5	11	2	9
Position	66.9-66.9 cm	87.8-87.8 cm	100-100c m	131.6-134.9 cm	49.5-49.5 cm	49.1-49.1 cm	45-45 cm
Co- localized markers	RG1	RG910	RG437	RM26,K M31	RZ35	RG139	RG662
Neighbouring markers	RG978- RZ66	RG96,RG48 2,RZ393	RG157,RG5 54	C246,RG43 5	RG1094,RG 118	RG102,RZ1 03	RG570,RZ4 04
QTL accession id	AQHP073	AQHP088	AQAN00 2	AQHP076	AQHP063	AQHP057	AQHP077
linkage group	7	11	12	10	9	1	11
Position	11.4-11.4 cm	49.5-49.5 cm	69.1-93.6 cm	12.4-12.4 cm	45-45 cm	60.1-60.1 cm	36.4-36.4 cm
Co- localized markers	CD6533	RZ35	RZ457,G1 184,RG36 1,Y12817	RZ561	RG662	RG462	CD0365
Neighbouring markers	RZ351,CD0 405	RG1094,RG 118	G2140,G13 91	RZ241,RZ8 11	RG570,RZ4 04	RG101,CD0 920	RG16,RG21 1

QTL accession id	AQHP081	AQHP079	AQAN00 5	AQAN00 4	AQHP066	AQHP080
linkage group	4	3	8	4	2	4
Position	82.9-82.9 cm	59.2-59.2 cm	96.5-96.5 cm	162.6-162.6 cm	19.1-19.1 cm	11.2-11.2 cm
Co- localized markers	RG190	RG394	G192	G318	RG324	RG214
Neighbouring markers	RG449,RZ6 9	RZ284,RZ5 76	G93,RM42	RG190,Y34 L RG322, RG73	RG73	RG620,RG4 75
QTL accession id	AQHP061	AQHP075	AQHP074	AQHP060	AQHP087	AQHP064
linkage group	6	9	8	4	11	11
Position	2.3-2.3 cm	45-45 cm	66.9-66.9 cm	82.9-82.9 cm	30.4-30.4 cm	30.4-30.4 cm
Co- localized markers	RZ516	RZ662	RG1	RZ190	CD0365	CD0365
Neighbouring markers	CD0475,RZ 398	RG570,RZ4 04	RG978,RZ6 6	RG449,RZ6 9	RG16,RG21 1	RG16,RG21 1
QTL accession id	AQAN00 3	AQAN00 6	AQHP086	AQHP071	AQHP072	
linkage group	1	3	10	4	6	
Position	97.3-97.3 cm	11.3-11.3 cm	6.3-6.3 cm	82.9-82.9 cm	2.3-2.3 cm	
Co- localized markers	RG541	RM231	RZ892	RG190	RZ516	
Neighbouring markers	RM212,G10 85,G396	CT121,G17 5	RZ257,RZ4 66	RG449,RZ6 9	CD0475,RZ 398	

Conclusion

The attainment of a complete and accurate map-based sequence for rice is compelling. We now have a blueprint for all of the rice chromosomes. We know, with a high level of confidence, the distribution and location of all the main components the genes, repetitive sequences and centromeres. Substantial portions of the map-based sequence have been in public databases for some time, and the availability of provisional rice pseudomolecules based on this sequence has provided the scientific community with numerous opportunities to evaluate the genome, as indicated by the number of publications in rice biology and genetics over the past few years. Furthermore, the wealth of SNP and SSR information provided here and elsewhere will accelerate marker-assisted breeding and positional cloning, facilitating advances in rice improvement. The syntenic relationships between rice and the cereal grasses have long been recognized for comparing genome organization, genes and intergenic regions between cereal species will permit identification of regions that are highly conserved or rapidly evolving. Such regions are expected to yield crucial insights into genome evolution, speciation and domestication. Global gene expression analysis showed that about 55 per cent of genes differentially expressed in roots of rice in response to drought stress treatments. In contrast to the drought-induced genes, many of which are functionally annotated, the genes with the highest repression factors were mostly of unknown or putative function. These genes could be as relevant for drought-tolerance as the highly induced genes, however, they are obviously much more difficult to interpret and much more time consuming to study functionally. The candidate list could be further narrowed down by checking candidate gene expression for their contrasting drought tolerance in the QTL region of interest. This strategy could also include genes of unknown function and thus open up the chance to discover truly unknown genes that are relevant for drought stress tolerance.

References

Alford, J. and Duguid, N., 2003. *Seductions of rice*. Artisan Books.

Chaudhary, R.C., 2001. Speciality rice of the world: a prologue. *Speciality rices of the world. Breeding, production and marketing*.

Chopra, V.L. and Prakash, S., 2002. *Evolution and adaptation of cereal crops*. Science Publishers, Inc.

International Rice Research Institute, 2001. Rice Research and Production in the 21st Century. (Gramene Reference ID 8380)

International Rice Research Institute, 2002. Rice Almanac, 3rd Edition. (Gramene Reference ID 8379)

International Year of Rice, 2004. Aquatic biodiversity in rice fields factsheet. (Gramene Reference ID 8373)

International Year of Rice, 2004. "Gender and rice" factsheet. (Gramene Reference ID 8371)

International Year of Rice, 2004. Rice and water: a long and diversified story factsheet. (Gramene Reference ID 8372)

International Year of Rice, (2004). Rice and human nutrition factsheet. (Gramene Reference ID 8370)

10

Application of Different Pretreatment Strategies on Grass Biomass for Enhanced Delignification and Cellulose Exposure for Bioethanol Production

S. Mohapatra[1*], S.S. Mishra[1] and H.N. Thatoi[2]

[1]Department of Biotechnology, College of Engineering and Technology, Ghatikia, Bhubaneswar – 751 003, Odisha

[2]Department of Biotechnology, North Orissa University, Baripada – 757 003, Odisha

****E-mail: sonu0625@gmail.com***

ABSTRACT

Bioethanol production from renewable sources like lignocellulosic biomass holds a remarkable future as a substitute for the depleting fossil fuels. Lignocellulosic substrates such as grass biomass consist of complex cell wall components which require various processes such as pretreatment, enzymatic hydrolysis and fermentation for bioethanol conversion among which pretreatment is most important one. This paper highlights the action of pretreatment strategies on individual grass cell wall components. The major forms of pretreatment include alkali, acid, ultrasono-assisted alkali/acid, microwave and biological pretreatments. Further, the application of ultrasono assisted alkali/acid pretreatment on Denannath and Hybrid napier grass (CO-3) varieties for enhanced delignification and cellulose exposure for bioethanol production has been discussed.

Keywords: *Delignification, Fermentation, Pennisetum sp., Pretreatment, Ultrasonication.*

Introduction

The fossil fuels which are the major source of energy supply are depleting at a faster rate which has become a global concern (Teramura *et al.*, 2015). Besides, the use of fossil fuels leads to environmental pollution and global warming. Thus it has necessitated the search for alternative source of energy to meet the growing energy demand. Thus current energy policies are focusing on utilization of eco-friendly, bio-degradable and economical alternatives such as biofuel which can help in reduction of global warming and simultaneously improve energy security (Tao *et al.*, 2017). At present the renewable sources supply only about 20 per cent of the global energy consumed, which is expected to rise to 80 per cent by 2050. Biofuels can be produced from different sources such as sugary, starchy, cellulosic biomass and algal biofuels.

Among bio-fuels, bioethanol production through fermentation of lignocellulosic biomass (*i.e.* agricultural wastes, energy crops) has several advantages over other biologically produced fuels. However, ethanol production from lignocellulosic biomass requires a variety of steps such as pretreatment, enzymatic hydrolysis and fermentation. Among these processes, pretreatment plays a significant role in destructuring the cell wall components of biomass for easy accessibility of enzymes for efficient hydrolysis. However, the type of pretreatment for a biomass greatly depends on the inter-linking units of the cell wall. The underlying cause is the dissimilar effects that occur on the individual cell wall components through pretreatment. Apart from that, the result of the pretreatment must also validate its impact on the economics of downstream processing steps and must maintain a balance mainly between capital investments and biomass and operating costs (**Lynd, 1996**). Therefore, in the present paper a brief discussion about the most effective pretreatments that have been observed for grass biomass *i.e.* alkali, acid, ultrasono assisted alkali and acid pretretment and biological pretreatments are presented. The effects of these pretreatments on the major cell wall linkages that exist between cellulose, hemicellulose and lignin have also been elucidated. Besides, the application of pretreatment of delignification and cellulose exposure for bioethanol production from grass biomass (Dennanath) has been discussed.

Composition of Grasses

The composition of grass biomass plays an important role for bioethanol production. The grass cell wall generally consists of cellulose, hemicellulose and lignin as the major cell wall components. Grass cell wall composition is important for determination of pretreatment. The greater the amount of lignin present in the cell wall, the harsher is the mode of pretreatment. Table 10.1 illustrates the composition of different grasses which have the potential to be used as energy feed stocks.

Pretreatment Strategies for Enhanced Delignification of Grass Biomass

Delignification of lignocellulosic biomass is required for easy accessibility of saccharifing enzymes to cellulose and hemicellulose for enhanced bioethanol production. However the process of delignification is substrate specific, as the composition and arrangement of lignin varies among different substrates. For grass biomass the efficient pretreatment technologies that have reported are chemical pretreatments which comprises of acids and alkalis and physiochemical pretreatm ents (Mohapatra *et al.*, 2016).

Table 10.1: Composition of different Grass Biomass Used for

Sl. No.	*Lignocellulosic Biomass*	*Cellulose (wt per cent)*	*Hemicellulose (wt per cent)*	*Lignin (wt per cent)*	*Reference*
1	Bamboo	26-43	15-26	21-31	Kumar *et al.*, 2016
2	Coastal Bermuda grass	25	35.7	6.4	Prasad *et al.*, 2007
3	Elephant grass	22	24	24	Olukemi, 2015
4	Esparto grass	33-38	27-32	17-19	Akchiche *et al.*, 2007
5	Esparto grass	33-38	27 - 32	17 – 19	Rowell *et al.*, 1992
6	King grass	50	23	21	Cardona *et al.*, 2016
7	Grasses (average)	25-40	25-50	10-30	Sun and Cheng, 2002
8	Orchard grass (medium maturity)	32	40	4.7	Howard *et al.*, 2003
9	Rice straw	32.1	24	18	Li *et al.*, 2009
10	Rye straw	37.6	30.5	19.0	Nigam *et al.*, 2009
11	Sabai grass	49.90	23.72	20.88	Tang *et al.*, 2013
12	Sugarcane bagasse	38.1	26.9	18.4	Lee, 1997
13	Miscanthus	28–49	24–32	15–28	Zhang *et al.*, 2012
14	Switch grass	45	31	12	Nigam *et al.*, 2009
15	Timothy grass	34	30	18	Nanda *et al.*, 2012
16	Napier grass	30-33	35-37	3-4	Nanda *et al.*, 2012
17	Dwarf napier	35.64	34.19	3.66	Wongwata-napaiboon *et al.*, 2012
18	King napier	32.01	31.13	3.10	Wongwata-napaiboon *et al.*, 2012
20	Srilanka vetiver grass	37.54	38.33	3.99	Wongwata-napaiboon *et al.*, 2012
21	KamphengPhet 1	34.95	39.67	5.06	Wongwata-napaiboon *et al.*, 2012
22	Thatch grass	45.07	29.06	17.43	Das *et al.*, 2013
23	Wheat straw	36.6	24.8	14.5	Lee,1997
24	Reed canary straw	42.6	29.7	7.6	Bridgeman *et al.*, 2008
25	Rice straw (*Oryza sativa* sp.)	41	21.5	9.9	Lee, 1997
26	Verge grass	30.7	15.6	14.1	Dijkerman *et al.* (1997)
27	Rye grass (early leaf)	21.3	15.8	2.7	Howard *et al.* (2003)

Acid Pretreatment

Acid pretreatments traditionally use dilute hydrochloric acid (HCl), sulphuric acid (H_2SO_4), nitric acid (HNO_3) among others. The main benefit offered by this technique is that, it sometimes eliminates the use of subsequent enzymatic saccharification step, as it mostly targets the hydrolysis of hemicelluloses and lignin in the plant cell wall.

Effect of Acid Pretreatment on Lignin Content

Acid pretreatment, generally dilute acid (DA), leads to fragmentation of aryl ether linkages (primarily β-O-4 linkages) by acidolysis and acid catalysed recondensation while stabilizing the linkages such as resinol and phenyl-coumaran subunits (Cao *et al.*, 2012). However, as acid hydrolysis mostly targets the β-O-4 linkages in lignin, the pretreated biomass generally has a lower content of the same. DA pretreatment in switch grass was found to decrease the β-O-4 linkages by 36 per cent (Samuel *et al.*, 2010). The authors also observed a distinguished variation in the monolignol S/G ratio after the DA pretreatment. Moreover, they observed a reduced biomass, suggesting the loss of cellulose and hemicellulose fractions. Similarly, Moxley *et al.* (2012) compared free phenolic OH groups associated with S and G lignin in grass biomass and found an increased phenolic S unit content than G units which was mostly due to the liable cleavage of β-O-4 linkages in syringyl lignin. Acid pretreatments are mostly effective to decrease the ferulates which act as bridges between lignin and polysaccharides. The decrease in ferulate promotes enhanced enzymatic hydrolysis as it helps in creating proximity of polysaccharide moieties with enzymes. Apart from ferulates, DA pretreatments effectively decrease aliphatic linkages in grasses, such as 5-5/4-O-β (dibenzodioxin), β-O-4 (β-aryl ether), β-O-4-H/G (β-aryl ether p-hydroxyphenyl/guaiacyl), β-O-4-S (β-aryl ether syringyl) and β-5 (phenylcoumaran) linkages. However in rice straw, DA pretreatment is observed mostly to have little effect on aromatic regions of lignin such as *p*-coumarate, guaiacyl, syringyl, *p*-hydroxyphenyl and cinnamyl alcohol end group (Teramura *et al.*, 2015). Similar observations were also made by Tao *et al.* (2017) on acid, ultrasonication assisted acid and microwave assisted acid pretreated *Triarrhenalutarioriparia*. However, greater modifications in lignin were observed in the later two pretreatments. In order to achieve effective delignification, it is important not only to understand the inter unit lignin linkages but also the association of lignin with other cell wall components which are responsible for the recalcitrance nature of the cell wall.

The effect of acid pretreatment on hemicellulose is an important aspect, as grasses unlike woody biomass are mostly dominated by hemicellulose rather than lignin (Pattahil *et al.*, 2015). DA pretreatment hydrolyzes the components of hemicellulose (xylan, arabinoxylan, mannan, glucomannan *etc.*) thereby enhancing the saccharification efficiency of cellulose. Mild acid causes depolymerisation of hemicellulose by selectively hydrolysing the glycosidic linkages and thus liberating O-acetyl group with other acid moieties for formation of acetic and uronic acids. The hydrolysis of glycosidic linkages in hemicellulose have been observed for hybrid napier grass and denannath grass by the authors when treated with ultrasonication assisted dilute HCl (Mohapatra *et al.*, 2016).

Although the release of acetic and uronic acids aid to the breakdown of hemicelluloses and oligosaccharides in these pretreatment conditions yet acetyl groups that are produced as a by-product of this pretreatment, inhibit necessary bindings of cellulases to cellulose thereby protecting the accessibility of cellulase to cellulose (Pan *et al.*, 2005). Selig and co-workers (Selig *et al.*, 2009) also supported the inhibitory action of acetyl groups and stated that when acetyl groups bind to the xylan backbone, the accessibility of cellulase to the β-1,4glycosidic linkages is hindered. The dissolved xylo-oligomers which contain acetyl groups can be cleaved using severe pretreatment conditions such as longer residence time in a process of deacetylation.

Deacetylation by DA is not only helpful for providing more sites for enzymatic attacks, but is also helps in reduction of cell-wall recalcitrance by formation of xylo-oligomers which have few side branches and are easily hydrolysed (Chen *et al.*, 2012). However, increasing DA severity also increases the risk of xylan degradation to by-products such as furfural which is inhibitory to the ethanol fermenting microorganisms. Another interesting phenomenon of translocation and redistribution of lignin into lignin droplets which is seen to increase cell wall accessibility for enzymes have been found in corn stover and grass biomass during DA pretreatment (Pingali *et al.*, 2010). It was found that the lignin-hemicellulose complex underwent phase separation during DA pretreatment, which drove the lignin aggregation followed by collapse, leading to increasing the cell wall porosity (Langan *et al.*, 2014). The interaction of lignin with polysaccharides and the effect of DA pretreatment on the biomass have been recently emphasized by Pattahil *et al.* (2015). The authors used a novel glycome profiling technique in which cell wall glycan-directed monoclonal antibodies were employed to observe the structural changes involving major classes of polysaccharides and non-polysaccharide cell-wall components in switch grass during similar pretreatment conditions as that of acid hydrolysis. Glycome profile results indicated significant loss of pectic and arabinogalactan epitopes, xylans and xyloglucans along with disruption of lignin-polysaccharide interactions after pretreatment of grass biomass.

Further, cellulose crystallinity and degree of polymerisation (DP) have been also considered as major biomass recalcitrance features that affect enzymatic hydrolysis performance. Nonetheless acid pretreatments have been observed to have antagonistic effects on cellulose component of grass cell wall. Foston *et al.* (2010) have observed the para-crystalline content of cellulose in switch grass appeared to increase during the DA pretreatment. They suggested that the majority of increase in crystallinity and para-crystalline percentage of cellulose is primarily due to localized hydrolyzation and removal of cellulose from the amorphous regions. The cellulose consists of two different allomorphs *i.e.* cellulose Iα which is a single-chain triclinic phase and cellulose Iβ, a two-chain monoclinic phase with the latter being more stable than the former (Baker *et al.*, 2000). DA pretreatments have been seen to primarily convert cellulose Iα to para-crystalline cellulose followed by conversion of para-crystalline cellulose to cellulose Iβ.

Alkali Pretreatment

Alkaline pretreatment involes pretreatment of the biomass with alkaline compounds like sodium hydroxide (NaOH), potassium hydroxide (KOH), calcium hydroxide [Ca $(OH)_2$] ammonia (NH_3), *etc.* Alkaline pretreatments selectively remove lignin without strongly affecting cellulose and xylan of the cell wall (Swain and Krishnan, 2015). Subsequently, the removal of lignin from lignocellulosic biomass results in improved cellulose accessibility.

Effect of Alkali Pretreatment on Lignin Content

Alkaline pretreatments mostly lead to cleavage of aryl ether linkages in lignin and lignin-carbohydrate complexes (LCC) that aids the disruption of cell wall resulting in increased pore volume and cellulose accessibility. It has been reported that alkali pretreatment of lignocellulosic biomass with high S/G ratio leads to the cleavage of labile β-O-42 linkages resulting in depolymerization and higher occurrence of β-β2 bonds subsequently leading to delignification of biomass. Lapierre *et al.* (1989) pretreated wheat straw with 1N NaOH and indicated aryl ether linkages or β-O-42 linkages as the main cleavages within the lignin bulk. Additionally, peroxide treatments of grasses are reported to cleave alkyl-aryl ether linkages at pH values around 11 (Li *et al.*, 2012). The positive co-relation of the presence of S/G ratio was well documented by the researchers like Yu *et al.* (2014) who found that high S/G ratio of lignin in the alkaline pretreated biomass positively affected the enzymatic digestibility

It is also noteworthy that the influence of H-lignin on recalcitrance of grasses biomass is comparatively less investigated than that of S and G units of lignin. However, in a study conducted by Li *et al.* (2014) it was revealed that the hexose yield in alkali pretreated *Miscanthus* increased upon increased occurrence of H lignin content. Moreover, the H/G ratio also plays a major role as measured in the KOH-extractable lignin of wheat and rice which revealed that it had a strong positive correlation with enhanced glucose yield in pretreated biomass (Wu *et al.*, 2013).

Furthermore, the higher concentrations of alkali are also witnessed for cleavage of two types of aryl ether bonds: $C_{aliphatic}$–O–$C_{aromatic}$ and $C_{aromatic}$–O–$C_{aromatic}$ that produces non-core lignin (NCL) phenolics, such as ferulates and *p*-coumarates. As a consequence, xylan and NCL-lignin depletion in the remaining residues after pretreatment should be observed. In an experiment by Li *et al.* (2012), the pCA (p-coumarates) content of grass biomass was observed to be substantially less after pretreatment with alkaline hydrogen peroxide.

Alkaline pretreatment cleaves ferulate cross-linkages and promotes lignin degradation to enhance the digestibility of the grass biomass (Yoo *et al.*, 2016). Alkaline pretreatment disrupts the lignin structure in biomass by saponification of intermolecular ester bond as well as acetyl groups and various uronic acid substitutions that cross links xylan and lignin, thereby improving the susceptibility of the remaining polysaccharides (cellulose and hemicellulose) for other treatment (Karp *et al.*, 2013).

Dilute alkali and modified alkali pretreatments have been observed to solubilise more than 60 per cent of grass lignin which have relatively low lignin contents of 10-18 per cent (Bjerre *et al.*, 1996). Iyer *et al.* (1996) used ammonium hydroxide of different concentration for the pretreatment of a switch grass and found the efficiency of delignification was 65-85 per cent for the grass biomass. Alkaline hydrogen peroxide (AHP) pretreatment has also been studied as a chemical pretreatment explicitly as delignifying post-treatment (Kumar *et al.*, 2011), which is based on the action of hydrogen peroxide on biomass at alkaline pH (optimally at pH 11.5) at ambient or near-ambient temperatures and pressures. H_2O_2 reacts strongly with the aromatic rings of lignin and generates aromatic carboxylic acids that act as inhibitors in the later stages of transformations (Bhaumik and Dhepe, 2015). AHP pretreatment attributes to the destruction of ferulate cross links as well as mild oxidation and solubilisation of lignin. These outcomes have the net effect of improving the overall hydrophilicity of cell wall matrix that allows water and hydrolytic enzyme penetration.

Modifications of alkaline pretreatment such as using alkali pretreatment in combination with ambient and elevated temperatures [*e.g.* Ammonium fibre explosion (AFEX), Ammonia recycle percolation (ARP) and soaking aqueous ammonia (SAA)], microwave and ultrasonication process have been observed to have more positive results both for delignification and sugar release than that of only alkali pretreatment. While, AFEX is conducted at ambient temperature, ARP is carried out at high temperatures and SAA is a form of AFEX which treats biomass through aqueous ammonia in a batch reactor at 30–60 °C (Kumar and Sharma, 2017). Among these three types of ammonia pretreatment, AFEX treatment have been extensively experimented for herbaceous feedstock like alfalfa, wheat straw, elephant grass, rice hulls, bermuda grass and switch grass (Dien, Iten and Skory, 2017). Ultrasonication in amalgamation with alkaline pretreatment such as NaOH increases the delignification considerably. This effect of ultrasonication assisted NaOH (UA-NaOH) treatment on lignin, cellulose and hemicellulose of hybrid napier grass and denannath grass was extensively studied by the author (Mohapatra *et al.*, 2017). In the study by the author, it was observed that the effect of UA-NaOH treatment was more effective for dennanath grass rather than hybrid Napier grass. Wang *et al.* (2017) pretreated grass clippings (the lawn waste that remains after mowing) using ultrasound assisted calcium hydroxide treatment and achieved higher saccharification yields. Using microwave treatments in combination with NaOH has also been experimented for grass feedstocks. In one such study, Hu *et al.* (2008) soaked switch grass in NaOH solutions of different concentrations and then treated the samples by microwave or conventional heating and found 90 per cent of maximum potential sugars at a concentration of 0.1 g/g. These results suggest that microwave-assisted alkali treatment to be an effective pretreatment technique which can be utilised to improve the enzymatic digestibility of switch grass. Park *et a.* (2010) conducted lime pretreatment on rice straw with slight alterations and observed that this helped in elimination of the solid–liquid separation step resulting in 89 per cent glucose recovery.

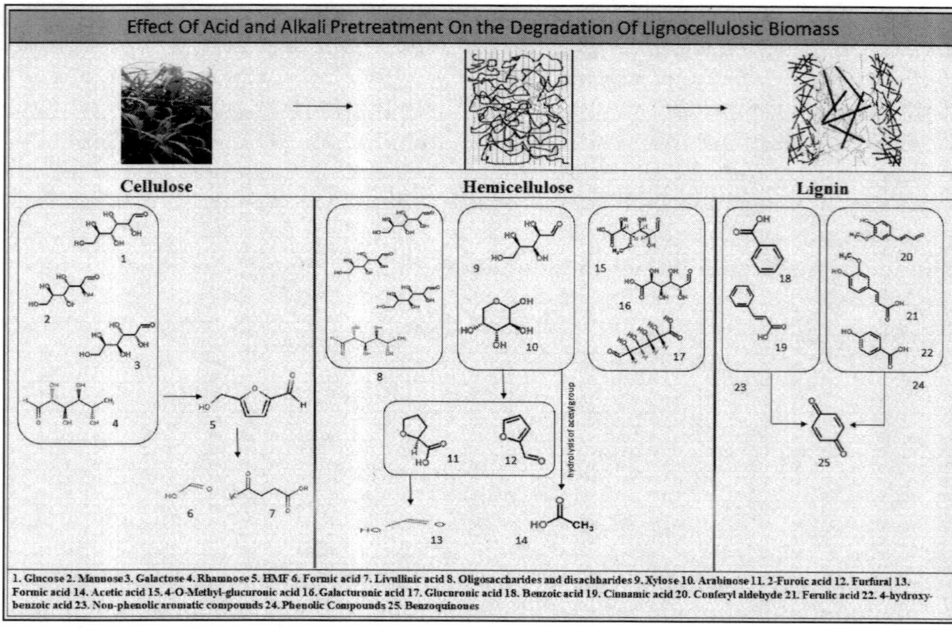

Figure 10.1: Mechanism of Acid and Alkali Pretreatment on Components of Grass Cell Wall.

Pretreatment Techniques for Delignification and Maximum Cellulose Exposure from Dennanath Grass and Hybrid Napier Grass

The pretreatment techniques that are mentioned above have been used by the authors for enhanced delignification and cellulose exposure from Denannath and Hybrid napier grass (CO-3), the two varieties of *Pennisetum* sp. for bioethanol production. Denannath grass is a tropical to sub-tropical species which is a native to India and is tolerant to salinity. It also has very strong recovery ability after watering even under severe drought conditions with an annual productivity of 65- 71 tons per hectare. Similarly, hybrid napier grass which is a hybrid developed by Tamil Nadu agricultural university in Coimbatore is a very high yielding grass variety with yields of up to 150 tonnes per year per acre. Further, the two *Pennisetum* varieties have a cellulose content of 45-50 per cent and low lignin concentration of 18-21 per cent, which makes them very attractive biomass for bioethanol production.

The pretreatment strategy that has been used by the authors is alkali, acid and ultrasonication assisted alkali/acid pretreatment techniques (Mohapatra *et al.*, 2016 and 2017). Different alkalis such as sodium hydroxide (NaOH), potassium hydroxide (KOH), calcium hydroxide (Ca_2OH) and sodium carbonate (Na_2CO_3) and acids such as sulphuric acid (H_2SO_4), hydrochloric acid (HCl), phosphoric acid (H_3PO_4) and nitric acid (H_2NO_3) were initially used for screening of the best pretreatment technique. All the alkali and acids were diluted with distilled water prior to pretreatment process in a range of 1 to 2 per cent concentration. The optimisation

Table 10.2: Comparative Data of Pretreatment, Enzymatic Hydrolysis and Ethanol Production by Separate Hydrolysis and Co-fermentation (SHCF) and Simultaneous Saccharification and Co-fermentation (SSCF) from two *Pennisetum* sp. by NaOH and Ultrasono Assisted NaOH Pretreatment

Pretreatment				*Enzymatic Hydrolysis*				*Fermentation using E1 Saccharified Biomass (270 g)*	
Delignification (Per cent)		*Reducing Sugar (mg/g)*		*Reducing Sugar (E1)*(mg/g)*		*Reducing Sugar (E2)*(mg/g)*		*Ethanol (Per cent)*	*Ethanol (Per cent)*
NaOH	*UA-NaOH*	*NaOH*	*UA-NaOH*	*Glucose*	*Xylose*	*Glucose*	*Xylose*	*SHCF*	*SSCF*
Dennanath Grass									
32.2	89.3	76.8	227.2	504.4	157.6	254.8	223.9	62.1	77.6
Hybrid Napier Grass									
43.2	86.7	86.8	242.8	323.4	107.6	301.6	181.7	43.6	49.6

* E1 and E2 refers to two different set of enzymes *i.e.* Palkonal MBW and mixture of isolated cellulase and xylanase respectively.

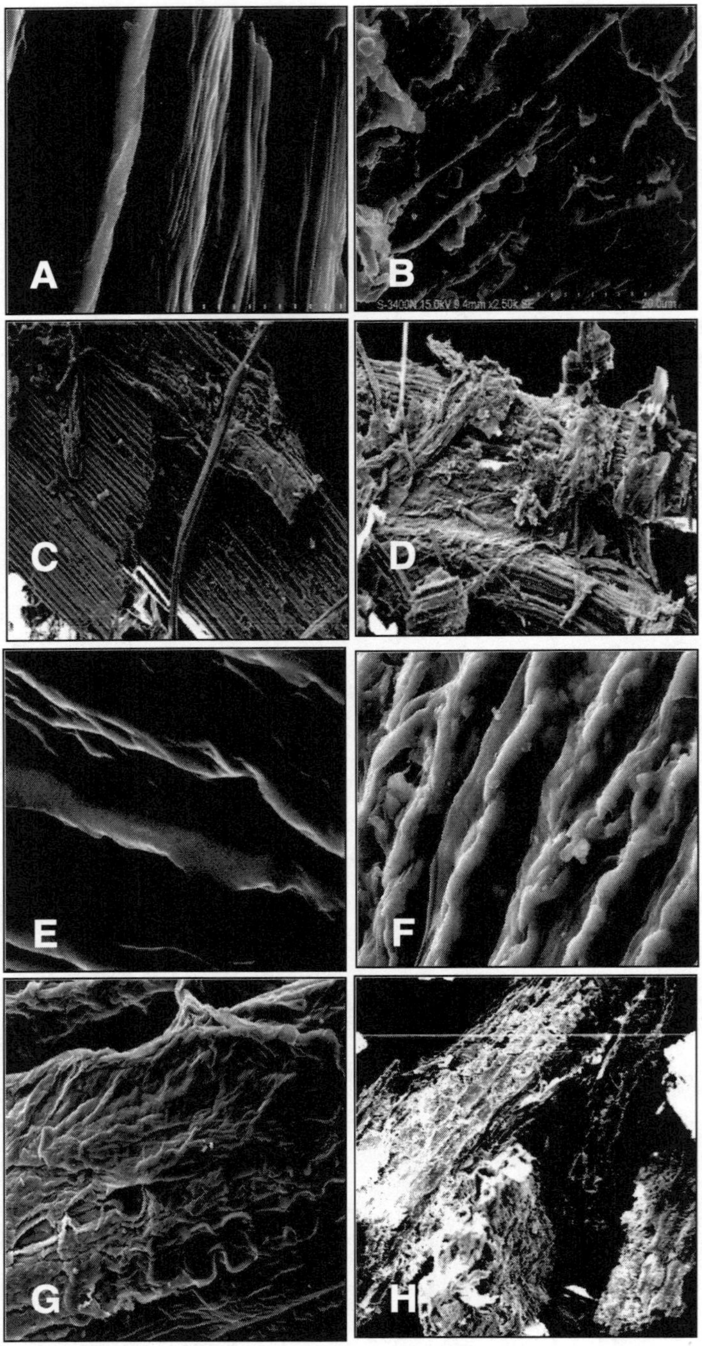

Figure 10.2: SEM Analysis-A and F-untreated DG and HNG, B and G- 1 per cent NaOH-treated DG and 1 per cent NaOH treated HNG, C and H- UA-NaOH DG and UA-NaOH HNG, D and I- Palkonal MBW hydrolysed DG and Palkonal MBW hydrolysed HNG.

process was conducted using taguchi optimisation method using various parameters such as concentration of alkali/acid, temperature and incubation time.

The best alkali and acid were found to be NaOH and HCl which were further subjected to ultrasonication pretreatment. The ultrasono assisted alkali and acid pretreatments were further optimised for different parameters. The best delignification results were obtained from ultrasonication assisted NaOH pretreatment were 89.3 per cent and 86.7 per cent for DG and HNG respectively which was greater as compared to ultrasonication assisted HCl pretreatment. Similarly, the maximum reducing sugar that was obtained using ultrasonication assisted NaOH pretreatment was also higher, as compared to that of ultrasonication assisted HCl pretreatment. Further enzymatic hydrolysis of the UA-NaOH DG and HNG were conducted using two different set of enzymes *i.e.* 1) a commercial enzyme Palkonal MBW and 2) a mixture of isolated cellulase enzyme and commercial xylanase enzyme. The experiments for enzymatic hydrolysis were carried out using response surface methodology (RSM). Alike taguchi method various parameters like temperature, enzyme concentration, substrate concentration and incubation time were considered for designing the experiments for enzymatic hydrolysis. Further fermentation experiments were conducted for the grass biomass using fed-batch separate hydrolysis and co-fermentation (SHCF) and simultaneous saccharification and co-fermentation (SSCF) process. The results obtained in pretreatment, enzymatic hydrolysis and ethanol production are given in Table 10.2.

The structural variations in the biomass were revealed by scanning electron microscopy (SEM) as given in Figure 10.2. There was a remarkable variation in the untreated, pretreated and Palkonal MBW saccharified grass biomass for both the *Pennisetum* sp.

Conclusion

The emerging concern for energy fulfilment promotes the development of bioethanol production from lignocellulosic feed stocks. Among the lignocellulosic biomass, grass has emerged as a promising feedstock due to their high yields, low costs and low environmental impacts. This paper highlights the various prospects of pretreatment techniques and their corresponding mechanisms on degradation of lignin in grass cell wall that is necessary for enzymatic accessibility and bioethanol production. This paper also delivers the information about the interaction of the chemicals, ultrasound/microwave waves and organisms that are used during pretreatment process with the cell wall components of plant cell and the application the techniques for delignification of dennanath grass and hybrid napier grass in an effort for bioethanol production.

References

Akchiche, O. and Boureghda, K.M. (2007). Esparto grass (*Stipa tenacissima L.*) raw material of paper making. The Chemistry of Plant Raw Materials, 4: 25-30

Baker, A.A., Helbert, W., Sugiyama, J. and Miles, M.J. (2000). New insight into cellulose structure by atomic force microscopy shows the Iα crystal phase at near-atomic resolution. Biophysical Journal, 79(2): 1139-1145.

Bhaumik, P. and Dhepe, P.L. (2015). Conversion of Biomass into Sugars. RSC Green Chemistry, 44: 1-53.

Bjerre, A.B., Olesen, A.B. and Fernqvist, T. (1996). Pretreatment of wheat straw using combined wet oxidation and alkaline hydrolysis resulting in convertible cellulose and hemicellulose. Biotechnology and Bioengineering, 49: 568– 577.

Bridgeman, T.G., Jones, J.M., Shield, I. and Williams, P.T. (2008). Torrefaction of reed canary grass, wheat straw and willow to enhance solid fuel qualities and combustion properties. Fuel, 87: 844-856.

Buranov, A.U. and Mazza, G. (2008). Lignin in straw of herbaceous crops. Industrial Crops and Products. 28: 237-259.

Cao, S., Pu, Y., Studer, M., Wyman, C.L. and Ragauskas, A.J. (2012). Chemical transformations of *Populus trichocarpa* during dilute acid pretreatment. RSC Advances, 2: 10925–10936.

Cardona, E.M., Luis A.R., and Juan D.P. (2012). Disponibilidad de Variedades de Pastos Y Forrajes Como Potenciales Materiales Lignocelulósicos Para La Producción de Bioetanol En Colombia. Informacion Tecnologica, 23(6): 87–96.

Chen, W.H., Ye, S.C. and Sheen, H.K. (2011). Hydrolysis characteristics of sugarcane bagasse pretreated by dilute acid solution in a microwave irradiation environment. Applied Energy, 93: 237-244.

Chen, X., Shekiro, J., Elander, R. and Tucker, M. (2012b). Improved xylan hydrolysis of corn stover by deacetylation with high solids dilute acid pretreatment. Industrial and Engineering Chemistry Research, 51: 70–76.

Chen, X., Shekiro, J., Franden, M.A., Wang, W., Zhang, M., Kuhn, E., Johnson, D.K. and Tucker, M.P. (2012a). The impacts of deacetylation prior to dilute acid pretreatment on the bioethanol process. Biotechnology for Biofuels, 5: 8–21.

Das, S.P., Deka, D., Ghosh, A., Das, D., Jawed, M. and Goyal, A. (2013). Scale up and efficient bioethanol production involving recombinant cellulase (Glycoside hydrolase family 5) from *Clostridium thermocellum*. Sustainable Chemical Processes, 1: 19.

Dien, B.S., Iten, L.B. and Skorey, C.D. (2017). Handbook of Industrial Biocatalysis: Chapter 23, Converting herbaceous energy crops to bioethanol; A review with emphasis to pretreatment processes. CRC publications. Edited by Hou, C.T.

Dijkerman, R., Bhansing, D.C., Op den Camp, H.J., van der Drift, C. and Vogels, G.D. (1997). Degradation of structural polysaccharides by the plant cell-wall degrading enzyme system from anaerobic fungi: and application study. Enzyme and Microbial Technology, 21: 130-136.

Foston, M. and Ragauskas, A.J. (2010). Changes in lignocellulosic supramolecular and ultrastructure during dilute acid pretreatment of populus and switch grass. Biomass and Bioenergy, 34: 1885–1895.

Howard, R.L., Abotis, E., van Rensburg, E.L. and Howard, S. (2003). Lignocellulose biotechnology: Issues of bioconversion and enzyme production. African Journal of Biotechnology, 2: 602-619.

Hu, Z., Wang, Y. and Wen, Z. (2008). Alkali (NaOH) Pretreatment of switch grass by radio frequency-based dielectric heating. Applied Biochemistry and Biotechnology, 148: 71–81.

Iyer, V., Wu, Z.W., Kim, S.B. and Lee, Y.Y. (1996). Ammonia recycled percolation process for pretreatment of herbaceous biomass. Applied Biochemistry and Biotechnology, 57-58: 121–132.

Jiang, B., Wang, W., Gu, F., Cao, T. and Jin, Y. (2016). Comparison of the substrate enzymatic digestibility and lignin structure of wheat straw stems and leaves pretreated by green liquor. Bioresource Technology, 199: 181–187.

Karp, S.G., Woiciechowski, A.L., Soccol, V.T. and Soccol, C.R. (2013). Pretreatment strategies for delignification of sugarcane bagasse: a review. Brazilian Archives of Biology and Technology, 56: 4.

Kim, T.H., Kim, J.S., Sunwoo, C. and Lee, Y.Y. (2003). Pretreatment of corn stover by aqueous ammonia. Bioresource Technology, 90: 39-47.

Kosikova, B., Joniak, D. and Kosakova, L. (1979). The properties of benzyl ether bonds in the ligninsaccharidic complex isolated from spruce. Holzforschung, 33: 11-14.

Kumar, A., Gautam, A. and Dutt, D. (2016). Biotechnological transformation of lignocellulosic biomass in to industrial products: An overview. Advances in Bioscience and Biotechnology, 7: 149-168.

Kumar, A.K. and Sharma, S. (2017). Recent updates on different methods of pretreatment of lignocellulosic feedstocks: a review. Bioresources and Bioprocessing, 4(1): 7.

Kumar, L., Chandra, R. and Saddler, J. (2011). Influence of steam pretreatment severity on post-treatments used to enhance the enzymatic hydrolysis of pretreated softwoods at low enzyme loadings. Biotechnology Bioengineering, 108: 2300-2311.

Langan, P., Petridis, L., O'Neill, H.M., Pingali, S.V., Foston, M. and Nishiyama, Y. (2014). Common processes drive the thermochemical pretreatment of lignocellulosic biomass. Green Chemistry, 16: 63–68.

Lapierre, C., Jouin, D. and Monties, B. (1989). On the molecular origin of the alkali solubility of Gramineae lignins. Phytochemistry, 28: 1401–1403.

Lee, J. (1997). Biological conversion of lignocellulosic biomass to ethanol. Journal of Biotechnology, 56: 1-24.

Li, H., Kim, N.J., Jiang, M., Kang, J.W. and Chang, H.N. (2009). Simultaneous saccharification and fermentation of lignocellulosic residues pretreated with phosphoric acid-acetone for bioethanol production. Bioresource Technology, 100: 3245-3251.

Li, M., Foster, C., Kelkar, S., Pu, Y., Holmes, D., Ragauskas, A., Saffron, C. and Hodge, D. (2012). Structural characterization of alkaline hydrogen peroxide pre-treated grasses exhibiting diverse lignin phenotypes. Biotechnology for Biofuels, 5: 38.

Li, M., Si, S., Hao, B., Zha, Y., Wan, C., Hong, S., Kang, Y., Jia, J., Zhang, J., Li, M., Zhao, C., Tu, Y., Zhou, S. and Peng, L. (2014). Mild alkali-pretreatment effectively extracts guaiacyl-rich lignin for high lingocellulose digestibility coupled with largely diminishing yeast fermentation inhibitors in *Miscanthus*. Bioresource Technology, 169: 447-454.

Lynd, L.R. (1996). Overview and evaluation of fuel ethanol from cellulosic biomass: technology, economics, the environment, and policy. Annual Review of Environment and Resources, 21: 403–465.

Mohapatra, S., Dandapat, S.J. and Thatoi, H.N. (2016). Physicochemical characterization, modelling and optimization of ultrasono-assisted acid pretreatment of two *Pennisetum* sp. using Taguchi and artificial neural networking for enhanced delignification. Journal of Environmental Management, 187: 537-549.

Mohapatra, S., Pattathil, S. and Thatoi, H. (2017b). Structural and functional characterization of two *Pennisetum* sp. biomass during ultrasono-assisted alkali pretreatment and enzymatic hydrolysis for understanding the mechanism of targeted delignification and enhanced saccharification. ACS Sustainable Chemistry and Engineering, 5: 8: 6486–6497.

Moxley, G., Gaspar, A.R., Higgins, D. and Xu, H. (2012). Structural changes of corn stover lignin during acid pretreatment. Journal of Industrial Microbiology and Biotechnology, 39(9): 1289–1299.

Nanda, S., Mohanty, P., Pant, K., Naik, S., Kozinski, J. and Dalai, A. (2012). Characterization of North American lignocellulosic biomass and biochars in terms of their candidacy for alternate renewable fuels. Bioenergy Research, 6: 663-677.

Nigam, P.S., Gupta, N. and Anthwal, A. (2009). Pre-treatment of agro-industrial residues. Netherlands: Springer. 2009: 13-33.

Okeke, B.C. and Obi, S.K.C. (1994). Lignocellulose and sugar compositions of some agro-waste materials. Bioresource Technology, 47: 283-284.

Olukemi, M. (2015). Production of bioethanol from elephant grass (*Pennisetum purpureum*) stem by aiyejagbara. Department of Chemical Engineering, Faculty of Engineering, Ahmadu Bello University, Zaria, Nigeria.

Pan, X. J., Xie, D., Gilkes, N., Gregg, D. J. and Saddler, J. N. (2005). Strategies to enhance the enzymatic hydrolysis of pretreated softwood with high residual lignin content. Applied Biochemistry and Biotechnology, 121: 1069-1079.

Park, J.Y., Shiroma, R., Al-Haq, M.I., Zhang, Y., Ike, M., Arai-Sanoh, Y., Ida, A., Kondo, M. and Tokuyasu, K. (2010). A novel lime pretreatment for subsequent bioethanol production from rice straw–calcium capturing by carbonation (CaCCO) process. Bioresource Technology, 101: 6805–6811.

Pattathil, S., Hahn, M.G., Dale, B.E. and Chundawat, S.P.S. (2015). Insights into plant cell wall structure, architecture, and integrity using glycome profiling of native and AFEX™-pre-treated biomass. Journal of Experimental Botany, 66: 4279–4294.

Pedersen, M. and Meyer, A.S. (2010). Lignocellulose pretreatment severity–relating pH to biomatrix opening. New Biotechnology, 27(6): 739–750.

Pingali, S.V., Urban, V.S. and Heller, W.T. (2010). Breakdown of cell wall nanostructure in dilute acid pretreated biomass. Biomacromolecules, 11: 2329–2335.

Prasad, S., Singh, A. and Joshi, H.C. (2007). Ethanol as an alternative fuel from agricultural, industrial and urban residues. Resources, Conservation and Recycling, 50: 1-39.

Ralph J. (2010). Hydroxycinnamates in lignification. Phytochemistry. Rev., 9: 65–83.

Raveendran, K., Ganesh, A. and Khilar, K.C. (1995). Influence of mineral matter on biomass pyrolysis characteristics. Fuel, 74: 1812-1822.

Rowell, R.M. (1992). Opportunities for lignocellulosic materials and composites. Emerging Technologies for Materials and Chemicals from Biomass, American Chemical Society, Washington DC, pp. 12-27.

Samuel, R., Pu, Y., Raman, B. and Ragauskas, A.J. (2010). Structural characterization and comparison of switch grass ball-milled lignin before and after dilute acid pretreatment. Applied Biochemistry and Biotechnology, 162: 62–74.

Selig, M.J., Adney, W.S., Himmel, M.E. and Decker, S.R. (2009). The impact of cell wall acetylation on corn stover hydrolysis by cellulolytic and xylanolytic enzymes. Cellulose, 16: 711–722.

Sun, Y. and Cheng, J. (2002). Hydrolysis of lignocellulosic materials for ethanol production: a review. Bioresource Technology, 83: 1–11.

Sundaram, V. (2017). Understanding the impacts of ammonia fiber expansion (AFEX™) pretreatment and densification on densified products quality and the bioproducts yield through enzymatic hydrolysis and fast pyrolysis. *Theses and Dissertations*, p. 1146.

Swain, M. R. and Krishnan, C. (2015). Improved conversion of rice straw to ethanol and xylitol by combination of moderate temperature ammonia pretreatment and sequential fermentation using *Candida tropicalis*. Industrial Crops and Products, 77: 1039–1046.

Taboada, A., Novo-Uzal, E., Flores, G., Loureda, M., Ros Barceló, A. and Masa, A. (2010). Digestibility of silages in relation to their hydroxycinnamic acid content and lignin composition. Journal of the Science of Food and Agriculture, 90: 1155–1162.

Tang, J., Chen, K., Huang, F., Xu, J. and Li, J. (2013). Characterization of the pretreatment liquor of biomass from the perennial grass, *Eulaliopsis binata*, for the production of dissolving pulp. Bioresource Technology, 129: 548–552.

Tao, X., Li, J., Zhang, P., Nabi, M., Jin, S., Li, F., Wang, S. and Ye, J. (2017). Reinforced acid-pretreatment of *Triarrhena lutarioriparia* to accelerate its enzymatic hydrolysis. International Journal of Hydrogen Energy, 42(29): 18301-18308.

Teramura, H., Sasaki, K., Oshima, T., Aikawa, S., Matsuda, F., Okamoto, M., Shirai, T., Kawaguchi, H., Ogino, C., Yamasaki, M., Kikuchi, J. and Kondo, A. (2015). Changes in lignin and polysaccharide components in 13 cultivars of rice straw following dilute acid pretreatment as studied by solution-state 2D 1H-13C NMR. PLoS One, 10(6): 0128417.

Wang, S., Li, F., Zhang, P., Jin, S., Tao, X., Tang, X., Ye, J., Nabi, M. and Wang, H. (2017). Ultrasound assisted alkaline pretreatment to enhance enzymatic saccharification of grass clipping. Energy Conversion and Management, 149: 409-415.

Wongwatanapaiboon, J., Kangvansaichol, K., Burapatana, V., Inochanon, R., Winayanuwattikun, P., Yongvanich, T. and Chulalaksananukul, W. (2012). The potential of cellulosic ethanol production from grasses in Thailand. Hindawi Publishing Corporation. Journal of Biomedicine and Biotechnology, 10 pages.

Wu, Z., Zhang, M., Wang, L., Tu, Y., Zhang, J. and Xie, G. (2013). Biomass digestibility is predominantly affected by three factors of wall polymer features distinctive in wheat accessions and rice mutants. Biotechnology for Biofuels, 6: 183.

Xu, N., Zhang, W., Ren, S., Liu, F., Zhao, C., Liao, H., Xu, Z., Huang, J., Li, Q., Tu, Y., Yu, B., Wang, Y., Jiang, J., Qin, J. and Peng, L. (2012). Hemicelluloses negatively affect lignocellulose crystallinity for high biomass digestibility under NaOH and H_2SO_4 pretreatments in *Miscanthus*. Biotechnology for Biofuels, 5(1): 58.

Yeh, T. F., Chang, M. J. and Chang, W. J. (2014). Comparison of dilute acid and sulfite pretreatments on *acacia confusa* for biofuel application and the influence of its extractives. Journal of Agricultural and Food Chemistry, 62: 10768–10775.

Yoo, C. G., Kim, H., Lu, F., Azarpira, A., Pan, X. and Oh, K. K. (2016a). Understanding the physicochemical characteristics and the improved enzymatic saccharification of corn stover pretreated with aqueous and gaseous ammonia. Bioenergy Research, 9: 67–76.

Yu, Z., Gwak, K. S., Treasure, T., Jameel, H., Chang, H. M. and Park, S. (2014). Effect of lignin chemistry on the enzymatic hydrolysis of woody biomass. Chem Sus Chem, 7: 1942–1950.

Zhang, T., Wyman, C., Jakob, K. and Yang, B. (2012). Rapid selection and identification of Miscanthus genotypes with enhanced glucan and xylan yields from hydrothermal pretreatment followed by enzymatic hydrolysis. Biotechnology for Biofuels, 5: 56.

11

Review of Machine Learning Applications in Bioinformatics

S. Dash

Department of Computer Application,
North Orissa University, Baripada – 757 003, Odisha
E-mail: Sujata238dash@gmail.com

ABSTRACT

The machine learning methods specially supervised learning method designed for bioinformatics are reviewed in this article. It presents different modelling methods namely, supervised classification and clustering model for knowledge discovery/extraction, as well as deterministic and stochastic heuristics for optimization process. A brief overview of the above models is highlighted in this article.

Keywords: *Machine learning, Bioinformatics, Supervised classification, Clustering, Optimisation, Heuristic.*

Introduction

The exponential growth of the biological data raises two problems: (i) efficient information storage and management (ii) extraction of useful information from these data. The second problem is one of the main challenges in computational biology, which requires the development of tools and methods capable of transforming all these heterogeneous data into biological knowledge about the underlying mechanism. These tools and methods should allow us to go beyond a mere description of the data and provide knowledge in the form of models. By this simplifying abstraction that constitutes a model, we will be able to obtain predictions of the system. There are several biological domains where machine-learning techniques are applied for knowledge extraction from data. Figure 11.1 shows a schema of the main biological problems where computational methods are

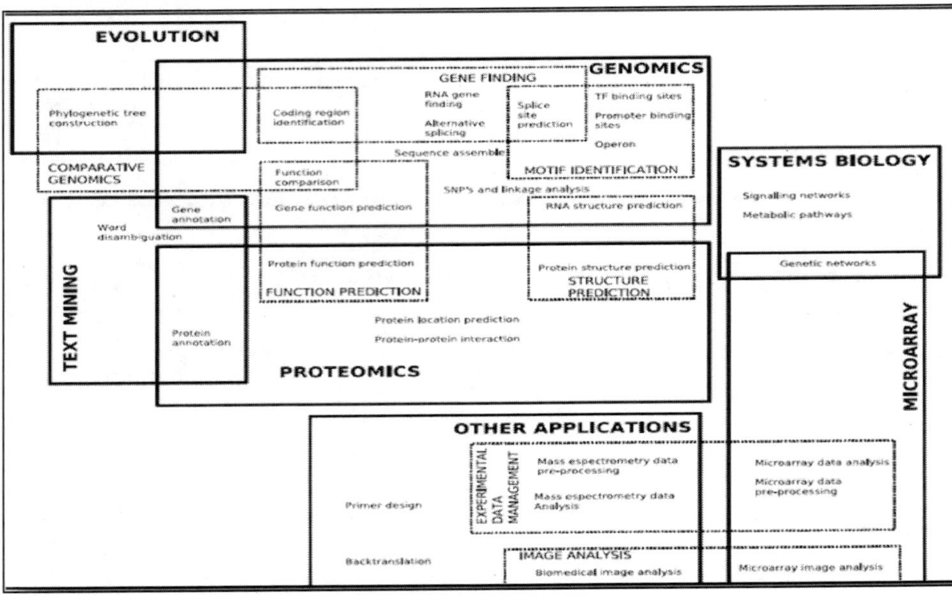

Figure 11.1: Application of Machine Learning Methods.

being applied. The problems are classified into six different domains: i) genomics, ii) proteomics, iii) microarrays, iv) systems biology, v) evolution and vi) text mining. These categories should be understood in a very general way, especially genomics and proteomics. One of the most important domains in bioinformatics is Genomics. The number of sequences available is increasing exponentially that need to be processed to derive useful information. We can extract the location and structure of the genes (Mathe *et. al.*, 2002) from the sequence. Recently, the identification of regulatory elements (Aetrs *et al.*, 2004) and non-coding RNA genes (Carter *et al.*, 2001) from the sequence are made applying computational model. Sequence information is also used for gene function and RNA secondary structure prediction.

Proteins also play a very important role in the life process, and their three-dimensional (3D) structure is a key feature in their functionality. In the proteomic domain, the main application of computational methods is protein structure prediction. Proteins are very complex macromolecules with thousands of atoms and bounds. Hence, the number of possible structures is huge. This makes protein structure prediction a very complicated combinatorial problem where optimization techniques are required. In proteomics, as in the case of genomics, machine learning techniques are applied for protein function prediction. Another interesting application of computational methods in biology is the management of complex experimental data. Microarray datasets are the best-known domain where this kind of data is collected. Complex experimental data raise two different issues. First, data need to be pre-processed *i.e.,* modified to be suitably used by machine learning algorithms and second, the analysis of the data *i.e.,* what we are looking for. In the case of microarray data, the most typical applications are expression pattern identification, classification, and genetic network induction.

Systems biology is another domain where biology and machine learning work together. It is very complex to model the life processes that take place inside the cell. Thus, computational techniques are extremely helpful when modelling biological networks (Bower *et al.*, 2004), especially genetic networks, signal transduction networks, and metabolic pathways. Evolution and, especially phylogenetic tree reconstruction also take advantage of machine learning techniques. Phylogenetic trees are schematic representations of the evolution of organism. Traditionally, they were constructed according to different features but, nowadays, with the great amount of genome sequences available, phylogenetic tree construction algorithms are based on the comparison between different genomes (Baldi *et al.*, 2001). This comparison is made by means of multiple sequence alignment algorithms, where optimization techniques are very useful.

A side effect of the application of computational techniques to the increasing amount of data is an increase in available publications. This provides a new source of valuable information, where text-mining techniques are required for the knowledge extraction. Thus, text- mining is becoming more and more interesting in computational biology, and it is being applied in functional annotation, cellular location prediction, and protein interaction analysis (Krallinger *et al.*, 2005).

Machine learning consists of algorithms to optimize the performance criterion by using past data. The optimized criterion can be the accuracy provided by a predictive model–in a modelling problem and the value of a fitness function–in an optimization problem. In a modelling problem, the 'learning' term refers to running a computer program to induce a model by using training data or past experience. Machine learning uses statistical theory when build in gcomputational models since the objective is to make inferences from a sample. It must be noticed that the efficiency of the learning and inference algorithms, as well as their space and time complexity and their transparency and interpretability, can be as important as their predictive accuracy. The process of transforming data into knowledge is both iterative and interactive. In the iterative step, we need to integrate and merge the different sources of information into one format. By using data warehouse techniques, the detection and resolution of outliers and inconsistencies are solved. In the second step, it is necessary to select, clean, and transform the data. To carry out this step, we need to eliminate or correct the data, as well as decide the strategy to impute missing data. This step also selects the relevant and non-redundant variables; this selection could also be done with respect to the instances. In the third step, called data mining, we take the objectives of the study into account in orderto choose the most appropriate analysis for the data. In this step, the type of paradigm for supervised or unsupervised classification should be selected and the model will be induced from the data. Once the model is obtained, it should be evaluated and interpreted–both from statistical and biological point of view–and, if necessary, we should return to the previous steps for a new iteration. This includes the solution of conflicts with the current knowledge in the domain.

Optimization problems are the task of finding an optimal solution in a space of multiple sometimes exponentially sized possible solutions. The choice of the optimization method to be used is crucial for the problem solution. Optimization approaches to biological problems can be classified, according to the type of solutions found, into exact and approximate methods. Exact methods output the optimal solutions when convergence is achieved. However, they do not necessarily converge for every instance. Approximate algorithms always output a candidate solution, but it is not guaranteed to be the optimal one. Optimization is also a fundamental task when modelling from data. In fact, the process of learning from data can be regarded as searching for the model that gives the data the best fitting. In this search, in the space of models any type of heuristic can be used. Therefore, optimization methods can also be seen as an ingredient at modelling.

The goal of this article is to serve as an insightful categorization and classification of the machine learning methods in bioinformatics including a listing of their applications. This article is organized as follows. 'Supervised classification' section presents the supervised classification problem, techniques for assessing and comparing classification algorithms, feature subset selection and several classification paradigms. 'Clustering' section shows different types of clustering–partition clustering, hierarchical clustering and clustering based on mixture models. 'Optimization' section shows heuristic optimization methods that have been proposed in bioinformatics to solve some hard computational problems. Final section explains the conclusions of this revision on machine learning methods in bioinformatics.

Supervised Classification

In a classification problem, we have a set of elements divided into classes. Given an element or instance of the set, a class is assigned according to some of the element's features and a set of classification rules. In many real-life situations, this set of rules is not known, and the only information available is a set of labelled examples (*i.e.* a set of instances associated with a class). Supervised classification paradigms are algorithms that induce the classification rules from the data. As an example, we will see a possible way to tackle splice site prediction as a supervised classification problem. The instances to be classified will be DNA sequences of a given size for example, 22 base pairs, 10 upstream and 10 downstream the and 2 bp splice site. The attributes of a given instance will be the nucleotide at each position in the sequence. Here, we are looking for donor sites, so the possible values for the class will be true donor site or false donor site. As we are approaching the problem as supervised classification, we need a set of labelled examples, *i.e.* a set of sequences of true and false donor sites along with their label. At this point, we can use this training set to build up a classifier. Once the classifier has been trained, we can use it to label new sequences, using the nucleotide present at each position as an input to the classifier and getting the assigned label as an output shown in Table 11.1.

Evaluation Metrics

Error rate and ROC curve: The error calculations are based on the confusion matrix shown in Table 11.2. In this case, we can define the error rate as (|FN| +

Table 11.1: Data in a Supervised Classification Problem

	X_1	...	X_n	C
$(\mathbf{x}^{(1)}; c^{(1)})$	$x_1^{(1)}$	...	$x_n^{(1)}$	$c^{(1)}$
$(\mathbf{x}^{(2)}; c^{(2)})$	$x_1^{(2)}$	...	$x_n^{(2)}$	$c^{(2)}$
$(\mathbf{x}^{(N)}; c^{(N)})$	$x_1^{(N)}$	...	$x_n^{(N)}$	$c^{(N)}$
$\mathbf{x}^{(N+1)}$	$x_1^{(N+1)}$	...	$x_n^{(N+1)}$	???

Table 11.2: Confusion Matrix for a Two-Class Problem

		Predicted class	
		Positive	**Negative**
True class	Positive	TP: True positive	FN: False negative
	Negative	FP: False positive	TN: True negative

|FP|)/N, where N = |TP| + |FP| + |TN| + |FN| is the total number of instances in the validation set. To fine-tune a classifier, another approach is to draw the receiver operating characteristics (ROCs) curve (Kamarudin *et al.*, 2017), which shows hit rate versus false alarm rate, namely, 1-specificity = |FP| / (|FP| + |TN|) versus Sensitivity = |TP| / (|TP| + |FN|), and has a form shown in Figure 11.2.

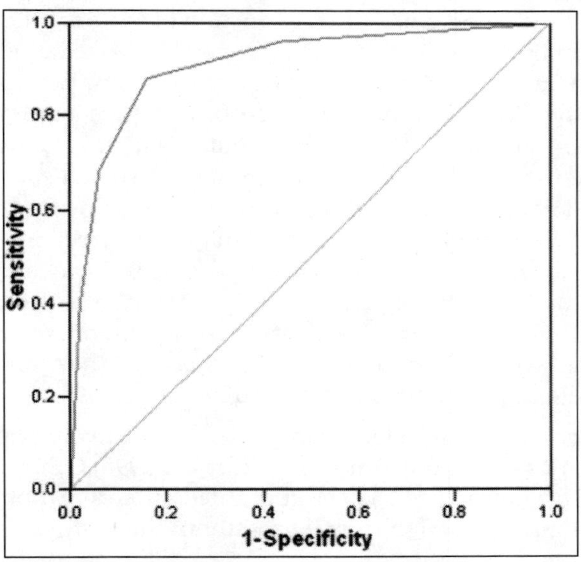

Figure 11.2: An Example of ROC Curve.

Estimating the Classification Error

An important issue related to a designed classifier is how to estimate its (expected) error rate when using this model for classifying unseen (new) instances. The simplest and fastest way to estimate the error of a designed classifier in the absence of test data is to compute its error on the sample data itself. This re-substitution estimator is very fast to compute and a usually optimistic (*i.e.* low-biased) estimator of the true error. In k-fold cross-validation (Basavanhally *et al.*, 2017), dataset is partitioned intok folds. Each fold is left out of the design process and used as a testing set. The estimate of the error is the overall proportion of the errors committed on all folds. In leave-one-out cross-validation, a single observation is left out each time, which corresponds to N-fold cross-validation. The bootstrap methodology is a general resampling strategy that can be applied to error estimation (Nieminen *et al.*, 2017). It is based on the notion of an 'empirical distribution, which puts mass 1/N on each of the N data points. A 'bootstrap sample' obtained from this 'empirical distribution' consists of N equally likely draws with replacement from the original data set DN. The bootstrap zero estimator and the 0.632 bootstrap estimator are used.

Feature Subset Selection (FSS)

One question that is common to all supervised classification paradigms is whether all the n descriptive features are useful when learning the classification rule. In trying to respond to this question, the so-called FSS problem appears, which can be reformulated as follows: given a set of candidate features, select the best subset under some learning algorithm. This dimensionality reduction made by an FSS process can bring several advantages to a supervised classification system, such as a decrease in the cost of data acquisition, an improvement in the understanding of the final classification model, a faster induction of the final classification model and an increase in the classifier accuracy. FSS can be viewed as a search problem, with each state in the search space specifying a subset of the possible features of the task. An exhaustive evaluation of possible feature subsets is usually unfeasible in practice because of the large amount of computational effort required. Four basic issues determine the nature of the search process: a search space starting point, a search organization, a feature subset evaluation function and a search-halting criterion. The search space starting point determines the direction of the search. One might start with no features and successively add them, or one might start with all the features and successively remove them. One might also select an initial state somewhere in the middle of the search space. The search organization determines the strategy of the search in a space of size 2^n, where n is the number of features in the problem. The search strategies can be optimal or heuristic.

Two classic optimal search algorithms which exhaustively evaluate all possible subsets are depth-first and breadth-first (Liu *et al.*, 1998).Otherwise, branch and bound search (Narendra *et al.*, 1977) guarantees the detection of the optimal subset for monotonic evaluation functions without the systematic examination of all subsets. When monotonicity cannot be satisfied, depending on the number of features and the evaluation function used, an exhaustive search can be impractical.

In this situation, heuristic search is interesting because it can find near optimal solutions, if not optimal. Among heuristic methods, there are deterministic and stochastic algorithms. On one hand, classic deterministic heuristic FSS algorithms are sequential forward and backward selection (Dash *et al.*, 2012), floating selection methods or best-first search. They are deterministic in the sense that all runs always obtain the same solution and, due to their hill-climbing nature, they tend to get trapped on local peaks caused by interdependencies among features. On the other hand, stochastic heuristic FSS algorithms use randomness to escape from local maxima, which implies that one should not expect the same solution from different runs. Genetic algorithms (Dash *et al.*, 2016) and estimation of distribution algorithms have been applied to the FSS problem.

The evaluation function measures the effectiveness of a particular subset of features after the search algorithm has chosen it for examination. Each subset of features suggested by the search algorithm is evaluated by means of a criterion (accuracy, area under the ROC curve, mutual information with respect to the class variable, *etc.*) that should be optimized during the search. In the so-called wrapper approach to the FSS problem, the algorithm conducts a search for a good subset of features using the error reported by a classifier as the feature subset evaluation as the feature subset evaluation criterion. However, if the learning algorithm is not used in the evaluation function, the goodness of a feature subset can be assessed by only regarding the intrinsic properties of the data. The learning algorithm only appears in the final part of the FSS process to construct the final classifier using the set of selected features. The statistics literature proposes many measures to assess the goodness of a candidate feature subset (Patra *et al.*, 2011). This approach to the FSS is called filter in the machine-learning field. Regarding the search-halting criterion, an intuitive approach is the non-improvement of the evaluation function value of alternative subsets. The applications of FSS methodology to microarray data try to obtain robust identification of differentially expressed genes. The most usual approach to FSS in this domain is the filter approach, because of the huge number of features from which we obtain information. Wrapper approaches have been proposed in Dash *et al.*, 2016) sequential wrapper and in genetic algorithms. Hybrid combinations of filter and wrapper approaches have also been proposed (Dash *et al.*, 2012).

Classification Paradigms

In this section, we introduce the main characteristics of some of the most representative classification paradigms. Each supervised classification paradigm has an associated decision surface that determines the type of problems the classifier is able to solve. In this sense, a version of the no free-lunch theorem introduced in optimization is also valid for classification–there is no best classifier for all possible training sets.

Bayesian Classifiers

Bayesian classifiers minimize the total mis classification cost using the following assignment:

$$\gamma(x) = arg\ \min_k \sum_{c=1}^{r0} cost\ (k, c)p(c|x_1, x_2, \ldots, x_n)$$

where cost(k, c) denotes the cost for a bad classification. In the case of a 0/1 loss function, the Bayesian classifier assigns the most probable a posterioriclass to a given instance, that is: γ (x) = argmaxcp (c | x_1, x_2,., x_n) = argmaxc p(c)p(x_1, x_2,., x_n | c). Depending on the way p(x_1, x_2,., x_n | c) is approximated, different Bayesian classifiers of different complexity are obtained.

Naive Bayes (Langarizadeh *et al.*, 2016) is the simplest Bayesian classifier. It is built upon the assumption of conditional independence of the predictive variables given the class shown in Figure 11.3. Although this assumption is violated in numerous occasions in real domains, the paradigm still performs well in many situations. The mostprobable a posteriori assignment of the class variable is calculated as:

$$c^* = arg\ \max_c p(c|x_1, \ldots, x_n) = arg\ \max_c p(c) \prod_{i=1}^{n} p(x_i|c).$$

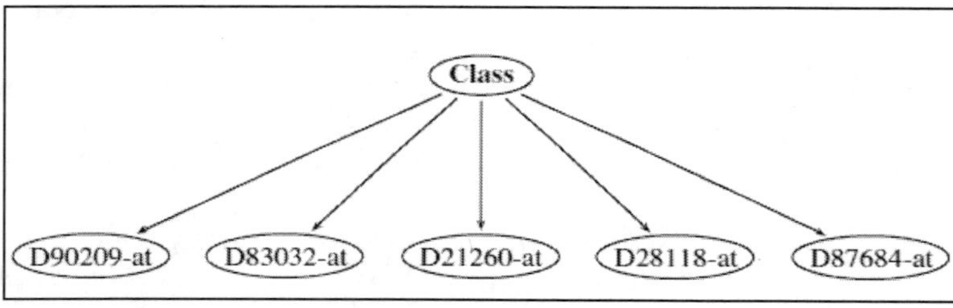

Figure 11.3: Representation of Naïve Bayes Algorithm.

The semi-naive Bayes classifier (Langarizadeh *et al.*, 2016) tries to avoid the assumptions of conditional independence of the predictive variables, given the class variable, by taking into account new variables. These new variables consist of the values of the Cartesian product of domain variables which overcome a condition. The condition is related to the independence concept and the reliability on the conditional probability estimations.

The tree augmented naive Bayes (Kim *et al.*, 2012) classifier also takes into account relationships between the predictive variables by extending a naive Bayes structure with a tree structure among the predictive variables. This tree structure is obtained adapting the algorithm proposed by Chow and Liu (1968) and calculating the conditional mutual information for each pair of predictive variables, given the class. The tree augmented naive Bayes classification model is limited by the number of parents of the predictive variables. In it, a predictive variable can have a maximum of two parents: the class and another predictive variable. The k dependence Bayesian (kDB) classifier (Kim *et al.*, 2012) avoids this restriction by allowing a predictive variable to have up to k parents aside from the class.

Classification Trees

It is natural to classify a pattern through a sequence of questions in which the next question asked depends on the answer to the current question. It is also usual to display the sequence of questions in a directed classification tree (Song *et al.*, 2015) also called classification tree where the root node is located at the top, connected by successive and directional links or branches to other nodes. These are similarly connected until we reach terminal or leaf nodes, which have no further links. The classification of a particular pattern begins at the root node, which asks for the value of a particular property of the pattern. The different links from the root node correspond to the different possible values. Based on the answer, we follow the appropriate link to a subsequent or descendant node. In classification trees, the links must be mutually distinct and exhaustive, *i.e.* one and only one link will be followed. The next step is to make the decision at the appropriate subsequent node, which can be considered the root of a subtree. We continue this way until we reach a leaf node, which has no further questions. Each leaf node bears a category label, and the test pattern is assigned to the category of the leaf node. In order to select the appropriate variable at each level of the tree, different impurity measures such as Gini index and gain ratio have been used.

Nearest Neighbour

The nearest-neighbour rule (Zhang, 2016) to classify x is to assign it to the label associated with the prototype nearest to the test point. An obvious extension of the nearest-neighbour rule is the k-nearest-neighbour rule. This rule classifies x by assigning it to the label most frequently represented among the k nearest samples. In other words, a decision is made by examining the labels on the k-nearest-neighbours and voting.

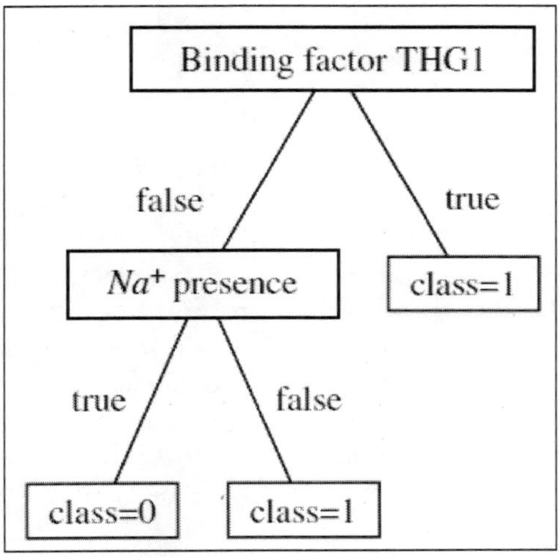

Figure 11.4: Example of Classification Tree.

Clustering

Clustering use to make partitioning of the set of elements into subsets according to the differences between them. In other words, it is the process of grouping similar elements together. The main difference from the supervised classification is that, in clustering, we have no information about the class values. The most typical example of clustering in bioinformatics is the clustering of genes in expression data. In microarray datasets, we obtain the expression value for thousands of genes in a few samples. An interesting information we can extract from these data is which genes are co-expressed in the different samples. This is a clustering problem where genes with similar expression level in all samples are grouped into a cluster. Cluster analysis, also called data segmentation, has a variety of goals. All relate to grouping or segmenting a collection of objects into subsets or' clusters', such that those within each cluster are more closely related to one another than objects assigned to different clusters. Sometimes the goal is to arrange the clusters into a natural hierarchy. This involves successively grouping the clusters so that, at each level of the hierarchy, clusters within the same group are more similar to each other than those in different groups. Clustering method is classified into: partition clustering, hierarchical clustering and clustering based on mixture models

Optimization

Optimization approaches to bioinformatics problems can be classified, according to the type of solutions found, into **exact and approximate methods**. Exact methods output the exact solutions when convergence is achieved. However, they do not necessarily converge at every instance. Approximate algorithms always output a candidate solution, but not necessarily the optimal one. **Exact optimization** approaches include exhaustive search methods. However, these algorithms are feasible only for small search domains and are not relevant to our review. Some methods are able to use knowledge about the problem to reduce the search space. This can be done by enforcing some constraints which the optimal solution has to fulfil (Dash *et al.*, 2017).

Approximate algorithms can be further classified into **deterministic and stochastic** (Dash *et al.*, 2017) according to the way solutions are found. Given a set of input parameters, a deterministic method will converge to the same solution. Stochastic methods use a random component that may cause them to obtain different solutions when running with the same input parameters. Stochastic algorithms can be divided into **local search method and population-based search methods** (Dash *et al.*, 2017). Local search algorithms visit one point of the search space at each iteration. Population-based search methods use a set or population of points instead of a single point. Examples of local search methods are MonteCarlo-based search, simulated annealing and tabu search and population-based methods are genetic algorithm, genetic programming, and estimation of distribution algorithm.

Conclusion

Now days, one of the most challenging problems in computational biology is to transform the huge volume of data, provided by newly developed technologies, into knowledge. Machine learning has become an important tool to carry out this transformation. The article can serve as a gateway to some of the most representative works in the field and as an insightful categorization and classification of the machine learning methods in bioinformatics.

References

Aerts, S., Loo, P.V. and Moreau, Y. (2004). A genetic algorithm for the detection of new cis-regulatory modules in sets of co-regulated genes. Bioinformatics, 20(12): 1974–1976.

Baldi, P. and Brunak, S. (2001). Bioinformatics. The Machine Learning Approach. MIT Press.

Basavanhally, A., Viswanath, S. and Madabhushi, A. (2015). Predicting Classifier Performance with Limited Training Data: Applications to Computer-Aided Diagnosis in Breast and Prostate Cancer. PLOS ONE 10(5): e0117900

Bower, J.M. and Bolouri, H. (2004). Computational Modelling of Genetic and Biochemical Networks. MIT Press.

Carter, R.J., Dubchak, I. and Holbrook, S.R. (2001). A computational approach to identify genes for functional RNAs in genomic sequence. Nucleic Acids Research, 29(19): 3928–3938.

Chow, C. and Liu, C. (1968). Approximating discrete probability distributions with dependence trees. IEEE Transactions on Information Theory, 14: 462–467.

Dash, S. and Patra, B. (2016). Genetic diagnosis of cancer by evolutionary fuzzy-based neural network ensemble, International Journal of Knowledge Discovery in Bioinformatics (IJKDB), IGI-Global, 6: 1.

Dash, S. (2017). Meta-heuristic based ensemble for feature selection and classification of gene-expression datasets. In Handbook of Research on the Modelling, Analysis and Application on Nature-Inspired Metaheuristic Algorithms, IGI-Global, USA, pp. 1-22.

Dash, S., Patra, B. and Tripathy BK. (2012). "A Hybrid Data Mining Technique for Improving the Classification Accuracy of Microarray Data Set", International Journal of Information Engineering and Electronic Business (IJIEEB), 2: 43-50.

Green, D.M. and Swets, J.A. (1974). Signal Detection Theory and Psychophysics. Wiley.

Kamarudin, A N., Cox, T. and Kolamunnage-Dona, R. (2017). Time-dependent ROC curve analysis in medical research: current methods and applications. BMC Medical Research Methodology, 17, 53.

Kim, D.C., Wang, X., Yang, C.-R., and Gao, J.X. (2012). A framework for personalized medicine: prediction of drug sensitivity in cancer by proteomic profiling. Proteome Science, 10(Suppl 1), S13.

Krallinger, M., Erhardt, R.A. and Valencia, A. (2005). Text mining approaches in molecular biology and biomedicine. Drug Discovery Today, 10(6): 439–445.

Langarizadeh, M., and Moghbeli, F. (2016). Applying Naive Bayesian Networks to Disease Prediction: a Systematic Review. ActaInformatica Medica, 24(5): 364–369.

Liu, H. and Motoda, H. (1998). Feature Selection for Knowledge Discovery and Data Mining. Kluwer Academic.

Mathe, C., Sagot, M.-F. and Schlex. (2002). Current methods of gene prediction, their strengths, and weaknesses. Nucleic Acids Research, 30(19): 4103–4117.

Narendra, P. and Fukunaga, K. (1977). A branch and bound algorithm for feature subset selection. IEEE Transactions on Computation, 26(9): 917–922.

Nieminen, P., Virtanen, J.I. and Vähänikkilä, H. (2017). An instrument to assess the statistical intensity of medical research papers. PLOS ONE 12(10): e0186882.

Patra, B., Singh, S.P. and Dash, S. (2011). "A different technique for gene selection and classification of gene expression data", published in data. Journal of Computer and Information Technology, 2(1): 15-24.

Song, Y., and Lu, Y. (2015). Decision tree methods: applications for classification and prediction. Shanghai Archives of Psychiatry, 27(2): 130–135.

Zhang, Z. (2016). Introduction to machine learning: k-nearest neighbours. Annals of Translational Medicine, 4(11): 218.

12

Rural Livelihood Through Aquaculture Development

P. Mishra

Department of Zoology,
Gangadhar Meher University, Sambalpur – 768 004, Odisha
E-mail: mishra.prasanti@gmail.com

ABSTRACT

Freshwater prawn culture, a profitable and remunerative business for the local fish farmers which can be adopted very easily at low cost and with little culture technologies. With growing demand for freshwater prawn, an emergent need is felt for its growth and culture in local ponds in rural villages. Aquaculture has an important role in the development of many national economies and plays a key role in rural development. Three clusters cluster I, cluster II and cluster III of WSHGs (Women Self Help group) in three villages namely Prataprudrapur, Bolamara and Nagapur, respectively were selected from Balipatna Block, Khurda, Odisha. They were doing composite fish farming along with mrigal in six Grampanchyat ponds. All the six ponds were stocked with seed of Macrobrachium rosenbergii with average weight 0.15gm at 1,000 Nos/pond along with carps, rohu (Labeo rohita) and catla (Catla catla) of size 5-7 g and 6-8 g, respectively, were stocked by the women farmers at 3,000 nos/ha. The selected WSHGs were provided hands on training on different management protocol which include pond preparation, liming, fertilization and stocking of seed. The results indicated that freshwater prawn culture along with rohu and catla gave encouraging production in both prawn and carps which convinced the women of the self-help groups. It may be an ideal example for other rural farmers to motivate them for adoption of this management practice with adequate training, infrastructure, marketing and financial credit.

Keywords: *Freshwater prawn, Fish, Clusters, Farmers, Indian major carps.*

Introduction

It is widely recognized that aquaculture play an important role in rural development. It is also recognized that aquaculture's potential for contributing to rural development has not been fully realized in many parts of the world. On the other hand, the integration of aquaculture into rural development to date has been associated with both beneficial and inauspicious trends. Effective rural development comes through sound governance. With participation at all levels, sustainable development will be people-oriented, integrated, and have a multi-sectoral agenda. There are significant challenges in developing holistic, people-centered rural development plans where the role for aquaculture is determined by an understanding of people's livelihoods. There are opportunities and challenges for effectively integrating aquaculture into rural development. Rural development, the process of sustained growth of the rural economy and improvement of well-being of rural men, women and children, has various dimensions, but in many cases it is particularly the development of the agricultural sector which is widely believed to provide the main impetus not only for reducing poverty and hunger, but also for ensuring food security for all. If more rapid agricultural growth takes place in countries with impoverished rural populations, rural farm and non-farm incomes can rise sufficiently to enable the rural poor to become more food secure.

Aquaculture–Definition, Current Status and Potential

Aquaculture may be defined simply as farming fish and other aquatic organisms. Fish is used here generically to include all farmed aquatic organisms. Land-base systems are commonly integrated with agriculture by stocking fish in rice fields and ponds. Water-based systems involve stocking fish directly in enclosures or attaching them to substrates in water bodies such as rivers, lakes, reservoirs or bays. Water-based systems may provide an entry point for landless people and poor fishers to farm fish. 'Rural aquaculture' implies low-cost production with extensive and semi-intensive technologies most appropriate for the limited resource base of small-scale households (Edwards and Demaine, 1997). It is based on natural food for fish such as plankton, which is occasionally fertilized and/or supplemented by other feed. Fertilizer and feed may be derived from on-farm by-products, at least in the initial stage of intensification, although even here formulated and pelleted feed from agro-industry are increasingly used. By contrast, intensive systems invariably depend on relatively high-cost, nutritionally complete diets. The annual growth of aquaculture has averaged some 11 per cent for nearly two decades. Although salmon and shrimp receive most publicity, they comprise less than 10 per cent of global aquaculture production by weight compared with 50 per cent for carps and tilapias which contribute most to the domestic food supply in developing countries. Aquaculture production is also skewed geographically with Asia producing over 90 per cent of global products, dwarfing Africa and Latin America at less than 0.5 per cent and 2 per cent, respectively. In China, with 67 per cent of global aquaculture production, inland aquaculture production has increased at least fivefold in the past decade; it has only doubled in the rest of the world, implying large potential in other developing countries if constraints to its expansion were removed.

In rural areas, there are diversified freshwater resources with immense potentialities for aquaculture development. Large proportion of freshwater resources available in rural areas falls under community ponds (over 60 per cent) vested with community or local level institutions, religious bodies without and/or with resources and local self-government for multiple use and multiple users (Sinha, 1999; Radheyshyam *et al.*, 2011 and 2013). Out of the total freshwater ponds and tanks (2.214 million ha) about 44 per cent are in use for aquaculture, producing 3.38 million tones (t) of fish at national average productivity of 2.9 t/ha/yr. However, little is known about the rest 56 per cent of freshwater ponds and tanks (mostly community ponds and tanks). Fish produced in these ponds and tanks are unreported, locally consumed and ignored in the policy domain.

The community based aquaculture is a subset of the rural aquaculture in which governance and management are community centric with the central role played by the community institutions. The typical nature of aquaculture in rural areas in scattered distributed manner is termed as rural aquaculture and it includes farming under traditional and extensive system to meet the needs of small scale farming households, fitting the resources available (De and Saha, 1999; Edwards *et al.*, 2002; Littlem, 2003, Radheyshyam *et al.*, 2009) Such system is characterised by locally based small scale farming, using extensive technology without government involvement. The Gram Panchayat ponds in the village are being leased out to the women self-help groups (WSHGs) on priority basis for a short term period (3 to 5 years) and the women are encouraged for pisciculture with provision of adequate technical input and capacity building. Many WSHGs are successfully implementing aqua-farming and their success stories may be of different kinds. Present success story deals with freshwater prawn culture by women self- help groups in Balipatna block of Khurda district, Odisha – a successful venture.

Materials and Methods

Three WSHGs namely Cluster- I, Cluster – II and Cluster – III were selected from Balipatna Block, Khurda, Odisha. WSHGs comprise of 15, 18 and 13 number of members with age group of 24-48 years. More than 80 per cent of the members have agriculture as their main occupation and about 20 per cent of the members depend on service for their household income. These three WSHGs diverted to pisciculture from agriculture as their group activity since 2009. They were doing composite fish farming along with mrigal in these six Grampanchayat ponds. In Sept, 2010; DST project was implemented in these six ponds for adoption of prawn culture along with carps (rohu, *Labeo rohita* and catla, *Catla catla*). The size of the ponds were 0.18 and 0.16 ha (Cluster I), 0.17 and 0.21 ha (Cluster – II) and 0.25 and 0.19 ha (Cluster – III). All the six ponds were stocked with seed of *M. rosenbergii* with average weight 0.15gm at 1,000 nos/pond along with carps (rohu, and catla). The carp fingerlings comprising of rohu and catla of size 5-7 g and 6 - 8g respectively, were stocked by the women farmers at 3,000 nos/ha. The water quality parameters of ponds were monitored every month by standard procedures. The growth rate was checked at periodic intervals (fortnightly). At the end of six months, final harvesting of prawns and carps were made after draining out the water from the pond.

Results and Discussion

During culture, water temperature of the ponds varied from 28.2°C to 30.1°C, water pH was slightly alkaline in reaction, while other important parameters like dissolved oxygen (DO), total alkalinity, total hardness and dissolved ammonia (NH_3) were within the acceptable limits with very little variations (Table 12.1). Including major river systems in the Punjab and Bangladesh, has concentrated in recent times on achieving self-sufficiency in food grains through agricultural intensification and floodwater management. However, this has been at the expense of aquatic animal production, which has declined due to drying out of fish habitat and blocking of migration routes (Haylor and Bhutta, 1997; Barr and Haylor, 2001). Earlier studies on the ponds indicated that the village community ponds are characterised by anaerobic benthic sediments (Olah and Sinha, 1984; Olah *et al.*, 1987; Kumar, 1992). Agricultural growth at the expense of fish production in rural development in societies where culture and food security are based on fish and rice. The average growth of *M. rosenbergii* was maximum at 65.85g in cluster - I followed by cluster - II (61.05g) and cluster – III (60.65g) from the initial stocking size of 0.15 g. On the other hand the survival in three clusters was 46 per cent, 42.5 per cent and 43 per cent respectively. Expenditure on different heads in prawn culture with carp in six selected ponds is given at Table 12.2. From this activity WSHGs harvested 75kg, 78 kg and 84 kg of prawn, 151kg, 90kg of rohu and 120 kg and 122 kg, 140 and 152kg of catla in cluster I, II and III respectively. Prawn was sold in a range of 180 - 200 rupees and carp 65-70 rupees at local markets (Table 12.3). Sale price of prawn and carps varied from cluster to cluster. In cluster I, unit sale price of prawn was Rs. 200, higher than other two Clusters and quantity of rohu and catla produced were also more than these two Clusters I and II that has resulted in more Gross Income in Cluster I. Women form about 48 per cent of the total population in India Radheyshyam (1997). About 78 per cent of them are economically active and are engaged in agriculture and allied fields. Moreover, total expenditure was lower than these two Clusters, I and II, which has brought out highest net return on expenditure and ultimately lowest Cost Benefit ratio and highest Profitability Index. The vital role of small-scale yet widespread systems in family nutrition, food security and income generation, is now beginning to gain recognition (UNICEF, 1994; Gregory and Guttman, 1997; Ahmed *et al.*, 1998; Haylor *et al.*, 1999). Taking rural development in the lower Mekong basin as an example, 80 percent of the 60 million people living in rural areas are rice farmers with 1-2 ha plots and a per capita income of US$186-400; rice and aquatic resources from paddies and nearby wetlands are the basis of their food security.

Table 12.1: Mean Water Quality Parameters in Three Clusters of Village Ponds

Parameters	*Cluster I*		*Cluster II*		*Cluster III*	
	Range	*Mean*	*Range*	*Mean*	*Range*	*Mean*
Temp.	28.2-30.1	29.09	28.3-29.9	29.1	28.3-30.1	28.9
pH	7.2-7.6	7.3	6.9-7.8	7.3	7.1-7.8	7.39
DO	5.3-6.6	4.9	5.8-6.4	5.12	5.8-6.4	6.2
Total alkalinity	135-160	146.9	130-140	135.8	122-143	133.8
Total hardness	120-135	126.8	120-140	129.3	120-140	128
Ammonia	0.01-0.06	0.022	0.03-0.07	0.039	0.03-0.08	0.046

Table 12.2: Economics of Prawn (*M. rosenbergii*) with Carp Culture (Rohu and Catla)

Sl.No.	*Expenditure Head*	*Rate*	*Cluster I*	*Cluster II*	*Cluster III*	*Total*
1	Lease value (Rs/ha/yr)	5000	1700	1900	2200	5800
2	**Pond Preparation**					
	Diesel(Rs/litre)	47	188	116	235	539
	Bleaching powder (Rs/kg)	10	306	342	396	1044
	Lime (Rs/kg)	12	1020	1368	1320	3708
3	**Manure and Fertilizer**					
	*Cow dung (Rs/tonne)	5000	340	380	440	1160
	Urea (Rs/kg)	9	184	205	238	627
4	**Seed Prawn**					
	(*M.rosenbergii*) (Rs/PL)	0.5	1040	1040	1040	3120
	Carp (Rohu + Catla) (Rs/fingerling)	1	1020	1140	1320	3480
5	**Feed**					
	GNOC (Rs/kg)	25	6375	9500	11000	26875
	Rice bran (Rs/kg)	7	1700	2660	3080	7440
6	Miscellaneous (labour charge for netting only) **	120	1440	1440	1560	4440
7	Subtotal (Rs)		15313	20091	22829	58233
8	Interest on finance (12 per cent per annum for 6 months) (Rs)		900	1200	1320	3420
	Total (Rs)		16213	21291	24149	61653

Table 12.3: Gross Return and Net Income from Sale of Prawn and Carp

Sl. No.	*Item*	*Cluster I*			*Cluster II*			*Cluster III*		
		Prawn	*Rohu*	*Catla*	*Prawn*	*Rohu*	*Catla*	*Prawn*	*Rohu*	*Catla*
1	Quantity produced (kg)	75	151	122	78	90	140	84	120	152
2	Unit sale price (Rs/kg)	200	65	65	180	70	70	195	65	65
3	Sale price (Rs)	15000	9815	7930	14040	6300	9800	16380	7800	9880
4	Gross income (Rs)	32745			30140			34060		
5	Total expenditure (Rs)	16213			21291			24149		
6	Net income (Rs) (iv-v)	16532			8849			9911		
7	Net return on expenditure (per cent)	101.97			41.56			41.04		
8	Profitability index	1.02			0.42			0.41		
9	Cost benefit ratio (C:B)	0.98			2.41			2.44		

Conclusion

The results indicated that freshwater prawn culture along with rohu and catla gives encouraging production in both prawn and carps which convinced the women of the self- help groups. Although the ponds used were seasonal with water level varying from 0.8-1.2 meter during culture period, the achievement in production has drawn the attention of local farmers for adoption of freshwater prawn culture along with rohu and catla. It may be an ideal example for other rural farmers to motivate them for adoption of this management practice with adequate training, infrastructure, marketing and financial credit. The major gaps in knowledge include the social and environment/resource aspects of rural aquaculture; the actual and potential contribution of aquaculture to sustainable livelihoods of the poor; and the trade-offs between capture of wild fish and culture of farmed fish, and between agriculture and aquaculture

References

Ahmed, M., Hap, N., Ly, V. and Tiengco, M. (1998). Socio- economic assessment of freshwater capture fisheries of Cambodia. A report on a household survey. mrc/dof/danida project for the management of the freshwater capture fisheries of Cambodia. Phnom penh, Mekong river commission, 185 pp.

Barr, J.J.F. and Haylor, G.S. (2001). Experiences of applying the sustainable livelihoods framework on Bangladesh flood plains. World development. Integrated coastal management: south Asia. Department for marine science and coastal management, University of New Castle.

De, H. K and Saha, G S. (1999). Rural aquaculture prospects and potentials. Indian Farming, 49(4): 17.

Edwards, P. and Demaine, H. (1997). Rural aquaculture: overview and framework for country reviews. FAO Regional Office for Asia and the Pacific (RAP), RAP Publ. 1997/36 RAP/FAO B

Edwards, P., and Demaine, H. (1997). Rural aquaculture: Overview and framework for country reviews. Bangkok: RAP/FAO.

Edwards, P., Demaine, H., Innes-Taylor, N. and Turonguang, D. (1996). 'Sustainable aquaculture for small-scale farmers: Need for a balanced model'. Outlook on Agriculture, 25(1): 19–26.

Edwards, P., Little, D.C. and Demaine, H. (2002). Rural Aquaculture, CABI, Wallingford, 358 ppangkok. 61 pp

Gregory, R. and Guttman, H. (1997). Poor in all but fish. A study of the collection of rice field foods from three villages in Svay Theap District, Cambodia. Asian Institute of Technology, AIT Aquaculture Outreach, Draft Working Paper 4, 27.

Haylor, G.S. and Bhutta, M.S. (1997). The role of aquaculture in the sustainable development of irrigated farming systems in Punjab. Aquaculture Resarch, 28: 691-705.

Kumar, D. (1992). Fish culture in un-drainable ponds: A manual for extension, FAO Fisheries Technical paper, 325: 239 pp

Little, D. (2003). Defining Peri-urban Aquaculture in Asia, PAPUSSA, (Production in Aquatic Peri-Urban Systems in Southeast Asia), EC-IncoDEV.ICA-2001- 10072

Little, D.C., Innes-Taylor, N.L. Turongruang, D. and Komolmarl, S. (1991). 'Large seed for small-scale aquaculture'. Aquabyte, 4(2): 2–3.

Olah, J. and Sinha, V.R.P. (1984). Principles and methods of monitoring of perennial undrainable pond ecosystems in tropical monsoon lands. Aquaculture Hungeria, 4: 103–110.

Olah, J., Sinha V.R.P., Ayyappan, S., Purushothaman, C.S. and Radheyshyam. (1987). Detritus associated respiration during macrophyte decomposition. Archeological Hydrobiology, 111(2): 309-315.

Radheyshyam (1997). Rural Aquaculture in Odisha: Sarkana Success story, CIFA Bhubaneswar.

Radheyshyam, De, H. K. and Saha, G.S. (2009). Role of community in production and supply of larger quality fingerlings. AquacultureAsia, 14 (1): 16-17.

Radheyshyam, Saha, G.S., Barik, N.K., Eknath, A.E, De, H.K., Safui, L., Adhikari, S. and Chandra, S. (2011a). Constraints to fish production in community ponds in Odisha, India. Aquaculture Asia, 16(1): 25-30.

Radheyshyam, Saha, G.S., Safui, L., Eknath, A.E., Adhikari, S., De, H.K., Barik, N.K. and Chandra, S. (2013). Status and economy of community fish farming in rural Odisha. Indian Journal of Fisheries, 60(4): 59-67

Sinha, V. R. P. (1999). Rural aquaculture in India. Rap Publication, FAO of the UN, Bangkok, Thailand, 84 pp.

UNICEF. (1994). Report on food and nutrition survey's 1993-4 UNICEF-FFP Phnom Penh Cambodia, 19 pp.

13

Medicinal Plants Diversity with Special Reference to Phytosociological Aspect of Kuldiha Wildlife Sanctuary, Balasore, Odisha

S.D. Rout[1], S.K. Panda[1] and T. Panda[2]*

[1]Department of Wildlife and Biodiversity Conservation, North Orissa University, Baripada – 757 003, Odisha
[2]Department of Botany, Chandbali College, Chandbali – 756 133, Odisha
**E-mail: srusti_d_rout@rediffmail.com.*

ABSTRACT

The present study provides an assessment of medicinal plant species diversity of Kuldiha Wildlife Sanctuary, Odisha which is located in Balasore district. Various small plots nested inside 100 hector inside the sanctuary area near Jodachua and Purunapani Reserve forest were surveyed during December, 2015-16. The density, frequency, basal area and RVI (Relative Value Index) along with diversity indices viz. Shannon index, species richness, equitability and species dominance were computed. In this study 38 medicinal plants belonging to 23 families has been extensively studied. Proper conservation and management plans are needed to save the natural resources especially medicinal plants of the Kuldiha range.

***Keywords**: Phytosociology, Medicinal plants, Diversity index, Kuldiha.*

Introduction

Medicinal plants are the "backbone" of traditional medicine, which means more than 3.3 billion people in the less developed countries utilize medicinal plants on a regular basis (Hunt, 2000). The Indian sub-continent has a very rich diversity of plant species in a wide range of ecosystems. There are about 17,000 species of higher plants, of which approximately 8,000 species, are considered medicinal and used by village communities, particularly tribal communities, or in traditional medicinal systems, such as the Ayurveda.

Tropical forests are the most complex of all the terrestrial ecosystems. They provide many goods and ecosystem services, such as prevention of soil erosion and preservation of habitats for plants and animals (Anbarashan and Parthasarathy, 2013). Although tropical forests occupy only 7 per cent of the earth's land surface (Wilson, 1988), they harbors about two thirds of all biological populations (Hughes *et al.*, 1997). Phytogeographically, these forests are characterized by a large biological diversity and enriched with economically important species (Hare *et al.*, 1997; Khare *et al.*, 1997). On a global basis, 52 per cent of total forests are tropical and over 42 per cent of tropical forests have been classified as dry forests (Holdridge, 1967) and these currently are disappearing at an alarming rate, ranging between 0.8-2 per cent per year (May and Stumpf, 2000; Sagar *et al.*, 2003). Many tropical forests are under great anthropogenic pressure and require management interventions to maintain the overall biodiversity, productivity and sustainability (Kumar *et al.*, 2006). Tropical dry deciduous forest constitutes about 38 percent of total forest area in India (Dixit, 1997). This is correct in a sense that these forests produce huge amount of renewable resource that are being utilized by global people. So, economically the forests are significant, as it is productive in a sense and reducing the pollutants of the atmosphere which could be the safeguard of our toxic materials that lead to death of organisms. Therefore, all the forests of the same kind are the most productive and variable in ecosystem processes in the globe.

Plant sociology or Phytosociology is defined as the discipline which concerns itself with the study of vegetation as such, with its floristic composition, structure, development and distribution. The key component of vegetation study is phytosociological analysis, with the aid of sampling techniques. Phytosociological information serves as prerequisite for understanding the structure and function of the vegetation. Some region of India has been quantitatively explored in terms of phytosociological investigation (Negi and Nautiyal, 2005: Khare *et al.*, 1985: Singh and Yadav, 2006; Ahmed *et al.*, 2009). While ethnobotanical studies have been done in Odisha (Chaudhuri, 1975; Mudgal and Pal, 1980; Das and Misra,1987; Girach *et al.*, 1998; Pattanail, 2003; Behera and Mishra, 2005; Pattanaik *et al.*, 2006) The medicinal plants available in this remote area have not been explored thoroughly, and no previous records on ethnomedical knowledge from within the study area are available. Only Pattanaik and Reddy (2014) have studied on medicinal plant wealth of local communities in Kuldiha Wildlife Sanctuary (KWS), Odisha, India. Similarly, several researchers have studied on the ethnomedicinal plants in Similipal Biosphere Reserve and other parts of Mayurbhanj district (Pandey *et al.*, 2002; Kumar and Rout, 2005; Rout 2007a&b; Rout and Pandey, 2007; Rout and Thatoi, 2008; Rout *et al.*, 2009; Mishra *et al.*, 2011).

Due to the study areas heterogeneity the chosen approach is to distribute a maximum conductible number of sample plots covering the whole study area as well as representing the existing vegetation patterns. Therefore a preliminary landscape classification using the ERDAS IMAGINE 8.4 Knowledge Classifier was performed. In this diversity assessment, we endeavored to document the availability as much of the flora fauna of the study area. Some interviews were taken with the local people, forest guards and officials which helped for the initial assessment, evaluating and defining priority areas for biodiversity conservation. Based on these initial results it was planned for multi-taxa survey effort not only to help draw the boundaries but also to develop an operational plan of the Kuldiha Wildlife Sanctuary (KWS). Hence, this study was done to document medicinal plant species available and their phytosociological aspect of Kuldiha Wildlife Sanctuary (KWS), Balasore, Odisha.

Materials and Methods

Study Area

Located in the southern part of Balasore district of Odisha and situated between 21° 20′ 31″ to 21° 29′ 08″ N latitude and 86° 25′ 23″ to 86° 44′ 50″ E longitude, the Kuldiha Wildlife Sanctuary (KWS) extends over 272.75 sq. km and is linked with the Similipal Biosphere Reserve (SBR) of Mayurbhanj District through the Sakhuapada and Nato Hill ranges. The altitude within the sanctuary varies from 80 meters to 750 meters. The vegetation of the sanctuary is mostly tropical moist deciduous forest harboring many valuable economic and medicinal plant species. The climate is wet and humid with an average rainfall of 1,568 mm, favors a diversified flora.

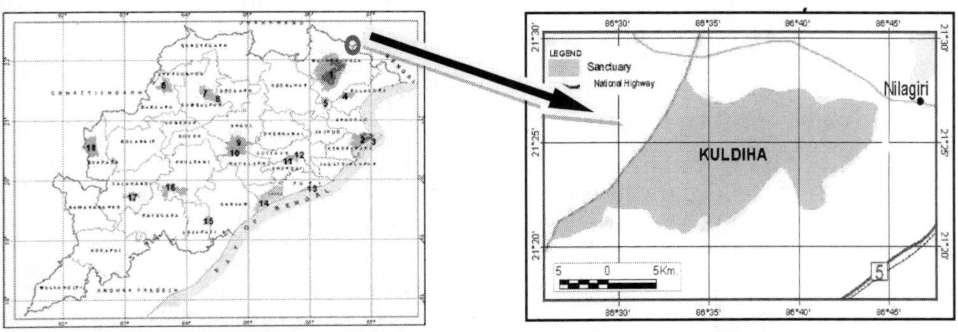

Figure 13.1: Map of the Kuldhia Wildlife Sanctuary (KWS), Balasore.

Survey

An extensive survey has been conducted to study the medicinal plants during 2015-16 which covered post monsoon periods and overall spectrum of vegetation of the study area. All the plots sampled were representative of most common types, sampling for herb species. Quadrate method was used (1m × 1m sized) to study the medicinal plants laying 10 quadrates randomly at different sites in Kuldiha range. These observations were used to calculate frequency, relative frequency, density, relative density, abundance, relative dominance, basal cover and RVI of each species by methods obtained as given below.

$$\textbf{Density} = \frac{\text{Total number of Individuals in all sampling units}}{\text{Total number of sampling units studied}}$$

$$\textbf{Frequency} = \frac{\text{Number of sampling units in which species occur}}{\text{Total number of sampling units}} \times 100$$

Basal area = $Cbh^2/4\pi$

$$\textbf{Relative density} = \frac{\text{Density value of species}}{\text{Sum of density value of all species}} \times 100$$

$$\textbf{Relative frequency} = \frac{\text{Frequency value of species}}{\text{Sum of frequency value of all species}} \times 100$$

$$\textbf{Relative dominance} = \frac{\text{Total basal area of the species}}{\text{Total basal area of all species}} \times 100$$

The medicinal plants species richness were identified with the help of Flora of Odisha (Saxena and Brahamam, 1994-96).The main purpose of the phytosociological analysis is to understand floristic vegetation characteristics to estimate the species richness and diversity which is existing in the study area, the standard protocols of Curtis (1950) have been adapted to analyze the density,frequency and abundance for calculation of frequency,density and abundance the following formulas have to be used.

Results and Discussion

The floristic composition of the reserve forests was remarkable in its diversity and luxuriance. The study revealed that there are 32 species in the ground flora representing 23 families of vascular plants. With regard to the number of species found within families, Rubiaceae were the most diverse families in the samples being represented by 3 species; they were followed by Apocynaceae, Asteraceae, Dioscoreaceae, Dipterocarpaceae, Euphorbiaceae, Liliaceae, Mimosaceae 2 species each, and 15 families were represented only by single species (Table 13.2).

The study revealed that in the herb layer the most dominant species was *Croton roxburghii* of RVI (17.11) which contributed with 15.6 per cent of total RVI followed by *Shorea robusta* (15.91) accounting for 14.4 per cent of total RVI. *Combretum decandrum* (5.87) which accounting 5.33 per cent of total RVI. Plant *Urginea indica, Mimosa himalayana, Hollarhena antydesenterica* and *Haldinia cordifolia* has minimum value of RVI (0.661) (Table 13.1). Besides these some species of herb plant were unidentified in the study area. A huge number of potential medicinal plant species were depleting rapidly because of unsustainable harvesting, lack of awareness and unrestricted grazing by domestic animals from nearby villages. Major medicinal plant species such as *Asparagus recemosus, Costus speciosus* and *Curculigo orchioides* were harvested in bulk for preparation of medicines by the local people. Unsustainable collection of above medicinal plants has placed them in threatened and vulnerable categories

Table 13.1: Phytosociology of Medicinal Plant Species of Kuldiha Wildlife Sanctuary, Balasore

Sl. No.	*Botanical Name*	*Local Name*	*Number of Individuals*	*Present in how Many Plots*	*Density*	*Relative Density*	*Frequency*	*Relative Frequency*	*Abundance*	*Relative Value Index (RVI)*	*A/F*
1.	*Hemidesmus indicus*	Anantamula	5	2	0.5	2.688	20	0.246	2.5	2.935	0.125
2.	*Combretum roxburghii*	Atundi	10	4	1	5.376	40	0.493	2.5	5.870	0.062
3.	*Urginea indica*	Banapiaja	1	1	0.1	0.537	10	0.123	1	0.661	0.1
4.	*Flacourtia jangomas*	Boinchakoli	6	3	0.6	3.225	30	0.370	2	3.596	0.067
5.	*Buchanania lanzan*	Char	6	3	0.6	3.225	30	0.370	2	3.596	0.067
6.	*Ageratum conyzoides*	Dahanimari	7	3	0.7	3.763	30	0.370	2.333	4.133	0.078
7.	*Elephantopus scaber*	Mayurachulia	10	1	1	5.376	10	0.123	10	5.499	1
8.	*Dioscoea bulbifera*	Pita alu	2	1	0.2	1.075	10	0.123	2	1.198	0.2
9.	*Icnocarpus frutescens*	Dudhilata	6	3	0.6	3.225	30	0.370	2	3.596	0.067
10.	*Andrographis paniculata*	Bhuinimba	2	1	0.2	1.075	10	0.123	2	1.198	0.2
11.	*Indigofera cassioides*	Gileri	3	3	0.3	1.612	30	0.370	1	1.983	0.033
12.	*Lygodium flexuosum*	Indrajal	4	4	0.4	2.150	40	0.493	1	2.644	0.025
13.	*Imperata cylindrica*	Juna	8	4	0.8	4.301	40	0.493	2	4.794	0.05
14.	*Ziziphus oenopila*	Kankoli	6	2	0.6	3.225	20	0.246	3	3.472	0.15
15.	*Diospyros melanoxylon*	Kendu	4	3	0.4	2.150	30	0.370	1.333	2.520	0.044
16.	*Mimosa himalayana*	Khirkhichi	1	1	0.1	0.537	10	0.123	1	0.661	0.1
17.	*Hollarhena antydesenterica*	Kuluchi	1	1	0.1	0.537	10	0.123	1	0.661	0.1
18.	*Haldinia cordifolia*	Kurma	1	1	0.1	0.537	10	0.123	1	0.661	0.1
19.	*Scleichera oleosa*	Kusuma	8	2	0.8	4.301	20	0.246	4	4.547	0.2
20.	*Ixora parviflora*	Luhajangi	2	2	0.2	1.075	20	0.246	1	1.322	0.05
21.	*Dioscorea pentaphylla*	Bana alu	3	2	0.3	1.612	20	0.246	1.5	1.859	0.075
22.	*Helicteres isora*	Modimudica	4	2	0.4	2.150	20	0.246	2	2.397	0.1

Sl. No.	Botanical Name	Local Name	Number of Indivi-duals	Present in how Many Plots	Density	Relative Density	Fre-quency	Relative Fre-quency	Abun-dance	Relative Value Index (RVI)	A/F
23.	*Cleistanthus collinus*	Parasi	1	1	0.1	0.537	10	0.123	1	0.661	0.1
24.	*Pterocarpus marsupium*	Piasal	2	1	0.2	1.075	10	0.123	2	1.198	0.2
25.	*Catunaregam spinosa* (Thunb.) Tirveng.	Potua	3	2	0.3	1.612	20	0.246	1.5	1.859	0.075
26.	*Croton roxburghii Balak.*	Putuli	30	8	3	16.129	80	0.987	3.75	17.11	0.046
27.	*Dillenia pentagyna Roxb.*	Rai	7	5	0.7	3.763	50	0.617	1.4	4.380	0.028
28.	*Smilax zeylanica*	Ramdantuni	3	2	0.3	1.612	20	0.246	1.5	1.859	0.075
29.	*Shorea robusta*	Sal	28	7	2.8	15.053	70	0.864	4	15.91	0.057
30.	*Asparagus racemosus*	Satabari	2	2	0.2	1.075	20	0.246	1	1.322	0.05
31.	*Albizzia lebbek*	Sirisha	5	3	0.5	2.688	30	0.370	1.667	3.058	0.05
32.	*Curculigo orchioides* Gaertn.	Talmuli	5	1	0.5	2.688	10	0.123	5	2.811	0.5

in Conservation Assessment and Management Plan (CAMP) of Odisha (Reddy and Pattanaik, 2009).

Table 13.2: Distribution of Medicinal Plants (Ground Species) in Kuldiha Wildlife Sanctuary, Balasore

Family	*Botanical Name*	*Local name*
Acanthaceae	*Andrographis paniculata*	Bhuinimba
Anacardiaceae	*Buchanania lanzan*	Char
Apocynaceae	*Icnocarpus frutescens*	Dudhilata
Apocynaceae	*Hollarhena antydesenterica*	Kuluchi
Asclepiadaceae	*Hemidesmus indicus*	Anantamula
Asteraceae	*Ageratum conyzoides*	Dahanimari
Asteraceae	*Elephantopus scaber*	Mayurachulia
Combretaceae	*Combretum roxburghii*	Atundi
Dilleniaceae	*Dillenia pentagyna*	Rai
Dioscoreaceae	*Dioscoea bulbifera*	Pita alu
Dioscoreaceae	*Dioscorea pentaphylla*	Bana alu
Dipterocarpaceae	*Pterocarpus marsupium*	Piasal
Dipterocarpaceae	*Shorea robusta*	Sal
Ebenaceae	*Diospyros melanoxylon*	Kendu
Euphorbiaceae	*Cleistanthus collinus*	Parasi
Euphorbiaceae	*Croton roxburghii*	Putuli
Fabaceae	*Indigofera cassioides*	Gileri
Flacourtiaceae	*Flacourtia jangomas*	Boinchakoli
Hypoxidaceae	*Curculigo orchioides*	Talmuli
Liliaceae	*Urginea indica*	Banapiaja
Liliaceae	*Asparagus racemosus*	Satabari
Lygodiaceae	*Lygodium flexuosum*	Indrajal
Mimosaceae	*Mimosa himalayana*	Khirkhichi
Mimosaceae	*Albizzia lebbek*	Sirisha
Poaceae	*Imperata cylindrical*	Juna
Rhamnaceae	*Ziziphus oenopila*	Kankoli
Rubiaceae	*Haldinia cordifolia*	Kurma
Rubiaceae	*Ixora parviflora*	Luhajangi
Rubiaceae	*Catunaregam spinosa*	Potua
Sapindaceae	*Scleichera oleosa*	Kusuma
Smilacaceae	*Smilax zeylanica*	Ramdantuni
Sterculiaceae	*Helicteres isora*	Modimudica

In the prevailing situation, conservation of plant resources is very important. Therefore sustainable utilization of medicinal plant is an urgent demand of the hour. Sustainable wild collection with fair trade would help to conserve the natural resources of the Kuldiha preservation plots. The state Government should promote Village Management Committee and Conservation Area Management Committee to protect the forest from denudation. It is necessary to improve the socio economic conditions of people living around the Kuldiha sanctuary to minimize the anthropogenic activities in order to prevent depletion of plant diversity.

Conclusion

The quantitative characters with references to density, frequency, dominance and their relative values distribution could well act as indicators of anthropogenic disturbances that are affecting the various forests types and such studies would help in understanding the threats that are being faced by the tropical forests and would help in deriving conservation policies. There is an urgent need for recognizing these traditionally valued natural systems at various levels and planning for their better management, ultimately aiming to conserve biodiversity.

Acknowledgment

The authors are grateful to the Forest Deportment of Kuldiha Wildife Sanctuary, Balasore, Odisha for providing their help during our field survey.

References

Ahmed, S.A., Kadam, J.A., Patil S.S. and Baig, M.M.V. (2009). Biological efficiency and nutritional contents of *Pleurotus florida* Singer cultivated on different agro-wastes. Nature and Science, (1): 44-48.

Anbarashan, M. and Parthasarathy, N. (2013). Tree diversity of tropical dry evergreen forests dominated by single or mixed species on the Coromandel coast of India. Tropical Ecology, 54(2): 179- 190.

Behera, S.K. and Mishra, M.K. (2005). Indigenous phytotherapy for genitourinary diseases used by the Kandha tribe of Odisha, India. Journal of Ethnopharmacology, 102(3): 319–325.

Chaudhuri, H.N., Pal, D.C. and Tarafder, C.R. (1975). Less known uses of some plants from the tribal areas of Odisha (India). Plant Taxonomy Geography, 17(1-4): 132–136.

Curtis, J.T. (1950). The Interrelation of certain analytic and synthetic phytosociological characters. Ecology, 31: 434-455.

Das, P.K. and Misra. M.K. (1987). Some medicinal plants used by the tribals of Deomali and adjacent areas of Koraput district, Odisha. Indian Journal of Forestry, 10(4): 301–303.

Davidson-Hunt, I. (2000). Ecological ethnobotany: stumbling toward new practices and paradigms. MASAJ, 16: 1.

Dixit, A.M. (1997). Ecological evaluation of dry tropical forest vegetation: An approach to environmental impact assessment. Tropical Ecology, 38: 87-99.

Girach, R.D., Singh S., Ahmed M., Brahmam, M. and Misra, M.K. (1998). Euphorbiaceae in native health practices of district Bhadrak, Odisha, India. Fitoterapia, 69(1): 24–28.

Hare, M.A., Lantagne, D.O, Murphy, P.G. and Chero, H. (1997). Structure and tree species composition in a sub tropical dry forest in Dominican Republic: comparison with a dry forest in Puerto Rico. Tropical Ecology, 38: 1-17.

Holdridge, L.K. (1967). Life zone ecology. San Jose, Coasta Rica, Tropical Science Centre, p. 206.

Hughes, J.B, Daily, G.C, and Ehrlich, P.R. (1997). Population diversity: its extent and extinction. Science, 278: 689-692.

Khare, P.K., Yadav, V.K. and Mishra G.P. (1985). Phytosociological structure of some forest communities in Central India. Journal of Tropical Forestr, (4): 321-326.

Kumar, A., Bruce, G.M. and Ajai, S. (2006). Tree species diversity and distribution patterns in tropical forests of Garo Hills. Current Science, 91: 1370-1381.

Kumar, J. and Rout, S.D. (2005). Ethnomedicinal study of Mayurbhanj district of North Odisha. Step towards participatory documentation of indigenous knowledge of Medicinal Plants. Revival of Local Health Traditions and Conservation of Medicinal Plants. Pp. 51-60 (ed. Bibhu Kalyan Mohanty). SAMBADH, Bhubaneswar, Odisha, India.

May, R.M. and Stumpf, M.P.H. (2000). Species area relations in tropical forests. Science, 290: 2084-2086.

Mishra, N., Rout S.D. and Panda, T. (2011). Ethno-zoological studies and medicinal values of Similipal Biosphere Reserve, Odisha, India. African Journal of Pharmacy and Pharmacology, 5(1): 6-11.

Mudgal, V., and D.C. Pal. (1980). Medicinal plants used by tribal of Mayurbhanj (Odisha). Botanical Survice India, 22(1-4): 59–62.

Negi, C.S. and Nautiyal, S. (2005). Phytosociological studies of a traditional reserve forest Thal Ke Dhar, Pithoragarh, Central Himalaya, India. Indian Forestry, p. 519-534.

Pandey, A.K. and Rout, S.D. (2006). Ethnobotanical uses of plants by tribals of Similipal Biosphere Reserve, Odisha. Ethnobotany, 18: 102-106.

Pattanaik, C. (2003). Studies pertaining to medicinal use of plants and plant parts available in South Odisha. M.Phil. Thesis submitted to Berhampur University, Berhampur, Odisha.

Pattanaik, C., and Reddy, C.S. (2008). Medicinal plant wealth of local communities in Kuldiha wildlife sanctuary, Odisha, India. Journal of Herbs Spices and Medicinal Plants Spices and Medicinal Plants, (3): 175-184.

Pattanaik, C., Reddy C.S., Dhal N.K. and Das. R. (2006). Some phytotherapeutic claims by tribals of Rayagada district, Odisha, India. Journal of Ethnobotany, 10: 189–197.

Pattanaik, C., Reddy, C.S., Murthy, M.S.R. and Manikya, R.P. (2006). Ethnomedicinal observations among the tribal people of Koraput district, Odisha, India. Research Journal Botany, 1(3): 125–128.

Reddy, C.S. and Pattnaik, C. (2009). An assessment of floristic diversity of Gandhamardan hill range, Odisha, India. Bangaladesh Journal of Plant Taxonomy, 16(1): 29-36.

Rout, S.D. (2007a). Ethnobotany of diversified wild edible fruit plants in Similipal Biosphere Reserve, Odisha. Ethnobotany, 19: 137-139.

Rout, S.D. (2007b). Medicinal plants of Similipal Biosphere Reserve and our health care system. Medicinal plants (eds: Dr. R.C. Rout and J.K. Behera). S.R. Mahavidyalaya, Baliapal, Balasore, Odisha. pp. 25-31.

Rout, S.D. and Pandey, A.K. (2007a). Some wild edible and other useful plants of Similipal Biosphere Reserve, Odisha. In: Advances in Ethnobotany (eds: A.P. Das and A.K. Pandey) Bishen Singh Mahendra Pal Singh, Dehera Dun. pp 61-72.

Rout, S.D. and Thatoi, H.N. (2008). Ethnomedicinal practices of Kol tribes in Similipal Biosphere Reserve, Odisha, India. Ethnobotanical Leaflets, 13: 379-387.

Rout, S.D., Panda, T. and Mishra, N. (2009). Ethno-medicinal plants used to cure different diseases by tribals of Mayurbhanj district of North Odisha. Ethno-Medicine, 3(1): 27-32.

Sagar, R., Raghubansi, A.S. and Singh, J.S. (2003). Tree species composition, dispersion and diversity along a disturbance gradient in a dry tropical forest region of India. Forest Ecological Managazine, 186: 61-71.

Saxena, H.O. and Brahmam, M. (1994-96). The flora of orissa. Regional Research Laboratory, Bhubaneswar, Vol. (1-4).

Shannon, C.E. and Wiener, W. (1963). The Mathematical Theory of Communication. University of Illinois press, Urbana. 117 pp.

Singh, J.S., Singh, S.P. and Gupta, S.R. (2006). Ecology, Environmental and Resource Conservation, Anamaya Publishers, New Delhi, 668.

Wilson, E.O, (1988). The current state of biological diversity. In: Biodiversity ed. Wilson EO, Peter FM, p. 3-18. Washington, DC, USA: National Academy Press.

14

Ethnomedicinal Uses of Plants from Lulung, Similipal Biosphere Reserve, Odisha

*D. Nayak and S.C. Sahu**

Department of Botany, North Orissa University,
Baripada – 757 003, Odisha
**E-mail: sudamsahu.bdk@gmail.com*

ABSTRACT

Ethnobotanical study was carried out in Lulung, Similipal Biosphere Reserve (SBR), Odisha during January- March, 2018 to document the ethnomedicinal uses of plants. The present study reports 44 plant species belong to 43genera and 29 families including their botanical name, vernacular name, locality, parts used and diseases/ailments. The reported medicinal plants were mostly used for different diseases such as skin diseases, stomach problem, leucorrhoea and gonorrhoea. Out of 44 plant species; 12 plant species are used by making juice, 22 are used by making paste, 5 by making powder, 1 are used applying their latex, 1 are used by making oil and 4 in the form of decoction. Majority of remedies were prepared as paste followed by juice and powder, either for oral or for external application.

Keywords: *Ethno-medicines, Tribal community, Medicinal plants, Diseases.*

Introduction

Ethnomedicine is sometimes used as a synonym for traditional medicine. Ethnomedicine is a study or comparison of the traditional medicine practiced by various ethnic groups and especially by indigenous peoples. Herbal drugs from ethnomedicinal plants have gained considerable importance in the recent past not only in India but also around the World. Ethnomedicines are much in demand as they are affordable and have much less side effects. The indigenous traditional knowledge of medicinal plants of various ethnic communities, where it has been transmitted orally for centuries is fast disappearing from the earth due to advent of modern technology and transformation of traditional culture (Ganesan *et al.*, 2004). However, during the last few decades there has been an increasing interest in the study of medicinal plants and their traditional uses in different parts of the world. It is estimated that about 64 per cent of the total global population depend on traditional medicines (Farnsworth, 1994; Sindiga, 1994). Nearly 8000 species of plants have been recognized as of ethnobotanical importance (Anon., 1994).India is one of the eight major centres of origin and diversification of domesticated taxa due to its glorious part of traditional medical system and used pattern of different plants, having rich biodiversity and is one of the world's twelve mega diversity countries. Odisha has dense forests with natural plant resources. The tribals residing deep inside the forest are still depending on medicinal plants for their primary health care and treatment of diseases.

There are about 357 ethnic communities found inside the SBR (Rout *et al.*, 2004). They have acquired good knowledge on medicinal uses of plants in the course of their lifetime. The SBR harbours a rich diversity of ethnobotanical species, which generate considerable benefits from social and economic perspectives. Till now, in the core and buffer areas of SBR people are preparing medicines from their available species of plants which are used to treat common diseases. There are so many studies are available on medicinal plants and its indigenous uses in different parts of SBR by different workers (Behera *et al.*, 2006; Behera, 2003; Duraipandiyan *et al.*, 2006; Jain, 1987; Singh, 2002; Naik, 1998; Rout, 2004). However, till date Lulung, a part of SBR is not explored properly for this purpose. Therefore, in this study an attempt has been taken to document various indigenous uses of medicinal plants by the tribes, living in and around Lulung, SBR.

Materials and Methods

Study Area

The SBR, one of the mega biodiversity zones of the country is situated in the central part of the Mayurbhanj district of Odisha lying between 21° 10′ to 22° 12′ N latitude and 85° 58′ to 86° 42′ E longitude and covers an area of 5569 km^2, within Mahanadian east-coastal bio-geographic region of tropical eastern India (Figure 14.1). In particular, Lulung lying between 21° 57′ N latitude and 086° 34′ longitude. There are four villages inside the core area, sixty one villages in the buffer zone and twelve hundred villages in the transitional zone of SBR with a population of about 4.5 lakhs. Tribes occupy a big chunk of the population constituting 73 per cent of the total population of Mayurbhanj. Agriculture is not well developed and therefore most of them depend solely on forests for catering their daily and perpetual need from food to medicine. Since time immemorial the intimate association and dependence of the tribal communities on the local natural resources have enriched them with invaluable knowledge on bio-resource utilization and consequently they have developed extensive knowledge on various plants.

Figure 14.1: Map of Study Area.

Vegetation and Climate

The Similipal forests occupy a group of hills known as Similipal hills. The hills rising very precipitously from the plains of Baripada and Udala in the east and south-east, extending as far as Jashipur in the west, Bisoi in the north and Thakurmunda in the south-west, exhibit an altitude range from 300m to 1200m. Due to varied topography, edaphic condition and climate Similipal harbours rich and varied flora and fauna. The climate of the Similipal is warm and humid. Three distinct seasons are felt during the year: rainy season (mid-June till October), winter (mid-October to February) and summer (March to mid-June). The annual rainfall is varying from 1200 to 2000 mm. The temperature ranges from 9.8 °C to 33.5 °C. The southern and western aspects are cooler and north eastern aspects are warmer. Periodic earth tremors, thunder storms in the rains and dust storms in late May and early June are further characteristic features of SBR.

Data Collection

The work is based on the explorations of medicinal plants in SBR during January- March, 2018. Field trips were carried out at Lulung and its adjacent areas for collection of medicinal plants. Questionnaire survey method was employed to record the traditional ethnomedicinal knowledge of local communities (Santal, Bathudi, Gonda, Kolha, Bhumija, *etc.*) in Similipal Biosphere Reserve. Study was carried out in different villages near to Similipal forest. Based on the close proximity to the forest and tribal dominance, villages were selected for survey.

The local tribal vaidyas and tribal old man were interviewed to record different plants used for various diseases. They were asked for advice for recording local name; plant parts used methods of medicine preparation and recommended doses. The age of the interviewed vaidyas ranged from 42 to 65 years. Regarding common diseases the respondents informed about the frequent occurrence of diseases like dysentery, diarrhoea, jaundice, malaria, leucorrhoea, gonorrhoea, headache and cold fever. After gathering the information from the field, data were analysed by statistical computer programs. The identification of plants was done using the available literatures (Haines 1921-25; Saxena and Brahmam 1989; Saxena and Brahmam 1994-96). Identification and specimen deposition of these medicinal plants were done at Biosystematic Laboratory, North Orissa University, Baripada. To establish the uses of these medicinal plants, the earlier published scientific literature sources were referred (Jain, 1991; Kirtikar and Basu, 1991; Ambasta *et al.*, 1992; Chopra *et al.*, 1996).

Results

The data on medicinal plants, which were collected from tribal people of Lulung, SBR were analysed. The investigation revealed that a total of 44 medicinal plant species belonging to 43 genera and 29 families, which are commonly used for various ailments by various tribes (Santal, Bathudi, Gonda,Kolha, Bhumija, *etc.*). In the enumeration, plant botanical names have been arranged alphabetically following family, local names, parts used and folk uses.

Table 17.1: Ethnomedicinal Plants in Lulung, Similipal Biosphere Reserve (SBR)

Botanical Name	*Family*	*Vernacular Name*	*Locality*	*Parts Used*	*Diseases/Ailments*
Abuliton indicum (Link) Sweet	Malvaceae	Pedipedika	Kuchulahati	Leaf	Jaundice
Asparagus racemosus Willd.	Asparagaceae	Satavari, Gaichira	Lulung	Root	Piles and leucorrhoea
Aegle marmelos (L.) Corr.	Rutaceae	Bela	Kuchulahati	Fruit	Digestive disorder
Aerva Lanata (L.) Juss.	Amaranthaceae	Paunsia	Kuchulahati	Whole plant	Gastric problem and urinary tract infection
Aloe vera (L.)Burm.f.	Xanthorrhoeaceae	Ghi-kumari	Kuchulahati	Leaf	White patches
Anogeissus latifolia (Roxb. ex DC.) Wall.ex Guillem and Perr.	Combretaceae	Dhaura	Lulung	Bark	Intestinal worm in children
Argyreia nervosa (Burm.f.) Bojer	Convolvulaceae	Brudha Jaraka	Pithabata	Root	Ageing
Atylosia scarabaeoides (L.) Benth.	Fabaceae	Bana-kulutha	Lulung	Seed, Root	Cure tape-worm and nose bleedings
Azadirachta indica A.Juss.	Meliaceae	Nimba	Kuchulahati	Leaf	Intestinal worms, skin disease and blood purifier
Bryophyllum pinnatum (Lam.) Oken	Crassulaceae	Amarpoi	Kuchulahati	Leaf	Dysentery
Butea monosperma (Lam.) Taub.	Fabaceae	Palas	Pithabata	Root, seed, latex	Tuberculosis, skin disease, piles and stomach problem
Chrysopogon zizanioides (L.) Roberty	Poaceae	Bena	Lulung	Root	Diabetes
Cissampelos pareira L.	Menispermaceae	Akalabindu	Pithabata	Leaf	Paralysis and fits
Croton roxburghii Wall.	Euphorbiaceae	Putudi	Lulung	Stem	Toothache
Curcuma angustifolia Roxb.	Zingiberaceae	Palua	Kuchulahati	Root	Dysentery
Cuscuta reflexa Roxb.	Convolvulaceae	Nirmuli	Kuchulahati	Whole plant	Poison affected goat and cow
Cynodon dactylon (L.) Pers.	Poaceae	Duba ghasa	Pithabata	Leaf	Malaria
Dendrocalamus strictus (Roxb.) Nees	Poaceae	Baunsa	Lulung	Stem	Leucorrhoea and gonorrhoea

Botanical Name	*Family*	*Vernacular Name*	*Locality*	*Parts Used*	*Diseases/Ailments*
Haldina cordifolia (Roxb.) Ridsdale	Rubiaceae	Kuruma	Pithabata	Bark	Redness in eyes
Hemidesmus indicus (L.) R.Br.	Apocynaceae	Anantamula	Lulung	Root	Joint pains and blood purification
Holarrhena pubiscens Wall.ex G.Don	Apocyanaceae	Kuduchi	Pithabata	Root	Paralysis, knee pains, arthralgia, and fits
Justicia adhatoda L.	Acanthaceae	Basanga	Kuchulahati	Leaf	Bronchitis
Lawsonia inermis L.	Lytharaceae	Manjuati	Pithabata	Root, Leaf	Jaundice
Mangifera indica L.	Anacardiaceae	Amba	Pithhabata	Bark	Dysentery and diarrhoea
Marsileaquadrifolia L.	Marsileaceae	Sunsuni	Pithhabata	Leaf	Leucorrhoea and gonorrhoea
Mimosa himalayana Lam.	Fabaceae	Kirkiti	Lulung	Root	Induces abortion and prevent pregnancy
Moringa oleifera Lam.	Moringaceae	Sajana	Kuchulahati	Bark	Control hair fall and better hair growth
Ocimum tenuiflorum L.	Lamiaceae	Tulasi	Pithhabata	Leaf	Cold fever and leucorrhoea
Papaver somniferum L.	Papaveraceae	Afima	Lulung	Seed	Leucorrhoea and gonorrhoea
Phoenix acaulis Buch-Ham. ex Roxb.	Arecaceae	Bana Khajuri	Pithabata	Root	Toothache
Piper longum L.	Piperaceae	Pippai	Kuchulahati	Fruit	Seizures, paralysis and knee pain.
Pipernigrum L.	Piperaceae	Golmaricha	Pithabata	Seed	Control high blood pressure
Puerariatuberosa (Willd.) DC.	Fabaceae	Bandunadi	Kuchulahati	Root, Leaf	Haemorrhoids
Rauvolfia serpentina (L.) Benth. Ex Kurz	Apocyanaceae	Patala garuda	Lulung	Root	Diabetes
Sapindus trifoliatus L.	Sapindaceae	Rithhaphala	Lulung	Fruit	Chronic dysentery, cholera and tubercular glands
Shorea robusta Gaertn.	Dipterocarpaceae	Sal	Lulung	Bark, Seed	Wound and Carbuncle
Smilax macrophylla Willd.	Smilacaceae	Ramadantuna	Lulung	Root	Leucorrhoea and gonorrhoea

Botanical Name	*Family*	*Vernacular Name*	*Locality*	*Parts Used*	*Diseases/Ailments*
Spondias pinnata (L.f.) Kurz	Anacardiaceae	Ambada	Kuchulahati	Bark	Dysentery and diarrhoea
Swertiachirata Buch.-Ham. ex Wall.	Gentianaceae	Chirata	Lulung	Root	Widening of chest in Children
Terminalia tomentosa Wight and Arn.	Combretaceae	Asana	Pithabata	Bark	Dysentery and diarrhoea
Tridax procumbens (L.) L.	Asteraceae	Bishalyakarani	Pithabata	Whole Plant	Healing of cut wounds
Urginea indica (Roxb) Kunth	Asperagaceae	Banapiyaja	Lulung	Bulb	Asthma, heart problem, paralytic affection and promote urination
Vitex negundo L.	Lamiaceae	Begunia	Pithabata	Stem	Deafness
Wendlandia tinctoria (Roxb.) DC.	Rubiaceae	Chelai	Lulung	Bark	Sunken part of head of baby

These medicinal plants total of44 medicinal plant species belonging to 24 families from Lulung were documented used for different purposes by tribal people; 16 (36.36 per cent) were trees, 14 (31.81 per cent) herbs, 7 (15.9 per cent) shrubs, 7 (15.9 per cent) climbers.

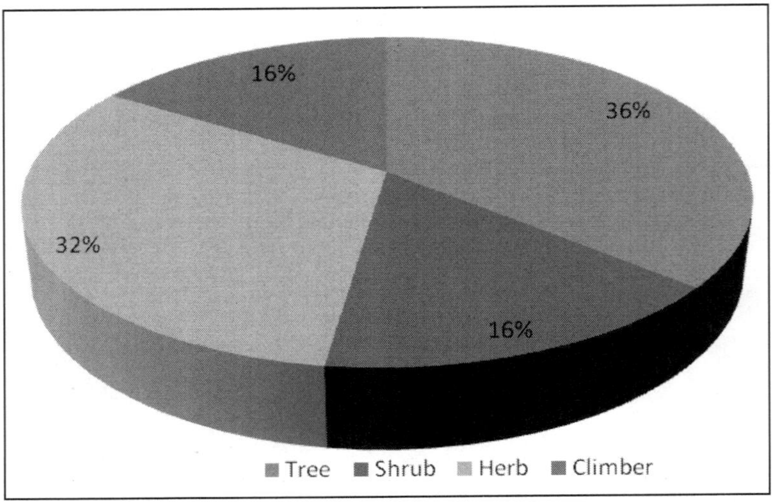

Figure 14.2: Life Form of Identified Plant Species at Lulung, Similipal Biosphere Reserve.

Local users make use of either part or entire plant for medicinal purpose to cure their different diseases. This study revealed that out of 40 plant species, 8 have medicinal properties in their bark, 14 have medicinal properties in their root, 11 have medicinal properties in their leaf, 2 have medicinal properties in whole plant, 6 have medicinal properties in their seed, 1 each have medicinal properties in latex, 3 have medicinal properties in their stem, 3 have medicinal properties in fruit and 1

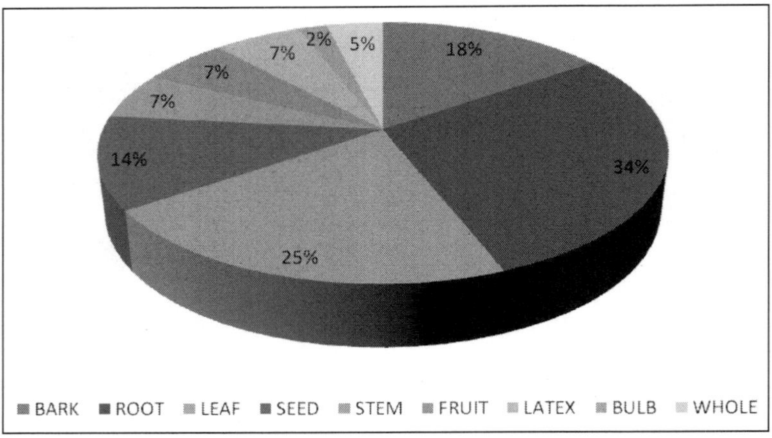

Figure 14.3: Plant Parts Used for Medicine Purpose by the Local Tribes of Lulung, Similipal Biosphere Reserve.

have the medicinal properties in their bulb. Additionally plants possess medicinal properties in their resin, heart wood and fronds. The most frequently utilized plant part was the leaf followed by the root, bark, fruits and whole plant.

People depend on plant-based medicines for primary health care. For the identified 44 plant species, respondents were requested to share their knowledge of each plant and its medicinal properties to cure different ailments. However, some plants having ethnomedicinal uses are not documented due to improper identification and lack of common language. The present study showed that out of 40 plant species; 10 plant species are used for stomach-ache (4 for diarrhoea and dysentery, 1 digestion problem, 1 for gastric, 1 stomach pain, 3 intestinal worm); 2 species are used for rheumatic problems; 7 are used to treat respiratory problems (1 in cold, 1 in bronchitis, 1 in asthma, 1 in tuberculosis and 3 in throat infection); 2 species are used to cure fever (1 in malaria and 1 in normal fever); 5 are used to cure skin diseases (1 for each small spot and white spot, 2 for septic wounds, 1 dandruff, 1 for urinary tract infection); 8 for urinary problem (5 for gonorrhoea, 1 for diabetes, 1 for urinary tract infection and 1 for urination); 1 species are used for antidote for poison; 9 are used for blood disease (2 for jaundice, 6 for piles, 1 each as anticoagulant and blood purifier). Additionally 4 species are used for treating different types of headache problems and ailments; 2 for toothache); 1 plant species are used for eye problem; 1 for aging; 1 for brain nerve cell disorder; 1 plant species are used for gynaecological problems after traditional knowledge on medicinal plants.

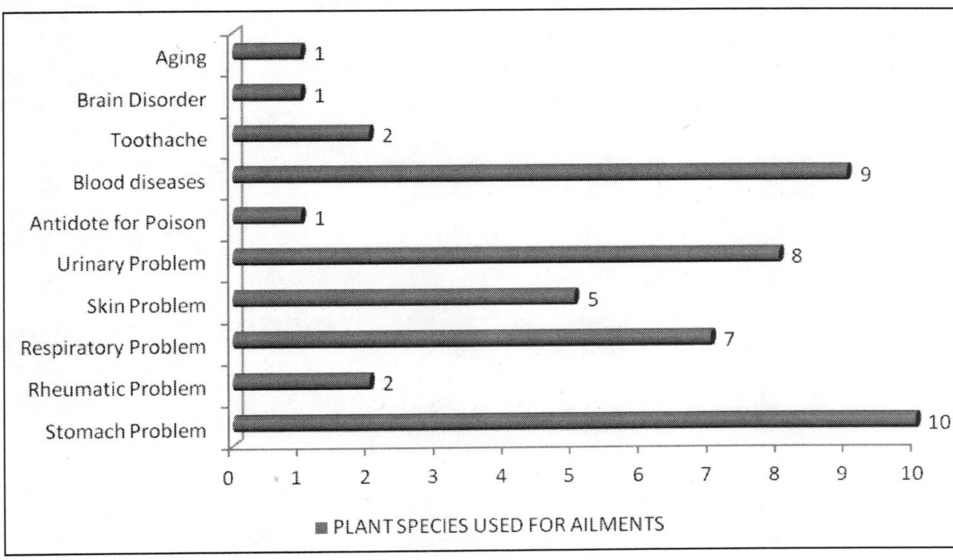

Figure 14.4: Plant Species Used for different Ailments by the Local Tribes of Lulung, Similipal Biosphere Reserve.

The tribals of SBR have been found using specific plant parts for the treatment of different ailments. Out of 44 plant species; 12 plant species are used by making juice, 22 are used by making paste, 5 by making powder, 1 are used applying their latex, 1 are used by making oil and 4 in the form of decoction. Majority of remedies were prepared as paste followed by juice and powder, either for oral or for external application.

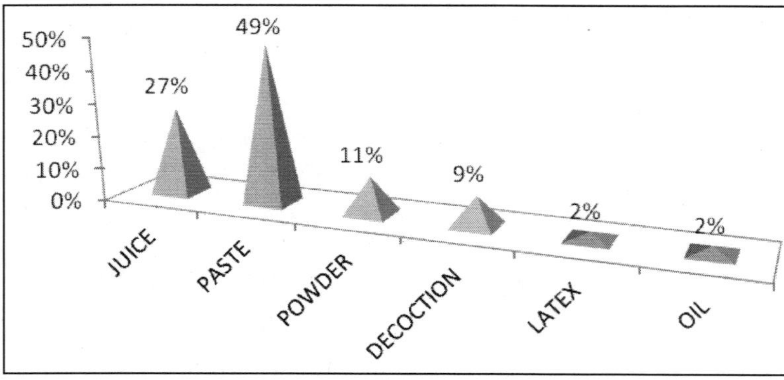

Figure 14.5: Different Forms of Medicine Prepared from Plant Parts by the Local Tribes of Lulung, Similipal Biosphere Reserve.

Local users rely on subsistence agriculture; with plants performing vital roles as crops, fodder, fruit and vegetables, fuel wood, building materials and medicines. This study found that out of known plant species 44 are being used for medicine, 6 for fodder and 3 for fuel wood. 3 out of 44 plants have religious value, 7 are used for construction as timber, 6 provide edible fruits, 7 are important for vegetable and 2 for rope making, 4 as spices, 6 for leaves, 3 are used as dye, 6 are used as insect repellent, 2 for making beverages and 6 are used for cosmetics.

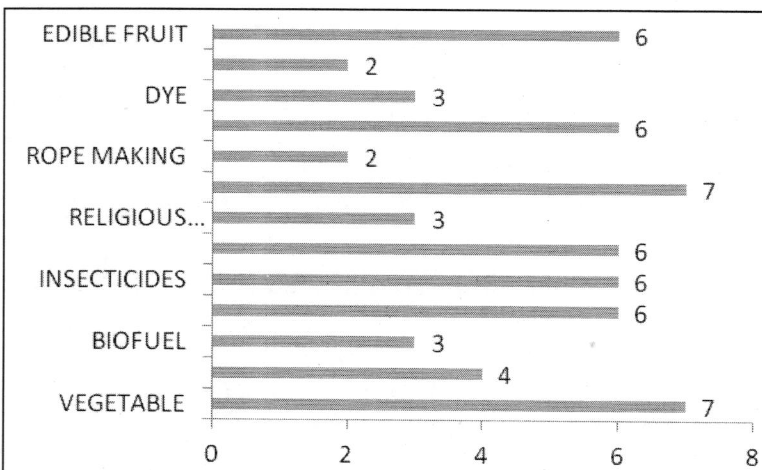

Figure 14.6: Different Uses of Plant Species by the Local Tribes of Lulung, Similipal Biosphere Reserve.

Discussion

The ethnomedicinal uses of 44 plant species were reported from the different tribes of Lulung, Similipal Biosphere Reserve. It is evident from the present study that the tribals are dependent on many medicinal plants for treatment of various diseases. Some experienced tribals have shared their knowledge with the authors about the cure of some important diseases like diarrhoea, dysentery, piles, rheumatism, diabetes and fever. This information is being reported by us for the first time (not reported by earlier investigators *i.e.*, Saxena *et al.* (1988) as well as Jain (1991).This information was also checked with available literature of Karuppusamy *et al.* (2001), Girach and Aminuddin (1989), Mishra *et al.* (2001) and Singh *et al.* (1989). Girach *et al.* (1999) studied the medicinal flora of Similipahar by collecting 270 plant species, belonging to 217 genera and 86 plant families. They carried out a regression analysis by analyzing the number of medicinal species per family against the total number of species in each family for the 138 flowering plant families in Similipahar. Among the studies families, Fabaceae, Euphorbiaceae, Rubiacae and Verbenaceae were dominant having maximum number of medicinal plants. Panda *et al.* (2014) studied the flora of SBR by collecting 187 species. Among the studies families, Fabaceae, Euphorbiaceae, Asteraceae were dominant having maximum number of medicinal plants. Further, 186 species were collected by Rout *et al.* (2012) in SBR. In the present survey, we found Fabaceae having maximum number of medicinal plants *i.e.* 4 followed by Apocyanaceae and Poaceae (3 in each family). Sahu *et al.* (2014) studied Ethnomedicinal plants used in the Healthcare systems of Tribes of Dantewada, Chhattisgarh by collecting 44 species and 23 families. Rajadurai *et al.* (2009) studied Ethnomedicinal Plants of Panchamalai Hills, Tamil Nadu and collected 12 species. Some information recorded in the study particularly for *Pueraria tuberosa, Abuliton indicum, Argyreia nervosa, Chrysopogon zizanioides* were found to be either not known or little known, whereas *Andrographis paniculata, Asparagus racemosus, Holarrhena pubescens, Butea monosperma, Azadirachta indica, Atylosia scaraboides* and *Aervalanata* were very commonly used by other tribes indicating the authenticity of their usefulness as evidenced by earlier researchers (Saxena *et al.*, 1988; Sarkar *et al.*, 1999). Some common medicinal plants have different medicinal uses for different diseases. *Hemidesmus indicus* used for cooling body (Rajadurai *et al.*, 2009), used as antidote to snakebite, scorpion sting and wound healing (Sahu *et al.*, 2014); *Cuscuta reflexa* used for joint pain, headache (Sahu *et al.*, 2014), for fever and malaria (Rout *et al.*, 2012) and *Cynodon dactylon* used for Urinary tract infection, diabetes fever, hypertension, cancer and headache (Sahu *et al.*, 2014), for dysentry (Panda, 2013), for diarrhoea, vomiting, gastric problem (Rout *et al.*, 2012). In the present study, the plant species such as *Hemidesmus indicus* was used for blood purification, gout and joint pain; *Cuscuta reflexa*for antidote of poison; *Cynodondactylon* for malaria.

Traditionally, the rural women prefer plant medicines rather than modern medicines for abortion, menstrual trouble, conception disorders, sterility, delivery problem *etc.* Some ethnomedicinal observations made here, revealed valuable phytotherapeutic information on the various gynecological disorders. The important plant species like *Pueraria tuberosa, Curcuma angustifolia,* and *Smilax macrophylla* are used in various gynecological disorders. Similarly many of the herbs used by the tribal people for the treatment of various joint diseases, like gout, arthritis and rheumatism are very common. Their modes of administration are also simple and convenient. The common man can easily afford to take the treatment without any side effect. The species *Vitex negundo* root used for rheumatic problem. These species play a vital role for families with many children without income, women (gynecological problem), old aged parents (always rheumatic problem) and snake biting problem in rainy season.

Tribal users have found using specific plant parts to treat different diseases and ailments. The most frequently utilized plant parts are leaf, followed by roots, and bark. Additionally, some have medicinal value in their flowers, rhizomes, tubers. In some cases, the whole plant including the roots is utilized. It is important to note that the removal of bark and roots can have significant detrimental effects on plant survival and regeneration (Dhillion and Amundsen 2000). Although the exploitation of some plant parts (*e.g.* fruits, seeds and latex) is less damaging than others (*e.g.* bark, leaf, stem and roots), almost any form of over harvesting has an impact on the structure and function of plant population. The data collected show that majority of the remedies are taken orally. People use these plants in different forms such as juice, extract, decoction, paste, infusion, powder *etc.* Most of the reported preparations are drawn from a mixture of plant; single plants are used rarely. Our study has shown that majority of the plant parts are prepared in the form of paste followed by juice and powder. Similarly, internal administration of these substances predominates over external. For current use, the most important methods used were direct application of the paste or oil. The mode of preparation is based upon the nature of the ailments. For the bitterness of the herbal medicine they use ghee and honey (its sweet taste makes the bitter and unpalatable taste of some of the herbs) and cow milk. The knowledge and practices are not directly related to any organized system but has come entirely through oral tradition and personal experience.

Conclusion

In this study we have enlisted a record of 44 number of plant species which are used by the tribals of SBR for various ailments. These include plants whose medicinal importance has not been recorded and reported so far, or about which new information on medicinal uses and mode of administration has been obtained. The observation revealed that the tribes used leaf and root as the most common crude drug in their preparations to cure different diseases. Although the information given by tribals of Similipal is not comparable to the modern medicine, their efficacy is claimed to be high. In this study, we have observed that the younger generation is not interested to inherit the traditional knowledge from their elders. Presently, very few elders in the tribal community practice herbal cure, while the young and current generation knows little or nothing about the traditional herbal medicines. If this trend continues, a few years from now, there will not be even a single elder member in tribal community who knows traditional knowledge on medicinal plants. This is particularly serious consequences as the knowledgeable people are very old and the young have not shown much interest in this accumulated knowledge. This will surely affect the resilience of the local traditional knowledge. There is great danger of extinction of such healing practices. Further, the demand for medicinal plants is increasing; their survival in their natural habitat is under threat. This is due to rapid degradation and loss of natural habitats and in certain cases due to over-harvesting of species. The existing deforestation and habitat fragmentation would pose a serious threat to the growth of wild plants. The declination of traditional ethno-botanical knowledge with over exploitation and habitat destruction amounts to threatening the survival of wild plants. Hence, documentation of traditional knowledge needs to be given top priority to help conservation resources and preservation of the disappearing knowledge. Ethno-botanical knowledge if recognized, valued and implemented properly can help the environment and society in a holistic way.

References

Ambasta, S.P., Chandran, K.R., Kashyappa, K. and Chand, R. (1992). The Useful Plants of India. Publications and Information Directorate, CSIR, New Delhi.

Behera, K.K. (2003). M. phil thesis. Ethnobotanical studies of Kaptipada subdivision, Utkal University, Bhubaneswar.

Behera, K.K. (2006). Potential Medicinal Plants and Possible Threats for their Survival in the District of Nuapada, Odisha, India, National Seminar on forest Biodiversity, OUAT, Bhubaneswar, From Feb 12 to 15.

Chopra, R.N., Nayar, S.L. and Chopra, I.R. (1996). Glossary of Indian Medicinal Plants (Rep. Ed.). National Institute of Science Communication, CSIR, New Delhi.

Cunningham, A.B. (2001). Applied ethnobotany, people wild plant use and conservation. Earthsean Publishing Ltd. London and Sterling VA.

Duraipandiyan, V. and Ignacimuthu, S. (2007). Antibacterial and antifungal activity of *Casia fistula* L.: An ethnomedicinal plant. Journal of Ethnopharmacology. 112: 590-594.

Farnworth, N. (1994). Ethnopharmacology and drug development. pp. 42-51. In: Chadwick, D.J. and J. Marsh (Eds.). *Bioactive Compounds from Plants.* Cifa Foundation Symposium, 185, Wiley, Chichectar.

Ganesan, S. and Kesavan, L. (2003). Ethnomedicinal plants used by the ethnic group of Valaiyans of Vellimalai hills (Reserved forest), Tamil nadu, India, Journal of Economic and Taxonomic Botany, 27: 754-760.

Girach, R.D., Aminuddin, C. (1989). Ethnopteridological Notes on *Lygodium. Flexuosum* (Linn.) Sw. Journal of Economic and Taxonomic Botany, 13(2): 255-257.

Girach, R.D., Shaik, A.A., Singh, S.S. and Ahmad, M. (1999). The medicinal flora of Similipahar forests, Odisha State, India. Journal of Ethnopharmacology, 65: 165–172.

Haines, H.H. (1921-25). The Botany of Bihar and Odisha. London. (Reprint, 1961, Calcutta).

Jain, S.K. (1987). A manual of ethnobotany. Scientific Publishers, Jodhpur, India.

Jain, S.K. (1987). Glimpses of Indian Ethnobotany. New Delhi: Oxford and IBH Publishing Co.

Jain, S.K. (1991). Dictionary of Indian Folk Medicine and Ethnobotany. Deep Publications, New Delhi.

Jain, S.K., Rao, R.R. (1976). A handbook of field and Herbarium Methods.

Karuppusamy, S., Karmegam, N. and Rajasekaran, K.M. (2001). Enumeration, Ecology and Ethnobotany of Ferns of Sirumalai hills, South India. Journal of Economic and Taxonomic Botany, 25: 631-634.

Kirtikar, K.R. and Basu, B.D. (1991). Indian Medicinal Plants. 4 Vols. (Repn. Edn). Lalit Mohan Basu, Allahabad.

Martin, G.J. (1995). Ethnobotany: A Methods Manual. Chapman and Hill, London.

Mishra, R.C., Panda, P.C. and Das, P.A. (2001). Taxonomic study of the ferns and ferns allies of Gandhamardan hills, Odisha. Journal of Economic and Taxonomic Botany, 25(3): 577- 579.

Naik, D. (1998). Tribal culture in the context of Similipal. Workshop Journal on Different Problematic Aspects of Similipal Protection and its Solutary Action.

Panda, S.K., Rout, S.D., Mishra, N. and Panda, T. (2011). Phytotherapy and traditional knowledge of tribal communities of Mayyurbhanj district, Odisha, India. Journal of Pharmacognosy and Phytotherapy, 3: 101–113.

Panda, S.K., Rout, S.D., Mishra, N. and Panda, T. (2011). Folk uses of some medicinal plants by Kol tribes of Similipal Biosphere Reserve, Odisha, India. International Journal of Biological Technology, 2: 16–20.

Rajadurai, M., Vidhya, V. G., Ramya, M. and Bhaskar, A. (2009). Ethnomedicinal plants used by traditional healers of Pachamalai hills, Tamil Nadu, India. Ethnomedicine, 3(1): 39-41.

Rout, S.D. and Pandey, A.K. (2007). Ethnomedicobiology of Similipal Biosphere Reserve, Odisha. In: A. P. Das and AK, Pandey, eds. Advance Ethnobotany, p. 247-252.

Rout, S.D., Panda, T. and Mishra, N. (2009). Ethno medicinal plants used to cure different diseases by tribals of Mayurbhanj district of North Odisha. Ethnomedicine, 3(1): 27-32.

Sahu, P.K., Masih, V., Gupta, S., Sen D.K. and Tiwari, A. (2014). Ethnomedicinal plants used in the healthcare systems of tribes of Dantewada, Chhattisgarh, India. American Journal of Plant Sciences, 5: 1632-1643.

Sarkar, N., Rudra, S. and Basu, S.K. (1999). Ethnobotany of Bangiriposi, Mayurbhanj, Odisha. Journal of Economic and Taxonomic Botany, 23(2): 509-514.

Saxena, H.O. and Brahmam, M. (1989). The flora of Similipahar (Similipal), Odisha, Regional Research Laboratory (CSIR); Bhubaneswar.

Saxena, H.O. and Brahmam, M. (1994-96). The flora of Orissa, vol. I-IV. Regional Research laboratory (CSIR), Bhubaneswar and Odisha forest Development Corporation Ltd. Bhubaneswar

Saxena, H.O. and Dutta, P.K. (1975). Studies on the ethnobotany of Odisha. Bulletin of Botanical Survey of India, 17: 124-131.

Saxena, H.O., Brahmam, M. and Dutta, P.K. (1988). Ethnobotanical studies in Similipal Forests of Mayurbhanj District (Odisha) Bulletin of Botanical Survey of India, 30: 83- 89.

Singh, J.S. (2002) The biodiversity crisis. A multifaceted review, Current Science, 82(6): 638.

Singh, K.K., Saha, S. and Maheswari, B. (1989). Ethnomedicinal uses of some ferns among the tribals of Uttar Pradesh. Indian Fern Journal, 6: 63-67.

15

Status of Wild Tasar Silkworm Populations in Mayurbhanj Revealed Through GIS and Remote Sensing

S.N. Achary[1], *Debabrata Nandi*[1*] and *Prasanta Kumar Kar*[2]

[1]*Department of Remote Sensing and GIS, North Orissa University, Takatpur, Baripada – 757 003, Odisha*
[2]*Regional Tasar Research Station, Central Silk Board, Prafulla Nagar, Baripada – 757 001, Odisha*
**E-mail: debabrata.gis@gmail.com*

ABSTRACT

Wild Tasar silk worm are endemic to the Indian subcontinent. As a result of continued habitat loss and degradation over the past century, Wild Tasar silkworm populations have been in steady decline and now exist only in isolated or fragmented habitat across the entire range. So, the conservation program should be provided to overcome the extinction of the species of interest. In Similipal by using the Maximum Entropy model we develop a distribution pattern with reference to presence only data the present study included the wild Tasar silkworm populations in different zones of Mayurbhanj, including Similipal Biosphere Reserve. The MaxEnt model was used to ascertain the contribution of environmental variables with an aim to develop a predictive mapping. It was clear from the study that at higher and medium altitude tasar silkworm quantitative traits are well expressed. This may be due the fact that the habitat there is more adaptive for the silkworm. On the other hand, the landscape or habitat in majority of Mayurbhanj district, Odisha is not suitable for wild tasar conservation. There are sporadic distribution of forests with tasar silkworm host plants like Shorea robusta and Terminalia tomentosa. But the canopy along with other environmental variables may not be sufficient to bear the load of carrying capacity for the perpetuation of tasar silkworm in wild conditions.

Keywords: *Antheraea mylitta, GIS, Maximum entropy model, Wild Tasar silkworm*

Introduction

Silk has been used by human beings for various purposes since ancient times. Pure silk is one of the finest and most beautiful natural fibers of the world and is said to be the queen of fibers (Ahsan and Sinha, 2000). The silk was known about 2200 B.C. in China. After its cultivation, it passed an adventurous course of journey and from time to time it has established itself across different nations through the famous "Silk Route" (Dutta, 1988; Sahu *et al.*, 1997). Due to silk's great value and usefulness, there have been many attempts in various parts of the world for the large scale production of silk. One of the methods is the rearing of the silkworms on large scale with great care in natural and controlled conditions. Different rearing techniques are applied in different parts of the world for large scale production of silk threads of fine quality. This process or method of practice is known as "Sericulture" (Ahsan and Sinha, 2000). Sericulture in India is practiced both in temperate and tropical zones, where seasons and rearing conditions vary, hence, the silkworm breeds differ (Rangaswamy *et al.*, 1992).The natural silks are broadly classified as mulberry and non-mulberry silks. Non-mulberry sericulture is universally known as forest based or wild sericulture (Jolly, 1976; Jolly *et al.*, 1968, 1974 and 1975). Of late these silks are also known as *Vanya* Silk. Tropical and temperate Tasar, Eri, Muga are the non-mulberry silks. About 95 per cent of the global production of non-mulberry silk is tasar (Zhuang *et al.*, 1994). Tropical tasar silkworm, *Antheraea mylitta* Drury is exploited in country for commercial *vanya* silk production and thousands of poor tribal families earn their livelihood through tasar culture. Tasar silkworm rearing for production of cocoons has also been a tradition of tribal people inhabiting the states of Bihar, Jharkhand, Chhattisgarh, Madhya Pradesh Odisha, Andhra Pradesh, Maharashtra, West Bengal, Telengana and Uttar Pradesh. Nature grown cocoons are collected on large scale in some of the tasar producing areas. These states are famous for traditional and unique tasar silk weaving.

Tasar silk industry has directly and indirectly been providing gainful employment to approximately 2.5 lakh family in rural and semi-urban areas. Mostly tribal people are engaged in the production of silk. Out of the total tribal population of about 38 million nearly 30 million (77 per cent) are located in tasar cultivating states of tropical (25.67 million) and temperate (4.23 million) belts. However, 13.2 million and 3.2 million tribals are living in tropical and temperate tasar cultivating districts, respectively. About 1.23 lakh families are engaged in tasar silkworm rearing in tropical zone.

Wild silkworms are polyphagous in nature and reared outdoor on various food plants. The polyphagous nature of tasar silkworms is an interesting and intriguing aspect of sericulture. Food plants of tropical tasar have been categorized as primary, secondary and tertiary on the basis of preferential feeding and/or adaptability of these silkworms on them (Singh and Srivastava, 1997). Tasar silkworm *Antheraea mylitta* Drury is reared outdoor on *Terminalia arjuna* and *T. tomentosa*. The silkworm populations are also found in natural *Shorea robusta* forests. Daba, Sarihan, Bhandara and Sukinda ecoraces of *A. mylitta* have been adapted to *T. arjuna* and *T. tomentosa*. Laria, Barharwa, Raily, Nalia and Modal ecoraces have been adapted to *Shorea robusta*. Tira, Jiribam and Belgaum ecoraces have the first choice for *Lagerstroemia* and *Zizyphus* species of plants. Hence these host plants may serve as a primary host plants to these ecoraces. The most important secondary host plants as recorded by Jolly *et al.* (1979) are *T. chebula, T. belerica, T. catapa, T. paniculata, Anogeissus latifolia* and *Syzygium cumini*. Apart from the primary and secondary host plants discussed above, there is a long list of secondary and tertiary host plants of tropical tasar silkworms but they are less significant. Therefore the current investigation was carried out to understand the distribution of wild tasar silkworm species using environmental factors and other discrete factors such as elevation, slope *etc.* in Mayurbhanj district of Odisha. To focus on fitting a probability distribution for occurrence of tasar silkworm to set of pixels across the study area. To understand the application of species distribution modelling using MaxEnt model. To produce the habitat suitability zone or area map of the particular or group or given species.

Materials and Methods

Data Source: Tasar Silkworm

Through survey of wild tasar in Mayurbhanj districts, Odisha, samples were collected from 36 places. The geographical coordinates along with elevation are presented in Table 15.1. The cocoon quantitative traits of wild cocoons *viz.* cocoon volume, cocoon weight, shell weight, pupal weight and shell ratio were recorded. The data were statistically analysed for mean, standard deviation, coefficient of variation and ANOVA using MS-Excel.

Data Source: GIS and RS

Archive data

IRS P6 LISS III Image

Land use/Land Cover map of year 2011-12

Cartosat Image

Cartosat DEM

Ancillary Data

Soil data

Meteorological Data

Administrative Atlas

Software

ArcGIS 10.2

ERDAS imagine 2015

Microsoft Word, Excel 2007

MaxEnt Model v3.3.3e Tutorial (ArcGIS v10)

Study Area

The study was conducted in Mayurbhanj district of Odisha. Mayurbhanj lies between north latitudes 21°17′ and 22°34′ and east longitudes 85°40′ and 87°10′ falling in the survey of India toposheet nos 73F, 73G, 73J and 73K. It is bounded on the north by the district Singbhum of Jharkhand and Midnapore district of West Bengal, on the south by Balasore and Keonjhar districts and on the west by Keonjhar and Singhbhum districts and on the east Midnapore and Balasore districts. The district comprises of 4 subdivisions namely Sadar sub-division with headquarter at Baripada, Bamanghati sub-division with headquarter at Rairangpur, Panchpir with headquarter at Karanjia and Kaptipda subdivision with headquarter at Udala. The district has a total geographical area of 10418 sq km. There are 26 blocks and 3945 villages out of which 227 villages are uninhabited. The elevation above mean sea level varies from a maximum of 1165 meter at Meghasani peak of Similipal to a minimum of 40 meter in the plains.

Methodology

Continuous data environmental variables

Discrete data soil types, forest types, slope, aspect, elevation *etc.*

MaxEnt software for species habitat modeling (version 3.3.3k) was used. The specific settings used to run the MaxEnt models are as follows:A random test percentage of 25 were used. Replicates -5, Replicate type-sub sample, No output grids, Maximum Iterations - 5000. This withheld 20 per cent of species occurrences and tested the validity and accuracy of the training model. Response curves and jackknife measures of variable importance were created to provide additional descriptive data and graphs to help interpret the MaxEnt model. The logistic output format was selected. Logistic output gives an estimate between 0 and 1. The cutoff value (to determine that typical presence localities have probability of presence) was set at 0.5 (Phillips 2011).

MaxEnt output includes a suite of statistical results. The receiver operator characteristic (ROC) curve or AUC gives an estimate of how well the model distinguishes species occurrences from random locations. The AUC estimate has a range from 0 to 1, and values close to 1 indicate that the model does very well distinguishing species occurrences. An AUC of 0.5 indicates that the model predicted species occurrence no better than by random chance. In addition to the statistical results, MaxEnt creates a map of the logistic habitat predictions in ASCII format. The MaxEnt ASCII output was converted into an integer file geodatabase raster by multiplying each ASCII layer by 100 and converting the result to an integer dataset. Storing the model output as a 0 to 100 value integer raster dataset resulted in significant file size reduction from the original ASCII. The final model raster datasets have a data range from 0 to 100, with 100 equaling 1 in the original ASCII output. Binary maps of species occurrence were created by reclassifying the integer raster datasets, where all values from 0 to 50 were given a value of 0 and all values from 50 to 100 were given a value of 1. A value of 1indicates predicted species occurrence; a value of 0 indicates unsuitable habitat.

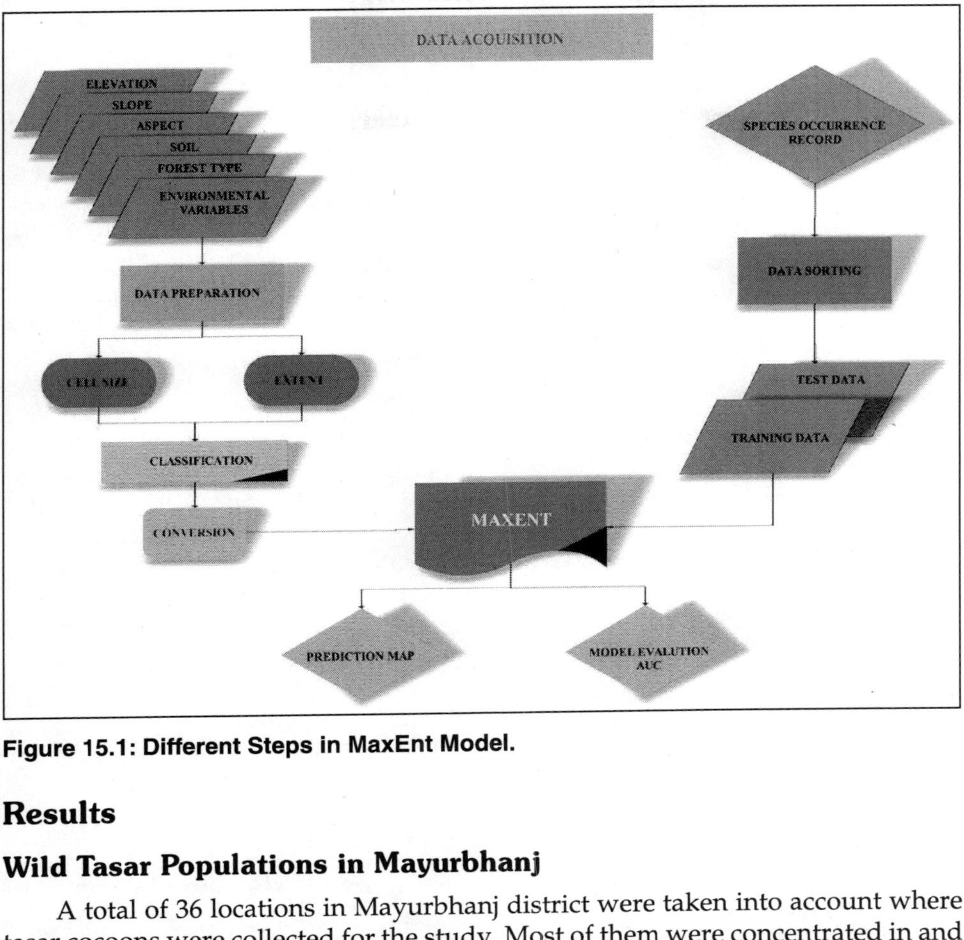

Figure 15.1: Different Steps in MaxEnt Model.

Results

Wild Tasar Populations in Mayurbhanj

A total of 36 locations in Mayurbhanj district were taken into account where tasar cocoons were collected for the study. Most of them were concentrated in and around Similipal Biosphere Reserve. For convenience in analysing the data they were grouped into Central, Western, Southern, Nothern and Eastern zones. The details of locations, their geographical coordinates and altitude are indicate that there was a well defined difference in the altitude of collection site, the lowest was Suliapada (77 mASL) while highest was at Patbil (971 mASL). Suliapada is an isolated patch of forest situated towards the extreme eastern corner of Mayurbhanj detached from original Similipal. Like Patbil, the other places at high altitude were Gudgudia (843 mASL), Bahaghar (848 mASL), Chandikhaman (877 mASL) and Nuagaon (887 mASL). These are situated at the central part of Similipal or towards the western zone.

In case of tasar silkworm cocoon traits are very important not only from commercial viewpoint but also on the aspects of adaptation of the species in its natural habitat. Perusal of data in revealed that there exist wide differences in almost all the traits in male cocoons as the F value ranged from 2.912 ($P < 0.01$) to 21.871 ($P < 0.001$). When the CD values are taken into account, the wild tasar populations in Central and Western zones came in one group while southern populations also occupied position closure to them. On the other hand the populations from northern and eastern zone were separated from them for all the five traits. In case of female cocoons the trend remained almost similar to that of male. But, in case of shell ratio the difference was not significant. There was great variation in the size of the cocoons. In the cenytral and western zone, the cocoon volume was 37.2 and 34.55 cc respectively while lowest was in Eastern zone (21.9cc) and the difference was highly significant ($F = 225.7$, $P < 0.001$). From Table 2 and 3 it is also seen that there was significant difference of cocoon traits between male and female. The females had always higher values than the males except the shell ratio.

MaxEnt Model Outputs

Analysis of Omission/Commission

Figure 15.2 shows the test omission rate and predicted area as a function of the cumulative threshold, averaged over the replicate runs. The omission rate should be close to the predicted omission, because of the definition of the cumulative threshold. Similarly, Figure 15.3 is the receiver operating characteristic (ROC) curve for the same data, again averaged over the replicate runs. Note that the specificity is defined using predicted area, rather than true commission (Phillips, Anderson and Schapire). The average test AUC for the replicate runs is 0.858, and the standard deviation is 0.050.

Pictures of the Model

The Figure 15.4 show the positional distribution pattern of the host plant and there percentage of accuracy.

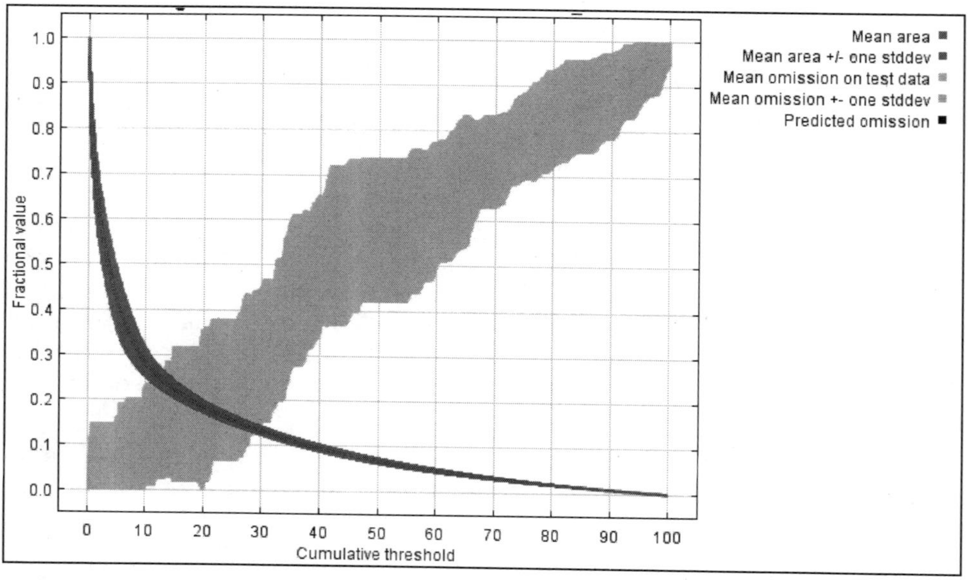

Figure 15.2: Average Omission and Predicted Area.

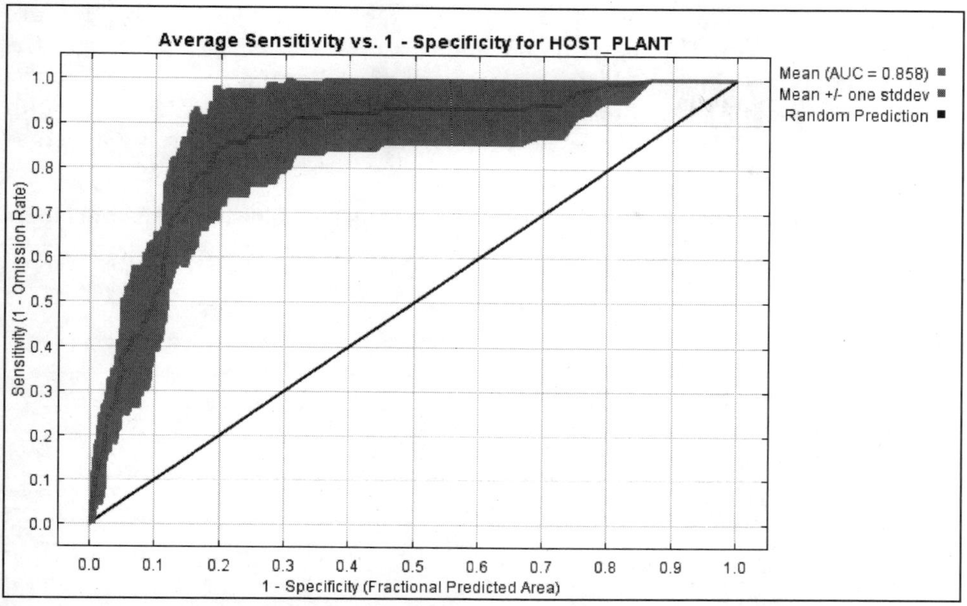

Figure 15.3: Average Sensitivity vs Specificity.

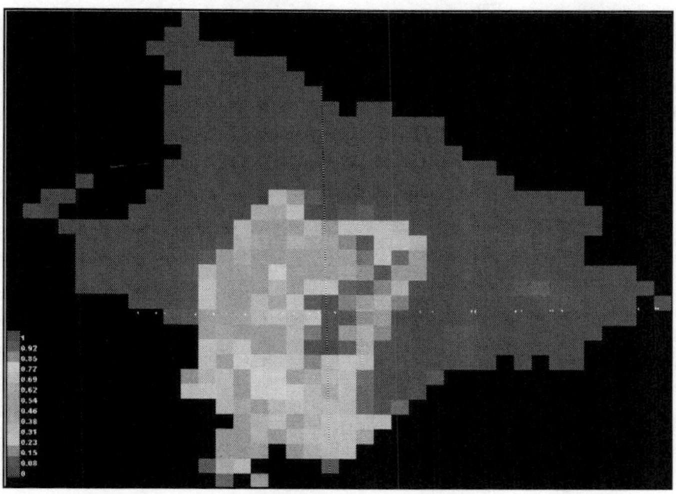

Figure 15.4: Prediction Map of Mayurbhanj for Wild Tasar Silkworm.

Analysis of Variable Contributions

Table 15.1 gives estimates of relative contributions of the environmental variables to the Maxent model. To determine the first estimate, in each iteration of the training algorithm, the increase in regularized gain is added to the contribution of the corresponding variable, or subtracted from it if the change to the absolute value of lambda is negative. For the second estimate, for each environmental variable in turn, the values of that variable on training presence and background data are randomly permuted. The model is reevaluated on the permuted data, and the resulting drop in training AUC is shown in the table, normalized to percentages. As with the variable jackknife, variable contributions should be interpreted with caution when the predictor variables are correlated. Values shown are averages over replicate runs.

Table 15.1: Relative Contributions of the Environmental Variables to the MaxEnt Model

Raster BIOs	*Variable*	*Per cent Contribution*	*Permutation Importance*
BIO 4	Temperature seasonality	20.6	25.8
BIO14	Precipitation of driest month	20.3	1.7
BIO 8	Mean temperature of wettest quarter	15.6	9
BIO10	Mean temperature of warmest quarter	12.2	0
BIO 2	Mean diurnal range	8.9	4.9
BIO19	Precipitation of coldest quarter	7.5	7.2
BIO 3	Isothermality	6.4	9.1
BIO11	Mean temperature of coldest quarter	2.3	15.4
BIO 1	Annual mean temperature	1.5	0
BIO 7	Temperature annual range	1.3	15.6
BIO16	Precipitation of wettest quarter	1.2	0
BIO18	Precipitation of warmest quarter	0.9	4.5
BIO12	Annual precipitation	0.5	0
BIO15	Precipitation seasonality	0.2	1
BIO17	Precipitation of driest quarter	0.2	1.6
BIO 9	Mean temperature of driest quarter	0.2	3.5
BIO 6	Min temperature of coldest month	0.1	0.3
BIO 5	Max temperature of warmest month	0	0.3
BIO13	Precipitation of wettest month	0	0

Figure 15.5 shows the results of the jackknife test of variable importance. The environmental variable with highest gain when used in isolation is Bio16, which therefore appears to have the most useful information by itself. The environmental variable that decreases the gain the most when it is omitted is Bio 3, which therefore appears to have the most information that isn't present in the other variables. Values shown are averages over replicate runs.Note that conclusions about which variables are most important can change, now that we're looking at test data. The jackknife test for area under curve (AUC)

Discussion

In the present study the species under consideration is *A. mylitta*, the tropical tasar silkworm which has immense commercial value as it also provides bread and butter to the thousands of poor tribal families through its lucrative and invaluable silk. Once upon a time about 44 ecoraces or ecological races were available in India

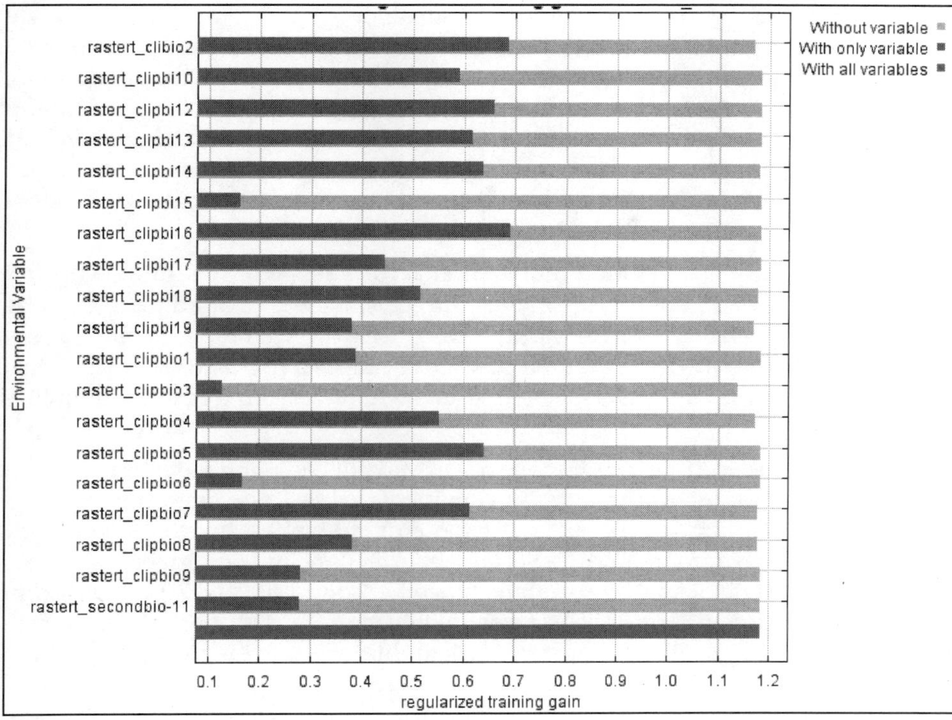

Figure 15.5: Jackknifing of Regularized Training Gain for Tasar Silkworm.

(Suryanarayan and Srivastava, 2005) but due to deforestation, habitat loss, habitat degradation due to anthropogenic reasons many of them are either lost or existing in a dwindling condition (Kar *et al.*, 2010). Although there have been efforts for their conservation, systematic scientific approach is the need of the hour. Mayurbhanj district of Odisha is of prime importance as Similipal Biospeher Reserve is known to be the habitat of three important ecoraces *viz.* Modal, Nalia and Jata. Although the extent of their distribution covered to a quite large area n and beyond Mayurbhanj, there has been shrinkage in recent years. So it was pertinent to ascertain the habitat pattern and its distribution in the district.

In the present study the Maximum entropy (MaxEnt) method was used because of its predictive abilities and wide usage. Among the well-established methods, maximum entropy (Maxent) is one of the most reliable (Phillips *et al.*, 2004, 2006; Elith *et al.*, 2006; Wisz *et al.*, 2008). Precise prediction of the distribution of endemic and endangered species is useful for decision makers, especially for those whose conservation and management activities involve large areas but constrained by resources to carry out detail exploration/investigation. Potential distribution of species prioritizes the favorable biogeographic areas to lead the conservation or management activity in a more focused way with precise scientific inputs. The advantages of distribution modeling is manifold like, explaining basic ecological phenomenon behind species distribution, understanding biogeography and dispersal barriers, verification of the earlier presence records, explored the yet uncovered regions, assessment of impacts of environmental changes over species distribution, conservation planning and reserve system design (Peterson, 2006; Guisan and Thuiller, 2005; Johnson, 2005). Considering the extent of plant distribution and diversity in India, available data related to ecology and environmental preference still represents a small fraction of this vast field especially distribution modeling (Ganeshaiah *et al.*, 2003; Irfan Ullah *et al.*, 2007; Giriraj *et al.*, 2008).

From the study it was found that tasar silkworm has most favourable habitat in Central, Wester and Southern zones of Mayurbhanj where Similipal biosphere lies. It is also interesting to note that the eastern part of Similipal is climatically different as it is also evident from the predictive map. At high altitudes silkworm cocoon quantitative traits are also very high which indicates the habitat suitability for this species.

The jackknife test of variable importance in MaxEnt has identified four important variables. Temperature seasonality (bioclimatic variable 4), precipitation of driest month (bioclimatic variable 14), mean temperature of wettest quarter (bioclimatic variable 8) and mean temperature of warmest quarter (bioclimatic variable 10) are the most important variables as they combined together contributed 68.7 for the MaxEnt prediction regarding distribution of tasar silkworm in the district. Soil parameters were also studied, however, in Mayurbhanj condition except magnesium, zinc and boron other micronutrients seemed to contribute negligibly.

Lepidoptera are among the few insect groups so far studied using predictive distribution modeling (Hill *et al.*, 2003; Romo *et al.*, 2006; Jiménez-Valverde *et al.*, 2008; Newbold *et al.*, 2009; Lozier and Mills 2011; Gutierrez *et al.*, 2012; Lv *et al.*, 2012; Romo *et al.*, 2013). In India so far there has been no report on the Lepidopteran wild silkworm. When the known distribution of this important and endemic species was studied in relation to the its conservation values, it became clear that there is the need of conservation plans with appropriate scientific approach. Simultaneously, there should be effort for the conservation of its habitat in true sense. Because, the loss and degradation of the habitat is one of the main threats to Lepidoptera (Hill *et al.*, 2001; Warren *et al.*, 2001; New 2014), the habitat in Mayurbhanj district must be restored for tasar silkworm along with its host plants.

From the study another important aspect was discernible that at higher altitude and dense canopy the tasar silkworm fitness traits in the form of cocoon size, cocoon and shell weight as well as pupal weight showed maximum values. Earlier,(Nayak *et al.* (2000), Dash *et al.* (1991) have mentioned about univoltinism and higher quantitative trait gain for tasar silkworm at high elevation areas of Similipal. From this study, MaxEnt model also derived at the same conclusion with contributions from selected environmental variables. Hill *et al.* (2001) have mentioned that for expansion of range of any Lepidopteran species the landscape structure is very important. Hence, the conservation plans for tasar silkworm must be reoriented with emphasis on habitat structure.

Because the distribution of any species depends on variables related to climate, it is likely that the species could rapidly respond to climatic change (Luoto *et al.*, 2006) and responding to the global warming may also be a challenge for the Lepidopteran insect *Antheraea mylitta*. Therefore, ultimately a good knowledge of the biological and ecological characteristics of the species is required as also advocated by (Lv *et al.*, 2012.)

Thus, the predictive map of tasar silkworm and the information generated from this study will be helpful in conserving this precious genetic resource for posterity. However, further studies are required to know the changes in habitat over years so as to predict the future scenario under the era of global warming.

Conclusion

The present study included the wild tasar silkworm populations in different zones of Mayurbhanj, including Similipal Biosphere Reserve, their cocoon quantitative traits, soil micronutrients and fitting the MaxEnt model to ascertain the contribution of environmental variables with an aim to develop a predictive mapping. It was clear from the study that at higher and medium altitude tasar silkworm quantitative traits are well expressed. This may be due the fact that the habitat there is more adaptive for the silkworm. On the other hand, the landscape or habitat in majority of Mayurbhanj district is not suitable for wild tasar conservation. This may be due to non-suitability of habitat structure. In recent times there has been fragmentation of forest landscapes. There are sporadic distribution of forests with tasar silkworm host plants like *Shorea robusta* and *Terminalia tomentosa*. But the canopy along with other environmental variables may not be sufficient to bear the load of carrying capacity for the perpetuation of tasar silkworm in wild conditions. On the other hand, the areas in central, western and eastern zones must be judiciously used for the conservation of this precious genetic resource.

References

Ahsan, J. and Sinha, S.P. (2000). A Handbook of Economic Zoology, S. Chand and Company Ltd., New Delhi. pp. 48-62.

Dutta, R.K. (1988). Indian Sericulture: Past, Present and Future. International Congress of Tropical Sericulture Practices, pp. 217-240.

Elith, J., Graham, C.H., Anderson, R.P., Dudik, M., Ferrier, S., Guisan, A., Hijmans, R.J., Huettmann, F., Leathwick, J.R., Lehmann, A., Li, J., Lohmann, L.G., Loiselle, B.A., Manion, G., Moritz, C., Nakamura, M., Nakazawa, Y., Mac, M., Overton, J., Peterson, A.T., Phillips, S.J., Richardson, K., Scachetti-Pereira, R., Schapire, R.E., Soberon, J., Williams, S., Wisz, M.S. and Zimmermann, N.E.. (2006). Novel methods improve prediction of species' distributions from occurrence data. Ecography, 29: 129–151.

Ganeshaiah, K.N., N. Barve, N. Nath, K. Chandrashekara, M. Swamy and Uma Shaanker, R. (2003). Predicting the potential geographical distribution of the sugarcane woolly aphid using GARP and DIVA-GIS. Current Science, 85, 1526-1528.

Giriraj, A., Irfan-Ullah, M., Ramesh, B.R., Karunakaran, P.V., Jentsch, A. and Murthy, M.S.R. (2008). Mapping the potential distribution of *Rhododendron arboreum* Sm. ssp. *Nilagiricum* (Zenker) Tagg (Ericaceae), an endemic plant using ecological niche modelling. Current Science, 94: 1605-1612.

Guisan, A. and Thuiller, W. (2005). Predicting species distribution: Offering more than simple habitat models. Ecological Letters, 8: 993-1009

Gutierrez, A.P., Ponti, L., Cooper, M.L., Gilioli, G., Baumgärtner, J. and Duso, C. (2012). Prospective analysis of the invasive potential of the European grapevine moth *Lobesia botrana* (Den. and Schiff.) in California. Agricultural and Forest Entomology, 14: 225–238.

Hill JK, Collingham YC, Thomas CD, Blakeley DS, Fox R, Moss D, Huntley B. (2001) Impacts of landscape structure on butterfly range expansion. Ecology Letters 4: 313–321.

Hill JK, Thomas CD, Huntley B. (2003) Modeling present and potential future ranges of European butterflies using climate response surfaces, pp. 149–167.

Irfan-Ullah, M., A. Giriraj, M.S.R. Murthy and A.T. Peterson: Mapping the geographic distribution of Aglaia bourdillonii Gamble (Meliaceae) an endemic and threatened plant, using ecological niche modelling.

Jiménez-Valverde A, Gómez JF, Lobo JM, Baselga A, Hortal J. (2008). Challenging species distribution models: the case of *Maculinea nausithous* in the Iberian Peninsula. Annales Zoologici Fennici 45: 200–210.

Jolly, M.S. (1976). Package of practices for tropical tasar culture. Central Tasar Research Station, Ranchi, India. p.32.

Jolly, M.S., Chaturvedi, S.N. and Prasad, S.A. (1968). A survey of tasar crops in India. Indian Journal of Sericulture, 1: 50-58.

Jolly, M.S., Chowdhury, S.N. and Sen, S.K. (1975) Non-mulberry Sericulture in India. Central Silk Board, Bombay. pp.1-89.

Jolly, M.S., Sen, S.K, and Ahsan, M.M. (1974) *Tasar Culture*, Ambika Press, Bombay. pp. 1-166.

Jolly, M.S., Sen, S.K., Sonwalker, T.N. and Prasad, G.K. (1979) Non-mulberry Silk. F.A.O. Agricultural Bulletin, Rome 29, 1- 178.

Lozier J, Mills N. (2011) Predicting the potential invasive range of light brown apple moth (*Epiphyas postvittana*) using biologically informed and correlative species distribution models. Biological Invasions 13: 2409–2421.

Luoto M, Heikkinen RK, Pöyry J, Saarinen K. (2006). Determinants of the biogeographical distribution of butterflies in boreal regions. Journal of Biogeography 33: 1764–1778.

Lv W, Li Z, Wu X, Ni W, Qv W, Li D, Chen Y. (2012) Maximum entropy nichebased modeling (Maxent) of potential geographical distributions of *Lobesia botrana* (Lepidoptera: Tortricidae) in China. IFIP Advances in Information and Communication Technology 370: 239–246.

Nayak, B.K. and Dash, M.C. (1998) Studies on biology, ecology and culture practices of the wild tasar silk moth *Antheraea paphia* Linn. of Similipal biosphere reserve, Odisha, Bulletin of Indian Academy of Sericulture, 2, 1-13.

New, T.R. (2014). Lepidoptera and Conservation. John Wiley and Sons, Ltd., United Kingdom.

Newbold T, Gilbert F, Zalat S, El-Gabbas A, Reader T. (2009). Climate-based models of spatial patterns of species richness in Egypt's butterfly and mammal fauna. Journal of Biogeography 36: 2085–2095.

Peterson, A.T. (2006).: Uses and requirements of ecological niche models and related distributional models. Biodiversity Information, 3: 59-72.

Phillips SJ, Anderson RP, Schapire RE. (2006). Maximum entropy modeling of species geographic distributions. Ecological Modeling 190: 231–259.

Phillips SJ, Dudík M, Schapire RE. (2004). A maximum entropy approach to species distribution modeling, pp. 655–662 *In* Greiner R, Schuurmans D. [eds.], Proceedings of the 21st International Conference on Machine Learning, Alberta, Canada.

Rangaswamy, G., Narsimahanna, M.N., Kasiviswanathan, Sastry, C.R. and Jolly, M.S. (1992). Manual on Sericulture. FAO Agriculture Service Bulletin, Vol. IV. Non-mulberry Silk, pp. 38-48.

Romo H, Camero-R E, García-Barros E, Munguira ML, Martín Cano J. (2014). Recorded and potential distributions on the Iberian Peninsula of species of Lepidoptera listed in the habitats directive. European Journal of Entomology 111: 407–415.

Sahu, P.K., Das, K.K. and Das, N.K. (1997). Growth of Indian Tasar Culture: An overview,

Suryanarayana N. and Srivastava, A.K. (2005) Monograph on tropical tasar silkworm. Central Tasar Research and Training Institute, Ranchi, India pp. 134.

Wisz MS, Hijmans RJ, Li J, Peterson AT, Graham CH, Guisan A. (2008). Effects of sample size on the performance of species distribution models. Diversity and Distributions 14: 763–773.

Zhuang, Ga-Huan, Liu, Shi-Xian and Long, L. (1994). Sericultural production strategies in the 21st century. *Global Silk Scenario-2001*. pp. 121-130.

16

Cryptogamic Flora: A Prospect in Environmental Toxicity Remediation with Physiological Interventions

M.K. Adak, C. Mandal, K. Das and A.K. De*

Crop Physiology and Biotechnology Laboratory
Department of Botany, University of Kalyani,
Kalyani – 741 235, West Bengal
**E-mail: mkadak09@gmail.com*

ABSTRACT

The lower groups of plants mainly of that cryptogam are in general less explored with their physiology and biochemistry for stress tolerance. This paper entails some information regarding the cellular activities of pteridophytic species with reference to aquatic ferns for their in-built efficacy to hyper-accumulate xenobiotic compounds. The emphasis would be given on anti-oxidation cascade for those plant materials which would show both by enzymatic and non enzymatic paths to follow and their over-expression according to the gradient of pollutant. Water fern is a common name of several plant species and those may refer to species of Salvaniales, Marsiales. Aquatic fern species might have prospect to quench xenobiotics and related other hazardous materials from industrial toxicity. Phytochelatin like compounds, glutathione like anti-oxidants as well as other cell wall bound energy rich compound are likely to be possible biomarkers in this fern mediated environment decontamination.

***Keywords**: Aquatic ferns, Remediation, Heavy metals.*

Introduction

The lower groups of plants mainly of that cryptogam are in general less explored with their physiology and biochemistry for stress tolerance (Abercrombie *et al.*, 2008). These species have originated in different geographical time scales when the environment was mostly harsh as well as anoxygenic gases. Therefore, it would be a highlighting feature for those species surviving in that environment with their cellular activities as well as modification of life processes. The aquatic Pteridophytes are those classes which are in a transition of both vascular and flowering mode of sexual reproduction. Pteridophytic species are limited with their two stages life cycles, however, a plenty diversity exists for their morphological variation in leaf, sporangia *etc.* (Alloway, 1995). These variants of fern species with almost 1200 species of ferns, *Lycopodes, Selaginella* are omnipresent in each ecological niche like deserts, saline land, water bodies and understory and epiphytic on angiospermic as well as conifers. The seedless ferns species have their wider adaptability for developing primary xylems had reached them for efficient water translocation and thereby makes a bridge for seedless non flowering plants and seed bearing flowering plants. The development of secondary xylem occurs from early Devonian period so the modern land plants have the advantage to capitalize the well developed water conducts that gives mileage for water stress recovery (Bani *et al.*, 2009). Still, the ancestors as ferns with their primary xylem and other cellular attribute my pose mild to moderate water deficit stress and the commonly associated secondary effects of water stress. These factors may contribute the unrelenting success of pteridophytic species. So, there may be some active principle to alleviate the toxicity or environmental hazards by these fern groups that may be exciting in environmental cleanup (Cunningham *et al.*, 1995). With these back ground this paper entails some information regarding the cellular activities of pteridophytic species with reference to aquatic ferns for their in-built efficacy to hyper-accumulate xenobiotic compounds. The emphasis would be given on anti-oxidation cascade for those plant materials which would show both by enzymatic and non enzymatic paths to follow and their over-expression according to the gradient of pollutant.

Criteria for Selection of Pteridophytic Species

The Aquatic Ferns

Water fern is a common name of several plan species and those may refer to species of Salvaniales, Marsiales and few others like *Azolla, Blecknan, Bolbitis, Histiopteris, Osmunda, Regnellidium etc.* The fern with their adaptability are categorized into two: Leptosporangiate and Eusporangiate (Deng *et al.*, 2010). The pteridophytic nature of fern primarily denominates the seedless vascular plant. The appearance of fern as by fossil record was derived as in the Devonian period. The development of primary xylem is the most promising image for initial vascularisation and thus, shows a wider variety of habitats from hilly alleviation to sandy rocks, to water or a open field (Davis *et al.*, 2006). Ferns generally thought as special species with various limiting environmental factors constrains for flowering plant. Few species like *Pteridium, Azolla* are also regarded as weeds and destructive in nature. Four major habitats are found in fern: moist, sedum forest, faces of rocks, epiphytes and acidic wet lands including bogs and swamps. With this diverse habitat the different environmental factors particularly on pH ranges, moist acidic soil and even a limestone the fern develops a wider genetic plasticity through their adapted physiological attributes (Gupta *et al.*, 2004).

More than 50 fern species have been found to absorb metals or metalloids from soil and accumulated in high level of leaf tissues. This is hyper-accumulation with profound ecological significant for energy flow by the metals at varying trophic levels, as well as ecological; distribution of the soil chemical nature. The general recommendation for assignment of a hyper-accumulating plant species varying significantly, still fern species are a good source with metal interaction by physiological, genetic and evolutionary point of view. The use of artificially amendment in control condition of hyper accumulation is undesirable for true sense of hyper-accumulating species (Ehsan *et al.*, 2009). However, the fern species have also face the criticism that those could be induced to take up more metal under high soil concentration. As a result they fail to sustained and even enter in next and consequent growth phases. It is also believed that the soil amendment do not mimic the condition of natural soil. Only under condition of high soil concentration, not only these fern species can hyper accumulate (passive hyper accumulation), even under lower salt concentration in soil plants can absorb more salt. Therefore, the minimum concentration at which the hyper accumulation can be achieved and maximum concentration for tolerance both could be done by the fern species (Kotrba *et al.*, 1999). The amendment of soil can identify the order soil characters like pH, CEC *etc.* on hyper-accumulation and determine whether hyper-accumulator of one metal has the ability to take up the other metals. In this way depending on natural soil metal concentration fern species cannot be detected as true hyper-accumulator. If plants are continued with reproduction that confirms its tolerance therefore, fern species may be one material to determine the minimum and maximum metal concentration for accumulation and maximum level of tolerance either through a particular metal or collective salt concentration of the soil (Li *et al.*, 2003).

Physiological Attributes in Supports for Hyper Accumulation

On the ecological perspective a plant species would recognize as hyper accumulator beyond the threshold leaf metal concentration for tolerance from natural population in the field. This also ignores the natural definition to the hyper accumulating phenomenon at physiological level and is not consistent with physiological trait (Pal *et al.*, 2006). Likewise, stress tolerant species could also survive in absence of those. However, the energetic activities may result the loss of tolerance due to non-exercise of the physiological character. In addition, hyper-accumulation of metal may be done by inadvertent metal uptake as a byproduct of affinity for other elements as soil nutrients (Rajkumar *et al.*, 2010). Therefore, in the fern specie also specific metal uptake can evolve without a species that a significant amount of specific metal absorption. In case of fern also, example *Pteris vitata*, a genetic drift that cause some transporter proteins for their over expression without any tolerance. Species exhibit threshold leaves metal concentration in natural population is inappropriate in the sense that plant has unique ability to sequester the metal in other plant parts. On that basis four categories of accumulating species are recognize: high accumulation but low tolerance; low accumulation low tolerance; high accumulation high tolerance; low accumulation high tolerance (Robinson *et al.*, 1998). Therefore, the first two in serial are hyper-accumulator and last two are low accumulator. It is interesting to note the few non flowering plants including algae, mosses and ferns have their unique ability to grow and reproduce upon natural metali-ferous soil and soil contaminated under anthropogenic activities. To categories of such species are recognize hyper tolerant but non hyper accumulator and hyper tolerant as well as hyper accumulator. The hyper tolerance and heavy metals distribution is an excluder non hyper accumulator and a hyper accumulator. For hyper tolerant but not hyper accumulator, the concentrations of particular metals/metalloids are less in concentration in leaf tissue.

Physiological-Genetical Significance for Aquatic Fern of Hyper-Accumulation of Metal

In fern sp the development of primary xylem and the variety of stele give an access for more water translocation along with solutes. The membrane function of roots for different transporter protein is important. Both anionic and cationic uptakes are facilitated by these transporters. As for example Cd and Al, the two toxic metals over a critical range depend on proportionate metal concentration in the soil. The decrease uptakes of Cd by *Marsilea* root are hindered with increasing Al concentration (Suzuki *et al.*, 2001). Therefore, metal competition for a particular transporter is also useful. After translocation root to shoot movement is important factor for hyper-accumulator species. For hyper accumulator plants not only detoxify the metal by chelation in the cytoplasm or storing into vacuole and efficient upward movement of the metals towards shoot via vascular system (Yadav *et al.*, 2011). Therefore, it is possible to adopt the strategy of these fern sp. (*Salvinia, Marsilea*) not to toxify the root tissue with the metals and thus, low sequestration in the root. By this way roots remain active for more accumulation of the metal followed by shoot translocation.

Energy Yielding Metabolism for Root to Shoot Translocation

The different membrane bound protein like ATPase (H^+/ATPase) are important in this context. Hydrolysis of ATP by the H^+ ATPase pump is another source for ion channel to function in an active manner. Plants loose a significant amount of cations like K^+ when a heavy metal is absorbed in addition a significant amount of solutes can function both metal binding ligands as well as provide compatible solutes like activities. This includes the TCA cycle acids like citric and malic which decarboxylated are from can bind the metals (Verma *et al.*, 2006). In fact, these organic acids are more prevalent in leaf cell where a high concentration heavy metal are sequester in vacuole. In root the concentration of ligand chelated with heavy metals are much lower in vacuoles that facilitates more free form of metals to be transported over the membrane chanel following uploading into the xylem. Besides this, the key role in heavy metals sequestration may be played by free amino acids and nicotinamide which remain more stable complexes likewise free histidine is one of the important of those. The biosynthesis of histidine with few regulatory genes is over expressed in such hyper accumulatory species (Patra and Sharma, 2006). On the other hand genes for nicotinamide biosynthesis are also important particulars for Zn hyper accumulation. Following xylem uploading of the heavy metals there are few transporters up regulated. Heavy metals transporting ATPase are important to be constitutively over express in such hyper accumulator species. Both these genes are required for metal uploading as well as concentration homeostasis for tolerance. Therefore, ongoing expressions of these genes are implicated the root-shoot translocation through driving force of the heavy metal accumulation.

Detoxification and Sequestering System

Therefore, it is the plant potential that can down regulate the ROS, both of its generation and lysis. The generation of ROS resourced by three cellular compartments; cell wall or apoplastic spaces, cell membrane and cellular compartment like chloroplast and mitochondria (Martins *et al.*, 2011). NADPH oxidase is another important enzyme that is the chief source of ROS, particularly the super oxide (O_2^-). Other enzyme protein like xanthin, oxidase and flavin mono-oxigenase are also important to generate ROS. Detection of the ROS is mostly achieved by either electrochemical reaction or histo-chemical stain. Both NBT and DAB are quite common to locate the tissue specific activity of super oxide (O_2^-) and H_2O_2 accumulation in different cellular compartments (Gratao *et al.*, 2005). Therefore, the metal induced activity of ROS is the most predominant to deliver its impact initially within a threshold concentration. H_2O_2 is granted as a chemical eliciter often a secondary messenger for cellular responses to heavy metals. Thus, fern species in this experiment have shown to mediate tolerance against toxic and heavy metals. Moreover, the activity of phytochelatins, a compound with sulphur linked oligopeptide serve as nucleophilic bonding of metals followed by sequestering in vacuole (Freeman *et al.*, 2009). In other cases the hyper accumulators also do not rely on prohibitive metabolic cost that a huge synthesis of this kind of chelators that require. So, anti-oxidizing cascade and its related genes like glutathione can supplement the non enzymatic anti-oxidation pool.

Gene Expression in Metal Hyper-accumulation

For the expression potential of gene, plants are focused with two broad aspects: transporter of metal over the membrane and second, minimize the oxidation stress by some over express proteins. For the first, it is the different kinds of transporters and channel proteins those are effective in transport of metal over the membrane (Cook *et al.*, 2009). The most important gene for such transport has already been cloned from *Arabidopsis* and from other crop species. For antiporter and symporter are responsible include or/and exclude the different metal ions. Likewise, the antiporter Na+/H+ can exchange a Na+ for hydrogen. Symporters are various including NO_3^-/H^+, PO_4^-/H^+, K^+/H^+, Na^+, sucrose/H^+, amino acid/H^+. For all the cases different solutes are incorporated into cytosol along with H^+ or any cation thus, increasing the solute concentration in the cytosol. On the other hand anti-porter can exchange one H^+ with Na^+. Few genes are there which are independent for solutes called efflux carrier causing release of sucrose outside the cytosol (Keeling *et al.*, 2003). In addition genes for H^+ pump are important which are energetic in reactions. ATP hydrolysis along with its energy can exchange out H^+ from cytosol. Membrane is also responsible for few genes to exchange ions through extra cellular spaces. Inward and outward rectifying channels for Ca^{2+} and Cl^- as well as K^+ for both are interesting (Cheng 2003). All the cases ions are exchanged through extra cellular space. For the heavy metal ions are exchange through tonoplast into vacuoles. On tonoplast ABC transporter are important. The ABC transporters with its ATP hydrolyzing activity can include Cd into vacuole. This is mediated by phyto-chelating Cd complex (Brooks *et al.*, 1997).

Another ABC transporter which is anthocyanin- GS (glutathione conjugate pumps) is important. This is typically covalently associated with glutathione. In plants such pump activities are also capable of transporting few non glutathionated compounds into vacuoles. ABC-GS pumps are mostly function with herbicide detoxification, pigment accumulation like anthocyanin are the examples. Few isozymic forms like glutathione-S-transferase are responsible for attaching glutathione to any solutes for transportation. In glutathione conjugation the C-C four electrons sharing attached to the element/functional groups which can withdraw electron (CH_2=CH_2/functional groups/chemical moiety like anthocyanin, IAA, phenolic residues *etc.* Phyto-chelatins synthesized from glutathione can also be transported into vacuoles along with their chelating heavy metals by ABC-GS pumps. From the *Arabidopsis* ABC transporters are represented by four genes (Kramer *et al.*, 2007). Each of the genes encodes a polypeptide of 170-180 kD containing 4-6 membrane spanning the cell membrane provides the pathway for solute transport and also molecular specificity of the transport is also recognized. On the tonoplast both fast vacuolar channel (FVC) and slow vacuolar channel (SVC) are also important. For SVC anions as well as cations are excluded out of vacuole whereas, with FVC anions like malate, chlorides and NO_3 are included the vacuoles. Hydrolysis of the ATP by H+ pump are also found on tonoplast membrane and thus called V-H+/ATP pump. Pyrophosphate dependent V-type PPi pump are also found. Few anti-porter like Na^+/H^+, $Ca^{+2}/3H^+$, Cd^+/H^+, hexose/H^+ and sucrose/H^+ are important (Janouskova *et al.*, 2005).

Conclusion

Therefore, this study may be a comprehensive one that connects the different modules for metal hyper accumulation along with its impact at different growth indices. Initially, different metals have varying affinities to consider its bioaccumulation at tissue level and thereby specific symptoms are also adhered to it. The properties of such metals also denote the ability for bioaccumulation, sequestering on cell wall or vacuoles and thereby poses the phyto toxicity. ROS could induce the oxidative stress with different effects on cellular levels. Mostly, gene concerned ion sequestration as well as anti-oxidation paths are the most targeted domains to probe for such fern species to exercise for phyto -remediation. With this concept these aquatic fern species might have prospect to quench xenobiotics and related other hazardous materials from industrial toxicity. Phytochelatin like compounds, glutathione like anti-oxidants as well as other cell wall bound energy rich compound are likely to be possible bio markers in this fern mediated environment decontamination.

References

Abercrombie, J.M., Halfhill, M.D., Ranjan, P., Rao, M.R., Saxton, A.M., Yuan, J.S. and Stewart, C.N. (2008). Transcriptional responses of Arabidopsis thaliana plants to As (V) stress. BMC Plant Biology. 8: 1–15.

Alloway, B.J. (1995). Heavy metals in soils, 2nd edn. Blackie edn. The University of Reading U.K, Glasgow.

Bani, A., Echevarria, G., Mullaj, A., Reeves, R., Morel, J.L. and Sulce, S. (2009). Nickel hyperaccumulation by brassicaceae in serpentine soils of Albania and Northwestern Greece. Northeast Nat., 16: 385–404.

Brooks, R.R., Lee, J., Reeves, R.D. and Jaffre, T. (1977). Detection of nickeliferous rocks by analysis of herbarium specimens of indicator plants. Journal of Geochem. Explor., 7: 49–57.

Cheng, S.P. (2003). Heavy metal pollution in China: origin, pattern and control. Environ. Sci. Pollut. Res., 10: 192–198.

Cook, L.L., Inouye, R.S. and McGonigle, T.P. (2009). Evaluation of four grasses for use in phytoremediation of Cs-contaminated arid land soil. Plant Soil, 324: 169–184.

Cunningham, S.D., Berti, W.R. and Huang, J.W.W. (1995). Phytoremediation of contaminated soils. Trends Biotechnol., 13: 393–397.

Davis, A.K., Hildebrand, M. and Palenik, B. (2006). Gene expression induced by copper stress in the diatom *Thalassiosira pseudonana*. Eukaryot. Cell, 5: 1157–1168.

Deng, H., Li, M.S., Chen, Y.X., Luo, Y.P. and Yu, F.M. (2010a). A new discovered manganese hyperaccumulator-Polygonum pubescens Blume. *Fresenius Environ. Bull.*, 19: 94–99.

Ehsan, M., Santamaria-Delgado, K., Vazquez-Alarcon, A., Alderete-Chavez, A., De la Cruz-Landero, N., Jaen-Contreras, D. and Molumeli, P.A. (2009). Phytostabilization of cadmium contaminated soils by *Lupinus uncinatus* Schldl. Span J. Agric. Res., 7: 390–397.

Freeman, J.L., Quinn, C.F., Lindblom, S.D., Klamper, E.M. and Pilon-Smits, E.A.H. (2009). Selenium protects the hyperaccumulator Stanleya Pinnata against black-tailed prairie dog herbivory in native seleniferous habitats. Am. J. Bot., 96: 1075–1085.

Gratao, P.L., Polle, A., Lea, P.J. and Azevedo, R.A. (2005). Making the life of heavy metal-stressed plants a little easier. Funct. Plant Biol., 32: 481–494.

Gupta, D.K., Tohoyama, H., Joho, M. and Inouhe, M. (2004). Changes in the levels of phytochelatins and related metal-binding peptides in chickpea seedlings exposed to arsenic and different heavy metal ions. J. Plant Res., 117: 253–256.

Janouskova, M., Pavlikova, D., Macek, T. and Vosatka, M. (2005b). Influence of arbuscular mycorrhiza on the growth and cadmium uptake of tobacco with inserted metallothionein gene. Appl. Soil Ecol., 29: 209–214.

Keeling, S.M., Stewart, R.B., Anderson, C.W.N. and Robinson, B.H. (2003). Nickel and cobalt phytoextraction by the hyperaccumulator *Berkheya coddii*: Implications for polymetallic phytomining and phytoremediation. Int. J. Phytorem., 5: 235–244.

Kotrba, P., Macek, T. and Ruml, T. (1999). Heavy metal-binding peptides and proteins in plants. A review. Collect Czech Chem. Commun., 64: 1057–1086.

Kramer, U., Talke, I.N. and Hanikenne, M. (2007). Transition metal transport. FEBS Lett., 581: 2263–2272.

Li, Y.M., Chaney, R.L., Brewer, E.P., Angle, J.S. and Nelkin, J. (2003). Phytoextraction of nickel and cobalt by hyperaccumulator Alyssum species grown on nickel-contaminated soils. Environ. Sci. Technol., 37: 1463–1468.

Martins, L.L., Mourato, M.P., Cardoso, A.I., Pinto, A.P., Mota, A.M., Goncalves, M.D.S. and de Varennes, A. (2011). Oxidative stress induced by cadmium in Nicotiana tabacum L.: effects on growth parameters, oxidative damage and antioxidant responses in different plant parts. Acta Physiol. Plant, 33: 1375–1383.

Pal, M., Horvath, E., Janda, T., Paldi, E. and Szalai, G. (2006). Physiological changes and defense mechanisms induced by cadmium stress in maize. J. Plant Nutr. Soil Sci., 169: 239–246.

Patra, M. and Sharma, A. (2000). Mercury toxicity in plants. Bot. Rev., 66: 379–422.

Rajkumar, M., Ae, N., Prasad, M.N.V. and Freitas, H. (2010). Potential of siderophore-producing bacteriafor improving heavy metal phytoextraction. Trends Biotechnol., 28: 142–149.

Robinson, B.H., Leblanc M., Petit, D., Brooks, R.R., Kirkman, J.H. and Gregg, P.E.H. (1998) The potential of Thlaspi caerulescens for phytoremediation of contaminated soils. Plant Soil, 203: 47–56.

Suzuki, N., Koizumi, N. and Sano, H. (2001). Screening of cadmium-responsive genes in Arabidopsis thaliana. Plant Cell Environ. 24: 1177–1188.

Verma, P., George, K.V., Singh, H.V., Singh, S.K., Juwarkar, A. and Singh, R.N. (2006). Modeling rhizofiltration: heavy-metal uptake by plant roots. Environ. Model. Assess., 11: 387–394.

Yadav, B.K., Siebel, M.A. and van Bruggen, J.J.A. (2011). Rhizofiltration of a heavy metal (lead) containing wastewater using the wetland plant *Carex pendula*. Clean Soil, Air, Water, 39: 467–474.

Animal Biotechnology

17

Anuran Tail Resorption as a Model to Study Lysosomal Enzyme Mediated Cell Death

C. Mahapatra

Department of Zoology, North Orissa University, Takatpur, Baripada – 757 003, Odisha
E-mail: cuckoomahapatra@gmail.com

ABSTRACT

Anuran tail resorption is a remarkable example of programmed cell death by apoptosis. Apoptosis has been well described and documented in model organisms as well as cell culture model systems; still many aspects of the apoptotic pathway remain elusive. A comprehensive understanding of apoptotic pathways require assessment of alternative pathways or additional molecules which may be involved in cell death. Lysosomal enzymes have emerged as an important molecule that may modulate some cell death programs. Lysosomal enzymes are known to cause degradation of tail tissues during anuran metamorphosis and so anuran tail resorption can provide important insights regarding the role of lysosomal enzymes during the cell death process. The present review evaluates the possibility of anuran tail resorption as a model to study lysosomal enzyme mediated cell death by apoptosis.

Keywords: *Anurans, Apoptosis, Lysosomal enzymes, Tail resorption.*

Introduction

Anurans, comprising frogs and toads, are tailless amphibians. They undergo a complex phenomenon, known as metamorphosis, during post-embryonic development where extensive morphological, biochemical and cellular changes occur in the aquatic anuran tadpoles to transform them into terrestrial adult frogs. Anuran metamorphosis is a complex programme mediated by hormones thyroxine

(T_4) and triiodo thyronine (T_3) where different tissues or group of cells within the same tissue can exhibit different hormonal responses, which range as widely as *de novo* metamorphosis, functional reprogramming and total tissue regression (Tata, 2003). Anuran metamorphosis is a well-studied phenomenon but has been very scarcely employed in biomedical research. Recently, Buccholz (2015) emphasised the importance of anuran model in medical understanding of endocrine disease, stress, and endocrine disruption affecting the perinatal period in humans. Anuran tail resorption has been projected as a model to study apoptosis (Ishizuya-Oka *et al.*, 2010) but till now it has not been employed for understanding disorders related to cell death regulation. A better understanding of cell death regulation would provide the basis for treating a variety of human disorders associated with abnormal cell death, such as cancer, auto-immunity, viral infection, AIDS, sepsis, ischemia, neuro-degeneration, impaired healing and tissue regeneration. There is very scanty information regarding the process of selection of cells for death *in vivo*, and integration of different signaling pathways in the decision between cell death and survival under physiological conditions (Fuchs and Stellar, 2011). The scarcity of information is primarily due to investigations being limited to only few model organisms. A complete understanding of cell death pathways requires elaborate studies in non-model organisms as well.

Tadpole tail atrophy is the most amazing event that occurs during anuran metamorphosis and is exclusively mediated by apoptosis. The tail, which accounts for one third of the body weight, is completely resorbed or regressed in a few days during metamorphic climax (Kerr *et al.*, 1974). The dramatic involution of the tadpole tail attracted the attention of pathologists a century ago as a model case of physiological tissue destruction (Weber, 1967). However, anuran tail resorption or regression began to be considered as a model to study programmed cell death after electron microscopic studies revealed that tail regression in dwarf tree frog, *Litoria glauerti* mainly occurs by apoptosis (Kerr *et al.*, 1974). Thus, it presents a unique *in vivo* model to study the molecular mechanisms of apoptotic cell death during vertebrate development (Nakajima *et al.*, 2005; Tata, 2006).

After the studies of Kerr *et al.* (1974), there was an explosion of studies elaborating the mechanisms of apoptosis during tail resorption. However, most of the studies were confined to established model, *Xenopus*. Anuran genera *Xenopus*, is relatively well understood metabolically and physiologically and thus continues to contribute in this new era of systems biology and "X-omics" (Burggrren and Warburton, 2007). It also has short generation time and genetic constructs that makes it desirable for transgenic and knockout technology (O'Rourke, 2007). In India, there is not much available literature regarding studies on tail resorption. Perhaps, unavailability of established anuran models is a major constraint and the main reason why many developmental biologists in India have not taken up studies on Indian anurans, instead, focus on other models like *hydra*, *drosophila*, zebrafish and mice. In the modern age of transcriptomics and RNA-seq technology, however, absence of genomic database is no more an impediment to carry out sophisticated molecular research. In the absence of a reference genome or when the reference is poorly reconstructed or annotated, *de novo* assembly of RNA-seq reads is an enabling technology to study the transcriptomes of non-model species. Transcriptomes provide information on complex biological processes, and give a picture on how the static genome behaves in dynamic environments (Birol *et al.*, 2015).

The mechanism of cell death during tail resorption has been clearly described in *Xenopus* only. For better understanding of cell death regulation, studies should be taken up across multiple taxa to evaluate and analyse the species specific characteristics of tail resorption. Indian sub-continent is home to 414 species of anurans (Dinesh *et al.*, 2015), with many still awaiting description, and thus provides ample opportunities to use them for experimental research. With modern age technology like transcriptomics and a rich diversity of anurans there is abundant possibility for undertaking more and more studies on anuran metamorphosis in India. The present review tries to consolidate the studies on anuran tail resorption worldwide and in India impressing the need for such studies in India. The review also discusses the future prospects of studying anuran tail resorption as a model.

Anuran Tail Resorption: The Process

The post-embryonic development of anurans can be separated into three specific periods: premetamorphosis, prometamorphosis and metamorphic climax (Shi, 2000). Premetamorphosis is primarily a period of tadpole growth that takes place without thyroid hormone (TH). This is followed by pro-metamorphosis which is characterized by overt changes in limb morphology. The climactic stages mark the surge of TH that triggers the final processes associated with metamorphosis, including forelimb development and resorption of the tail (Dodd and Dodd, 1976). The tail in prometamorphosis stages is referred to as pre-resorbing tail and in climactic stages as resorbing tail in the text.

During anuran metamorphosis, tail, which is about twice longer than the body is completely absorbed by programmed cell death (Nakajima and Yaoita, 2003). In early descriptions, increased activity of lysosomal enzymes released from preformed lysosomes and phagocytic macrophages were considered as the primary cause of degradation of tail tissues (Weber, 1967; Gilbert and Frieden, 1981). Proteolytic enzymes like cathepsins, phosphatases, ribonucleases, deoxyribonucleases and β-glucuronidase were reported to exhibit higher levels of activity in the resorbing tail tissues, all having an acid pH optimum (Weber, 1967). Recent reports of aryl sulphatase, N- acetyl-β-D-glucosaminidase and cathepsin activity in *X. laevis* (Das *et al.*, 2006) and cathepsin D activity in *R. catesbeiana* (Domanski and Helbing, 2007) also indicate the involvement of lysosomal enzymes during anuran tail resorption.

After, Kerr *et al.* (1974) described that apoptosis was the major cause of cell death in the resorbing tail, a series of studies in *X. laevis* described apoptosis to be the key mediator of cell death in tail muscles (Nishikawa and Hayashi, 1995; Nakajima and Yaoita, 2003) and spinal cord (Estabel *et al.*, 2003). Apoptosis related enzymes, caspases and regulatory mitochondrial factors such as BID, Bax, and xR11, a homologue of mammalian Bcl-XL, were also shown to be involved in TH-induced apoptosis in the tail of *Xenopus* tadpoles (Nakajima *et al.*, 2000; Shi *et al.*, 2001; Sachs *et al.*, 2004; Rowe *et al.*, 2005; DuPasquier *et al.*, 2006). Thus, it is now well perceived that tail tissues mainly die by apoptosis during anuran tail resorption (Ishizuya-Oka *et al.*, 2010; Ishizuya-Oka, 2011).

In addition, in the *X. laevis* tail muscles, it was demonstrated that TH can induce apoptosis cell-autonomously both *in vitro* (Yaoita and Nakajima, 1997) and in *vivo* (Das *et al.*, 2002). Noteworthy, in the tail muscles, over expression of matrix metalloproteinase, stromelysin 3 (ST3) caused apoptosis via a laminin receptor-independent pathway (Mathew *et al.*, 2009) while in the tail epidermis expressing laminin receptor, ST3 induced apoptosis via a laminin receptor-dependent pathway (Mathew *et al.*, 2010). Furthermore, a recent study using transgenic tadpoles indicated that Ouro proteins, which encode keratin-related proteins, function as larva-specific immune antigens (Mukaigasa *et al.*, 2009), suggesting a key role of the immune system in the tail regression. Thus, multiple apoptotic pathways seemed to work together to ensure massive apoptosis. The cell debris resulting out of apoptosis in the tail tissues are finally removed by the action of macrophages (Kinoshita *et al.*, 1985). Fibroblasts remove the collagenous components of tail like the notochord and the basement membrane (Berry *et al.*, 1998).

Although, programmed cell death by apoptosis cause the disappearance of the tadpole tail but the critical factors that trigger programmed cell death still remain an enigma (Hanada *et al.*, 1997). Lysosomal enzymes were among the first proteases to be associated with cell death in the tadpole tail during the 1960s but this line of research has not been actively perpetuated into the modern age of apoptosis. It is still not known whether they actually kill cells (*i.e.* participate in apoptosis) or only remove the cell debris resulting out of apoptosis (*i.e.* heterophagy). Hence, investigating the role of lysosomal enzymes during tail resorption was one of the major focal points in the studies carried out on Indian anuran tadpoles as discussed below.

Anuran Tail Resorption Studies in India

The first published reports on the tail resorption in an Indian anuran was probably by Mahapatra *et al.* (2001) where they reported the role of oxidative stress and acid phosphatase during tail resorption in an Indian frog, *Polypedates maculatus*. There was also a report on role of deiodotyrosine dehalogenase during tail resorption in *Microhyla ornata* (Gaupale *et al.*, 2009) but, it did not elucidate the factors involved in cell death during tail resorption.

Mahapatra *et al.* (2001) reported for the first time the involvement of acid phosphatase, a lysosomal enzyme, and its possible role in causing cell death during tail resorption in an Indian anura, *P. maculatus*. This preliminary work was followed by a series of studies on *P. maculatus* along with other locally available anuran species *Duttaphrynus melanostictus* and *Hoplobatrachus tigerinus*. Mahapatra (2007) showed the involvement of acid phosphatase and alkaline phosphatase in the resorbing tails of *D. melanostictus*. Both the enzymes showed higher levels of specific activity in the resorbing tails as compared to the pre-resorbing tails. The specific activity of various lysosomal enzymes (Acid phosphatase, Aryl sulphatase, N-acetyl-β-D-glucosaminidase, Cathepsin B, Cathepsin H, Cathepsin L and Cathepsin D) was found to be comparatively low in the pre-resorbing tail and showed a high rate of activity in the resorbing tails of *D. melanostictus*, *P. maculatus and H. tigerinus* (Mahapatra *et al.*, 2012; Mahapatra and Mahapatra, 2013).

The description of morphology of resorbing tail of *D. melanostictus* by Mahapatra (2007) and Mahapatra and Mahapatra (2007) was probably the first for any Indian anura. Later, detailed description of histology of the resorbing tails of *P. maculatus* and *D. melanostictus* gave a discrete idea of the morphology of regressing tail tissues. The pre-resorbing tails of *D. melanostictus* and *P. maculatus* had well-developed muscle bundles, intact epidermis, tail fins, notochord, spinal cord and blood vessels. But in climactic stages there was degradation of various tail tissues, which was highest in the distal half and lowest in the proximal half in both the species. Bending of spinal cord and blood vessels during tail regression was also reported for the first time. At any given point of time, muscles were the most degraded tissue in the tail and notochord was the least. The resorption occurred in a graded manner with distal end resorbing faster than proximal end (Mahapatra and Mahapatra, 2013). Species specific differences were also observed in the activity of lysosomal enzymes and tail histology implicating variations in the mechanisms of tail resorption (Mahapatra and Mahapatra, 2013).

The extensive work on the biochemical assay of lysosomal enzymes revealed that cathepsin D could be an important lysosomal enzyme as the activity of the enzyme increased significantly in the resorbing tails as compared to other lysosomal enzymes (Mahapatra and Mahapatra, 2013). Cathepsin D, a member of aspartic proteinases was earlier claimed to play a crucial role in the breakdown of larval tissues during anuran metamorphosis (Mukai *et al.*, 1995; Das *et al.*, 2006; Domanski and Helbing, 2007), but its exact role was not elaborated. Thus, immunohistochemical localization of cathepsin D was taken up in the pre-resorbing and resorbing tail tissues of *P. maculatus* and *D. melanostictus* to have an idea about the involvement of cathepsin D during cell death *in situ*. Cathepsin D was associated with the programmed cell death not only in the degrading epidermis but also in degenerating muscle fragments, spinal cord, blood cells and notochordal cells. The studies on localization of cathepsin D, however, gave enough indication that cathepsin D was at least involved in initiating the programmed cell death of epidermal cells, muscle cells, blood cells, neurons in spinal cord and notochordal cells (Mahapatra and Mahapatra 2011; 2012a and b;2013). Retractions of tail fins represented the first sign of tail atrophy (Weber, 1967) and presence of cathepsin D positive cells in the tip of tail fins in pre-resorbing tail, in both the species, suggested that proteolysis and subsequent degradation of tail tissues occurred much before visible tail resorption. Also, cathepsin D was found to be involved in the death of blood cells much before visible tail resorption (Mahapatra and Mahapatra, 2011; 2012a&b; 2013).

Presence of large melanocytes around the degrading tail tissues was an important finding in studies carried out by Mahapatra and Mahapatra (2013). Melanocytes are key components in the skin pigmentary system (Uong and Zon, 2010) and associated in maintaining epidermal homeostasis (Tsatmali *et al.*, 2002). They have been reported in resorbing tadpole tails of *Xenopus laevis* but have not been associated with cell death (Berry *et al.*, 1998; Jung *et al.*, 2002). Divya *et al.* (2010)

studied the involvement of melanocytes in an Indian anura *Clinotarsus curtipes* and proposed that the long processes of melanocytes pervade the skeletal muscle fibers concomitant with or preceding the breakdown of muscle fibers into sarcolytes and subsequently cause muscle disintegration by delivering melanin. They suggested a lysosome–like role of melanocytes in the fragmentation of muscle into sarcolytes. Melanocytes were also found around the skeletal muscle in pre-resorbing tail and were found surrounding the dying epidermal cells, sarcolytes, blood vessels and spinal cord in the resorbing tails in both *P. maculatus* and *D. melanostictus* (Mahapatra and Mahapatra, 2015a). The melanocytes were also shown to increase in size in the resorbing tails of both *P. maculatus* and *D. melanostictus* as compared to the pre - resorbing tails. Melanocytes have thus emerged as important cell types that mediate cell death, in addition to macrophages and fibroblasts (Mahapatra and Mahapatra, 2015b). Cathepsin L has been found to be involved in the proliferation, apoptosis, and differentiation of keratinocytes, and melanogenically active hair-bulb melanocytes (Tobin *et al.*, 2002). Significant increase in the levels of cathepsin L in the resorbing tails of anurans *P. maculatus* and *D. melanostictus* was correlated to the increase in size of melanocytes (Mahapatra and Mahapatra, 2015a&b).

Future Prospects

This review tries to highlight the importance of anuran metamorphosis as models in apoptotic research and also emphasises the need to take up such studies in non-model organisms. During anuran metamorphosis, tail undergoes apoptosis in a spatiotemporally-regulated manner and thus provides a unique opportunity to study the entire mechanisms underlying apoptosis at the molecular level (Tata, 2003). This model has been extensively utilized to study the role of matrix metalloproteinases during cell death and remodelling as well as studying the various aspects of apoptosis (Walsh *et al.*, 2007; Ishizuya-Oka *et al.*, 2010). However, there is very little information about the actual role of lysosomal enzymes during apoptotic death of different tissues in the resorbing tail.

Lysosomal enzymes comprise a variety of proteases, nucleases, glycosidases, sulphatases and lipases. While the importance of caspases in the apoptotic processes is firmly established, the role of lysosomal enzymes in cell death has been only recently unmasked by a number of *in vitro* and *in vivo* studies that indicate several lysosomal enzymes and mainly cathepsins to mediate or modulate some cell death programmes (Tardy *et al.*, 2006). Release of cathepsins from the lysosomes into the cytoplasm is a pre-condition for their participation in the regulation of apoptosis (Chwieralski *et al.*, 2006). Thus, alternative modes of programmed cell death have been proposed (Leist and Jaattela, 2001a) and the cysteine proteases cathepsins B and L and the aspartatic protease cathepsin D which are the most abundant proteases of the lysosomes have been described as good candidates to account for cell degradation during programmed cell death (Leist and Jaattela, 2001b). Especially, cathepsin D has been reported to trigger caspase-independent AIF (apoptosis-inducing factor) release in T lymphocytes committed to die (Bidere *et al.*, 2003). Additionally, cathepsins are also able to activate caspases, indicating that they can act in the early steps of apoptosis (Ishizaki *et al.*, 1998; Schotte *et al.*, 1998; Vancompernolle *et al.*, 1998). So far, the lysosomal pathway of apoptosis has been observed in cell death associated to pathologic processes, such as tumors (Foghsgaard *et al.*, 2001), neurodegenerative diseases (Houseweart *et al.*, 2003) or in cells treated with a variety of cytotoxic agents (Liaudet-Coopman *et al.*, 2005). Recently, lysosomal cathepsins have been found to cooperate with caspases in the execution of cell death in the developing limb of chick and mice embryos and expression of cathepsins B and D genes have been shown to be specific markers for programmed cell death in this system (Zuzarte-Luis *et al.*, 2007a;b). The studies on tail resorption in *Polypedates maculatus* and *Duttaphrynus melanostictus* indicate the involvement of cathepsin D in apoptosis (Mahapatra and Mahapatra, 2012a; b; 2013) and thus project anuran tadpoles as alternate models to study lysosomal enzyme mediated cell death.

Anuran tail resorption can thus be an effective *in vivo* model to study lysosomal enzyme mediated apoptosis. Additionally, studies on melanocytes, which are a proven source of lysosomal enzymes, is also warranted to discover their specific roles during cell death during anuran tail resorption. Recently, melanocytes of the epidermis have been suggested to contribute to the phagocytosis of invading pathogens and to present antigens to competent immune cells. Melanocytes produce key cytokines such as IL-1β, IL6 and TNF-α as well as chemokines which subsequently alert macrophages, neutrophils, fibroblasts and keratinocytes through unique crosstalk mechanisms (Gasque and Jaffar-Bandjee, 2015). Tail resorption studies also suggest that melanocytes along with macrophages and fibroblasts remove the cell debris in the resorbing tadpole tail (Mahapatra and Mahapatra, 2015a). Thus, explicit work is essential to discover the exact role of melanocytes and their contribution in cell death. This would also open a new area of research with potential to formulate alternate strategies of treatment of human disorders pertaining to uncontrolled cell proliferation. With the advent of modern technology in proteomics (Domanski, 2008) and transcriptomics (Zhao *et al.*, 2016), studies on molecular aspects in non-model organisms are no more a constraint. Anuran tadpoles are relatively easy to rear, maintain and hardy enough to sustain experimental manipulations. So this model system offers enough opportunities to be taken up for applications in biomedical research.

References

Berry, D.L., Schwartzman, R.A. and Brown, D.D. (1998). The expression pattern of thyroid hormone response genes in the tadpole tail identifies multiple resorption programs. Developmental Biology, 203: 12-23.

Bidere, N., Lorenzo, H.K., Carmona, S., Laforge, M., Harper, F., Dumont, C. and Senik, A. (2003). Cathepsin D triggers Bax activation, resulting in selective apoptosis inducing factor (AIF) relocation in T lymphocytes entering the early commitment phase to apoptosis. The Journal of Biological Chemistry, 278: 31401–31411.

Birol, I., Behsaz, B., Hammond, S.A., Kucuk, E., Veldhoen, N. and Helbing, C.C. (2015). *De novo* Transcriptome Assemblies of *Rana* (*Lithobates*) *catesbeiana* and *Xenopus laevis* tadpole livers for comparative genomics without reference genomes. PLoS ONE, 10(6): e0130720.

Buchholz, D.R. (2015). More similar than you think: Frog metamorphosis as a model of human perinatal endocrinology. Developmental Biology, 408: 188–195.

Burggren, W.W. and Warburton, S. (2007) Amphibians as Animal Models for Laboratory Research in Physiology. Institute for Laboratory Animal Research Journal, 48(3): 260-269.

Chwieralski, C.E., Welte, T. and Bühling, F. (2006). Cathepsin-regulated apoptosis. Apoptosis, 11 (2): 143-149

Das, B., Schreiber, A.M., Huang, H. and Brown, D.D. (2002). Multiple thyroid hormone-induced muscle growth and death programs during metamorphosis in *Xenopus laevis*. Proc. Natl Acad. Sci., USA. 99: 12230–12235.

Das, B., Cai, L., Carter, M.G., Piao, Yu.L., Sharov, A.A., Ko, M.S.H. and Brown, D.D. (2006). Gene expression changes at metamorphosis induced by thyroid hormone in *Xenopus laevis* tadpoles. Developmental Biology, 291: 342-355.

Dinesh, K.P., Radhakrishnan, C., Channakeshavamurthy, B.H. and Kulkarni, N.U. (2015). A Checklist of Amphibia of India, updated till January 2015 available at http: //mhadeiresearchcenter.org/resources (online only).

Divya, L., Beyo, R.S., Sreejith, P., Akbarsha, M.A. and Oommen, O.V. (2010). Skeletal muscle–melanocyte association during tadpole tail resorption in a tropical frog, *Clinotarsus curtipes* Jerdon (Anura, Ranoidea). Zoology (Jena), 113: 175-83.

Dodd, M.H.I. and Dodd, J.M. (1976). The biology of metamorphosis, in B Lofts (ed), Physiology of Amphibia. Academic Press, New York, 467-599.

Domanski, D. and Helbing, C.C. (2007). Analysis of the *Rana catesbeiana* tadpole tail fin proteome and phosphoproteome during T3-induced apoptosis: identification of a novel type I keratin. BMC Developmental Biology, 7: 94-122.

Domanski, D. (2008). Proteomic analyses of thyroid hormone-sensitive tissues during frog tadpole metamorphosis. Ph.D. thesis. University of Victoria, 1-208.

Du Pasquier, D., Rincheval, V., Sinzelle, L., Chesneau, A., Ballagny, C., Sachs, L.M., Demeneix, B. and Mazabraud, A. (2006). Developmental cell death during *Xenopus* metamorphosis involves BID cleavage and caspase 2 and 8 activation. Developmental Dynamics, 235: 2083–2094.

Estabel, J., Mercer, A., Konig, N. and Exbrayat, J.M. (2003). Programmed cell death in *Xenopus laevis* spinal cord, tail and other tissues, prior to, and during, metamorphosis. Life Sciences, 73: 3297-3306.

Foghsgaard, L., Wissing, D., Mauch, D., Lademann, U., Bastholm, L., Boes, M., Elling, F., Leist, M. and Jaattela, M. (2001). Cathepsin B acts as a dominant execution protease in tumor cell apoptosis induced by tumor necrosis factor. Journal of Cell Biology, 153: 999-1009.

Fuchs, Y. and Steller, H. (2011). Programmed cell death in animal development and disease. Cell, 147(4): 742-58.

Gaupale, T.C., Mathi, A.A., Ravikumar, A. and Bhargava, S. (2009). Localization and enzyme activity of Iodotyrosine dehalogenase1 (DEHAL1) during metamorphosis of frog, *Microhyla ornata*. Ann. N.Y. Acad.Sci. Trds. Comp. Endocrinology Neurobiology, 1163: 154-168.

Gasque, P. and Jaffar-Bandjee, M.C. (2015). The immunology and inflammatory responses of human melanocytes in infectious diseases. Journal of Infection, 71(4): 413-21.

Gilbert, L.I. and Freiden, E. (1981). Metamorphosis: A problem in Developmental Biology. Plenum Press, New York, 2nd edition, 139-176

Hanada, H., Kashiwagi, A., Takehara, Y., Kanno, T., Yabuki, M., Sasaki, J., Inoue, M. and Utsumi, K. (1997). Do reactive oxygen species underlie the mechanism of apoptosis in the tadpole tail? Free Radical Biology and Medicine, 23: 294-301.

Houseweart, M.K., Vilaythong, A., Yin, X. M., Turk, B., Noebels, J.L. and Myers, R.M. (2003). Apoptosis caused by cathepsins does not require Bid signaling in an *in vivo* model of progressive myoclonus epilepsy (EPM1). Cell Death and Differentiation, 10: 1329–1335.

Ishizaki, Y., Jacobson, M. D. and Raff, M. C. (1998). A role for caspases in lens fiber differentiation. The Journal of Cell Biology, 140: 153–158.

Ishizuya-Oka, A., Hasebe, T. and Shi, Y.B. (2010). Apoptosis in amphibian organs during metamorphosis. Apoptosis, 15(3): 350–364.

Ishizuya-Oka, A. (2011). Amphibian organ remodeling during metamorphosis: insight into thyroid hormone-induced apoptosis. Development, Growth and Differentiation, 53: 202–212.

Jung, J. C., Leco, K. J., Edwards, D. R. and Fini, M. E. (2002). Matrix metalloproteinases mediate the dismantling of mesenchymal structures in the tadpole tail during thyroid hormone-induced tail resorption. Developmental Dynamics, 223: 402-413.

Kerr, J.F., Harmon, B. and Searle, J. (1974). An electron microscope study of cell deletion in anuran tadpole tail during spontaneous metamorphosis with special reference to apoptosis of striated muscle fibres. Journal of Cell Science, 14: 571-585.

Kinoshita, T., Sasaki, F. and Watanabe, K. (1985). Autolysis and heterolysis of the epidermal cells in anuran tadpole tail regression. Journal of Morphology, 185(2): 269-275.

Leist, M. and Jaattela, M. (2001a). Four deaths and a funeral: from caspases to alternative mechanisms. Nature reviews. Molecular Cell Biology, 2: 589–598.

Leist, M. and Jaattela, M. (2001b). Triggering of apoptosis by cathepsins. Cell Death and Differentiation, 8: 324–326.

Liaudet-Coopman, E., Beaujouin, M., Derocq, D., Garcia, M., Glondu-Lassis, M., Laurent-Matha, V., Prebois, C., Rochefort, H. and Vignon, F. (2005). Cathepsin D: newly discovered functions of a long-standing aspartic protease in cancer and apoptosis. Cancer Letters, 273: 167–179.

Mahapatra, C. (2007). M.Phil dissertation, Utkal University, Bhubaneswar.

Mahapatra, C. and Mahapatra, P. K. (2007). Tail regression in tadpoles of *Bufo melanostictus*. Pranikee (J. Zoological Soc. Odisha), 24: 85-96.

Mahapatra, C. and Mahapatra, P. K. (2011). Involvement of cathepsin D during tail regression in tadpoles of the common Indian tree frog, *Polypedates maculatus* (Anura: Rhacophoridae). Acta Histochemica, 113: 756-762.

Mahapatra, C. and Mahapatra, P. K. (2012a). Anuran tail regression as a model to study cell death during development. In Khillare, Y. K., P. Natarajan and G. K. Kulkarni (eds.) Sustainable development of fisheries and livestock for food security Vol-I: Innovations in aquaculture and animal research. Narendra Publishing House Delhi, India, 69-76.

Mahapatra, C. and Mahapatra, P. K. (2012b). Immunohistochemical localization of cathepsin D and a possible role of melanocytes during tail resorption in tadpoles of a tropical toad. Journal of Morphology, 273: 791–800.

Mahapatra, C. and Mahapatra, P. K. (2013). Role of lysosomal enzymes during anuran tail regression. Scholar's Press, Saarbrucken, Germany.

Mahapatra, C. and Mahapatra, P. K. (2015a). Interspecific variations in duration of tail regression in two tropical anurans. Journal of Histology, 1-8.

Mahapatra, C. and Mahapatra, P. K. (2015b). Role of melanocytes during cell death in the resorbing tail of anuran tadpoles. Pranikee, 27: 70-94.

Mahapatra, P. K., Mohanty-Hejmadi, P. and Chainy, G. B. N. (2001). Changes in oxidative stress parameters and acid phosphatase activity in the pre-regressing and regressing tail of Indian Jumping frog *Polypedates maculatus* (Anura, Rhacophoridae). Comparative Biochemistry and Physiology, Toxicology and Pharmacology, 130: 281-288.

Mahapatra, P. K., Jamdar, S. N., Mahapatra, C., Dutta, S. K. and Harikumar, P. (2012). Interspecies variation in activity of lysosomal cathepsins during tail regression in the tadpoles of three sympatric anurans. Russian Journal of Herpetology, 19(3): 232-238.

Mathew, S., Fu, L., Fiorentino, M., Matsuda, H., Das, B. and Shi, Y.-B. (2009). Differential regulation of cell type-specific apoptosis by stromelysin-3: a potential mechanism via the cleavage of the laminin receptor during tail resorption in *Xenopus laevis*. Journal of Biological Chemistry, 284: 18545–18556.

Mathew, S., Fu, L., Hasebe, T., Ishizuya-Oka, A. and Shi, Y.-B. (2010) Tissue-dependent induction of apoptosis by matrix metalloproteinase stromelysin-3 during amphibian metamorphosis. Birth Defects Research Embryo Today, 90: 55–66.

Mukai, M., Obara, M. and Yoshizato, K. (1995). Characterization of gene of anuran cathepsin D as a metamorphosis-associated enzyme. Development, Growth and Differentiation. 37: 463-477.

Mukaigasa, K., Hanasaki, A., Maeno, M., Fujii, H., Hayashida, S., Itoh, M., Kobayashi, M., Tochinai, S., Hatta, M., Iwabuchi, K., Taira, M., Onoe, K. and Izutsu, Y.

(2009). The keratin-related Ouroboros proteins function as immune antigens mediating tail regression in Xenopus metamorphosis. Proceedings of National Academic of Science, USA, 106: 18309–18314.

Nakajima, K. and Yaoita, Y. (2003). Dual mechanisms governing muscle cell death in tadpole tail during amphibian metamorphosis. Developmental Dynamics, 227: 246-255.

Nakajima, K., Fujimoto, K. and Yaoita, Y. (2005). Programmed cell death during amphibian metamorphosis. Seminars in Cell and Developmental Biology, 16: 271–280.

Nakajima, K., Takahashi, A. and Yaiota, Y. (2000) Structure, expression, and function of the *Xenopus laevis* caspase family. The Journal of Biological Chemistry, 275: 10484–10491.

Nishikawa, A. and Hayashi, H. (1995) Spatial, temporal and hormonal regulation of programmed muscle death during metamorphosis of the frog *Xenopus laevis.* Differentiation, 59: 207-214.

O'Rourke, D. P. (2007) Amphibians used in research and teaching. ILAR Journal 48(3): 183-187.

Rowe, I., Le Blay, K., Du Pasquier, D., Palmier, K., Levi, G., Demeneix, B. and Coen, L. (2005) Apoptosis of tail muscle during amphibian metamorphosis involves a caspases 9-dependent mechanism. Developmental Dynamics, 233: 76–87.

Sachs, L. M., Le Mevel, S. and Demeneix, B. A. (2004) Implication of bax in *Xenopus laevis* tail regression at metamorphosis. Developmental Dynamics, 231: 671–682.

Schotte, P., Van Criekinge, W., Van de Craen, M., Van Loo, G., Desmedt, M., Grooten, J., Cornelissen, M., De Ridder, L., Vandekerckhove, J., Fiers, W., Vandenabeele, P. and Beyaert, R. (1998) Cathepsin B-mediated activation of the proinflammatory caspase-11. Biochemical and Biophysical Research Communications, 251: 379–387.

Shi, Y.-B., Fu, L., Hsia, S.C., Tomita, A. and Buchholz, D. (2001) Thyroid hormone regulation of apoptotic tissue remodeling during anuran metamorphosis. Cell Research, 11: 245-252.

Tardy, C., Codogno, P., Autefage, H., Levade, T. and Andrieu-Abadie, N. (2006) Lysosomes and lysosomal proteins in cancer cell death (new players of an old struggle). Biochimica et Biophysica Acta, 1765: 101-125.

Tata, J. R. (2003) Hormonal signaling during amphibian metamorphosis. Proceedings of the Indian National Science Academy - Part B. 69(5): 773-790.

Tata, J. R. (2006) Amphibian metamorphosis as a model for the developmental actions of thyroid hormone. Molecular and Cellular Endocrinology, 246: 10–20.

Tsatmali, M., Ancans, J. and Thody, A. J. (2002) Melanocyte function and its control by melanocortin peptides. The Journal of Histochemistry and Cytochemistry, 50(2): 125-133.

Tobin, D. J., Foitzik, K., Reinheckel, T., Mecklenburg, L., Botchkarev, V. A., Peters, C. and Paus, R. (2002) The Lysosomal Protease Cathepsin L Is an Important Regulator of Keratinocyte and Melanocyte Differentiation During Hair Follicle Morphogenesis and Cycling. The American Journal of Pathology, 160(5): 1807-1821.

Uong, A. and Zon, L. I. (2010). Melanocytes in development and cancer. Journal of Cellular Physiology, 222(1): 38–41.

Vancompernolle, K., Van Herreweghe, F., Pynaert, G., Van de Craen, M., De Vos, K., Totty, N., Sterling, A., Fiers, W., Vandenabeele, P. and Grooten, J. (1998) Atractyloside induced release of cathepsin B, a protease with caspase-processing activity. FEBS Letters, 438: 150–158.

Walsh, L. A., Carere, D. A., Cooper, C. A. and Damjanovski, S. (2007) Membrane type-1 matrix metalloproteinases and tissue inhibitor of metalloproteinases-2 RNA levels mimic each other during *Xenopus laevis* metamorphosis. PloS One, 2: e1000.

Weber, R. (1967) The Biochemistry of Animal Development. 2nd edition, Academic Press, New York.

Yaoita, Y. and Nakajima, K. (1997) Induction of apoptosis and CPP32 expression by thyroid hormone in a myoblastic cell line derived from tadpole tail. Journal of Biological Chemistry, 272: 5122– 5127.

Zhao, L., Liu, L., Wang, S., Wang, H. and Jiang, J. (2016) Transcriptome profiles of metamorphosis in the ornamented pygmy frog *Microhyla fissipes* clarify the functions of thyroid hormone receptors in metamorphosis. Scientific Reports, 6: 27310.

Zuzarte-Luis, V., Montero, J. A., Kawakami, Y., Izpisua-Belmonte, J. C. and Hurle, J.M. (2007a) Lysosomal cathepsins in embryonic programmed cell death. Developmental Biology, 301: 205–217.

Zuzarte-Luis, V., Montero, J. A., Torre-Perez, N., Garcia-Porrero, J. A. and Hurle, J.M. (2007b) Cathepsin D gene expression outlines the area of physiological cell death during embryonic development. Developmental Dynamics, 236(3): 880-885.

18

Characteristics Features of Mesenchymal Stem Cells

Seema Tripathy and Shikha Singh*

Centre for Biotechnology, Siksha 'O' Anusandhan (Deemed to be University), Kalinga Nagar, Ghatikia, Bhubaneswar – 751 030, Odisha

****E-mail: seema.tripathy@yahoo.com***

ABSTRACT

Currently, applications of mesenchymal stem cells (MSCs) to treat several chronic diseases are tend to be safe, feasible and promising. The therapeutic contribution of mesenchymal stem cells (MSCs) include their functional activity and interaction with host tissues after engraftment. Their unique biological characteristics which include self-renewal activity, differentiation potential, homing and migration ability and immune-tolerant capacity make them suitable candidates for both autologous and allogeneic cell based therapies. At current scenario MSCs associated clinical trials are gaining attention from common people to expert surgeon. The following communication describes the in vitro characteristics of MSCs which make them novel therapeutic agents to be used in different cell-based approaches and regenerative medicine.

***Keywords**: Mesenchymal stem cells, Migration, Immune-tolerent, Autologous and allogeneic.*

Introduction

Mesenchymal stem cells or mesenchymal stromal cells (MSCs) recently, have been given a new acronym "medicinal signaling cells" by Calpan (Caplan, 2010). This new acronym may specify the ability of MSCs to create and store drugs for patient-specific injured tissues. The existence of this cell population was first observed by Cohnheim more than 140 years ago (Cohnheim, 1867). However, its nature and *in vitro* characters was later confirmed by Friedenstein and their colleagues (Friedenstein *et al.*, 1976). This adherent non-haematopoietic cell population are multipotent in nature. These can differentiate into both mesenchymal and non-mesenchymal cell types (Nombela-Arrieta *et al.*, 2011). In addition to MSCs secrete many chemokines, growth factors and anti-inflammatory molecules which influence the survival ability, angiogenic capacity and immune modulating potential of the cells present in nearby environment (Pittenger *et al.*, 1999). This communication highlights *in vitro* characters of MSCs and its multi-dimensional features that recommend these as potent candidates for tissue regeneration and repair in current therapeutics.

Sources of Derivation

MSCs dwell in all through the adult organisms (Tripathy and Mohanty, 2016). Apart from bone marrow (BM), these cells can be derived from adipose tissue, periosteum, synovial membrane, synovial fluid (SF), muscles, dermis, deciduous teeth, pericytes, trabecular bone (TB), infrapatellar fat pad, articular cartilages, umbilical cord blood (UCB), peripheral blood, breastmilk, urine and menstrual blood (Figure 18.1). Interestingly, different methodologies are applied for isolation, expansion and characterization of MSCs from different sources. So, MSCs obtained from different sources are the same cells or different has long been debated and opinions are still conflicting.

Characterization of *In vitro* Cultured MSCs

An optimum protocol for *in vitro* culture of MSCs is highly necessary that reproduce sufficient number of MSCs for therapeutic purposes. In recent year's methodology of culturing MSCs are improved. Traditionally, MSCs were cultured in Dulbecco's modified eagle's medium-low glucose supplemented with 10 per cent fetal calf serum or new-born calf serum (NBCS) or fetal calf serum (FCS) at 37°C and 5 per cent CO_2 in ambient O_2 concentration for a week (Beyer Nardi and da Silva Meirelles, 2006). *In vitro* culture MSCs populations are highly heterogeneous (Phinney and Prockop, 2007. De Bari *et al.*, 2008). However, the two-dimensional (2 D) *in vitro* culture system for MSCs does not recapitulate the specialized micro-environments, or niches that reside in three dimensional (3 D) physiological system. So, currently by application of supportive scaffolds to culture MSCs mimic natural environment which may significantly improve condition of patients after engraftment and facilitate the recovery. *In vitro* culture, MSCs express unique stem cell markers and multilineage differentiation potential.

Immunophenotype Analysis of MSCs

According to guidelines of the international society for cellular therapy MSCs must express cluster of designation or classification determinant (CD) molecules such as CD105 (SH2), CD73 (SH3/4), and CD90 and must be negative for surface markers CD34, CD45, CD14, CD795 or CD19, and HLA-DR [9]. hMSCs are also negative for several other antigens like CD4, CD8, CD11a, CD14, CD15, CD16, CD25, CD31, CD33, CD49b, CD49d, CD49f, CD50, CD62E, CD62L, CD62P, CD80, CD86, CD106, (vascular cell adhesion molecule [VCAM]-1), CD117, cadherin V, and glycophorin A. Cell surface markers express by MSCs derived from different sources is shown in Table 18.1.

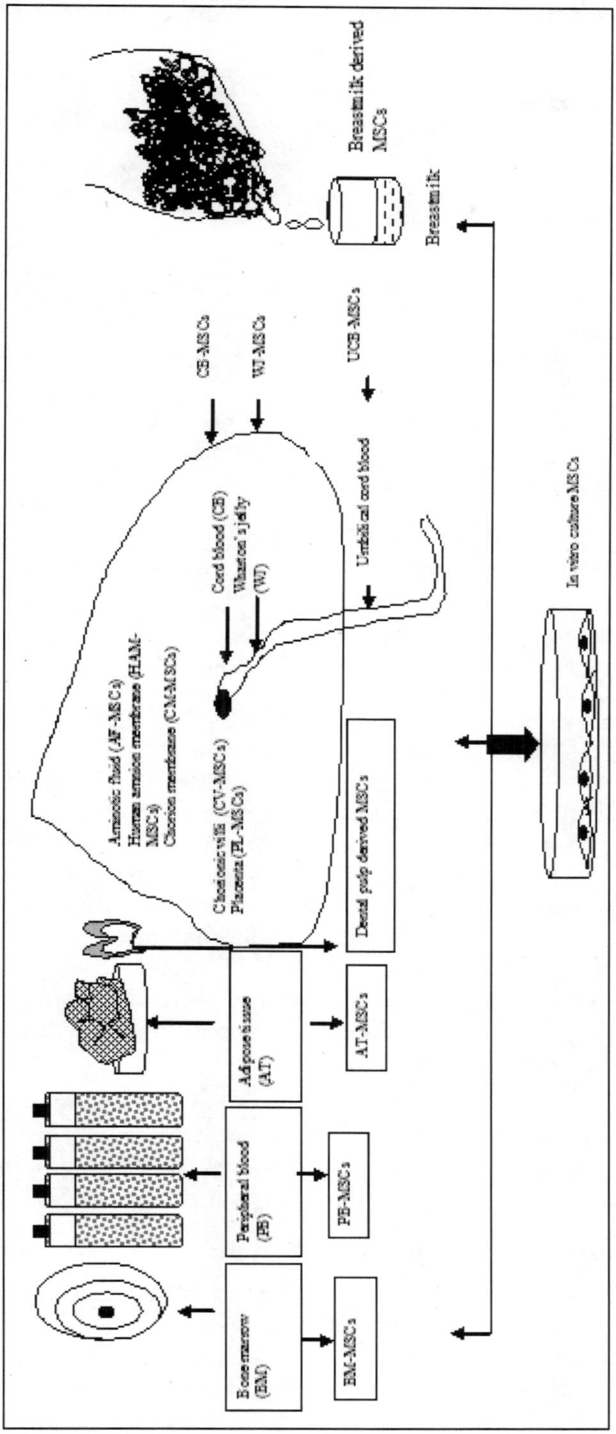

Figure 18.1: Major Sources to Obtain hMSCs for Therapeutic Interventions.

Table 18.1: Cell Surface Marker Expression Profile of MSCs Derived from different Sources

Sl.No.	Sources	CD Marker Expression in Derived MSCs	References
1	Bone marrow	SH2, SH3, CD105 (endoglin) and CD73 molecules $CD44^+$ $CD105^+$, $CD166^+$, $CD28^+$, $CD33^+$, $CD13^+$, CD9, CD10, CD13, CD73, CD105, CD166, HLA class I^+, SSEA-4, frizzled-9 (FZD-9), Oct-4 and nestin	Battula *et al.*, 2017
2	Adipose tissues	CD44, CD73, CD90, CD105, and HLA-ABC, and lack the expression of HLA-DR, CD14, and CD45	Bourin *et al.*, 2013
3	Mesenchymal stem cells of dental pulp (DPSCs)	CD13+, CD29+, CD44+, CD73+, CD90+, CD146+, CD166+, CD63+, CD45–, CD34–, CD105+, CD14–, CD34–,Stro-1+, CD271+, CD117+, OCT-3/4+, NANOG+, SSEA-3+, SSEA-4+, TRA-1-60+ and TRA-1-81+	Ledesma-Martínez *et al.*, 2016
4	Breastmilk	Positive expression of CD44, CD29, SCA-1, Oct-4, Rex-1 and Sox-2 and negative expression for CD33, CD34, CD45, CD73.	Patki *et al.*, 2008; Hassiotou *et al.*, 2012
5	Peripheral blood	Positive expression of CD105, CD166, and CD29 negative expression of CD34 and CD45.	Chong *et al.*, 2012

Differentiation Ability of MSCs

During initial period of investigation of MSCs, it is believe to show tri-lineage differentiation of mesoderm-type cells, for example, osteoblasts, adipocytes and chondrocytes. Extensive investigation during the years has altered this hypothesis. It is now confirm that MSCs can also differentiate into other ectodermal and endodermal cell types (Figure 18.2). These differentiation tendencies of MSCs are govern by different signalling pathways tightly regulated by signalling molecules and transcription factors closely linked to several controlling elements including type of inducers, adhesive contexts, mechanical signals and the physical responses of the cells (Tripathy and Mohanty, 2016). Specifically, the Wnt and transforming growth factor-β and bone morphogenetic protein signalling pathways are crucial players to affect MSCs nativity and help them to differentiate into cartilage and bone (Augello and De Bari, 2010).

In vitro differentiation of MSCs toward a particular lineage is govern by multiple factors such as cytokines, growth factors, extracellular matrix molecules, and transcription factors (TFs). This often achieved by addition of inducers to the culture medium (Table 18.2).

Immunomodulatory Properties of MSCs

The destiny, migration and commitment of MSCs and its precursors are regulated by various instructive signals from their immediate vicinity or micro-environment (Tripathy and Mohanty, 2016). MSCs always express low levels of major histocompatibility complex (MHC) class I molecules and are negative for MHC class II antigens. MSCs also do not express co-stimulatory molecules, such as

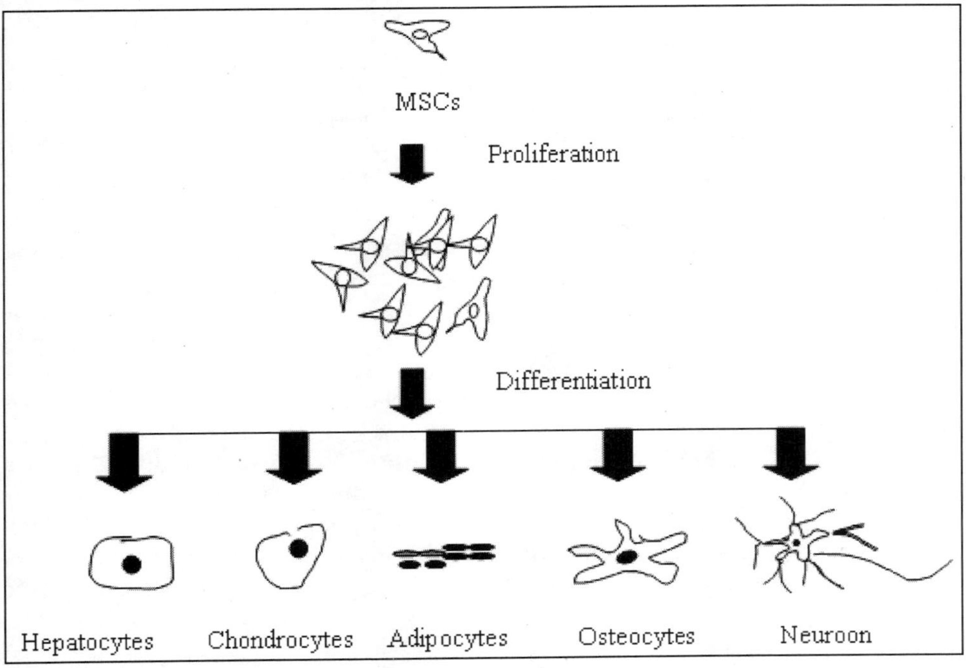

Figure 18.2: Differentiations of MSCs into different Cell Types in Presence of Appropriate Inducers.

B7-1, B7-2, CD80, CD86, CD40 and CD40L (Kode *et al.*, 2009). In response to injury MSCs secretes several bioactive molecules that have immunomodulatory properties. These bioactive molecules include interleukin (IL)-6, IL-7, IL-8, IL-11, IL-12, IL-14, IL-15, leukemia inhibitory factor, granulocyte colony stimulating factor (G-CSF),

Table 18.2: Factors Inducing Differentiation of MSCs Towards a Specific Cell Type

Inducers	*Concentration*	*Differentiated Cell Types*	*References*
Dexamethasone, (Dex) + 3-isobutyl-1-methylxanthine (IBMX)+ insulin (Ins)	1 μM Dex +0.5 mM IBMX +0.01 mg/ml Ins	Adipocytes	Yu *et al.*, 2012
Ascorbic acid (Asc acid) + Dex	50 μM Asc acid +100 nM Dex	Osteocytes	Yu *et al.*, 2012
TGF- β1 + asc acid + Dex	10 ng/ml TGF- β1 -50 μg/ml Asc acid +100 nM Dex	Chondrocytes	Vater *et al.*, 2011
Insulin, transferrin, selenium premix	6.25 μg/mL insulin, 6.25 μg/mL transferrin, 6.25 ng/mL selenious acid, 1.25 mg/mL BSA, and 5.35 mg/mL linoleic acid.	Hepatocytes	Lee *et al.*, 2004
5-azacytidine	3 μM-10 μM	Cardiomyocytes	Wei *et al.*, 2011
Retinoic acid (RA)+fibroblatic growth factor (FGF)	20 mM RA and 12.5 ng/ml bFGF	Neural cells	Salehinejad *et al.*, 2015

granulocyte macrophage colony-stimulating factor (GM-CSF), stem cell factor, macrophage colony-stimulating factor (M-CSF) and FMS like tyrosine kinase-3 ligand (flk-3L) implicated mainly in hematopoiesis. They also express cytokine receptors IL-1R, IL-3R, IL-4R, IL-6R and IL-7R. Cytokine-induced MSCs express very high level of several leukocyte chemokines, most notably CXCL9, CXCL10 and CXCL11. These cytokines and signalling molecules offer immune-modulatory, anti-inflammatory, antiapoptotic, pro-proliferative effects at site of injury.

The trophic migration of MSCs to the injured part might interfere differentiation, maturation and function of dendritic cells by mediating the soluble factors like IL-6 and macrophage colony-stimulating factor (M-CSF). Alongwith these functions MSCs are also markedly inhibit IL-2- induced NK cell proliferation and prevent the cytotoxic activity (Ren *et al.*, 2008). This is elicited by IL-10, TGF-ß, indoleamine 2, 3 dioxygenase (IDO) and prostaglandin E2 (PGE2). MSCs have not acquired immunosuppressive ability from birth it is later induced by pro-inflammatory cytokines such as IFN-g in combination with tumor necrosis factor α (TNF-α), IL-1α or IL-1β. Nitric oxide (NO is another important factor of MSCs-mediated immunosuppression mechanism. This rapidly diffusing gaseous bioactive molecule induces a notable up-regulation in inducible nitric oxide synthase (iNOS) and several leukocyte chemokines. High levels of NO can suppress immune cell function. (Sato *et al.*, 2007). Incredible, cytokine-induced chemokines and NO is the key to MSC-mediated immune-suppression (Kode *et al.*, 2009).

The toll-like receptor (TLR) proteins which play significant role in immunomodulation mechanisms are also expressed by MSCs. A study demonstrates that both TLR3 and TLR4 activation reduce the inhibitory activity of human BM-MSCs on T cell proliferation without influencing IDO activity or PGE2 levels. It is down-regulated the expression of Jagged1 which suggests that the Notch signaling pathway mediates contact-mediated immunosuppression by MSCs (Liotta *et al.*, 2008)·Whereas another study report that involvement of both TLR3 and TLR4 enhances the immunosuppressive properties of human BM-MSCs through the indirect induction of IDO1 (Opitz *et al.*, 2009). The biological functions and regenerative aptitudes of MSCs that make them novel therapeutic agents for cell-based therapies and regenerative medicine (Figure 18.3).

Conclusion

The multipotent MSCs have unique advantages over other contemporary cell sources to treat inflammatory diseases due to its immunosuppression property; However, MSC-mediated immunosuppression is dependent on the combined action of chemokines, inflammatory cytokines, and effector factors secreted by MSCs, as well as the micro-environment and potency of pro- inflammatory stimulus. Further, investigation is required to understand the molecular mechanisms behind immunomodulation mechanism shown by MSCs that will further reinforce its clinical application. Recent approach of establishment of MSCs cryobanks for future off-shelf therapy need detail knowledge regarding its nativity, *in vivo* and *in vitro* characteristics.

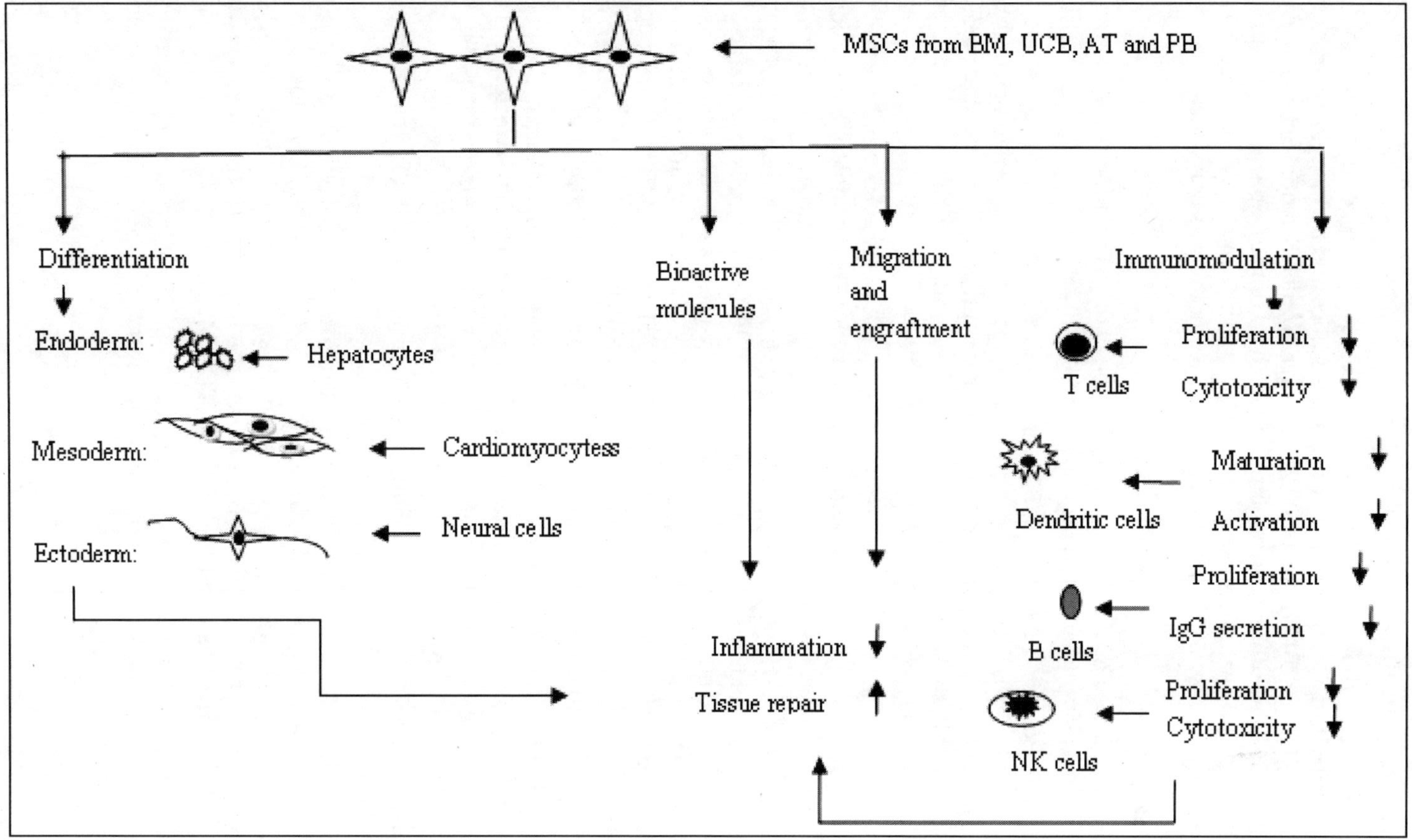

Figure 18.3: Strartegie Involve in MSCs Mediated Regenerative Medicine.

Acknowledgements

Authors are thankful to the Department of Science and Technology, Government of India, for providing financial assistance under vide reference number SR/WOS-A/LS-13/2016 dated 06.09.2016 under Women Scientist Scheme to carry out this work. The authors are indebted to the Head of the Department, Centre for Biotechnology, Siksha 'O' Anusandhan (Deemed to be University), Kalinga Nagar, Ghatikia, Bhubaneswar-751 030 India, for providing the necessary facilities.

References

Augello, A. and De Bari, C. (2010) The regulation of differentiation in mesenchymal stem cells. Human Gene Therapy, 21(10): 1226-1238.

Battula, V.L, Bareiss, P.M., Treml, S., Conrad, S., Albert, I., Hojak, S., Abele, H., Schewe, B., Just, L, Skutella, T. and Bühring, H.J. (2007) Human placenta and bone marrow derived MSC cultured in serum-free, b-FGF-containing medium express cell surface frizzled-9 and SSEA-4 and give rise to multilineage differentiation. Differentiation, 75(4): 279-291.

Beyer Nardi, N. and da Silva Meirelles, L. (2006). Mesenchymal stem cells: isolation, *in vitro* expansion and characterization. Handbook of Experimental Pharmacology, 174: 249-282.

Bourin, P., Bunnell, B.A., Casteilla, L., Dominici, M., Katz, A.J., March, K.L., Redl, H., Rubin, J.P., Yoshimura, K. and Gimble, J.M. (2013) Stromal cells from the adipose tissue-derived stromal vascular fraction and culture expanded adipose tissue-derived stromal/stem cells: a joint statement of the International Federation for Adipose Therapeutics and Science (IFATS) and the International Society for Cellular Therapy (ISCT). Cytotherapy, 15(6): 641-648.

Caplan, A. I. (2010) What's in a name? Tissue Engineering Part A, 16(8): 2415-2417.

Chong, P.P., Selvaratnam, L., Abbas, A.A. and Kamarul, T. (2012) Human peripheral blood derived mesenchymal stem cells demonstrate similar characteristics and chondrogenic differentiation potential to bone marrow derived mesenchymal stem cells. The Journal of Orthopaedic Research, 30(4): 634-642.

Cohnheim, J. (1867) Ueber entzundung und eiterung. Path Anat Physiol Klin Med. Berlin, p. 1-79.

De Bari, C., Dell'Accio, F., Karystinou, A., Guillot, P.V., Fisk, N.M., Jones, E.A., McGonagle, D., Khan, I.M., Archer, C.W., Mitsiadis, T.A., Donaldson, A.N., Luyten, F.P. and Pitzalis, C. (2008) A biomarker-based mathematical model to predict boneforming potency of human synovial and periosteal mesenchymal stem cells. Arthritis and Rheumatism, 58(1): 240-250.

Friedenstein, A.J., Gorskaja, J.F. and Kulagina, N.N. (1976) Fibroblast precursors in normal and irradiated mouse hematopoietic organs. Experimental Hematology, 4(5): 267-274.

Hassiotou, F., Beltran, A., Chetwynd, E., Stuebe, A.M., Twigger, A.-J., Metzger, P. and Hartmann, P.E. (2012) Breastmilk is a novel source of stem cells with multilineage differentiation potential. Stem Cells (Dayton, Ohio), 30(10): 2164-2174.

Kode, J.A., Mukherjee, S., Joglekar, M.V. and Hardikar, A.A. (2009) Mesenchymal stem cells: immunobiology and role in immunomodulation and tissue regeneration. Cytotherapy, 11(4): 377-391.

Ledesma-Martínez, E., Mendoza-Núñez, V.M. and Santiago-Osorio, E. (2016) mesenchymal stem cells derived from dental pulp: a review. Stem Cells International, vol. 2016, Article ID 4709572, 12 pages. https: //doi.org/10.1155/2016/4709572.

Lee, K.D., Kuo, T.K., Whang-Peng, J., Chung, Y.F., Lin, C.T., Chou, S.H., Chen, J.R., Chen, Y.P. and Lee, O.K. (2004) *In vitro* hepatic differentiation of human mesenchymal stem cells. Hepatology, 40(6): 1275-1284.

Liotta, F., Angeli, R., Cosmi, L., Filì, L., Manuelli, C., Frosali, F., Mazzinghi, B., Maggi, L., Pasini, A., Lisi, V., Santarlasci, V., Consoloni, L., Angelotti, M.L., Romagnani, P., Parronchi, P., Krampera, M., Maggi, E., Romagnani, S. and Annunziato, F. (2008) Toll-like receptors 3 and 4 are expressed by human bone marrow-derived mesenchymal stem cells and can inhibit their T-cell modulatory activity by impairing Notch signaling. Stem Cells, 26(1): 279-289.

Nombela-Arrieta, C., Ritz, J. and Silberstein, L. E. (2011) The elusive nature and function of mesenchymal stem cells. Nature Reviews. Molecular Cell Biology, 12(2): 126–131.

Opitz, C.A., Litzenburger, U.M., Lutz, C., Lanz, T.V., Tritschler, I., Köppel, A., Tolosa, E., Hoberg, M., Anderl, J., Aicher, W.K., Weller, M., Wick, W. and Platten, M. (2009) Toll-like receptor engagement enhances the immunosuppressive properties of human bone marrow-derived mesenchymal stem cells by inducing indoleamine-2, 3-dioxygenase-1 via interferon-beta and protein kinase R). Stem Cells, 27(4): 909-919.

Patki, S., Kadam, S., Chandra, V. and Bhonde, R. (2010) Human breastmilkis a rich source of multipotent mesenchymal stem cells. Human Cell, 23(2): 35-40.

Phinney, D.G. and Prockop, D.J. (2007) Concise review: Mesenchymal stem/multipotent stromal cells (MSCs): The state of transdifferentiation and modes of tissue repair - Current views. Stem Cells, 25(11): 2896-2902

Pittenger, M.F., Mackay, A.M., Beck, S.C., Jaiswal, R.K., Douglas, R., Mosca, J.D., Moorman, M.A., Simonetti, D.W., Craig, S. and Marshak, D.R. (1999) Multilineage potential of adult human mesenchymal stem cells. Science, 284(5411): 143-147.

Ren, G., Zhang, L., Zhao, X., Xu, G., Zhang, Y., Roberts, A.I., Zhao, R.C. and Shi, Y. (2008) Mesenchymal stem cell-mediated immunosuppression occurs via concerted action of chemokines and nitric oxide. Cell Stem Cell, 2(2): 141-150.

Salehinejad, P., Alitheen, N. B., Ali, A. M., Omar, A. R., Moshrefi, M., Motamedi, B. and Nematollahi-mahani, S. N. (2015). Neural differentiation of human umbilical cord matrix-derived mesenchymal cells under special culture conditions. Cytotechnology, 67(3): 449–460.

Sato, K., Ozaki, K., Oh, I., Meguro, A., Hatanaka, K., Nagai, T., Muroi, K. and Ozawa, K. (2007). Nitric oxide plays a critical role in suppression of T-cell proliferation by mesenchymal stem cells. Blood, 109(1): 228-234.

Tripathy, S. and Mohanty, P.K. (2016). Mesenchymal stem cells: a novel approach in regenerative medicine. World Journal of Pharmaceutical research, 2(5): 410-433.

Vater, C., Kasten, P. and Stiehler, M. (2011) Culture media for the differentiation of mesenchymal stromal cells. Acta biomaterialia, 7(2): 463-477.

Wei, F., Wang, T., Liu, J., Du, Y. and Ma, A. (2011). The subpopulation of mesenchymal stem cells that differentiate toward cardiomyocytes is cardiac progenitor cells. Experimental Cell Research, 317(18): 2661-2670.

Yu, W.H., Li, F.G, Chen, X.Y, Li, J.T, Wu, Y.H, Huang, L.H., Wang, Z., Li, P., Wang, T., Lahn, B.T. and Xiang, A.P. (2012). PPARgamma suppression inhibits adipogenesis but does not promote osteogenesis of human mesenchymal stem cells. The International Journal of Biochemistry and Cell Biology, 44(2): 377-384.

19

Bioprospecting of Horseshoe Crabs and Biotechnological Approaches Towards Sustainable Utilization

B.P. Dash

Department of Biosciences and Biotechnology, Fakir Mohan University, Nuapadhi, Balasore – 756 020, Odsha
E-mail: bisnubsbtfmu@gmail.com

ABSTRACT

Horseshoe crabs are unique type of arthropod having a lot of biomedical and evolutionary significance. Only four species of horseshoe crabs are presently available throughout the globe. In India two species of horseshoe crabs namely Tachypleus gigas and Carcinoscorpius rotundicauda have been reported from estuaries of Odisha and west Bengal coast. Besides the ecological significance they have immense biological, biomedical, economical and biotechnological prospects. The specific proteins present in the cells of blue blood of such animals have been used for detection of minute quantity of endotoxins. The chitin and chitosan of horseshoe crabs carapaces have been used for various biomedical purposes. The proteins and other compound present in perivitelline fluid of horseshoe crabs reported to have stem cell growth promoters, cardiogenesis enhancement capacity etc. But due to natural climatic change and excessive human exploitation, their population is decreasing. Keeping view of their importance, certain biotechnological approaches have been developed recently to increase and sustain the population of horseshoe on the earth.

Keywords: *Horseshoe crab, Chitin, Perivitelline fluid. Tachypleus gigas.*

Introduction

Horseshoe crabs or King crabs are a type of Arthropod found in marine waters. They are not true crabs; rather more closely related to Arachnids, a group that includes spiders and scorpions. It is also known by various names in Odia like:

Samudra Bichha, Ram Laxman Kankada, Kankua, Kanku *etc.* There are four extant species of horseshoe crabs found globally namely: *Limulus polyphemous, Tachypleus tridentatus, Tachypleus gigas* and *Carcinoscorpius rotundicauda.* They are regarded as an oldest creature in the earth and are surviving from the Mesozoic era, before the dinosaurs came into existence, since 425 million years ago. Because of their origin more than 450 million years ago, horseshoe crabs are also considered as "living fossils".

Bioresources

The blood and perivitelline fluid of eggs of horseshoe crabs have a number of biomedical important materials. Haemolymph or the blood of the Horseshoe crab is blue in colour, due to the presence of Haemocyanin. Plasma of the haemolymph contains the salt content of the sea water and has 3 abundant proteins i.e haemocyanin, the respiratory proteins and the C- reactive proteins which functions in the cytolytic destruction of bacterial cells. The protein isolated from the blood cells of American and Chinese horseshoe crabs have been used for development of very expensive and useful diagnostic kits (Lymulus Amoebocytes Lysate, LAL and Tachypleus Amoebocytes Lysate, TAL) for testing of the presence of bacterial endotoxins both Gram positive and Gram negative discovered by Levin *et al.* (1970). This test helps millions of human by providing safe and essential medical products. The America is getting huge revenue by supplying the product to different developed and developing countries. Indian has both the species of horseshoe crabs and therefore has enough scope to develop such kits for commercial uses. The lymph part of the horseshoe crab also contains some secondary metabolites and B_{12} binding proteins. Haemolymph contains some large granules such as Factor–B, C, D and G, Proclotting enzyme, coagulogen, Anti-LPS factor, Big Defensins, Tachylectin 1-5, Cystatin, Alpha–2 macroglobulin, LICI-123 as well as some small granules such as Tachyplesins, Big Defensins, Tachycitins and Tachystatins (Shigenaga *et al.*, 1993) The Perivitelline fluid of the horseshoe crab also contains some important primitive type of proteins, such as hemagglutinins and haemocyanin which plays an important role during embryogenesis by promoting cell proliferation. Lectin is the constituent of the peri vitelline fluid that helps in the proliferation of Chick embryonic heart cells (Ghaskadbi *et al.*, 2008). In wound healing activity and differentiation of human bone marrow stem cell into cardiomyocyte (Huma *et al.*, 2015a and b). The crude PVF has the ability to induce differentiation of dendritic cells from its bone marrow precursors (Chinnari *et al.*, 2015). The role of such molecule for growth stimulating effect can generate millions of rupees and enrich the global economy including India in future.

Potential Compounds for Anticancer Properties

The exploitation of new bio-active compounds from oceanic flora and fauna for its anti-carcinogenic properties has become an expanding biomedical field. This has lead to ever increasing quest to search for novel compounds with anti-cancer activity that could be used as drugs to cure the disease (Hoskin and Ramamoorthy, 2008). The naturally occurring antimicrobial peptides (AMPs), one of the efficient chemical defences of the eukaryotic cells against bacteria, fungi and viruses, have recently received more attention due to its potential anti-cancer activity (Zasloff, 2002).

AMPs that selectively targets cancer cells without harming the normal mammalian cells, are of great advantage in clinical treatment (Hoskin and Ramamoorthy 2008). Among the known AMPs, tachyplesin I obtained from hemocytes of horseshoe crab is currently the upcoming new bio-active compound known for its unique method of killing the cancer cells (Nakamura *et al.*, 1988). Recently, a report demonstrated that tachyplesin I selectively bind to hyaluronan on hyaluronan overexpressing human TSU prostate carcinoma cells, leading to lysis of tachyplesin I-coated cancer cells via activating the complement pathway (Adamia *et al.*, 2005). Like many tumor cells, hyaluronan is also present in higher amounts on the surface of endothelial cells involved in the process of neovascularisation that occur during tumor development. This suggests that tachyplesin I might also cause destruction of neovascularisation as well as neoplastic cells via classical complement pathway (Rooney *et al.*, 1995). Moreover, another report showed that heat inactivated serum complement leads to partial survival of the tachyplesin I coated TSU prostate cancer cells from lysis (Chen *et al.*, 2001). This result thus, suggests that tachyplesin I is involved in another mechanism that can cause cancer cell death. Also, it has been shown that a synthetic tachyplesin conjugated to the integrin homing domain RGD inhibited the proliferation of cultured tumor and endothelial cells and also reduced the colony formation of TSU prostate cancer cells.

The same group further emonstrated that RGD-tachyplesin activates caspase 9, caspase 8, and caspase 3 and increase the expression of Fas, FasL and caspase 7, and caspase 6, indicating the role of RGD-tachyplesin in promoting mitochondria and Fas/FasL mediated apoptosis. The potential mechanism of role of RGD-tachyplesin suggests that the overall positive charge of tachyplesin I allow internalization of RGD-tachyplesin I to target and this disrupt the negatively charged membranes of the mitochondria in the cancer cells, and thus promotes apoptosis (Chen *et al.*, 2001). Further, tachyplesin I is shown to inhibit the cancer cell growth via a non-cytolytic mechanism (Hoskin and Ramamoorthy 2008). Upon treatment of tachyplesin I of SMMC-7721 human hepatoma cell and BGC-823 human gastric adenocarcinoma cell cultures decreases the proliferative capacity of these cells (Shi *et al.*, 2006). Also, growth of SMMC-7721 cells decreases in the presence of tachyplesin I and this was shown to be associated with the reversal of malignant morphological and ultrastructural features (Ouyang *et al.*, 2002). At molecular level, it was shown to decrease the expression of tumor-associated antigens such as α-fetoprotein, modulation of differentiation-associated enzyme such as glutamyltransferase expression. Also, it leads to the decrease of expression of the *c-myc* oncogene, and an increase in the expression of the tumor suppressor gene p21WAF1/CIP1 (Ouyang *et al.*, 2002). This result was also demonstrated in BGC-823 cells in the presence of tachyplesin I where the morphology and ultrastructure of these cancer cells were altered (Li *et al.*, 2000). Moreover, tachyplesin I inhibited *in vitro* proliferation, decreased the expression of *c-myc*, *c-erbB-2* and *mtp53* oncogene expression, and increasing *p16* tumor suppressor gene expression. Taken together, the above findings suggest that tachyplesin I is capable of inducing tumor cell differentiation, and thereby reversing the malignant phenotype. Tachyplesin I might down-regulate cathepsins in lysosomes and up-regulate TOP2A to inhibit migration and promote apoptosis in glioma, thus contribute to its anti-tumor function. Our results suggest tachyplesin I is a potential candidate for treatment of glioma (Xuan *et al.*, 2017).

The Resources from Horseshoe Carapace

Chitin is a substance found in the shells, or exoskeleton, of horseshoe crabs. It is nontoxic and biodegradable and is processed to produce another substance called chitosan (aminopolysaccharide) that can be used to produce a variety of important products. Contact lenses, skin creams, and hair sprays can also be manufactured from chitin. Chitin can be used to remove lead and other harmful chemicals from wastewater. Chitosan also inhibits "bad" cholesterol uptake and boosts "good" cholesterol. Other chitosan uses include: promoting the healing of ulcers and lesions; serving in antibacterial action; acting as an antacid; helping to control high blood pressure; and treating and preventing irritable bowel syndrome (Pati *et al.*, 2018).

Conclusion

Besides their ecological and evolutionary significance the horseshoe crabs have enormous potential resources for biomedical, commercial usages of the human welfare. This has also regarded as a valuable biological indicator of marine and estuarine environment. But there is an urgent need to understand the biology of this organism properly to protect them from extinction and for sustainable utilisation also.

References

Adamia, S., Maxwell, C.A. and Pilarski, L.M. (2005). Hyaluronan and hyaluronan synthases: potential therapeutic targets in cancer. Current Drug Targets Cardiovascular and Haematological Disorders, 5(1): 3-14.

Chen, Y., Xu, X., Hong, S., Chen, J., Liu, N., Underhill, C.B., Creswell, K. and Zhang, L. (2001). RGD-Tachyplesin inhibits tumor growth. Cancer Research, 61(6): 2434-2438.

Chinnari, S., Pati, S., Dash, B.P. and Chatterji, A. (2015). A New Record on inducement of differentiation of dendritic cells by the peri-vitelline fluid of the fertilized eggs of Indian horseshoe crab (*Tachypleus gigas*, Müller). Biological -An International Journal, 7(2): 678-681.

Ghaskadbi, S., Patwardhan, V., Chakraborthy, M., Agrawal, S., Verma, M.K., Chatterjee, A., Lenka, N. and Parab, P.B. (2008). Enhancement of vertebrate cardiogenesis by a lectin from perivitelline fluid of horseshoe crab embryo. Cell and Molecular Life Sciences, 65 (20): 3312-3324.

Hoskin, D.W. and Ramamoorthy A. (2008). Studies on anticancer activities of antimicrobial peptides. Biochimica et Biophysica Acta, 1778(2): 357-375.

Huma, Alam., Chinnari, S., Pati, S., Dash, B.P. and Chatterji, A. (2015). A first record on the role of peri-vitelline fluid of the fertilized eggs of Indian horseshoe crab (*Tachypleus gigas*; Müller) promoting wound healing process *in vitro*. International Journal of Advanced Life Sciences, 8(4): 243-247.

Huma, Alam, Chinnari, S., Pati, S., Dash, B.P. and Chatterji, A. (2015). Peri-vitelline fluid obtained from the fertilized egg of *Tachypleus gigas* (Müller) triggers the differentiation of human bone marrow stem cell into cardiomyocyte *in vitro*. Cell and Developmental Biology, 4(3): 162.

Levin, J., Tomasulo, P.A., and Oser, R.S. (1970). Detection of endotoxin in human blood and demonstration of an inhibitor. Journal of Laboratory and Clinical Medicine, 75: 903.

Li, Q.F., Ou, Yang, G.L., Li, C.Y. and Hong, S.G. (2000). Effects of tachyplesin on the morphology and ultrastructure of human gastric carcinoma cell line BGC-823. World Journal of Gastroenterology, 6(5): 676-680.

Nakamura, T., Furunaka, H., Miyata, T., Tokunaga, F., Muta, T., Iwanaga, S., Niwa, M., Takao, T. and Shimonishi, Y. (1988). Tachyplesin: a class of antimicrobial peptide from the hemocytes of the horseshoe crab (*Tachypleus tridentatus*). Isolation and chemical structure. The Journal of Biological Chemistry, 263(32): 16709-16713.

Ouyang, G.L, Li, Q.F., Peng, X.X., Liu, Q.R. and Hong, S.G. (2002). Effects of tachyplesin on proliferation and differentiation of human hepatocellular carcinoma SMMC-7721 cells. World Journal of Gastroenterology, 8(6): 1053-1058.

Pati, S., Chatterji, A.and Dash, B.P. (2018). Chitosan from the carapace of Indian horseshoe crab (*Tachypleus gigas*, müller): Isolation and its characterization. Advances in Bioresources, 9(3): 120-132.

Rooney, P., Kumar, S., Ponting, J. and Wang, M. (1995). The role of hyaluronan in tumour neovascularization (review). International Journal of Cancer, 60(5): 632-636.

Shi, S.L., Wang, Y.Y., Liang, Y. and Li, Q.F (2006). Effects of tachyplesin and n-sodium butyrate on proliferation and gene expression of human gastric adenocarcinoma cell line BGC-823. World Journal of Gastroenterology, 12(11): 1694-1698.

Shigenaga, T., Takayenoki, Y., Kawasaki, S., Seki, N., Muta, T., Toh, Y., Ito, A. and Iwanaga, S. (1993). Separation of large and small granules from horseshoe crab (*Tachypleus tridentatus*) hemocytes and characterization of their components. Journal of Biochemistry, 114(3): 307-316

Xuan, Li., Jianguo, D., Yongjun, T., Lulu, L. and Gang, J. (2017). Quantitative proteomic profiling of tachyplesin I targets in U251 gliomaspheres. Marine Drugs, 15: 20-28.

Zasloff, M. (2002). Antimicrobial peptides of multi cellular organisms. Nature, 415 (6870): 389-395.

20

Biotechnology Intervention in Tropical Tasar Culture: The Need of the Hour

P.K. Kar

Regional Tasar Research Station, Central Silk Board, Baripada – 757 001, Odisha
E-mail: prasantakar@rediffmail.com

ABSTRACT

Tasar silk industry is rural based and provides employment to thousands of poor families in India. Tasar culture has certain associated constraints which are vulnerability of ecoraces for extinction, less amenability of wild ecoraces, population loss due to high temperature in summer, improper diapause induction and termination, asynchronous mating and prolonged grainage period, low mating propensity, susceptibility to diseases like microsporidiosis, virosis and bacteriosis, want of productive breeds with high yield and disease tolerance/resistance, domestication of tasar silkworm, want of dwarf variety, drought tolerant and high yielding host plants with low gestation period. During the last few decades works on the host plant augmentation, protection and silkworm rearing, grainage, conventional breeding, germplasm maintenance, post cocoon technology have been carried out at length. Quantitative and qualitative improvement in tasar is possible through genetic manipulation and simultaneously bringing improvement in post cocoon technology. Genetic manipulation can be made either through conventional breeding or through biotechnological means. Biotechnological tools are helpful in bringing out crop improvement, both in terms of quantity and quality. The progress on biotechnological application in tasar culture is at a slow pace. But many riddles could be solved in tasar culture and biotechnology has the answers. The present paper portrays the constraints associated with tasar culture and use of biotechnological means that can help to achieve horizontal and vertical growth in tasar silk production and productivity.

***Keywords**: Tasar silkworm, Antheraea mylitta, Cocoon.*

Introduction

Indian tasar silkworm, *Antheraea mylitta* Drury is exploited for *vanya* silk production in the country. This silkworm has tremendous commercial importance as it not only provides livelihood to thousands of poor tribal farmers but also earn sizeable foreign exchange. India is the only country in the world where all the five types of silkworm are found and tasar silkworm is unique among them. Tasar is represented by many ecoraces in India and majority of them are wild in nature but bestowed with superior quality traits. For example Raily and Modal ecoraces are known to have fecundity of 400-485 and one silkworm can spin cocoons with more than 1.5 km of silk filament. The exploitation of potentiality to the maximum level and the need for improved breeds/varieties require scientific research in order to enhance tasar silk production and productivity. Since, various components like soil, food plants, and silkworms are interlinked; research emphasis must be laid on each on the components in coherence and biotechnology holds the key.

Genetic variations in the natural population and diverse genes in the individuals of the populations of tasar silkworm are the prerequisites for evolving improved breeds/lines. With a resource of 44 known ecoraces, there is no dearth of genetic variability (Suryanarayana and Srivastava, 2005). The populations of wild tasar ecoraces are declining alarmingly in their natural habitat due to rapid deforestation and rampant cocoon collection. There is the need for scientific way of conservation of tasar silkworm and food plant diversity in different eco-geographical conditions so that inter as well as intra-population variability could be maintained for effective utilization and posterity.

Tasar silkworm and host plant genetic resources need to be characterized and evaluated for crop improvement, enhancement of yield traits and quality. Modern biotechnological tools hold the key for genetic characterization of the ecoraces for crop improvement.

Major Constraints Associated with Tasar Culture

Tasar culture has certain associated constraints which can be addressed scientifically. Rapid deforestation and habitat loss has imposed question on the survival of ecoraces and the populations. For sustainable utilization, conservation with genetics approach is a must so that there should not be allelic erosion. Wild ecoraces found mainly in Sal forests are least amenable to human handling. Being cold blooded, this insect has to face severe temperature during summer, resulting in the loss of populations. There is wide variation in tasar silkworm diapause induction and termination. During the reproductive period mating is asynchronous with prolonged emergence span which sometimes result in discontinuity of parental stocks in captive condition. There is low mating propensity under captivity. Silkworm is susceptible to diseases like microsporidiosis, virosis and bacteriosis. Moreover it is prone to pest and predator attack, as the rearing is mainly conducted outdoor. There is want of productive hybrids, disease tolerant or resistant breeds and lines. The host plant – tasar silkworm interaction is not adequately studied. Since, tasar cultivation is practiced in economic plantations, dwarf variety plants of Asan and Arjun with high leaf production are needed. There is need for large scale multiplication of true-to-type productive plants.

Some Molecular Studies in Tropical Tasar

Most of the molecular studies reported so far on tasar silkworm mainly focused on genetic diversity with the use of RAPD, ISSR or similar markers (Rao, 2000; Chatterjee *et al.*, 2004; Vijayan *et al.*, 2004, 2005; Mohandas *et al.*, 2004; Saha and Kundu, 2006; Mahendran *et al.*, 2005, 2006; Srivastava *et al.*, 2009). The studies were made sporadically with less population size, thus on a population genetic viewpoint are not adequate. Prasad *et al.* (2002) and Prasad and Nagraju (2003) have reported that the phylogeny and pattern of divergence of mariner like elements (MLEs) in silk moths were examined in relation to the phylogeny of the host species. Kar *et al.* (2005) studied the genetic diversity in the wild and semi-domesticated populations of Daba ecorace with Inter-Simple Sequence Repeat (ISSR) markers. They found allelic erosion in semi-domestic populations. The genetic variability present within wild Daba is of much importance for conservation and utilization. Ghosh *et al.* (2005) worked on molecular characterization of *Pao* like long terminal repeat retrotransposons in tasar silkworm. Mahendran *et al.* (2005) worked on molecular phylogeny of silkworm based on 16s ribosomal RNA and cytochrome-oxidase genes. Datta *et al.* (2000, 2001) have reported purification and characterization of fibroin from *A. mylitta* and further on differential expression of the fibroin gene in developmental stage of tasar silkworm. Qanungo *et al.* (2002) conducted molecular cloning and characterization of *A. mylitta* cytoplasm polyhedrosis virus genome segment-9. Datta *et al.* (2005) characterized A. *mylitta* CPV polyhedron gene and its variant forms.

Recently, there has been emphasis on use of silk as biomaterials. The team of workers from IIT, Kharagpur workers have established silk fibroin from Indian tropical tasar silkworm *A. mylitta* and shown that it has excellent properties as a biomaterial in terms of cell adhesion, migration, differentiation and proliferation of a number of different cell types (Talukdar *et al.*, 2011)]. Tasar silk is superior to mulberry silk in respect to mechanical properties like elasticity and tensile strength (Mandal and Kundu, 2008). These properties make it an ideal candidate material for the generation of contractile tissue like cardiac patches. Similarly, the other silk protein, sericin from tasar silkworm also holds the potential to be used as biomaterial.

Future Biotechnology Intervention

In order to improve the tasar silk production as well as sustainable utilization of genetic resources the intervention of biotechnology is required. Tasar silkworm genetic resources needs be characterized thoroughly at molecular level. Boundary between ecoraces/populations is not clearly understood and this has to be solved through molecular techniques. Studies of ecoraces with molecular markers will be more helpful to assess the conservation status and thus bringing a check in local extinction, migration and gene flow among populations, genetic drift and population bottleneck. Identification of quantitative trait loci (QTLs) associated with yield traits, disease resistance/tolerance *etc.* are required as done in recent agriculture research. Establishment of genetic map and linkage map of tasar silkworm is still a proposition. This will help in targeting the gene(s) for desirable crop improvement. Identification and characterization of genes, regulation of gene expression, role of receptors in switch on and switch off mechanism for the important physiological and metabolic activities in tasar silkworm is required. The area would include diapause induction and termination, silk moth emergence, mating, egg production and hatching, silk synthesis mechanism, quality, and quantity improvement of silk, mechanism of adaptation to different food plants and efficiency assessment, foraging behaviour/host plant interaction, immunological mechanism in tasar silkworm against different pathogens, genome-based approaches for the control of pest and parasite species, regulation in key metabolic pathways so as to reduce the level of feeding deterrents, control from pests and predators *etc.* To address the global warming issues with climate resilienceheat shock response needs to be studied. Evolution of transgenic races/breeds of tasar silkworm is needed with high productivity and tagged genes for defense against pathogens. Development of transgenic plants with disease resistance and high nutrition is also important so that the tasar silkworm could get adequate amino acid pool for silk synthesis.

Conclusion

From the above account it is also clear that the molecular aspect of Indian tasar silkworm is still at an infant stage and still a long way to go to harness the application of biotechnology in tasar culture. The molecular markers will be of immense help to assess the conservation status of tasar silkworm populations, their migration, gene flow, genetic shift, population bottleneck and the evolutionary consequence. Genome information provides powerful tools, where comparative genetics and genomics can work together to elucidate conserved evolutionary pathways and their diversification; identify new genes and gene systems as targets for transgenics, which can also be a reality in tasar silkworm. This will also help basic research leading towards new genome-based approaches for the control of pest and parasite species. The combination of conventional methods with modern biotechnological tools will certainly bring a change in the scenario of tasar culture. With emphasis on applications other than silk, more avenues could be created for revenue generation.

References

Chatterjee, S.N., Vijayan, K., Roy, G.C. and Nair, C.V. (2004). ISSR profiling of genetic variability in the ecotypes of *Antheraea mylitta* Drury, the tropical tasar silkworm. Russian Journal of Genetics, 40(2): 152-159.

Datta, A., Ghosh, A.K. and Kundu, S.C. (2001). Purification and characterization of fibroin from the tropical Saturniid silkworm, *Antheraea mylitta. Insect Biochem.* Molecular Bio*logy*, 31: 1013-1018 doi: 10.1016/S0965-1748(01)00049-2

Ghosh A.K., Datta A., Mahendran, B. and Kundu, S.C. (2005). Molecular characterization of *Pao* like long terminal repeat retrotranposon *Tamy* in Saturniid silkworm *Antheraea mylitta.* Current Science, 89(3): 539-543.

Kar, P.K., Vijayan, K., Mohandas, T.P., Nair, C.V., Saratchandra, B. and Thangavelu, K. (2005). Genetic variability and genetic structure of wild and semi-domesticated populations of tasar silkworm (*Antheraea mylitta*) ecorace Daba as revealed through ISSR markers. Genetica, 125: 173-183.

Mahendran, B., Padhi, B.K., Ghosh, A.K. and Kundu, S.C. (2005). Genetic variation in ecoraces of tropical tasar silkworm, *Antheraea mylitta* D. using RFLP technique. Current Science, 90(1): 100-102.

Mahendran, M., Acharya, C., Dash, R., Ghosh, A.K. and Kundu, S.C. (2006). Repetitive DNA in tropical tasar silkworm *Antheraea mylitta*. Genetics, 370: 51-57.

Mandal B.B. and Kundu S.C. (2008). Non-bioengineered silk gland fibroin protein: characterization and evaluation of matrices for potential tissue engineering applications. Biotechnology Bioengineering, 100: 1237-50.

Prasad M.D., Nurminsky, D.L., and Nagaraju, J. (2002). Characterisation and molecular phylogenetic analysis of mariner elements from wild and domesticated species of silk moths. Molecular Phylogentics. Evolution, 25: 210-217.

Qanungo K.R, Kundu, S.C., Mullins, J. and Ghosh, A.K. (2002). Molecular cloning and characterization of Antheraea *mylitta* cytoplasm polyhedrosis virus genome segment 9. Journal of Gentics Virology. 83: 1483-1491.

Rao, A.P. (2000). Some salient features of Andhra local ecorace of *Antheraea mylitta* Drury in relation to its conservation and multiplication. International Journal of Wild Silkmoth and Silk, 5: 356-358.

Saha, M. and Kundu, S.C. (2006). Molecular identification of tasar silkworm (*Antheraea mylitta*) ecoraces with RAPD, Biochemistry Genetics, 44: 72-85.

Srivastava, A.K., Kar, P.K., Sinha, R.B., Sinha, M.K. and Vijayaprakash, NB. (2009). Assessment of genetic diversity in different populations of Raily ecorace of Indian tasar silkworm, *Antheraea mylitta* Drury using ISSR primers. International Journal of Industrial Entomology, 4(3): 249-253.

Suryanarayana, N. and Srivastava, A.K. (2005). Monograph on tropical tasar silkworm. Central Tasar Research and Training Institute, Ranchi, India pp. 134.

Talukdar, S., Nguyen, Q.T., Chen, A.C., Sah, R.L. and Kundu, S.C. (2011). Effect of initial cell seeding density on 3D-engineered silk fibroin scaffolds for articular cartilage tissue engineering. Biomaterials, 32: 8927-37

Vijayan, K., Nair, C.V., Kar, P.K., Mohandas, T.P., Saratchandra, B. and Raje Urs, S. (2005). Genetic variability within and among three ecoraces of the tasar silkworm *Antheraea mylitta* Drury as revealed by ISSR and RAPD markers. International Journal of Industrial Entomology, 10(1): 51-59.

21

Study on the Anticancer Properties of Four Locally available Medicinal Plants in Human Glioma Cells (T98G)

S. Hansdah, S. Mohapatra and P.R. Debata*

Department of Zoology, North Orissa University,
Baripada – 757 003, Odisha
****E-mail: prdebata@gmail.com***

ABSTRACT

The anticancer properties of four locally available medicinal plants such as Catharanthus roseus, Nyctanthesar bortristis, Datura stromonium and Nerium oleander were studied in T98G glioma cells. The cells were treated with carrier and semi-purified extracts of doses 25µg/ml, 50µg/ml, 100µg/ml, 200µg/ml and 400µg/ml for 48 hrs. The data showed a dose dependent cytotoxic activity of the extracts against T98G glioma cells. Phytochemical analyses of these plant extracts revealed the presence of constituents like phenol, tannins, flavonoids, steroids, carbohydrates, proteins, terpenoids, glycosides, saponins and alkaloids. The results also showed that the plant extracts have antioxidant along with anticancer properties. Further studies are being carried out to know the mechanism of action of these plant extracts in relation to their anticancer property.

***Keywords**: Anticancer, Apoptosis, Medicinal plants, Phytochemicals, Glioma cells (T98G).*

Introduction

Glioma is a type of tumor that starts in brain or spine. It is called a Glioma because it arises from Glial cells. The most common site of the Gliomas is the brain. Glioma is the most common malignant tumors in the central nervous system of adults. The characteristic of these types of tumors are rapid proliferation, active invasiveness and rich angiogenesis. This is mainly due to highly regulated signaling pathways in the tumors (Nakada *et al.*, 2011). Recent advance in research techniques have given rise to wealth of new information regarding the distinct molecular subgroups of GBM with aberrant signaling pathways that drive the phenotype. According to the genomic analysis by The Cancer Genome Atlas (TCGA) network, GBM has four distinct molecular subtypes, proneural, neural, classical and mesenchymal (Van Meir *et al.*, 2010). There is at least one component of these pathways that occurs in many anaplastic astrocytoma and in the vast majority of GBMs (Sahreen *et al.*, 2015). Nearby all high grade tumors have impairments of this single critical cell cycle control pathways. It is likely as well that less profound defects in cell cycle regulation occurs in low grade gliomas, for invasive P53 gene mutation may affects both the G1-S and G2-M check point (Fan *et al.*, 2018).

The second event is over expression of growth factor and their receptors. A variety of growth factor such as epidermal growth factor (EPGF), platelet derived growth factor (PDGF), basic fibroblast growth factor (bFGF or FGF), transforming growth factor (TGF) – α and insulin like growth factor (IGF)-I are over expressed in GBM and thus provides a growth advantages to neoplastic cells. Generally glioma cells express both the EGFR and PDGF ligands and their receptor (Westermark *et al.*, 1995).The apoptosis of tumor cells is an important way through which anti-tumor drugs exert their therapeutic effects and the anti-tumor effect of the plant extracts and the induction of tumor cells apoptosis is currently a research hot spot. Apotosis is cell suicide phenomenon regulated by genes and significant in the occurrence and development of tumors and many chemotherapeutic drugs exerts their inhibitory effects on tumors by inducing the apoptosis of tumor cell (Xu *et al.*, 2014).

An optimistic approach against chemoprevention is by natural compounds, especially phytochemicals, minerals and vitamins to reduce the risk of cancer and is obtaining increasing attention because it is safe and cost effective alternative to cancer treatment (Khan *et al.*, 2015). The treatment includes chemotherapy, radiation therapy and surgery. Chemotherapy and radiation therapy not only kill the cancer cells but also damage the normal cells along with some side effects such as memory loss, headache and loss of hair. In spite of the current advance treatment it remains a challenge for oncologist. So they started treatment with adjuvant temozolamide therapy for development and testing of more effective therapeutic strategies is undertaken by several medical science communities (Cagle *et al.*, 2012). They have tested many dyes like both natural and synthetic compound known from antiquiety for their properties as colourants but also known for a variety of therapeutic properties which includes anti-inflammatory, antioxidant, antiseptic activity but also anti-proliferative, anti-metastatic, anti-angiogenic and anti-mutagenic (Desai *et al.*, 2017).

Cancer chemotherapeutic agents can often provide temporary relief from symptoms, prolongation of life and occasionally cures. A successful anticancer drug should kill or incapacitate cancer cells without causing excessive damage to normal cells. The life span of both normal and cancer cells is significantly affected by the rate of apoptosis. Synthesis or modification of known drugs continues as an important aspect of research. However, a vast amount of synthetic work has contributed relatively small improvements over the prototype drugs. There is a continued need for new prototypes – new templates to use in the design of potential chemotherapeutic agents. Significantly, natural products are providing such templates. Recent studies on tumor inhibitory compounds of plant origin have yielded an impressive array of novel structures. Besides, epidemiological studies suggest that consumption of diets containing fruits and vegetables, major sources of phytochemicals and micronutrients, may reduce the risk of developing cancer. Plants have a long history of use in the treatment of various disorders and provide an extensive reservoir of natural products, demonstrating outstanding structural diversity and offer a wide variety of novel and exciting chemical entities (Aung *et al.*, 2017). The importance of natural products in health care can be estimated by a report of World Health Organization that approximately 80 per cent of global population still relies on plant derived medicines for their primary health care. It is reported that 50 per cent of all drugs in clinical use and 74 per cent of the most important drugs are derived from natural sources (Lin *et al.*, 2012). At present, more than 60 per cent commercially available anticancer drugs are derived from natural sources including plants. Recent research has shown that more consumption of vegetables and fruits can prevent 20 per cent incident of cancer and about 200,000 cancer related deaths annually (Aguado *et al.*, 2007). Extracts of plants have long been used in the treatment of cancer and there are more than 3000 plant species that have reportedly been used in the treatment of cancer. There is a continued interest in the investigation of plant extracts for anticancer drug development because of four main reasons. First, plants often produce bioactive compounds that exceed the current capacity of synthetic chemistry. Second, natural anticancer compounds fit into mechanism based approach as perfectly as a hand fits into a glove. There is solid evidence that phytochemicals inhibit cancer by interfering with multiple mechanisms which are central to cancer progression. Third, there exists about 2.5 to 5.0 million plants species on earth and only about 10 per cent have been thoroughly studied for their potential values as source of drugs. Thus identification of novel structures and understanding their molecular mechanisms will contribute to the faster and more specific strategies for the development of successful therapies. Fourth, natural compounds are already being used in cancer treatment and their popularity in cancer treatment appears to be growing rather than declining (Fan *et al.*, 2018). In this study we have used the extracts from locally available medicinal plants to test anticancer properties in human glioma cells (T98G).

Materials and Methods

Collection of Plant Materials

Fresh parts of four medicinal plants, *Nyctanthes arbor-tristis* (Leaves, fruit, stem), *Datura stramonium* (Leaves, fruit, stem), *Nerium oleander* (Leaves, Flower, Stem), *Catharanthus roseus* (Leaves, Flower, Stem) were collected from in and around areas of North Orissa University campus. The plant materials were washed with water and shed dried until all the water molecules evaporated and plants became well dried for grinding. After drying, the plant materials were grinded by using mechanical blender into fine powder and transferred into airtight containers with proper labelling for future use.

Preparation of Plant Extraction

The plant materials were extracted through Soxhlet and Hot water extraction method using standard protocol.

Cell Lines and Culture

Human Glioma cells (T98G) were obtained as a gift from Institute of Life Sciences, Bhubaneswar, and cultured in Dulbecco's Modified Eagles Medium (DMEM) containing 10 per cent fetal bovine serum (FBS) and 1 per cent PS (Penicillin Streptomycin) in a humidified Incubator with 5 per cent CO2 and in atmospheric oxygen.

Plating of T98G Cells in 96 Well Plate

Cells were harvested from a culture of T98G cell line. The old media was aspirated and cells were washed with Phospahte buffered Saline (PBS, pH 7.2). Then 1ml of 1 × Trypsin-EDTA was added to the cell followed by gentle tapping at the bottom of the plate to detach the adherent cells. DMEM with 10 per cent FBS and 1 per cent PS was added to the detached cells to neutralize Trypsin. The cell suspension was then centrifuged and supernatant was discarded. Cell pellet was re-suspended in complete growth media to make uniform suspension. Cells were counted in haemocytometer and were plated in 96 well plate with cell density of 2000 cells per well. Cells were then allowed to adhere for overnight in CO_2 incubator.

Extract Treatment

One milligram of each plant extract in powder form of *Nyctanthes arbor-tristis, Datura stramonium, Nerium oleander* and *Catharanthus roseus* were taken and diluted in DMSO for stock. Then these prepared stocks were diluted in drug dilution media (with 1 per cent PS and 1 per cent Insulin-Transferrin Selenium (ITS) to achieve desired concentrations of (800, 400, 200, 100, 50, 25) µg/ml for treatment. 4 wells were treated for every drug and their different concentrations alongside a control with only DMSO treatment. Images of the cells were taken after 48 hrs and 72hrs of drug treatment.

Results

Collection and Identification of Plant Materials

The four medicinal plants such as *Datura stramonium, Nerium oleander, Nyctanthes arbor-tristis and Catharanthus roseus* were identified (Figure 21.1).

Extract Preparation, Column Chromatographic Separation, Thin Layer Chromatography and Phytochemical Analyses of Four Medicinal Plants

Chromatographic separation of all the four plants leave extracts were performed by using chloroform and hexane. A total of six fractions were collected and tested

Nyctanthes arbor-tristis **Datura stramonium**

Nerium oleander **Catharanthus roseus**

Figure 21.1: Four Medicinal Plants Used in the Study.

in the Thin layer Chromatography (TLC) for presence of phytocompound. The positive spot were tested for anticancer properties. For Phytocompound analysis hot water extract were used for the presence or absence of different biochemical and phytocompound such as carbohydrates, terpenoids, steroid, saponin, protein, flavonoid, glycosides, Alkaloids, Phenol and tannins (Figure 21.2 and Table 21.1). The results show that Protein, Steroid, Saponins and Alkaloid were present in all the plants. Carbohydrates are absent in the leaves of *D. stramonium,* Terpenoids were absent in *D. stramonium,* and *N. oleander.* Flavonoids were absent only in *N. arbotristis.*

Analysis of Anticancer Properties of Chloroform and Hexane Extracts of *N. arbor-tristis*

Both Chloroform and Hexane extract of *N. arbor-tristis* was used for growth inhibitory activities of glioma cells. T98 glioma cells (2000/well) were plated in 96 well plates for overnight, the dried extract were diluted in DMSO and different concentration of extract were prepared using DMEM and 1 per cent ITS. The cells were incubated with either carrier or 25µg/ml, 50µg/ml, 100µg/ml, 200µg/ml and

Table 21.1: Qualitative Analysis (+; Presence; - Absence) of the Phytochemicals of the Medicinal Plants

Plants	*Carbo-hydrate*	*Terpinoid*	*Steroid*	*Saponin*	*Protein and Amino Acid*	*Flavonoid*	*Glycoside*	*Alkaloid*	*Phenol and Tannin*
Nyctanthes Arbotritis	+	+	+	+	+	–	+	+	+
Datura stramonium	–	–	+	+	+	+	+	+	–
Nerium oleander	+	–	+	+	+	+	+	+	+
Cantharanthus roseus	+	+	+	+	+	+	+	+	+

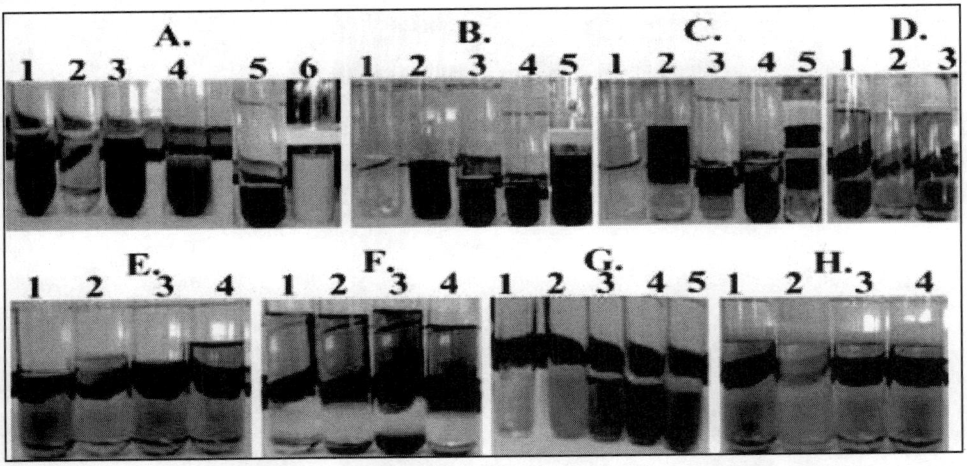

Figure 21.2: Qualitative Analysis of Four Plant Extracts for Presence of Carbohydrates, Terpenoids, Steroid, Saponin, Flavonoid, Glycosides, Alkaloids, Phenol and Tannins. A. Test for Carbohydrate, Tube 1 Positive control (Glucose), Tube 2 Negative control, Tube 3*N.arbor-tristis*, Tube 4 *D. stramonium*, Tube 5 *N. oleander*, Tube 6 *C. roseus*. B. Test for Terpenoid, Tube 1 Positive control, Tube 2 *N. arbor-tristis*, Tube 3 *D. stramonium*, Tube 4*N.oleander*, Tube 5 *C. roseus*. C. Test for Steroid, Tube 1 Positive control, Tube 2 *N. arbor-tristis*, Tube 3 *D. stramonium*, Tube 4 *N. oleander*, Tube 5 *C. roseus*.D. Test for Glycoside, Tube 1 *N. arbor-tristis*, Tube 2 *D. stramonium*, Tube 3 *N. oleander*. E. Test for Flavonoid, Tube 1 *N. arbor-tristis*, Tube 2 *D. stramonium*, Tube 3 *N. oleander*, Tube 4 *C. roseu*. F. Test for Phenol and tannin, Tube 1 *N. arbor-tristis*, Tube 2 *D. stramonium*, Tube 3 *N. oleander*, Tube 4 *C. roseus*. G. Test for Alkaloid, Tube 1 Positive control, Tube 2 *N. arbor-tristis*, Tube 3 *D. stramonium*, Tube 4 *N. oleander*, Tube 5 *C. roseus*. H. Test for Saponin, Tube 1 *N. arbor-tristis*,Tube 2 *D. stramonium*, Tube 3 *N. oleander*, Tube 4 *C. roseus*.

400μg/ml for 48 hrs. Bright field images were taken after 48 hrs. The Figure 21.3A represents the photo imaging of chloroform extract of *N. arbor-tristis* and Figure 21.3B represents the photo imaging of T98G cells treated with hexane extract of *N. arbor-tristis*.

Analysis of Anticancer Properties of Chloroform and Hexane Extracts of *D. stramonium*

Chloroform extract of *D. stramonium* was used for growth inhibitory activities of T98G cells. The T98G cells (2000/well) were plated in 96 well plates for overnight, the dried extract were diluted in DMSO and different concentration of extract were prepared using DMEM and 1 per cent ITS. The cells were incubated with Carrier or 25μg/ml, 50μg/ml, 100μg/ml, 200μg/ml and 400μg/ml for 48 hrs. Bright field images were taken after 48 hrs. The Figure 21.4 represent the photo imaging of T98G cells treated with Chloroform extracts of *D. stramonium*.

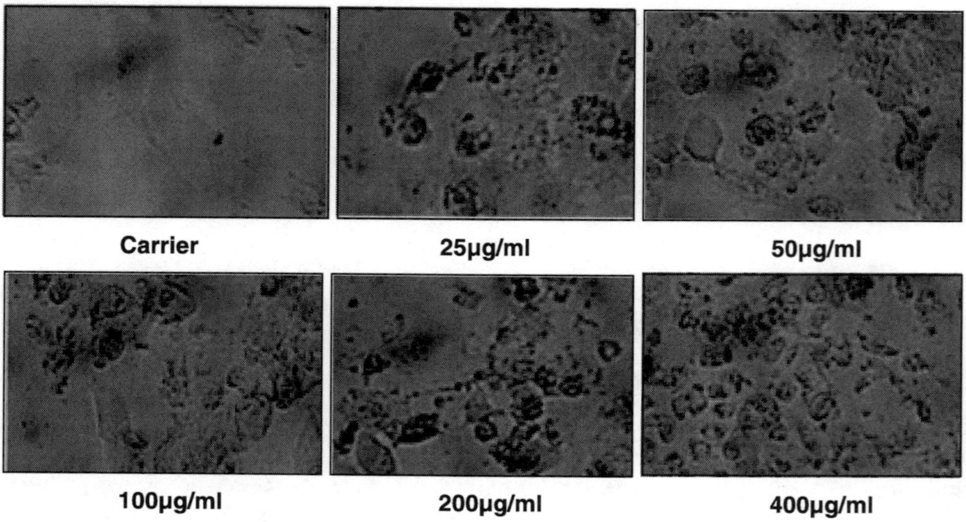

Figure 21.3A: Chloroform Extract of *N. arbotristis* Treated T98 Human Glioma Cells. The cells were treated with carrier, 25μg/ml, 50μg/ml, 100μg/ml, 200μg/ml and 400μg/ml. The Chloroform extract causes significant cell death of T98G cells.

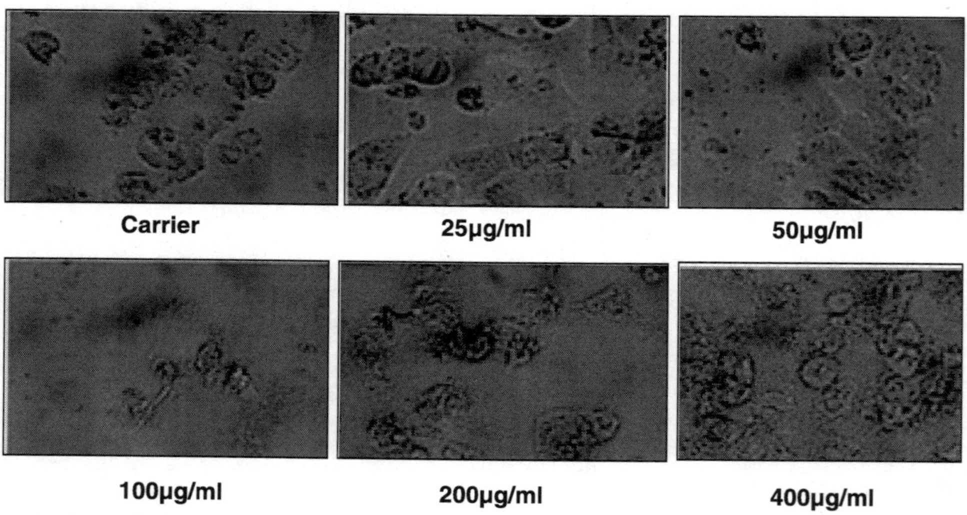

Figure 21.3B: Hexane Extract of *N. arbor-tristis* Treated T98 Human Glioma Cells. The cells were treated with carrier, 25μg/ml, 50μg/ml, 100μg/ml, 200μg/ml and 400μg/ml and images were taken. The Hexane extract causes significant cell death in T98G cells.

Analysis of Anticancer Properties of Chloroform and Hexane Extracts of *N. oleander*

Both Chloroform and Hexane extract of *N. oleander* was used for growth inhibitory activities of glioma cells. T98 glioma cells (2000/well) were plated in 96

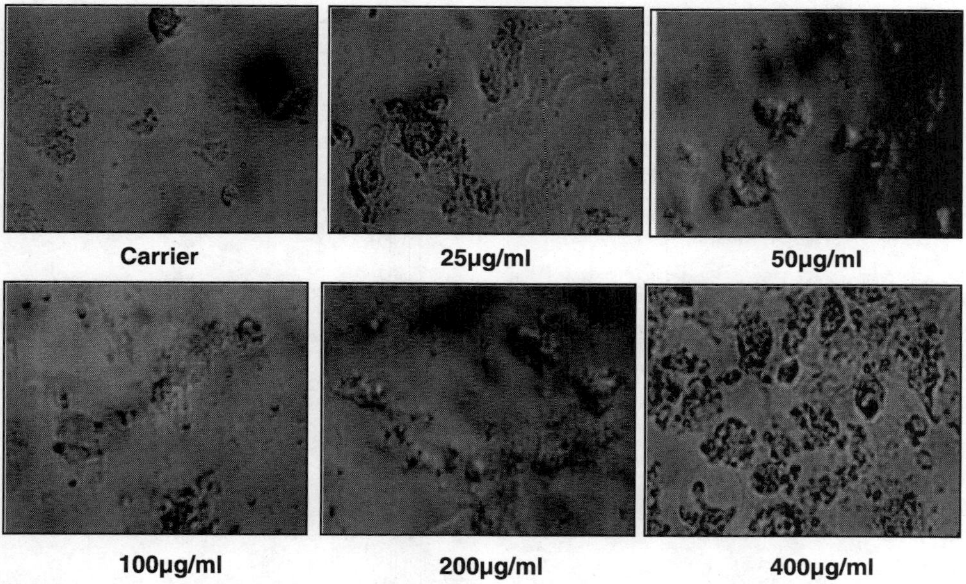

Figure 21.4: Chloroform Extract of *D. stramonium* Treated T98 Human Glioma Cells. The cells were treated with carrier, 25µg/ml, 50µg/ml, 100µg/ml, 200µg/ml and 400µg/ml and images were taken after 48h. The Chloroform extract causes significant apoptosis to T98G cells.

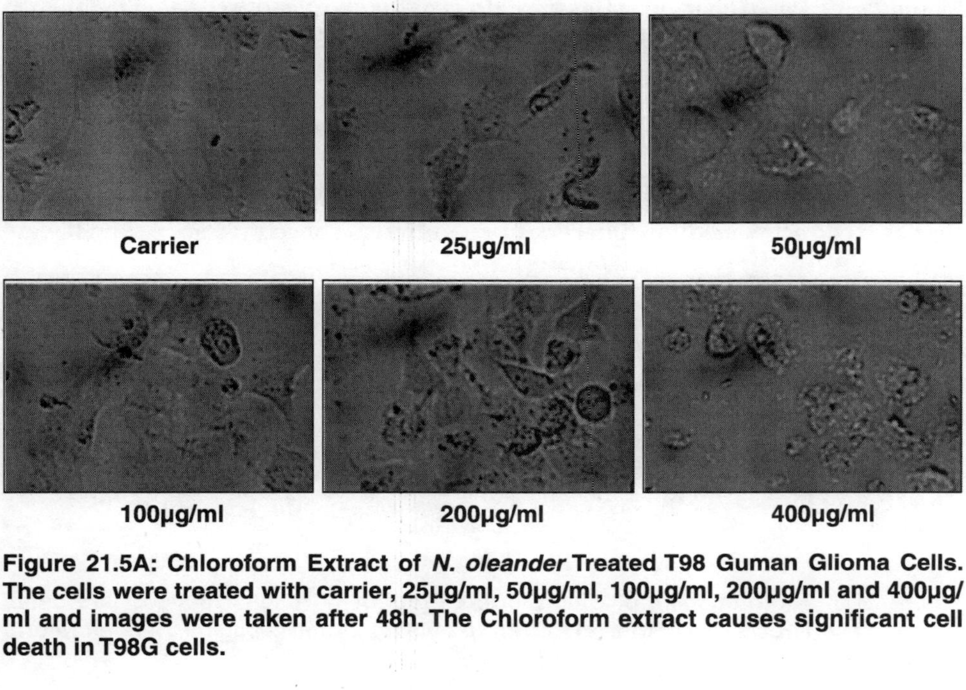

Figure 21.5A: Chloroform Extract of *N. oleander* Treated T98 Guman Glioma Cells. The cells were treated with carrier, 25µg/ml, 50µg/ml, 100µg/ml, 200µg/ml and 400µg/ml and images were taken after 48h. The Chloroform extract causes significant cell death in T98G cells.

well plates for overnight, the dried extract were diluted in DMSO and different concentration of extract were prepared using DMEM and 1 per cent ITS. The cells were incubated with Carrier or 25µg/ml, 50µg/ml, 100µg/ml, 200µg/ml and 400µg/ml for 48 hrs. Bright field images were taken after 48 hrs. The Figure 21.5A represents the photo imaging of chloroform extract of *N.oleander*and Figure 21.5B represents the photo imaging of T98G cells treated with hexane extractof*N. oleander*.

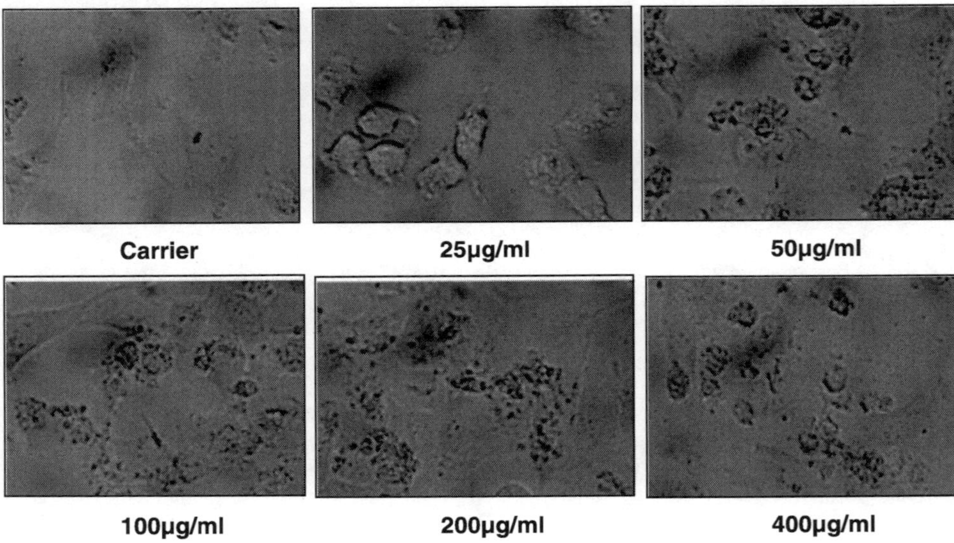

Figure 21.5B: Photo Imaging of Hexane Extract of *Nerium oleander*Treated T98 Human Glioma Cells. The cells were treated with carrier, 25µg/ml, 50µg/ml, 100µg/ml, 200µg/ml and 400µg/ml. The Hexane extract causes significant apoptosis to T98G cells.

Analysis of Anticancer Properties of Chloroform Extracts of *C. roseus*

Chloroform extract of *C. roseus* was used for growth inhibitory activities of T98G cells. The T98G cells (2000/well) were plated in 96 well plates for overnight, the dried extract were diluted in DMSO and different concentration of extract were

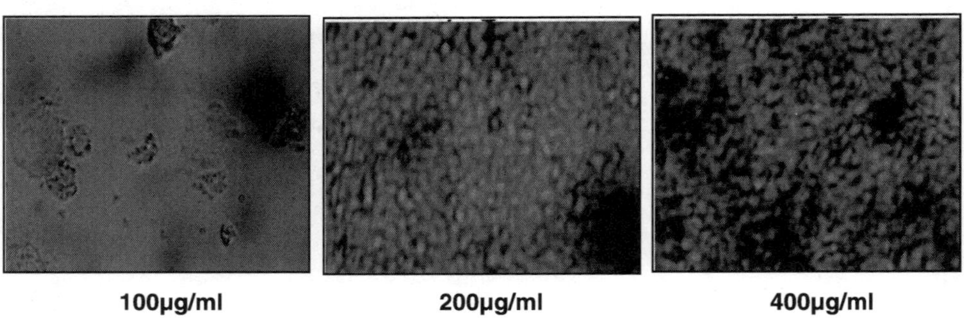

Figure 21.6: Photo Imaging of Chloroform Extract of *C. roseus* Treated T98 Human Glioma Cells. The cells were treated with carrier, 200µg/ml and 400µg/ml. The Chloroform extract causes significant apoptosis to T98G cells.

prepared using DMEM and 1 per cent ITS. The cells were incubated with carrier or 25μg/ml, 50μg/ml, 100μg/ml, 200μg/ml and 400μg/ml for 48 hrs. Bright field images were taken after 48 hrs. The Figure 21.6 represent the photo imaging of T98G cells treated with Chloroform extracts of *C. roseus.*

Discussion

In the present study, four locally available medicinal plants were tested for anticancer activites inT98 glioma cells. Both leave and stem extracts were prepared by Soxhlet method, partially purified by column chromatography and anticancer activities were tested for all four plants. A dose dependent cytotoxic activity was observed in leaves, stem mixture extracts of all the four plants. All four plants showed significant cytotoxic activity even act lower dilution concentration of extracts. The phytochemical analysis for all the four plants extract showed a variable degree of different phytocompound and biochemical tested.

Phytochemical analysis conducted on the plant extracts reveals the presence of constituent which are known to exhibit medicinal as well as the physiological activities. Analysis of the plant extracts revealed the presence of phytochemicals such as phenol and tannins, flavonoids, steroids, carbohydrates, protein, terpenoids, glycosides, saponins and alkaloids(Ramachandran *et al.,* 2014). The phenolic compounds are one of the largest and most ubiquitous groups of plant metabolites. Several studies have described the antioxidant properties of medicinal plants which are rich in phenolic compounds. Natural antioxidant mainly come from plants in the form of phenolic compounds such as flavonoid, phenolic acids, tocopherols *etc.,* Tannins bind to proline rich protein and interfere with protein synthesis(Parolaro *et al.,* 2008). Flavonoids are hydroxylated phenolic substances known to be synthesized by plants in response to microbial infection and they have been found to be antimicrobial substances against wide array of microorganisms *in vitro* . Their activity is probably due to their ability to complex with extracellular and soluble proteins and to complex with bacterial cell wall. They also are effective antioxidant and show strong anticancer activities (Vidak *et al.,* 2015).

The plant extracts were also revealed to contain saponins which are known to produce inhibitory effect on inflammation. Saponins has the property of precipitating and coagulating red blood cells. Some of the characteristics of saponins include formation of foams in aqueous solutions, hemolytic activity, cholesterol binding properties and bitterness(Xu *et al.,* 2016). Steroids have been reported to have antibacterial properties and they are very important compounds especially due to their relationship with compounds such as sex hormones (Dietrich *et al.,* 2011). Alkaloids have been associated with medicinal uses for centuries and one of their common biological properties is their cytotoxicity. Several workers have reported the analgesic, antispasmodic and antibacterial properties of alkaloids. Glycosides are known to lower the blood pressure according to many reports. The results obtained in this study thus suggest the identified phytochemical compounds may be the bioactive constituents and these plants are proving to be an increasingly valuable reservoir of bioactive compounds of substantial medicinal merit.

The phenolic compounds are one of the largest and most ubiquitous groups of the plant metabolites. They possess biological properties such as anti-apoptosis, anti-aging, anti-carcinogen, anti-inflamation, anti-atherosclerosis, cardiovascular protection and improvement of endothelial function as well as inhibition of angiogenesis and cell proliferation activities (Martos *et al.*, 2014). Several studies have described the antioxidant properties of medicinal plants which are rich in phenolic compounds. There are many potential chemopreventive agents for treatment against various types of cancer particularly as adjuvant therapy and in the prevention of recurrence. Natural bioactives mechanisms can be regulating the cell survival and proliferation which are important to development of new agents that can be utilized in low toxicity without the severe side effects accompanying chemotherapy (Abdul Rahman *et al.*, 2014).

Conclusion

The results revealed the presence of anticancer phytocompounds in the plants studied. Many evidences gathered in earlier studies which confirmed the identified phytochemicals to be bioactive. Several studies confirmed the presence of these phytochemicals contribute medicinal as well as physiological properties to the plants studied in the treatment of different ailments. Therefore, extracts from these plants could be seen as a good source for useful drugs. The traditional medicine practice is recommended strongly for these plants as well as it is suggested that further work should be carried out to isolate, purify, and characterize the active constituents responsible for the activity of these plants. Also additional work is encouraged to explain the possible mechanism of action of these extracts.

Acknowledgement

SH is supported by National Fellowship for Higher Education of ST Students and PRD is supported by Ramalingaswami Research fellowship, DBT, Govt. of India

References

Aguado, T., Carracedo, A., Julien, B., Velasco, G., Milman, G., Mechoulam, R., Alvarez, L., Guzmán, M. and Roperh, I.G. (2007). Cannabinoids induce glioma stem-like cell differentiation and inhibit gliomagenesis. Journal of Biological Chemistry (Papers in Press), 282: 6854-6862.

Aung, T.N., Qu, Z., Kortschak, R.D. and Adelson, D.L. (2017). Understanding the effectiveness of natural compound mixtures in cancer through their molecular mode of action. International Journal of Molecular Science, 18(3): 656.

Cagle, P., Idassi, O., Carpenter, J., Minor, R., Goktepe, I. and Martin, P. (2012). Effect of Rosehip (*Rosa canina*) extracts on human brain tumor cell proliferation and apoptosis. Journal of Cancer Therapy, 3: 534-545

Desai, V. and Bhushan, A. (2017). Natural bioactive compounds: Alternative approach to the treatment of glioblastoma multiforme. BioMed Research International, 9363040: 1-10.

Dietrich, J., Rao, K., Pastorino, S. and Kesari, S. (2011). Corticosteroids in brain cancer patients: benefits and pitfalls. Biochemistry, 4(2): 233–242.

Fan, H.C., Chi, C.S., Chang, Y.K., Tung, M.C., Lin, S.Z. and Harn H.J. (2018). The molecular mechanisms of plant-derived compounds targeting brain cancer. International Journal of Molecular Science, 19(2): 395.

Khan, M., Maryam, A., Qazi, J.I. and Ma, T. (2015). Targeting apoptosis and multiple signaling pathways with icariside II in cancer cells. International Journal of Biological Science, 11(9): 1100-1112.

Lin, P.C., Liu, P.Y., Lin, S.Z. and Harn. H.J. (2012). *Angelica sinensis*: A Chinese herb for brain cancer therapy. BioMedicine, 2(1): 30-35.

Martos, J.M.M., Mayas, M.D., Carrera, P., Saavedra, J.M.A.de., Agesta, R.S., Arrazola, M. and Expósito, M.J.R. (2014). Phenolic compounds oleuropein and hydroxytyrosol exert differential effects on glioma development via antioxidant defense systems. Journal of Functional Foods II, 221-234.

Nakada, M., Kita, D., Watanabe, T., Hayashi, Y., Teng, L., PykoI, V. and Hamada, J. (2011). Aberrant signaling pathways in Glioma. Cancers, 3: 3242-3278.

Parolaro, D. and Massi, P. (2008). Cannabinoids as potential new therapy for the treatment of Gliomas. Neurotherapeutics, 8: 1473-7175.

Rahman, A.A, Jamal A.R.A., Harun, R., Mokhtar, N.M. and Ngah W.Z.W. (2014). Gamma-tocotrienol and hydroxy-chavicol synergistically inhibits growth and induces apoptosis of human glioma cells. BMC Complementary and Alternative Medicine, 14(213): 1-11.

Ramachandran, H. D (2014). Role of phytochemical combating obesity. International Journal of Innovative Pharmaceutical Sciences and Research, 2: 1606-1624.

Sahreen, S., Khan, M.R., Khan, R.A. and Hadda, T.B. (2015). Evaluation of phytochemical content, antimicrobial, cytotoxic and antitumor activities of extract from *Rumex hastatus* D Don roots. BMC Complement Alternative Medicine, 15: 211.

Van Meir, E.G., Hadjipanayis, C.G., Norden, A.D., Shu, H.K., Wen, P.Y. and Olson, J.J. (2010). Exciting new advances in neuro-oncology: The avenue to a cure for malignant glioma. CA Cancer Jornal of Clinical Cancer, 60: 166-193.

Vidak, M., Rozman, D. and Komel R. (2015). Effects of flavonoids from food and dietary supplements on glial and glioblastoma multiforme cells. Cancer Cell Therapy, 20: 19406-19432.

Westermark, B., Heldin, C.H. andNistér, M. (1995). Platelet-derived growth factor in human glioma. Review, Glia, 3: 257-63.

Xu, C., Sun, G., Yuan, G., Wang, R. and Sun, X. (2014). Effects of platycodin D on proliferation, apoptosis and PI3K/Akt signal pathway of human glioma U251 cells. Molecules. 19: 21411-23.

Xu, X.H., Li T., Fong, C.M.V., Chen, X., Chen, X.J., Wang, Y.T., Huang, M.Q. and Lu, J.J (2016). Saponins from Chinese medicines as anticancer Agents. Biochemistry, 21: 1326.

22

Anti-Proliferative and Cytotoxic Effect of Chloroform Extract of *Shorea robusta* Resin on Cancer Cell Lines

G.Pradhan, N.S. Pal and R.N.Munda*

Department of Biotechnology, North Orissa University, Baripada – 757 003, Odisha
****E-mail: rudramunda@gmail.com***

ABSTRACT

Cancer is one of the major health problems that have widely affected the world's population. Current strategies for anticancer drug discovery involve less toxic, more selective, and highly effective agents to treat cancer has become more urgent especially from natural compounds. Shorea robusta (Sal) is one of the most important traditional Indian medicinal plants. The resin of the plant has been used in the treatment of different diseases. Present study investigates anti-proliferative and cytotoxic effect of chloroform extract ofShorea robustaplant resin on murine cell line. The air dried resin powder of was exhaustively extracted by cold percolation method in chloroform and column fractionate the chemical constituents into individual fractions. The phyto-chemical analysis of the resin shows major group of phytochemicals viz. alkaloids, glycosides, steroids, phenols, terpinoids, flavonoids etc., differing in concentration. The presence of compounds was confirmed using analytical TLC and preparative TLC with the aid of different solvent system. Cytotoxic effects of chloroform extract of Shorea robustaplant resin were investigated via monitoring of morphological changes, MTT assay and DAPI Test. Apoptotic or necrotic cell death by AO/EtBr double staining was investigated. In morphological analysis, apoptotic alterations related to apoptosis such as cytoplasmic blabbing, chromatin condensation, DNA fragmentation and irregularity in shape were detected in TC1 cell lines and significantly different from the PBMC normal cell

Keywords: *Anti-proliferation, Apoptosis, Cytotoxicity, Shorea robusta, Phytochemicals.*

Introduction

Medicinal plants are great importance the health of individual and communities in general. Ayurveda is the most ancient health care system and is practiced widely in India and other countries Charka (1000BC) and Sushruta (800BC) the two eminent physicians in India mentioned in their Somhitas about 700 plant species as therapeutic agents of which about 500 are being mentioned in Indian flora. Rigveda (4500-1600BC) and the Atharvaveda (2000-1500BC) have also mentioned some medicinal plants. Some important medicinal plants or phytoconstituents with anticancer activities are *Allium cepa, Aegle marmelos* (Cunningham, 1998).

Some important vegetable or phytoconstituents with anticancer activities are beet root, bitter gourd, brocoli, bottle gourd, cabbage, carrots, cauliflower, garlic, drumstick Ginger, neem, lady's finger and olive. Medicinal plants contents numerous bioactive phytochemicals or bionutrients. Natural products, such as plant extract either as pure compound or as standardized extract, provide unlimited opportunities for new drugs discoveries because of the unmatched ability of chemical diversity (Madhuri and Pandey, 2009). Plants use for traditional medicine content a wide range of substance that can be treat use chronic as well as infectious diseases. Many beneficial biological activity such as anticancer, antimicrobial, antioxidant, antidiarrheal, analgesic and wound healing activities were reported. However, clinical trials are necessary to demonstrate the effectiveness of bioactive compound to verify this traditional claim (Gupta, 1998). Clinical trials directed towards understanding the pharmacokinetics, bioavailability, efficacy, safety and drugs interaction of newly developed the bioactive compounds and their formulation require a carefully evaluation. The analysis of bioactive compounds present plant extracts involving the application of common phytochemical screening assay, chromatographic technique such as immunoassay and Fourier Transform Infra Red (FTIR)(Chopra and Doiphode, 2002).

Cancer is the second common cause of death in the developed countries next to cardiovascular diseases. According to WHO out of an estimated total of 50 million deaths annually in the world, more than 5 million are attributed to cancer and the number of deaths due to cancer throughout the world is increasing. The causes of cancer vary worldwide. Cancer case in Indian men is about half to one third of the incidence recorded in USA and Europe. A global comparison shows that India has high rate of cancer case in oral cavity, pharynx, and cervix (Singh and Khan 2000).

Cancer or neoplasm is thought to be developed from a cell in which the normal mechanisms for control of growth and proliferation are altered. Current evidences support the concept of carcinogenesis as a multi step process that is genetically regulated (Soliman *et al.*, 2006). The steps are as 1. Initiation 2.Promotion 3.Conversion or transformation and 4.Progression. Cancer is a disease caused by normal cells changing so that they grow in an uncontrolled way. The uncontrolled development causes to form a lump called tumour. If not treated, the tumour can cause problems either by scattering into normal tissues nearby or causing pressure on other body structures or scattering to other parts of the body through the lymphatic system or bloodstream. Formations of cancers depend mainly on cancer causing substances carcinogens), environment, viruses, *etc.* Host's genetic makeup, immune system, bodyweight, diet, physical activity, addiction in tobacco and bad food habit also play an important role in cancer development (Amy and Chow, 2010).

Shorea robusta or Sal tree is a large, deciduous tree found in India, Myanmar and Nepal. It is a medicinal tree and used in Ayurveda for thousands of years to treat variety of diseases including piles, leucorrhoea, gonorrhea, skin disorders, ulcers, wounds, diarrhea, dysentery, burning sensation, seminal weakness *etc.* The resin obtained from the tree is known as Sal damar, Bengal Dammar, Laldhuna Ral, Rhal or Dhoom. It has carminative, stomachic and astringent properties (Ames 1989). It gives relief in skin disorders, pain in nerves, burn injury, fever, enlargement of spleen, abnormal discharge from body (menorrhagia, leucorrhoea), diarrhea, and bleeding piles. It is also given in gonorrhea and for weak digestion. In Siddha, it is suggested for ulcers, wounds and menopausal disorders (Cullet, 2002).

Materials and Methods

Collection and Extraction of the Plant Samples

The plant samples are being collected from the vicinity of the North Orissa University and the surrounding areas of the town Baripada, and approximate amount of around 1kg of resin samples of *Shorea robusta* was collected and air dried in dark for around 2 weeks. The air dried samples are then grinded with the help of mechanical grinder and the resultant powder was then packed in the zip lock polythene bags to keep them moisture free and stored at cold temperature for further use (Lambert *et al.*, 1997).

Extraction of Crude Plant Extract

The air dried resin powder was exhaustively extracted by cold percolation method with different solvents of increasing order of polarity, starting with a highly nonpolar solvents *viz.*, Petroleum Ether followed by Chloroform, Ethyl Acetate, Methanol, Ethanol Chloroform and Water (Mc Chesney and James, 2001)). From that the Chloroform extract was taken for analysis.

Separation of Compound by Column Fractionation

Silica gel 100-200 mesh was made into a homogenous suspension by shaking with petroleum ether (first eluent). The bottom of the column was plugged with little cotton to prevent the adsorbent pass out, and then the silica gel suspension was poured into the column, set aside for 10 minutes and used (Houghton *et al.*, 2007).

Thin Layer Chromatography (TLC) Profile of the Crude Extracts

Thin Layer Chromatography is a very effective technique for the separation of chemical constituents of an extract and for their identification. A major breakthrough in this field was the commercial availability of convenient precoated plates in the early 70's Pharmacopoeias are increasingly employing this technique for assessing the quality and purity of compounds of both synthetic and natural origin. The movement of the active compound was expressed by its retention factor (Rf), values were calculated for different samples (Vashisht *et al.*, 2017).

$$R_f = \text{Distance Travelled by solute/}\\ \text{Distance Travelled by Solvent Front of TLC}$$

If the chemicals which are separated are colourless, several methods are available to visualize the spots. Iodine vapours are a common unspecific colour reagent. The TLC plate was dipped into the specific colour reagents or which were sprinkled on top of the plate, Potassium permanganate – oxidation, Iodine.

Phytochemical Analysis

The extracts were subjected to qualitative tests for detection of phytoconstituents present in it *viz.* alkaloids, carbohydrates, glycosides, phytosterols, fixed oils and fats, phenolic compounds and tannins, proteins and free amino acids, gums and mucilages, flavanoids, lignins and saponins.

Culture of PBMC and Cancer Cell Lines

The TC1 and PBMC cells were cultured in DMEM and supplemented with 10 per cent FBS and 1 per cent antibiotic (PSN) incubated at 37°C in a humidified atmosphere with 5 per cent CO_2. After 75-80 per cent confluence, cell were harvested with 0.025 per cent trypsin and 0.52mM EDTA in phosphate buffered saline (PBS) and were seeded at density to allow them to re-equilibrate a day before the start of experimentation. All experiment was conducted in DMEM supplemented with 10 per cent FBS and 1 per cent antibiotic and Antimycotic (ABAM) solution.

Assessment of Cell Morphology

The crude plant extract were dissolved in DMSO and mixed thoroughly by using a vortex rotator and then centrifuged at 1000 rpm for 1 min in a table centrifuge. The prepared drugs then were added to TC1 and PBMC cell cultured in the 96 well plates in duplicates for 24 h and 48 h. After the incubation the grown in 96-well tissue culture plates were observed for morphological changes were observed with an inverted phase contrast microscope. (Model: Eclipse Ti-U, Nikon, Japan).

MTT Assay

Measurement of cell viability and proliferation forms the basis for numerous *in vitro* assays of a cell population's response to external factors. The reduction of tetrazolium salts is now widely accepted as a reliable way to examine cell proliferation. The yellow tetrazolium MTT (3-(4, 5-dimethylthiazolyl-2)-2, 5-diphenyltetrazolium bromide) is reduced by metabolically active cells, in part by the action of dehydrogenase enzymes, to generate reducing equivalents such as NADH and NADPH. The resulting intracellular purple formazan can be solubilised and quantified by spectrophotometric means. The MTT Cell Proliferation Assay measures the cell proliferation rate and conversely, when metabolic events lead to apoptosis or necrosis, the reduction in cell viability. The number of assay steps has been minimized as much as possible to expedite sample processing (Patel *et al.*, 2009).

Results and Discussion

Phytomedicine is an ancient from health care known to human being. Phytoprinciples in medicinal plants such as steroids, tannins, alkaloids and

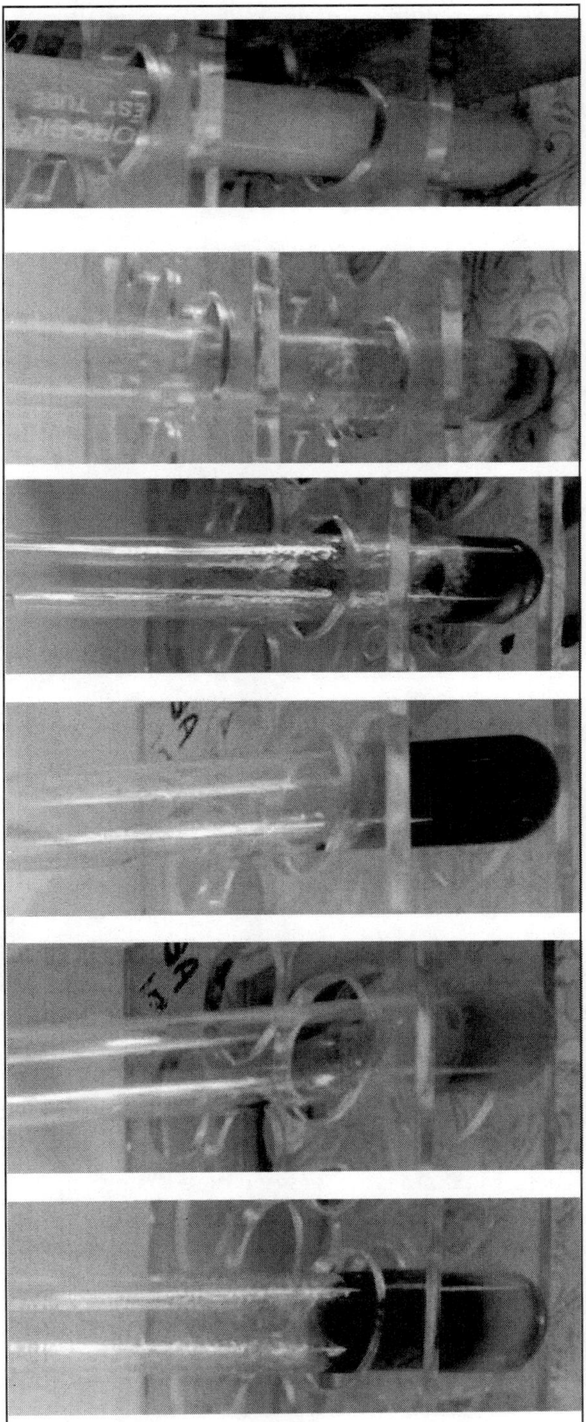

Alkaloids **Tannins** **Steroids** **Phenol** **Glycosides** **Flavonoids**

Figure 22.1: Phytoconstituents Presence in Phytochemical Tests of Chloroform Extract.

flavonoids a used to cure various human ailments like diabetes, obesity, inflammation associated diseases, cardiovascular diseases, cancer (Sathyanarayanan, 2009). The phytochemical analyses are only being the qualitative purpose. Phytochemical screening of the plants *Shorea robusta* resin confirmed the presence of indole, alkaloids, flavones, triterpenoids and fatty acids in the plants.

Table 22.1: Presence of Phyto-constituents in Chloroform Extract of Resin

Sl.No.	*Phytoconstituents*	*Chloroform Extract*
1	Alkaloids	+
2	Glycosides	+++
3	Steroids	+++
4	Tannins	-
5	Phenol	+
6	Terpenoids	++
7	Flavonoids	+
8	Gum and mucilage	++
9	Anthocyanin	-
10	Saponine	+

SPOT	Rf Value
1	0.64
2	0.58
3	0.54
4	0.5
5	0.46
6	0.42

Figure 22.2: TLC Profile and Rf Value.

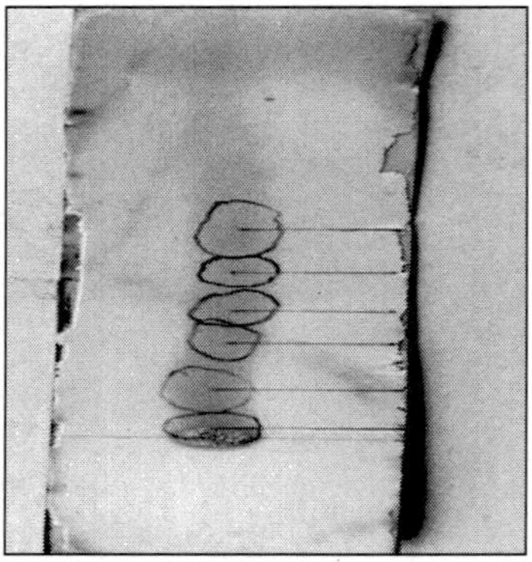

Figure 22.3: Spots on TCL.

The MTT assay of the chloroform extract was carried out using TC1 and normal blood cell lines. The extract was being treated with blood cells. The cells exhibited cell death *i.e.* apoptosis of the cell happen due to cytotoxicity of the drugs that we being treated to them the main drug range where drug concentration ranging from 5 μg/ml to 40 μg/ml showed low level of cytotoxicity, whereas drug level 50 μg/ml to 100 μg/ml showed medium level of cytotoxicity, and with drug level more than 100 μg/ml showed full cell population. The level of cytotoxicity can be observed from the staining where it is observed the formation apoptotic body and cell blabbing are main characteristics that are being observed.

In recent years, the cancer rate has increased, especially in less developed countries. Identification and development of new chemotherapy drugs has been

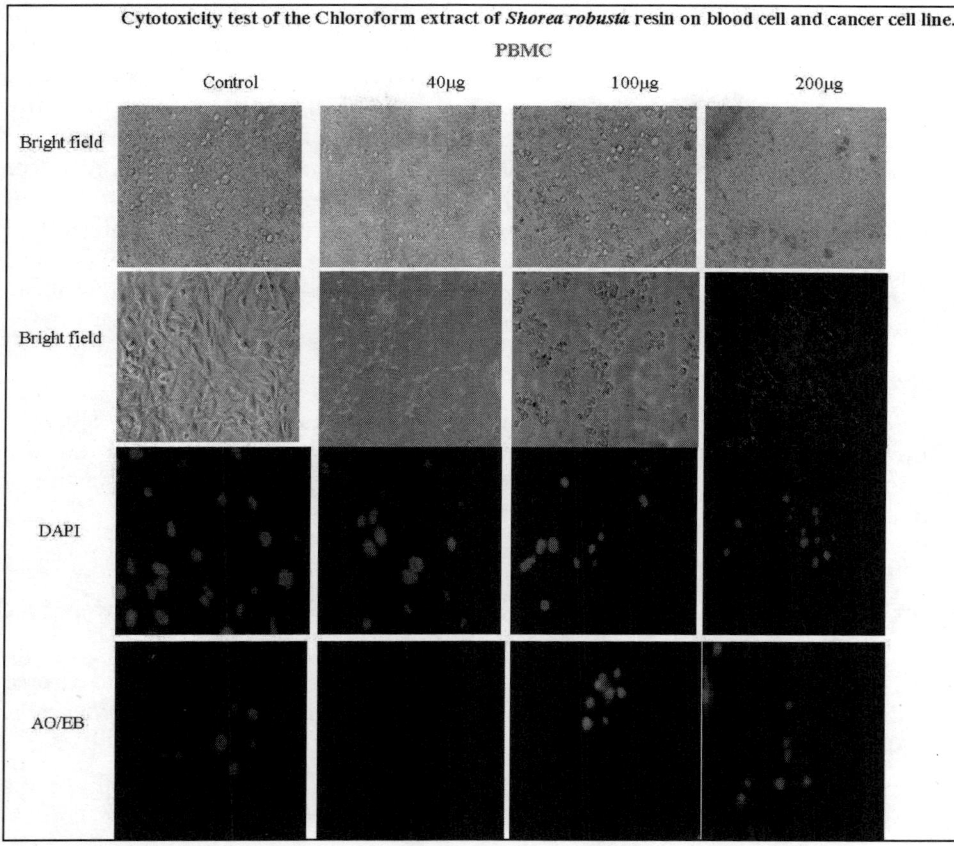

Figure 22.4: Morphological Data of PBMC and Tc1 Cell.

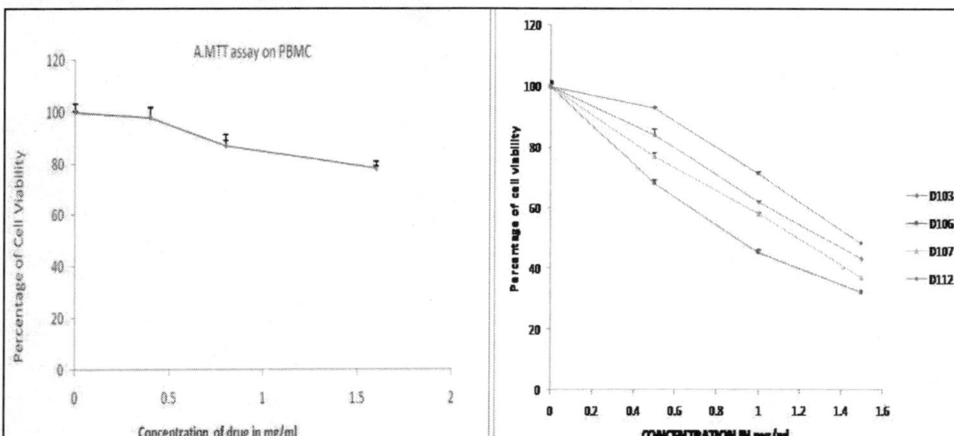

Growth inhibitory effect of Chloroform SR on different cancer cells and blood cell on (A) Blood cells and (B) Tc1. Cells were treated with different concentrations (0, 40, 100 and 200 μg) of the CSR for 24 h and viability was measured by MTT assay. The data are represented as mean ± SD from triplicate independent experiments. B. Morphological and nuclear changes in (A) Blood cell and (B) Tc1 treatment for 24 h. Upper row: Morphological changes seen under light microscope. Middle and lower row: Nuclear changes seen under fluorescence microscopy after DAPI and A.O./EtBr staining respectively.

critical for cancer treatment (Thanekar *et al.*, 2014). Plants have been a rich source of natural drugs. In addition, compounds derived from plants are diverse in structure and bioactivity and exhibit low toxicity, therefore they play an important role in pharmaceutical research (Saxena and Saxena, 2013). In cancer therapy, many plant-derived drugs such as vincristine, paclitaxel and resin have been identified and developed. As reported, resin of *Shorea robusta* showed promising anticancer properties *in vitro* (Misra and Ahmad, 1997). These phytoprinciples differ from plant to plant due to vast diversity. The cytotoxic activity of *Shorea robusta* was determined by using TC1 cell lines. The results indicate the presence of polyphenols and aqueous resin extract had shown dose depend effect on TC1 cell line proliferation, but minimal inhibitory effect was observed.

Conclusion

Present study report, indicates that *Shorea robusta* resin extract showed cytotoxicity activity on TC1 cell lines. Further work is essential to explore the compounds which are responsible for cytotoxicity action and their mechanism of actions.

References

Ames, B. (1989). Endogenous oxidative DNA damage, aging, and cancer. Free Radical Research Communication, 7: 121-128.

Amy, Y. and Chow, W. (2010). Cell cycle control by oncogenes and tumor suppressors: Driving the transformation of normal cells into cancerous cells. Nature Education, 3(7): 11-16

Chopra, A. and Doiphode, V. (2002). Ayurvedic medicine: Core concept, therapeutic principles and current relevance. Medical clinics of North America, 86(1): 75-89.

Cullet, P. (2002). Amended patents act and access to medicines after doha. Economic and Political Weekly, 37: 22-24.

Cunningham, A. (1998). Medicinal plants and sustainable trade. In International Conference on Medicinal Plants held in February at Bangalore, 25: 31-34.

Gupta, A. (1998). How cheap can a medicinal plant species be? Current Science, 74: 7: 565-566.

Houghton, P., Fang, R., Techatanawat, I., Steventon, G., Hylands, P. and Lee, C. (2007). The sulphorhodamine (SRB) assay and other approaches to testing plant extracts and derived compounds for activities related to reputed anticancer activity. Methods, 42(4): 377–87.

Lambert, J., Srivastava, J. and Vietmeyer, N. (1997). Medicinal plants: Resurrecting a global heritage. World Bank Technical Paper, 355: 254-258

Madhuri, S. and Pandey, G. (2009). Some anticancer medicinal plants of foreign origin. Current Science, 96: 779-783.

McChesney, K. and James, D. (2001). Development of herbal medicinals: A pharmaceutical perspective. Journal of Herbal Pharmacotherapy, 1: 51-79.

Misra, L., and Ahmad, A. (1997).Triterpenoids from *Shorea robusta* resin. Phytochemistry, 45(3): 575–578.

Patel, S., Gheewala, N., Suthar, A. and Shah, A. (2009). *In vitro* cytotoxicity activity of *Solanum nigrum* extract against HeLa cell line and Vero cell line. International Journal of Pharmacy and Pharmaceutical Sciences, 1(1): 38–47.

Sathyanarayanan, S. (2009). Preliminary phytochemical screening and study of antiviral activity and cytotoxicity of *Wrightia tinctoria*. International Journal of Chemical Science, 7(1): 1– 5.

Saxena, M, and Saxena, J. (2013). Phytochemistry of Medicinal Plants. Journal of Pharmacognosy and Phytochemistry, 1: 168-182.

Singh, A, and Khan S, (2000). India's competitive edge in international essential oil trade: Opportunities and challenges. Journal of Medicinal and Aromatic Plant Sciences, 22: 404-410.

Soliman, A.S., Samad, S., Banerjee, M., Chamberlain, R.M., Robert, M. and Aziz, Z. (2006). Brief continuing medical education (CME) module raises knowledge of developing country physicians international electronic. Journal of Health Education, 9: 31–41.

Thanekar, D., Tupe, P., Dhodi, J., and Juvekar, A. (2014). Evaluation of *in vitro* cytotoxic activity of petroleum ether and methanol extract of *Mentha arvensis* (Whole Plant) on human cancer cell lines. International Journal of Pharmacy and Pharmaceutical Sciences, 6: 169–172.

Vashisht, S., Singh, M. and Chawla, V. (2017). GC-MS analysis of phytocomponents in the various extracts of *Shorearo busta* Gaertn F. International Journal of Pharmacognosy and Phytochemical Research, 9: 783-788.

23

Effect of Protein, Lipid, Reduced Glutathione, Lipid Peroxidation, Superoxide Dismutase and Catalase Level in Earthworm Exposed to Nickel Chloride

P. Parida and S.P. Dhupal*

Post Graduate Department of Zoology, North Orissa University, Baripada – 757 003, Odisha
**E-mail: p_parida2000@yahoo.co.in*

ABSTRACT

Eudrilus eugeniae (Earthworm) cultured in laboratory condition were divided into four groups as C_{E0}, T_{E24}, T_{E48} and T_{E72}. Each group placed in culture tray containing 1 kg of soil.0.6 g of nickel chloride were mixed in each tray of experimental earthworms. The earthworms were sacrificed after different time intervals such as 0 hr (C_{E0}), 24hrs (T_{E24}), 48hrs (T_{E48}) and 72hrs (T_{E72}). The protein, lipid, reduced glutathione, lipid peroxidation, superoxide dismutase and catalase level at different time intervals were measured and compared.

Keywords: *Catalase, Eudrilus eugeniae, Glutathione, Lipid peroxidation, Nickel chloride, Superoxide dismutase.*

Introduction

Invertebrates are frequently used as biomarkers for the possible effect of xenobiotics applied to the environment. Earthworms have been successfully used as sentinel organisms for non-targeted toxicity of chemicals used in agricultural practices (Booth *et al.,* 2001). Earthworms have also been used to assess the bioaccumulation potential of xenobiotics because they are preferred prey species of multiple classes of vertebrates (Stephenson *et al.,* 1997). In biological systems, heavy metals have been reported to affect cellular organelles and components such as cell membrane, mitochondrial, lysosome, endoplasmic reticulum, nuclei, and some enzymes involved in metabolism, detoxification, and damage repair (Wang *et al.,* 2001). This study was designed to see the toxic effects of nickel chloride on *E. eugeniae* by measuring different biochemical parameters like protein, lipid, reduced glutathione, lipid peroxidation, superoxide dismutase and catalase at different time intervals.

Materials and Methods

Eudrilus eugeniae (Earthworm) were purchased from Soil Chemist Office, Takatpur, Baripada, Mayurbhanj. A culture of *E. eugeniae* was maintained in the laboratory. They were reared in buckets with perforated base, lined with gravel, coconut husk, farmyard manure on which worms are released. Bucket was covered with nylon net and gunny cloth and kept moist by sprinkling water. The cultured earthworms were divided into four groups as CE0, T_{E24}, T_{E48} and $T_{E72.}$ Each group placed in culture tray containing 1 kg of soil. To each tray (T_{E24}, T_{E48} and $T_{E72)}$ 0.6 g of nickel chloride were added and mixed properly as experimental group. The earthworms were sacrificed after exposure to nickel chloride for different time intervals, 24h (T_{E24}), 48h (T_{E48}) and 72h (T_{E72}). Another group containing 1 kg of soil without nickel chloride was used as control group (C_{E0}).

Earthworms (5-6 numbers) were picked up from the designated tray as time intervals and their pooled weight was measured in digital monopan balance (Shimadzu; ELB 300). A 20 per cent homogenate was prepared in ice-cold 50 mM phosphate buffer (pH 7.4) using pre-chilled porcelain mortar and pestle by up and down strokes at 4 °C. The homogenate was centrifuged at 4500 xpm for 10 minutes at 4 °C. The supernatant was taken for biochemical assay

Protein Estimation

Protein estimation of the samples was made according to the method of Lowry *et al.* (1951). Protein content was expressed as mg/g wet weight of the tissue and aqueous BSA (Bovine Serum Albumin) was taken as standard protein.

Lipid

Lipid was estimated by vanilline-phosphoric acid reagent according to Folch method (1957). The lipid content was expressed in **mg**/g tissue.

Superoxide Dismutase

Superoxide dismutase (SOD; EC 1.15.1.1) activity was determined according to the method of Das *et al.* (2000). SOD activity was expressed as units/mg protein

Estimation of Catalase

Catalase (CAT; EC 1.11.1.6) activity was estimated according to Beers and Sizer (1952). The activity of catalase was expressed as nkat/mg protein.

Reduced Glutathione

Reduced glutathione (GSH) of the sample were estimated by Ellman (1959) method. The data are expressed as mg/gm tissue.

Lipid Peroxidation

Lipid peroxidation (LPX) of the sample is estimated as thiobarbiturate acid reacting substance (TBARS) by thiobarbituric acid (TBA) according to the method of Ohkawa *et al.* (1979). The data are expressed as nmoles of TBARS/mg protein.

Statistical Methods

One-way ANOVA and Post Hoc analysis was carried out to find out the level of significance between *E. eugeniae* exposed over a period of 24 hrs, 48 hrs and 72 hrs of nickel chloride in comparison to the control. A difference was taken as significant when P was less than 0.05. Statistical analysis was done with the help of software SPSS package 16.0.

Results and Discussion

Protein Content

Protein content (µg/g tissue) in *E. eugeniae* treated with 0.6 g of nickel chloride were 3.778 ± 0.977 mg/g tissue, 3.444 ± 0.471 mg/g tissue, 2.800 ± 0.110 mg/g tissue, 4.857 ± 0.145 mg/g tissue after 0 hr, 24 hrs, 48 hrs and 72 hrs respectively. The protein content (mg/g tissue) gradually decreased at 48 hrs and slightly increased at 72 hrs in comparison to the control (Figure 23.1). One way ANOVA was performed in order to analyse the effect of nickel chloride on the protein content at different time intervals in *E. eugeniae*. One way ANOVA revealed that the protein content at different time intervals in *E. eugeniae* is significant [F (3, 11) = 7.339, P=0.11]. Post Hoc analysis revealed that the protein content at different time intervals when treated with nickel chloride in *E. eugeniae* was significant at 72 hrs, while, 24 hrs and 48 hrs are not significant with respect to control (P<0.05; LSD).

Lipid Content

Lipid content (mg/g tissue) in *E. eugeniae* treated with 0.6 g of nickel chloride were 0.006 ± 0.003 mg/g tissue, 0.033 ± 0.016 mg/g tissue, 0.002 ± 0.0005 mg/g tissue, 0.033 ± 0.016 mg/g tissue after 0 h, 24 hrs, 48 hrs and 72 hrs respectively. The Lipid content (mg/g tissue) gradually decreased at 0 hrs and 48 hrs. It was highest at 24 hrs and 72 hrs as comparison to control group (Figure 23.2). One way ANOVA was performed in order to analyse the effect of nickel chloride on the lipid content at different time intervals in *E. eugeniae*. One way ANOVA revealed that the lipid content at different time intervals in *E. eugeniae* is significant [F (3, 11) = 5.977, P=.019]. Post Hoc analysis revealed that the lipid content at different time

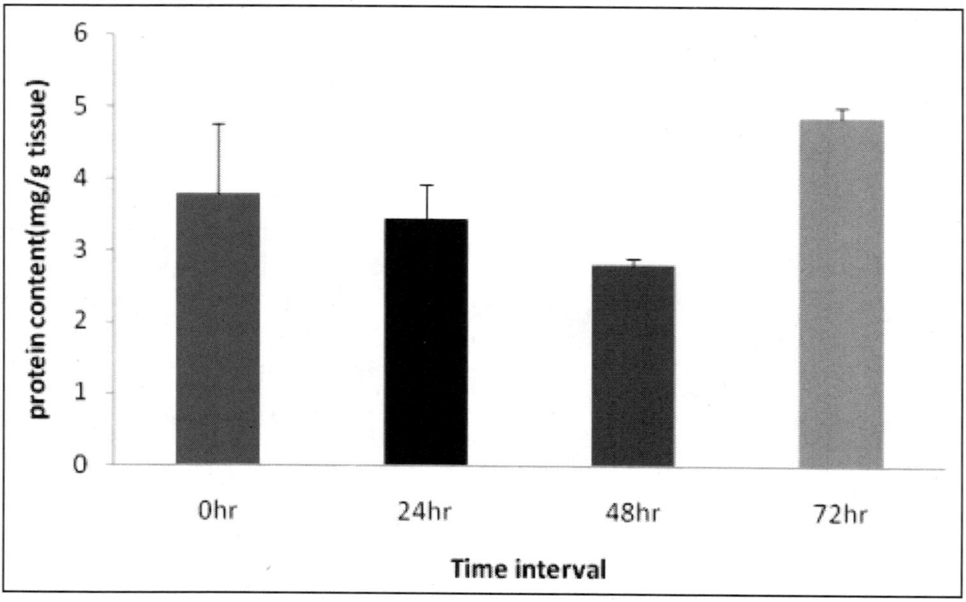

Figure 23.1: Comparison of Protein Content (mg/g tissue) of the *Eudrilus eugeniae* Exposed to Nickel Chloride at different Time Intervals.

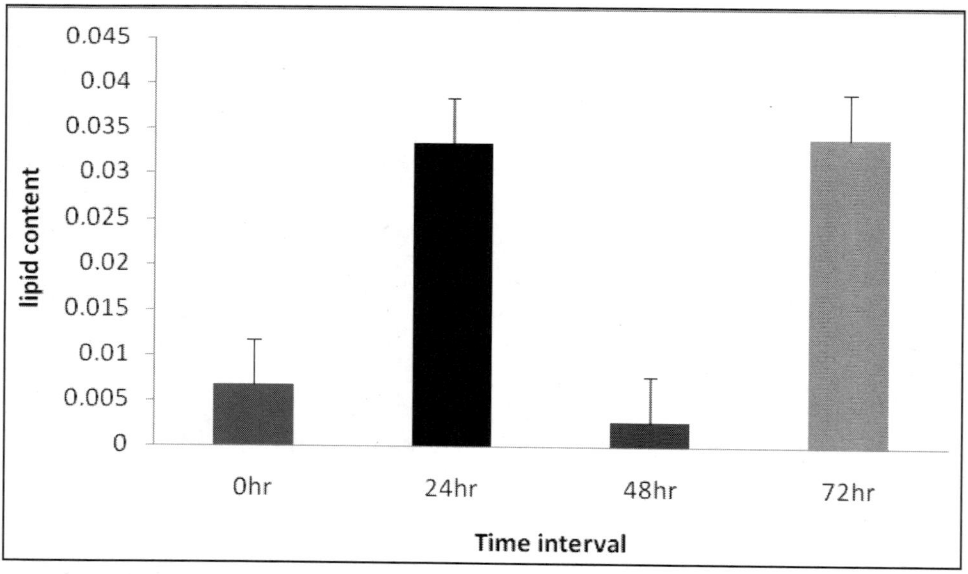

Figure 23.2: Comparison of Lipid Content (mg/g tissue) of the *Eudrilus eugeniae* Exposed to Nickel Chloride at different Time Intervals.

intervals when treated with nickel chloride in *E. eugeniae* was significant only at 24 hrs and 72 hrs ($P<0.05$; LSD). While, 48 hrs are not significantly different with respect to control.

Superoxide Dismutase (SOD)

Superoxide dismutase activity (Unit/mg protein) in *E. eugeniae* treated with 0.6 g of nickel chloride were 18.24 ± 3.94 mg/g tissue, 1.38 ± 0.66 mg/g tissue, 1.29 ± 0.69 mg/g tissue, 1.42 ± 0.67 mg/g tissue after 0 hr, 24 hrs, 48 hrs and 72 hrs respectively. The SOD level (Unit/mg protein) gradually decreased at 24 hrs, 48 hrs and 72 hrs. The SOD level was least at 72 hrs (Figure 23.3). One way ANOVA was performed in order to analyse the effect of nickel chloride on the protein content at different time intervals in *E. eugeniae*. One way ANOVA revealed that the protein content at different time intervals in *E. eugeniae* is significant [F (3, 11) =50.407, P=0.000]. Post Hoc analysis revealed that Superoxide dismutase activity (Unit/mg protein) at different time intervals when treated with nickel chloride in *E. eugeniae* was significant at 24 hrs, 48 hrs and 72 hrs (P<0.05; LSD) with respect to control group.

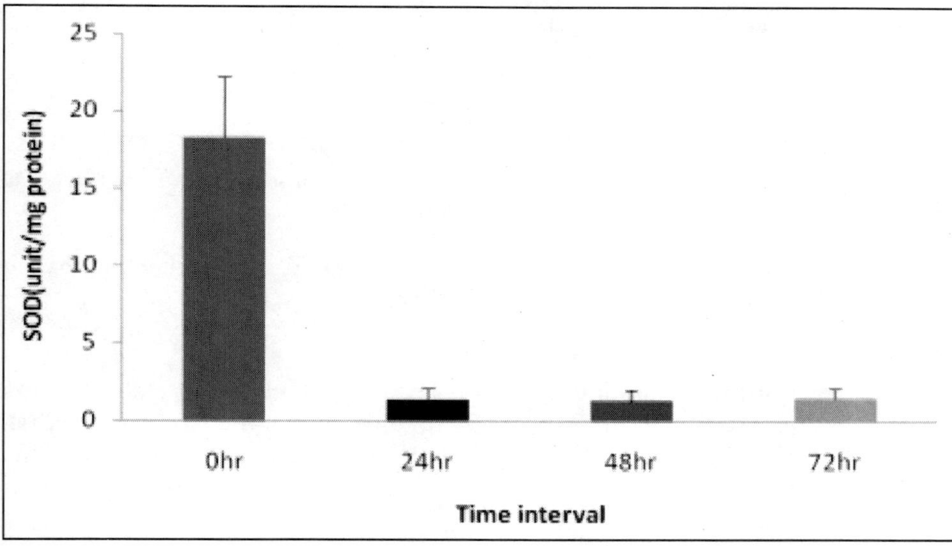

Figure 23.3: Comparison of SOD Activity (Unit/mg protein) of the *Eudrilus eugeniae* Exposed to Nickel Chloride at different Time Intervals.

Catalase (CAT)

Catalase activity (nkat/mg protein) in *E. eugeniae* treated with 0.6 g of nickel chloride were 41.31 ± 47.23 nkat/mg protein, 66.30 ± 12.34 nkat/mg protein, 59.52 ± 3.35 nkat/mg protein, 99.95 ± 43.19 nkat/mg protein, after 0 hr, 24 hrs, 48 hrs and 72 hrs respectively. The CAT level (nkat/mg protein) was lower in control in comparison to different time intervals. The CAT level (nkat/mg protein) was increased at 24 hrs and then decreased at 48 hrs but highest at 72 hrs (Figure 23.4). One way ANOVA was performed in order to analyse the effect of nickel chloride on the CAT activity (nkat/mg protein) at different time intervals in *E. eugeniae* is significant [F (3, 11) = 1,692, P=0.245]. Post Hoc analysis revealed that the at different

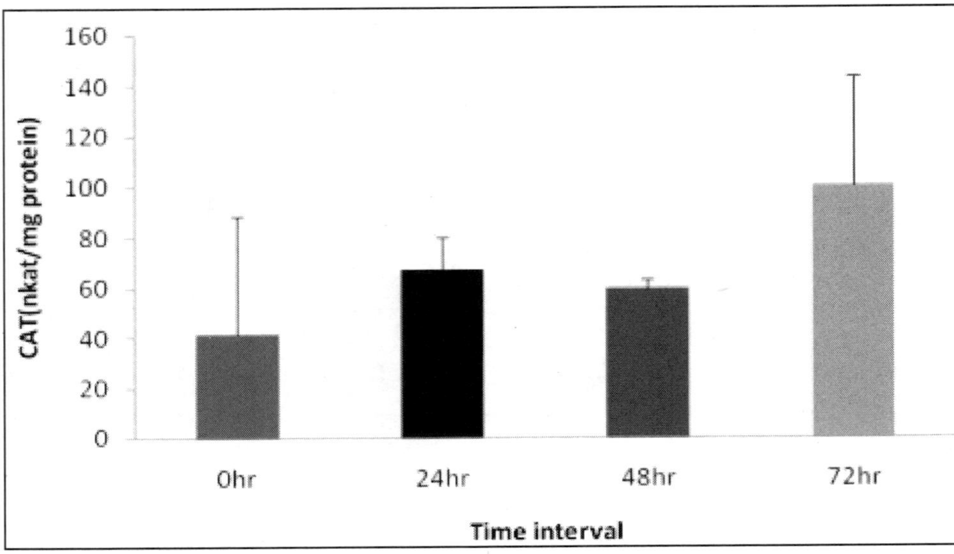

Figure 23.4: Comparison of CAT Activity (Unit/mg protein) of the *Eudrilus eugeniae* Exposed to Nickel Chloride at different Time Intervals.

time inter CAT activity (nkat/mg protein) when treated with nickel chloride in *Eudrilus eugeniae* ($P<0.05$; LSD) was not significant with respect to control group.

Reduced Glutathione (GSH)

GSH content (mg/g tissue) in *E. eugeniae* treated with nickel chloride were 183.49 ± 75.57 mg/g tissue, 261.94 ± 63.23 mg/g tissue, 95.04 ± 5.60 mg/g tissue, 130.18 ± 7.44 mg/g tissue after 0 h, 24 hrs, 48 hrs and 72 hrs respectively. The GSH

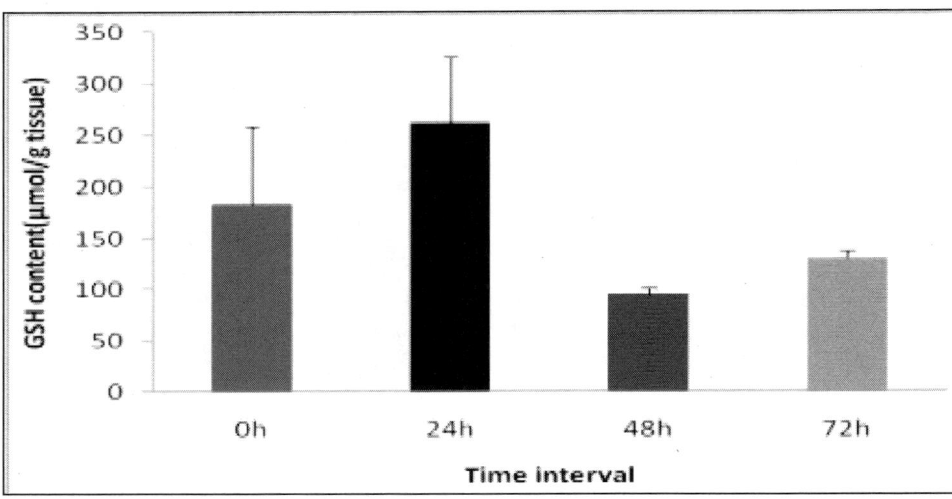

Figure 23.5: Comparison of GSH Content (µ mol/g tissue) of the *Eudrilus eugeniae* Exposed to Nickel Chloride at different Time Intervals.

content (mg/g tissue) gradually decreased at 0 hr, 48 hrs and 72 hrs and highest at 24 hrs (Figure 23.5). One way ANOVA was performed in order to analyse the effect of nickel chloride on the GSH content at different time intervals in *E. eugeniae*. One way ANOVA revealed that the protein content at different time intervals in *E. eugeniae* is significant [F (3, 11) = 32.814P=.000]. Post Hoc analysis revealed that the content at different time intervals when treated with nickel chloride in *E. eugeniae* was significant at 24 hrs, 48 hrs and 72 hrs (P<0.05; LSD) with respect to control group.

Lipid Peroxidation (LPX)

Lipid Peroxidation level (mg/g tissue) in *E. eugeniae* treated with nickel chloride were 2.65 ± 0.72 mg/g tissue, 1.31 ± 0.28 mg/g tissue, 6.63 ± 0.81 mg/g tissue, 1.09 ± 0.05 mg/g tissue after 0 hr, 24 hrs, 48 hrs and 72 hrs respectively. The LPX content (mg/g tissue) gradually decreased at 0 h, 24 h and 72 h and highest at 48 h in comparison to control (Figure 23.6). One way ANOVA was performed in order to analyse the effect of nickel chloride on the protein content at different time intervals in *E. eugeniae*. One way ANOVA revealed that the protein content at different time intervals in *E. eugeniae* is significant [F (3, 11) = 62.451, P=.000]. Post Hoc analysis revealed that the LPX content at different time intervals when treated with nickel chloride in *E. eugeniae* was significant at 24 hrs, 48 hrs and 72 hrs (P<0.05; LSD) with respect to control group.

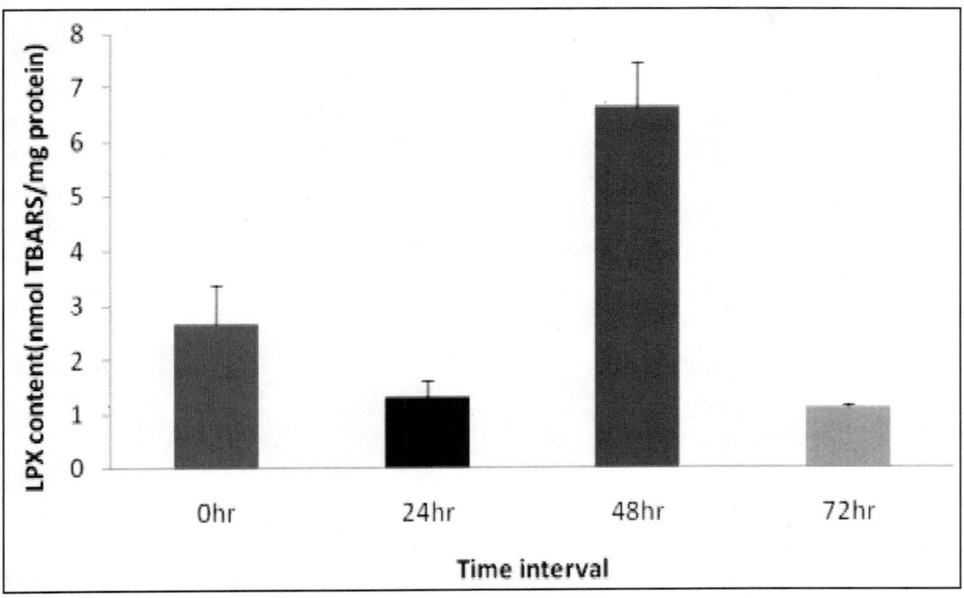

Figure 23.6: Comparison of LPX Level (n mol TBARS/mg protein) of the *Eudrilus eugeniae* Exposed to Nickel Chloride at different Time Intervals.

References

Beers, R.F. and Sizer, I.W. (1952). A spectrophotometric method for measuring the breakdown of hydrogen peroxide by catalase. Journal of Biological Chemistry, 195: 133-140.

Booth, L.H., Hodge, S. and O'Halloran, K. (2001). Use of biomarkers in earthworms to detect use and abuse of field applications of a model organophosphate pesticide. Archives of Environmental Contamination and Toxicology, 67: 633-640.

Das, K., Samanta, L. and Chainy, G.B.N. (2000). A modified spectrophotometric assay of superoxide dismutase using nitrite formation by superoxide radicals. Indian Journal of Biochemistry and Biophysics, 37: 201-204.

Ellman, G.L. (1959). Tissue sulphydryl groups. Archeal Biochemistry and Biophysics, 82: 70-77

Folch, J., Lees, M. and Stanley, G.H. (1957). A simple method for isolation and purification of total lipids from animal tissues. Journal of Biological Chemistry, 226: 497-509.

Lowry, O.H., Resbrough, N.J., Farr, A.L. and Randoll, R.J. (1951). Protein measurement with the Folin-phenol reagent. Journal of Biological Chemistry, 19: 265-275.

Ohkawa H., Ohishi, N. and Yagi, K. (1979). Assay of Lipid Peroxides in animal tissues by thiobarbituric acid reaction. Anals of Biochemistry and. Physiology, 118C (1): 33-37.

Stephenson, G.L., Wren, C.D. (1997). Exposure of the earthworm, *Lumbricus terrestris*, to diazinon, and the relative risk to passerine birds. Soil Biology and Biochemistry, 29: 717-720.

Wang S, Shi X (2001). Molecular mechanisms of metal toxicity and carcinogenesis. Molecular and Cell Biochemistry, 222: 3–9.

24

Colour Enhancement and Growth of Goldfish, *Carassius auratus* (Linnaeus) through Natural Carotenoids Feed Supplement

*T. Raj, K. Bisai, S. Bhattacharjee and M.K. Pati**

Department of Aquaculture Management and Technology, Vidyasagar University, Midnapore – 721 102, West Bengal
**E-mail: mkpati@mail.vidyasagar.ac.in*

ABSTRACT

A 60 days feeding experiment was carried out to investigate the effects of carrot (Daucus carota) meal, spirulina (Arthrospira platensis) meal and marigold (Tagetes erecta) petals meal as natural pigment material on growth and coloration of goldfish, Carassius auratus (L.). All fishes were divided into 4 different groups reared in duplicate and were fed with prepared basal feed, basal feed+carrot meal, basal feed+spirulina meal and basal feed+marigold petals meal respectively. Highest values of growth in terms of weight were recorded in group-C where the fish were fed with basal feed in combination with spirulina meal and minimum value in group-A where the fish were fed with basal feed (without the addition of natural carotenoids). Maximum colour was developed in group-B (fish were fed with basal feed in combination with carrot meal) and minimum in group-A where the fish were fed with the basal feed. So, it was demonstrated that natural pigment substances have an impact on growth and coloration in C. auratus.

***Keywords**: Carotenoids, Carrot, Spirulina, Marigold, Carassius auratus.*

Introduction

Ornamental fishes are rapidly gaining importance due to their commercial value in the export trade world over. The commercial value of the ornamental fish is determined by their attractive colour. Carotenoids are the primary source of pigmentation on the skin of fishes (Hata and Hata, 1973). In natural environment, the carotenoids requirements are compensated by eating aquatic plants or through their food chain as fish cannot make the carotenoid in their body (Meyers, 1994: Mukherjee *et al.*, 2009). Hence the carotenoid requirement of fish is included through the supplementary feed. Dietary carotenoid plays an important role in the regulation of skin and muscle colour in fish (Ezhil *et al.*, 2008).

Colour enhancement through the use of carotenoids in feed has been tried by various workers. Ako *et al.* (2000) have reported that feed diet containing carotenoid rich strain of *Spirulina platensis* and *Haematococcus pluvialis* resulted an intense colouration of freshwater red velvet swordtail (*Xiphophorus helleri*), rainbow fish (*Pseudomugil furcatus*) and topaz cichlids (*Cichlasana myrnae*). China rose (*Hibiscus rosasinensis*) petal was also found to enhance the colour of goldfish (*Carassius auratus*) (Sinha and Asimi, 2007). Dharmaraj and Dhevendaran, (2011) found that microbial carotenoid effectively enhance the pigmentation and growth in swordtail (*X. helleri*). Somnath and Jasmin (2013) tried dietary addition of China rose petals and spirulina and found enhanced total carotenoid content in the skin and muscle tissues of *C. auratus*. Yedier *et al.* (2014) also reported that skin colour was dependent on different pigment type and concentration in the ornamental red zebra cichlid, *Maylandia estherae*. Dananjaya *et al.* (2015) found that bixin-based diets significantly enhance the skin and fin colour in *C. auratus*. Viewing the above information, this study was designed to assess the effect of various sources of carotenoids namely carrot (*Daucus carota*) meal, spirulina (*Arthrospira platensis*) meal and marigold (*Tagetes erecta*) petals meal on the colour enhancement and growth rate of goldfish, *C. auratus*.

Materials and Methods

Collection and Acclimatization of Experimental Fish

For this experiment, advanced fingerlings of goldfish, *C. auratus* of average weight of 7.5±0.23 g were collected from local fish market of Kolkata, West Bengal, India. They were acclimatized in a 3000 litre capacity circular cemented cistern for 3 weeks. During this period the water parameters were maintained constantly and were fed with commercial feed [CP Aquaculture (India) Pvt. Ltd.] diet at the rate of 3 per cent body weight. After acclimatization, fishes were segregated into various groups for the experimental trials.

Experimental Design

Four different groups (A, B, C and D) consisting of glass aquarium (60 L) in duplicate were used to study the effect of various sources of carotenoids (carrot, spirulina and marigold petals meal) on the colour enhancement and growth rate in goldfish. Each group consists of 10 fishes. Group-A fed with prepared basal feed (without the addition of natural carotenoids). Group-B fed with basal feed + carrot meal. Similarly group-C fed with basal feed+spirulina meal and group-D fed with basal feed + marigold petals meal. All groups were fed with at the rate of 5 per cent of body weight once daily. The experiment was conducted for 60 days in aerated condition. During the experimental trial, water temperature was in the range of

29 °C and 31 °C. The average other water quality parameters like dissolved oxygen was 5.2±0.6 mgL^{-1}, total alkalinity and the pH of the water were 92.0 ± 1.8 mg L^{-1} and 7.3±0.5, respectively in the experimental aquarium.

Feed Formulation

Feed was formulated by using locally available ingredients like maize, wheat, rice bran, groundnut oil cake, mustard oil cake, tapioca flour, fish meal which were collected from local fish market of Midnapore, West Bengal, India. Vitamins and mineral mixture was added at 1 per cent. Common basal feed was prepared with the above mentioned ingredients. The proximate composition of this basal feed was crude protein 40.60±0.71 per cent, crude lipid 11.93±0.13 per cent, crude fibre 7.42±0.62 per cent, ash 8.30±0.17 per cent and moisture 7.20±0.11 per cent. For the experimental trials, carrot meal for group-B, spirulina meal for group-C and marigold petals meal for group-D were added at a rate of 10g/100g with the basal feed before pelletization.

Analysis of Feed Ingredients

The proximate composition of the fish feed ingredients (crude protein, crude lipid, crude fibre, ash and moisture) were evaluated by using standard methods from Analytical Food Testing Laboratory, Agricultural and Food Engineering Department, Indian Institute of Technology (IIT), Kharagpur, West Bengal, India.

Analysis of Growth Rate

The total body weights of fishes in each group were taken by using electronic balance. The specific growth rate (SGR) was calculated in every fortnight after inducing experimental feed by using given equation

$$\text{SGR (per cent)} = \frac{\text{The Natural logariitham of the final weight} - \text{The Natural logariitham of the initial weight}}{\text{Time (days)}} \times 100$$

Analysis of Colour Enhancement

The fishes were randomly sampled from each experimental group and the development of colouration of fish were assessed visually and recorded. The development of pigmentations of fin and skin for colouration of different experimental groups were compared under a microscope (Zeiss Axiocam).

Results

Growth Rate

All fishes were grown normally and no specific signs of disease were observed. All diets were accepted equally by the each group of fishes. Minimum growth was observed in group-A, whereas highest growth rate was found in group-C (basal feed+spirulina meal). The normal growth rate of each experimental group is given in Table 24.1 and the specific growth rates are given in Table 24.2.

Table 24.1: Shows the Normal Growth (in mean ± standard error) of Each Experimental Group Fed with different Diets after Every 15 Days Interval

Experimental Group	*0th Day (gm)*	*15th Day (gm)*	*30th Day (gm)*	*45th Day (gm)*	*60th Day (gm)*
Group-A	7.5±0.23	9.5±0.36	12.50±0.27	14.0±0.25	15.0±0.58
Group-B	7.5±0.23	11.0±0.28	14.0±0.46	16.5±0.42	18.5±0.73
Group-C	7.5±0.23	11.5±0.25	16.0±0.23	18.0±0.66	20.0±0.25
Group-D	7.5±0.23	10.0±0.54	13.5±0.12	16.0±0.38	17.0±0.58

Table 24.2: Specific Growth Rate (SGR) in Terms of Weight of each Experimental Group Fed with different Diets after Every 15 Days Interval

Experimental Group	*15th Day*	*30th Day*	*45th Day*	*60th Day*
Group-A	0.67	0.73	0.60	0.50
Group-B	1.07	0.87	0.73	0.65
Group-C	1.20	1.07	0.82	0.70
Group-D	0.80	0.83	0.71	0.58

Detection of Colour Enhancement

After the end of the experimental period, we found very less colour pigmentation (light orange colour in dorsal and silvery at the abdominal portion of fish) in group-A. Whereas in other experimental groups, deep colour pigmentation in whole body were noticed as shown in Figure 24.1. The visual observation of the colouration of experimental carotenoid fed groups was compared. Group-B and group-C fishes showed nearly same colouration but less colouration was observed in group-D. Under high magnification, the appearance of colour pigmentation of fin and skin of the experimental groups showed that group-B fishes develops maximum colour pigmentation as compared to group-C followed by group-D. Least colour pigmentation in fin and skin were observed in group-A fishes. The microscopic view of colour pigmentation of fin of each group is shown in Figure 24.2 and colouration of skin is shown in Figure 24.3.

Discussion

Carotenoids are important natural pigments generally produced by photosynthetic organisms that are accumulated by many animals through their diet (Goodwin, 1986). Carotenoids may play an important role in colour enhancement in animal as well as fish (Gupta *et al.*, 2007; Yedier *et al.*, 2014). Several workers have been tried for colour enhancement by using different natural carotenoids through diet in different fishes (Ako *et al.*, 2000; Sinha and Asimi, 2007; Ezhil *et al.*, 2008; Ramamoorthy *et al.*, 2010; Ho *et al.*, 2014). Harpaz and Padowicz (2007) reported that addition of 60 mg oleoresin paprika per kg diet is sufficient to obtain good colour in *Microgeophagus ramirezi*. Arulvasu *et al.* (2013) tried *Rosa rubiginosa* petals as natural sources of carotenoids and found growth and colour enhancement in *X. helleri*.

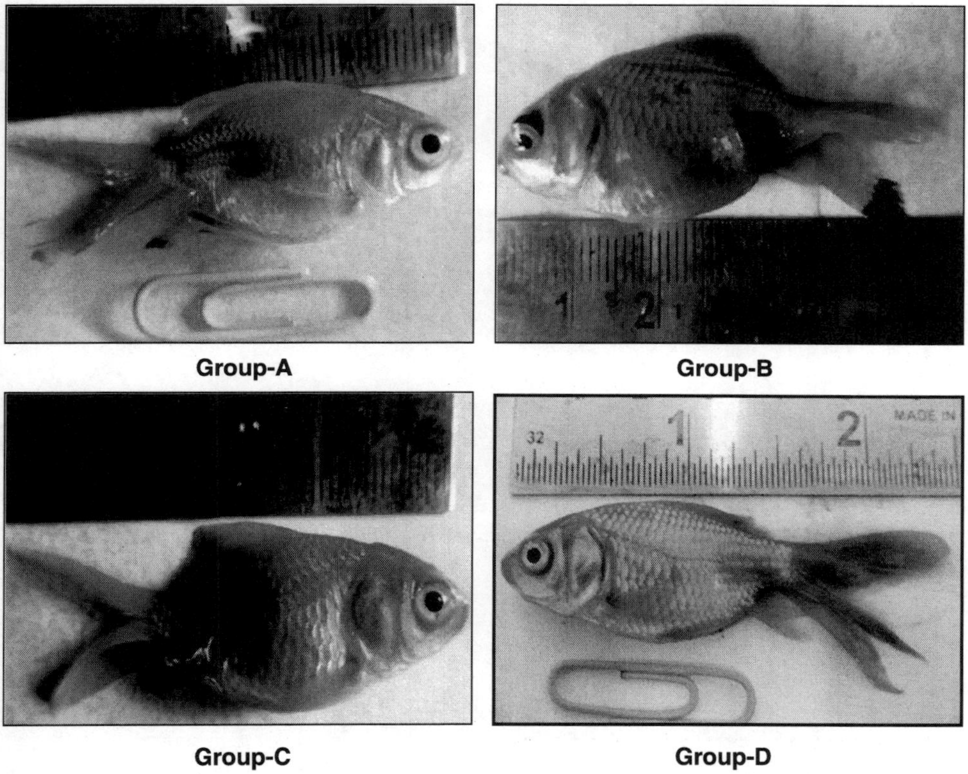

Figure 24.1: Colour Variation of Fishes in different Experimental Groups. In Group-A, light orange colouration appears in dorsal portion of body and silvery colouration in abdomen portion. Group-B fishes showed deep orange colouration throughout the body. Group-C fishes showed deep orange colouration but lighter than the group-B. Group-D possess light orange colouration in whole body.

The present study was carried out for a period of 60 days to determine the growth performance and colouration in *C. auratus* by using three different natural carotenoids through diet. The natural carotenoids used in this study were carrot meal, spirulina meal and marigold petals meal. Each of them was mixed with common basal feed and used in experimental group. After the experiment of 60 days, we found maximum colouration in group-B (carrot meal as carotenoids source) followed by group-C and group-D and least colour enhancement in group-A. Similar results have been reported by Ramamoorthy *et al.* (2010) where carrot meal was an effective colour enhancer. ß-carotene is the principal pigment in carrot, which gives a reddish orange appearance (Pintheiro-Santana *et al.*, 1998). Leza *et al.* (2013) reported that carrot has high nutritional value and also high concentration of bioactive constituents. Carrots are a good source of carbohydrates and minerals like Ca, P, Fe and Mg (Sharma *et al.*, 2012). In this study, fish in group-C and group-D showed good colour enhancement as compared to group-A. Golandaj *et al.* (2015)

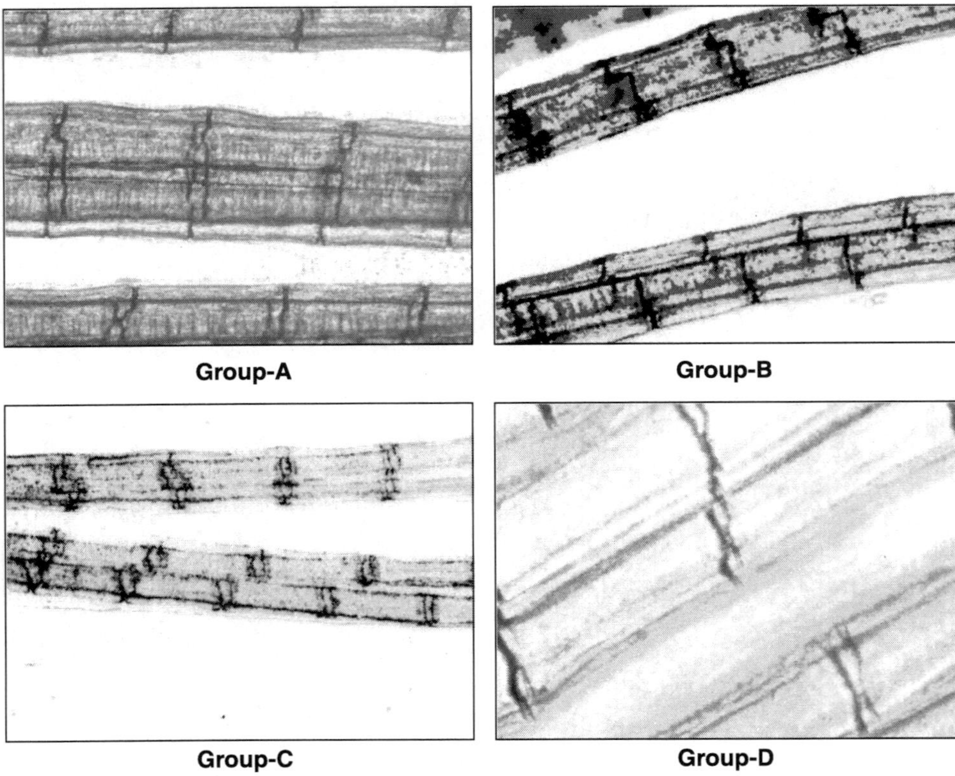

Figure 24.2: Magnified View (400X) of Fin of Fishes in different Experimental Groups. Group-A fishes showed weak colour pigments appearance in fin-rays with greyish-yellow colouration. In Group-B, fin-rays developed bright colour pigmentation with reddish-yellow colour combination. Appearance of moderate colour pigmentation with light reddish-yellow colour observed in Group-C fishes. The fin-rays of Group-D fishes showed yellow colouration with stripes of reddish pigmentation.

observed maximum colour intensity in *X. helleri* when fed with marigold based feed. Yedier *et al.*, 2014 recorded yellow and orange colour enhancement in skin of *Maylandia estherae*, when fed with spirulina in the diet.

Carotenoids are vital nutrients for healthy growth, metabolism and reproduction in animals (Miki, 1991). Spirulina is a rich source of protein (60-70 per cent), vitamins, essential amino acids, minerals, essential fatty acids and carotenoids (Rinna *et al.*, 2013). Jana *et al.* (2014) reported that spirulina as feed supplement of fish can raise its growth performance. Ezhil *et al.* (2008) fed marigold petal meal at 15g/100g diet for 60 days and found increased growth rate in *X. helleri*. In this experimental trial, we have demonstrated highest growth performance by group-C fed with spirulina based diet followed by group-B fed with carrot based and group-D fishes fed with marigold petal based feed. The least growth performance was recorded in group-A fed with basal diet.

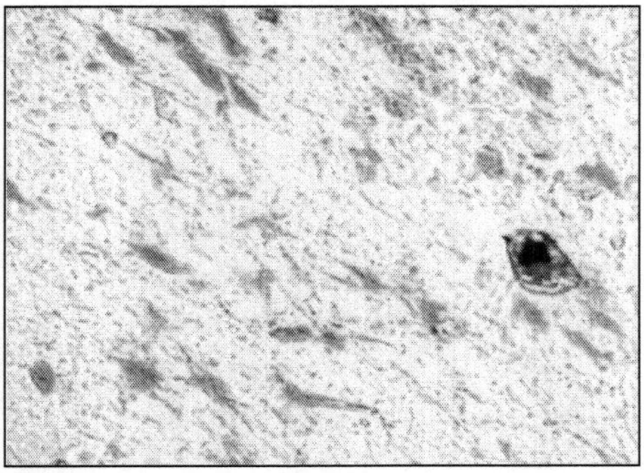

Group-A

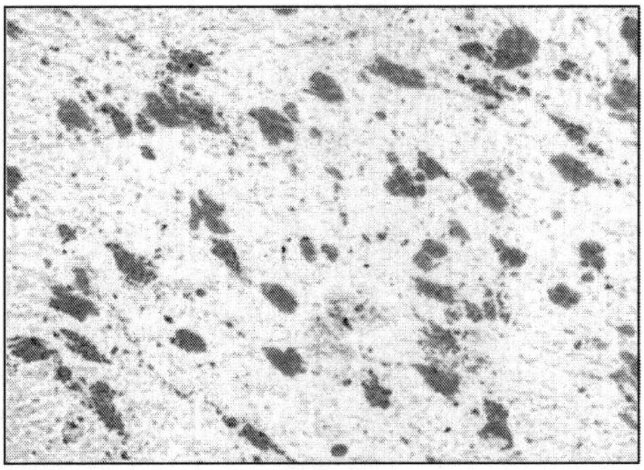

Group-B

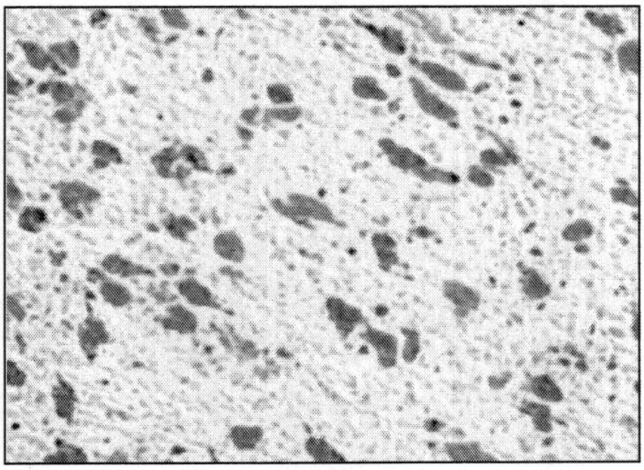

Group-C

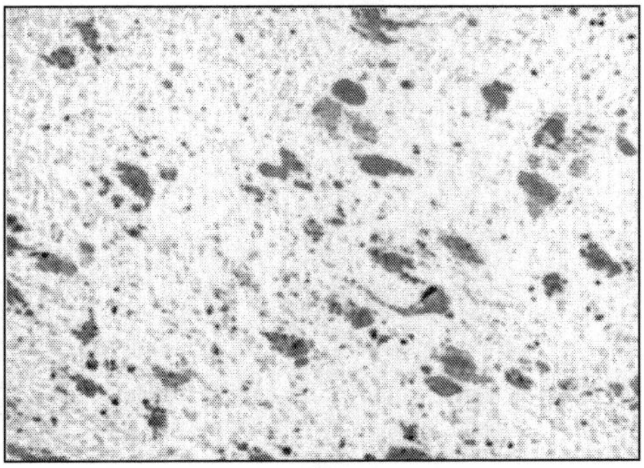

Group-D

Figure 24.3: Magnified View (400X) of Skin of Fishes in different Experimental Groups. Group-A showed lighter orange colouration with lesser pigmentation. Skin of Group-B fishes develop deep and bright reddish pigmentation. Intermediate pigmentation with moderate reddish colouration observed in the skin of Group-C fishes. Skin of Group-D fishes showed with lesser pigmentation with light reddish colouration.

The results of this study not only focus on skin colouration in goldfish (*C. auratus*) depending on types of carotenoids used but also suggest the important role of carotenoids in animal colour polymorphism, and speciation. In other words, carotenoids may help to integrate the empirical and theoretical perspectives underlying the bewildering diversity of colourful fishes. Based on the present study, it is concluded that the carrot meal as carotenoid source have best performance in terms of the appearance of colour pigmentation whereas the fish fed with spirulina as carotenoid source have highest growth rate. So it is suggest that supplementing the diet of farm fish with carotenoids can enhance skin colour, improve the growth factor and increase their market value.

Acknowledgements

The authors would like to thanks and acknowledge the authority of the Vidyasagar University, Midnapore, West Bengal, India for providing infrastructure to carry out the experiment under the Department of Aquaculture Management and Technology. We also thank the personnel of instrumentation facility of the Department of Agricultural and Food Engineering, Indian Institute of Technology, Kharagpur, and West Bengal, India for analysis of the samples.

References

Ako, H., Tamaru, C.S., Asona, L., Yuen, B. and Yamamoto, M. (2000). Achieving natural colouration in fish under culture. UJNR Technical Report, 28: 1-4.

Arulvasu, C., Meena, S.R., Chandhirasekar, D. and Sivaganam, S. (2013). Evaluation of natural sources of carotenoid pigments from *Rosa rubiginosa* on growth, survival and colouration of *Xiphophorus helleri* fish fry. European Journal of Biological Science, 5(2): 44-49.

Dananjaya, S.H.S., Munasinghe, D.M.S., Ariyaratne, H.B.S., Lee, J. and De Zoysa, M. (2015). Natural bixin as a potential carotenoid for enhancing pigmentation and colour in goldfish (*Carassius auratus*). Aquaculture Nutrition, 23(2): 255-263.

Dharmaraj, S. and Dhevendaran, K. (2011). Application of microbial carotenoids as a source of colouration and growth of ornamental fish *Xiphophorus helleri*. World Journal of Fish and Marine Sciences, 3(2): 137-144.

Ezhil, J., Jeyanthi, C. and Narayanan, M. (2008). Marigold as a carotenoid source on pigmentation and growth of red swordtail, *Xiphophorus helleri*. Turkish Journal of Fisheries and Aquatic Sciences, 8: 99-102.

Golandaj, A., Shyama, S., Dinesh, K., Sreenath, V.R. and Swain, S. (2015). Colour enhancement potential of selected local flowers in Sword tail, *Xiphophorus helleri* through dietary incorporation. Research Journal of Recent Sciences, 4: 37-43.

Goodwin, T.W. (1986). Metabolism, nutrition and function of carotenoids. Annual Review of Nutrition, 6: 273-297.

Gupta, S.K., Jha, A.K., Pal, A.K. and Venkateshwarlu, G. (2007). Use of natural carotenoids for pigmentation in fishes. Natural Product Radiance, 6(1): 46-49.

Harpaz, S. and Padowicz, D. (2007). Colour enhancement in the ornamental dwarf cichlid *Microgeophagus ramirezi* by addition of plant carotenoids to the fish diet. The Israeli Journal of Aquaculture - Bamidgeh, 59(4): 195-200.

Hata, M. and Hata, M. (1972). Carotenoid pigments in goldfish (*Carassius auratus*) - V. Conversion of zeaxanthin to astaxanthin. Bulletin of the Japanese Society of Scientific Fisheries, 38(4): 339-343.

Ho, A.L.F.C., Zong, S. and Lin, J. (2014). Skin colour retention after dietary carotenoid deprivation and dominance mediated skin coloration in clown anemone fish, *Amphiprion ocellaris*. AACL Bioflux, 7(2): 103-115.

Jana, A., Saroch, J.D. and Borana, K. (2014). Effect of spirulina as a feed supplement on survival and growth rate of *Pangasius sutchi*. International Journal of Fisheries and Aquatic Studies, 1(5): 77-79.

Leja, M., Kaminská, I., Kramer, M., Maksylewicz-Kaul, A., Kammerer, D., Carle, R. and Baranski, R. (2013). The Content of phenolic compounds and radical scavenging activity varies with carrot origin and root colour. Plant Foods for Human Nutrition, 68(2): 163-170.

Meyers, S.P. (1994) Developments in world aquaculture, feed formulation and role of carotenoids. Pure and Applied Chemistry, 66(5): 1069-1076.

Miki, W. (1991). Biological functions and activities of animal carotenoids. Pure and Applied Chemistry, 63(1): 141-146.

Mukherjee, A., Mandal, B. and Banerjee, S. (2009). Turmeric as a carotenoid source on pigmentation and growth of fantail guppy, *Poecilia reticulate*. Proceedings of the Zoological Society, 62(2): 119-123.

Pintheiro-Santana, H.M., Stringheta, P.C., Brandão, S.C.C., Páez, H.H. and Queiróz, V.M.V. de. (1998). Evaluation of total carotenoids, alpha- and beta-carotene in carrot (*Daucus carota* L) during home processing. Food Science and Technology, 18(1): 39-44.

Ramamoorthy, K., Bhuvaneswari, S., Sankar, G. and Sakkaravarthi, K. (2010). Proximate composition and carotenoid content of natural carotenoid sources and its colour enhancement on marine ornamental fish *Amphiprion ocellaris* (Cuveir 1880). World Journal of Fish and Marine Sciences, 2(6): 545-550.

Rinna, H.S., Jansi, M. and Vasudhevan, I. (2013). Effect of dietary *Spirulina plantensis* on feeding parameters of blue gourami, *Trichogaster trichopterus*. International Journal of Research in Fisheries and Aquaculture, 3(3): 103-106.

Sharma, K.D., Karki, S., Thakur, N.S. and Attri, S. (2012). Chemical composition, functional properties and processing of carrot - a review. Journal of Food Science and Technology, 49(1): 22-32.

Sinha, A. and Asimi, O.A. (2007). China rose (*Hibiscus rosasinensis*) petals: a potent natural carotenoid source for gold fish (*Carassius auratus* L.). Aquaculture Research, 38(11): 1123-1128.

Somanath, B. and Jasmin, K.J. (2013). Hibiscus petals and spirulina supplemented diet induced carotenoid changes in freshwater goldfish (*Carassius auratus*). International Journal of Pure Applied Zoology, 1(4): 352-362.

Yedier, S., Gümüs, E., Livengood, E.J. and Chapman, F.A. (2014). The relationship between carotenoid type and skin colour in the ornamental red zebra cichlid *Maylandia estherae*. AACL Bioflux, 7(3): 207-216.

25

Preliminary Studies on Determination of Efficacy of some Plant Extracts against Red Spider Mites (*Tetranychus urticae*) in Brinjal Plant (*Solanum melongena*)

S. Nayak[1], *A. Behera*[2], *M.K. Chand*[3] *and G. Mohanty*[1] *

[1]*Department of Zoology, North Orissa University,*
[2]*Department of Biochemistry, M.P.C (Autonomous) College,*
[3]*District Nodal Officer, Biological Control Field Station,*
Baripada – 757 003, Odisha
**E-mail: gargee.mohanty@gmail.com*

ABSTRACT

Instead of using chemical environmental risk pesticides we have tried to formulate biopesticide of plant origin. Leaf extract of ten different plants using different solvents for extraction were made and screened for the presence of different phytochemicals. The samples those showed presence of phenols and flavonoids in substantial amount were selected for pesticide efficacy experiments. Methanol leaf extracts of three plants namely, Bridelia retusa, Buchanania lanzan and Leucas aspera was prepared and applied at different concentrations (10 per cent, 20 per cent and 30 per cent) individually and in combinations against red spider mite (Tetranychus urticae) pest in brinjal plants (Solanum melongena). Counting of larva and adult mites was done prior and upto 96 hrs after exposure to these extracts. We have found leaf extracts of L. aspera and B. lanzan individually and in combination with a pesticidal effect against red spider mites.

Keywords: *Biopesticide, Bridelia retusa,Brinjal, Buchanania lanzan, Leucas aspera, Phytochemicals, Red spider mites.*

Introduction

Mites, the most diverse representatives of the phylum Arthropoda, belong to the subphylum Chelicerata and subclass Acari. Among the arachnids, Acari are the only group, which feeds on plants. Plant feeding mites play an important role as agricultural pests of fruits, vegetables, forage crops, ornamentals and other agricultural crops. Around 7,000 species of phytophagous mites are known worldwide which occur in five families namely Tetranychidae, Tenuipalpidae, Eriophyidae, Tarsonemidae and Tuckerillidae (Chhillar *et al.*, 2007). Tetranychidae, also known as spider mites, is a large family including about 1,200 species belonging to over 70 genera of worldwide distribution (Bolland, 1998).

Tetranychus urticae commonly called as red spider mite is a member of the family. The life cycle of *T. urticae* has been studied by several authors (Boudreaux, 1956; Hebert, 1981; Helle and Sabelis, 1985). It passes through five developmental stages: egg, larva, protonymph, deutonymph, and adult. Both nymphs and adults imbibe sap from between the veins from under surface of leaves. Red spider mite feeding on the underside of leaves generally results in the typical stippling damage, which is observed as white or grayish colored spots due to the punctures made by feeding (Brandenburg *et al.*, 1987).

Red spider mites affect a number vegetable plant that includes tomato, beans, brinjal plants and others. Brinjal, commonly known as eggplant or aubergine in the western world, is closely related to tomatoes and potatoes as it lies in the same *Solanum* family. Brinjal (*Solanum melongena* L.), a member of family Solanaceae, is a native of India. However, it is widely cultivated in both temperate and tropical regions of the globe mainly for its immature fruits as vegetable. Besides its use as a fresh vegetable, it is known to possess some medicinal properties in curing diabetes and a remedy for liver complaints.

Chemical pesticides have mostly been relied upon to control red spider mite populations; however, their continuous use has also resulted in pesticide resistance in them. Besides the problem of pest resistance, there is also concern about the high residues and toxicity of common miticides to humans. Concerns about the use of agrochemical control measures have led to search for alternative control measures to suppress the pest populations. Biopesticides are prepared from naturally occurring materials that controls pests by non-toxic mechanisms and in an eco-friendly manner. They may be derived from animals (*e.g.* nematodes), plants (*Chrysanthemum, Azadirachta*) and microorganisms (*e.g. Bacillus thuringiensis, Trichoderma, Nucleopolyhedrosis* virus) (Mazid *et al.*, 2011; Kumar, 2015). Therefore, biopesticides pose less risk to the environment and human health.

Phytochemicals are biologically active compounds found in vegetables, fruits, cereal grains, and plant based beverages such as tea and wine. Phytochemical consumption is associated with a decrease in risk of several types of chronic diseases because of their antioxidant properties and free radical scavenging effects. Other functions of the phytochemicals include; prevention against predators and pathogens, attraction and deterrence against pollinators, allelopathic action, attraction of symbionts, food for pollinators, symbionts, herbivores, pathogens and decomposers and ultraviolet (UV) rays protectants. Crude extracts and biochemical products of approximately 3000 plant species have worked out for insecticidal properties. Majority of plants showed antiherbivore mechanisms. These mechanisms are qualitatively allelochemicals, phenolic acids, tannins, flavonoids, alkaloids and terpenoids and quantitatively, carbohydrates, proteins, lipids and nitrogens. The present study thus projects on the preliminary evaluation of insecticidal effect of three methanol leaf extracts (crude) against red spider mite in brinjal plants were carried out. The results of the different combinations of the extracts showed promising results.

Materials and Methods

Incidence of Brinjal Red Spider Mites in Rabi Cropping Season

The seasonal incidence of the red spider mites was studied at District Biological Control Field station, Mayurbhanj district during *rabi* cropping season from September to March 2018 (86.75 longitude, 21.93 latitude with the GPS coordinates of 21 55′ 56.04″ N and 86 45′ 6.09″ E and has mild tropical rainy climate), Odisha, India. The observations on mite population were recorded on ten randomly selected plants. The mites were mounted on glass slides and identified as *Tetranychus urticae*.

Preparation of Crude Extracts

The initial stage is the preparation of plant sample for studies to preserve the biomolecules in plants prior to extraction is the pre extraction phase where fresh leaves of different plants are collected, washed thoroughly with running tap water and left for shade drying. 25g of each plant samples (powered leaves) were taken by weighing in electronic balance (Sartorious). Powered samples of each plant were transferred to conical flask. Two fifty milliliters of each solvent (cold water, hot water, methanol, ethanol, chloroform, benzene) was added to six different conical flasks carrying 25g of plant extract of individual plant. The mixture should always be in a ratio of 1:10 (plant extract: solvent quantity). Conical flasks were then placed in a rotary shaker for thorough mixing at 120-130 rotation per minute (rpm) for 1hr. The extract was collected in falcon tube for further use.

Screening of Plants

Ten plants having antioxidant properties were taken for the studies. Crude samples using different solvents (both polar and nonpolar) were obtained adapting the standard extraction process. These samples were screened for the presence of different types of phytochemicals. Three plants namely; *Bridelia retusa, Buchanania lanzan* and *Leucas aspera* were selected to find their efficacy as biopesticides against red spider mites in brinjal plants basing upon their phytochemcial properties especially the presence of phenols and falvanoids.

Efficacy Tests

To find out the efficacy of the crude extracts of three plant extracts and of their mixtures on red spider mites certain short term experiments were done. Methanol crude leaves extracts of three plants *B. retusa, B. lanzan* and *L. aspera* were taken for the experiments.

Experimental Design

The effect of the plant extract at varying concentration individually and in combinations was carried out on red spider mite population in brinjal plants. The leaves extracts of 10 per cent, 20 per cent and 30 per cent concentration from individual plant were made. Similarly a combination leaves extract from all three plants at a ratio of 1:1:1 crude extracts was also prepared and 10 per cent, 20 per cent and 30 per cent of the combination was considered for experiments. Red Mite population in all the plants used for the experiment was counted before the spray of the extracts. Three leaves from upper portion, middle and lower portion of the plants were taken and average leaf area of 2cm × 2cm was observed under a hand lens to count the number of mites. After counting, the treatments were done by spraying the extracts in late afternoon. One set of control group (only water spray), was also taken to see the difference between the treated groups and untreated groups. Reading (count of mites in three leaves of a particular plant in an area of 2 cm x 2 cm) of all plants was done at 24h, 48h, 72h and 96h of exposure.

Statistical Analysis

All the statistical analysis was done in Microsoft Office Excel Data Analysis Tool pack 2010. The mean percentage and standard error each concentration level and control was calculated using descriptive statistics of Microsoft Office Excel. One way (factor) ANOVA test was carried out to show the difference in the population of red spider mites between the treated and control group in each sampling day at $p < 0.05$. Student's t- test was carried out between the means of two extracts of particular doses of particular sampling time to compare the effect of different extracts of a particular dose on mites at a particular sampling period at $P < 0.05$.

Results

Phytochemical analysis showed the presence of phenols, flavanoids in some plants which are locally available in more quantities and can be grown in less time. Table 25.1 shows the results obtained by the treatment against red spider mite pest of brinjal plants with three different plant extracts (Extract A=*Bridelia retusa;* Extract B=*Buchanania lanzan;* Extract C=*Leucas aspera*) and in combination of all these extracts in 1:1:1 ratio. Three different doses were chosen (10 per cent, 20 per cent and 30 per cent) and sampling was done at four sampling periods (24 h, 48 h, 72 h and 96 h). Total numbers of larva and adult mites in the plants were counted both in the control and treated groups in each sampling hrs. Average number of mites with their standard deviation is shown in Tables 25.1 and 2. One way (factor) ANOVA test was carried out to show the difference in the population of red spider mites between the treated and control group in each sampling day at $p<0.05$.

The result of the test showed that there exists significant difference in between the control and treated groups in extract A in 48 h of exposure and thereafter in all doses. A similar type of trend was also observed with extract B. However, in case

Table 25.1: Effects of Leaves Extract of different Plants Individually and in Combination on the Larva Population of Red Spider Mites after Various Time Intervals

Hours of Exposure	*Treatment Groups*	*Extract A Bridalia retusa*	*Extract B Buchanania lanzan*	*Extract C Leucas aspera*	*Extract M Combination*
24 Hours	CONTROL	1.42 ± 0.4	0.42±0.40	1.42±0.40	1.42±0.40
	10 per cent	0.61±0.20*	0.89±0.37	0.66±0.16	0.80±0.35*
	20 per cent	0.57±0.24*	0.43±0.21	0.67±0.31	0.54±0.31*
	30 per cent	0.46±0.15*	0.38±0.18	0.52±0.16	0.70±0.45*
48 Hours	CONTROL	0.87±0.28	0.87±0.28	0.87±0.28	0.87±0.28
	10 per cent	0.93±1.05	0.65±0.16*	0.44±0.15*	0.61±0.20*
	20 per cent	0.72±1.08	0.40±0.18*	0.43±0.12*	0.69±0.08*
	30 per cent	0.54±0.20*	0.44±0.08*	0.47±0.07*	0.53±0.13*
72 Hours	CONTROL	1.21±0.12	1.21±0.12	1.21±0.12	1.21±0.12
	10 per cent	0.47±0.17*[b]	0.53±0.21*[b]	0.43±0.11*[a]	2.05±3.28
	20 per cent	0.28±0.06*[a]	0.35±0.05*[a]	0.45±0.18*[a]	0.60±0.2*
	30 per cent	0.41±0.20*[bd]	0.34±0.08*[cd]	0.39±0.26*[bd]	0.46±0.17*
96 Hours	CONTROL	1.11±0.042	1.11±0.42	1.11±0.42	1.11±0.42
	10 per cent	0.47±0.15*[a]	0.63±0.19*[a]	0.27±0.10*[a]	0.40±0.09*
	20 per cent	0.35±0.12*[b]	0.27±0.73*[b]	0.39±0.15*[b]	0.38±0.08*
	30 per cent	0.44±0.19*[cd]	0.25±0.03*[cd]	0.23±0.05*[cd]	0.38±0.14*[d]

* Significant difference from respective control in each extract group at $p< 0.05$ by one factor ANOVA analysis.

Different alphabets (a, b, c and d) in same groups refers to existence of significant difference between same alphabets across the different extract groups indicates the existence of significant difference between them at $p< 0.05$ in student's t – test.

of extract C and M significant difference in the three treated groups from that of the control were demonstrated after 24 hrs of treatment. In case of larva the effects of extracts have started from 72 h onwards. In case of larva the death were more significant in 20 per cent and 30 per cent doses of extract B and C (Table 25.1). Comparing the effect of different doses of a particular extract on red spider mites (larva and adult) in a particular sampling time; it was found that the 20 per cent and 30 per cent doses of extract B, C and M did significantly well in both adult and larva groups. Student's t- test was carried out between the means of two extracts of particular doses of particular sampling time to compare the effect of different extracts of a particular dose on mites at a particular sampling period. Results of the test showed existence of significant difference among the extract A with the three other extracts. Extract B and C at 20 per cent and 30 per cent doses showed more efficacy than the other two cases.

Discussion

Pest invasion leading to massive crop damage is most serious limiting factor in crop yield. In the course of evolution, many classes of proteins and secondary

Table 25.2: Effects of Leaves Extract of different Plants Individually and in Combination on the Adult Population of Red Spider Mites after Various Time Intervals

Hours of Exposure	*Treatment Groups*	*Extract A Bridalia retusa*	*Extract B Buchanania lanzan*	*Extract C Leucas aspera*	*Extract M Combination*
24HOURS	CONTROL	1.52±0.19	1.522±0.06	1.52±0.06	1.52±0.06
	10 per cent	0.22±0.06	0.39±0.03	0.25±0.04*	0.23±0.05*[a]
	20 per cent	0.32±0.20	0.16±0.03*	0.20±0.04*	0.14±0.02*
	30 per cent	0.25±0.12*[c]	0.15±0.01*[c]	0.16±0.02*[c]	0.02±0.06*[bc]
48Hours	CONTROL	0.92±0.25	0.92±0.08	0.92±0.08	0.92±0.08
	10 per cent	0.07±2.08*	0.28±0.29	0.18±0.01*	0.24±0.03
	20 per cent	0.15±0.06*	0.14±0.01*	0.09±0.01*	0.15±0.02*
	30 per cent	0.18±0.08*	0.17±0.01*	0.09±0.03*	0.13±0.02*
72 Hours	CONTROL	1.24±0.22	1.24±0.07	1.24±0.07	1.24±0.07
	10 per cent	0.18±0.07*	0.19±0.03*[a]	0.22±0.01*	2.85±0.56
	20 per cent	0.04±0.03*[a]	0.07±0.00*[b]	0.05±0.00*[a]	0.11±0.01*
	30 per cent	0.11±0.08*[b]	0.11±0.01*[ac]	0.07±0.02*[bc]	0.10±0.01*[c]
96Hours	CONTROL	1.20±0.21	1.20±0.07	1.20±0.07	1.20±0.07
	10 per cent	0.21±0.08*[b]	0.36±0.03*	0.22±0.01*	0.11±0.01*
	20 per cent	0.06±0.04*[a]	0.04±0.60*[a]	0.11±0.02*	0.09±0.00*
	30 per cent	0.23±0.10*[b]	0.05±0.00*[ac]	0.11±0.01*	0.09±0.00*[c]

* Significant difference from respective control in each extract group at $p < 0.05$ by one factor ANOVA analysis.

Different alphabets (a, b, c and d) in same groups refers to existence of significant difference between same alphabets across the different extract groups indicates the existence of significant difference between them at $p < 0.05$ in student's t – test.

metabolite substances have been expressed in plants as effective control measure against herbivory by pests and the application of these natural plants defense traits as candidates for genetic engineering for crop test pest control is now a reality. Plant secondary metabolites include glut of biochemical entities that are synthesized through multifarious pathways and therefore, the attempts to explore their biochemistry and physiology have always remained challenging. Broadly they belong to totally unrelated classes of compound like alkaloids, flavonoids, foliar enzymes (polyphenol oxidases and peroxidases) foliar phenolic acid esters (rutin and chlorogenic acid), non protein amino acids, peptide hormones, pyrethrins, steroids, and terpenoids.

The present study has been carried out to validate the efficacy of the crude leaves extracts of certain plants to control the population growth of red spider mite pest in the brinjal plants. Earlier, several researcher have successfully used various plant extract to control aphids, thrips, whitefly and jassid (leafhoppers) (Singh and Kumar, 2003; Binage *et al.*, 2004; Tiwari and Srivastava, 2004; Haq, 2006). Lowery and Isman (1994) have reported 50 per cent mortality of aphids by spraying neem based biopesticide while Kumar and Singh (2001) reported that neem seed kernel extract was effective in controlling insect pests with highest cost: benefit ratio. Singh and Kumar (2003) determined the efficacy of neem based pesticides against sucking complex and neem kernel extract was the most effective in controlling the aphid and jassid. Binage *et al.* (2004) found that 5 per cent neem seed extract showed the lowest infestation of aphids and maximum crop yield. In our study we have used three plants extracts namely, *Bridelia retusa, Buchanania lanzan, Leucas aspera.*

B. retusa has long been used in Indian folk medicine to treat broad spectrum of diseases, such as liver diseases, for removal of urinary concretion, inflammatory diseases *etc.* Phytochemical studies on *B. retusa* disclosed major chemical constituents including steroids, triterpenoids, tannins and flavonoids. A research report on *B. retusa* reveals for its analgesic activity, anti-inflammatory activity, hepatoprotective activity, antimicrobial activity, immunomodulatory activity and hypoglycemic effect. These effects may be attributed to the presence polyphenolic phytoconstituents in the extract. While studying the antimicrobial potentials of bark extract of *B. retusa* it shows significant activity against various pathogens. The total polyphenolic contents in the methanolic extract of *B. retusa* bark were 32.52 ± 0.24 mg/100 gm which was responsible for significant antioxidant activity. The above mentioned activities may be attributed to the presence of steroids, triterpenoids, tannins and flavonoids as major phytoconstituents. Since there are so many secondary metabolites present in the plant which can contribute to pesticidal effects and help the plant to withstand pest attack.

B. lanzan commonly called as Chironji indigenous to India is a medium-sized evergreen tree that grows up to 15m height. It is a member of the Anacardiaceae plant family and is also known as Almondette tree. Chironji is also called with other names such as 'Cuddapah almond, Charooli, and Chirolo'. The major therapeutic values of chironji oil are carminative, expectorant, diuretic, aphrodisiac, relaxant, antiseptic, anti-inflammatory and stimulant. The chemical components of chironji oil are camphene, a-pinene, y-terpinene, b-pinene, myrcene, 8-cineole, sabinene and triglycerides like triolein, dipalmitoolein and dioleopalmitin. *B. lanzan* seeds are rich in protein, fat, dietary fiber, and energy. They are also very good source of phosphorus, calcium, magnesium and iron. The nutritional and phytochemical composition of the seeds were demonstrated and found very rich source of bioactive components and can be a vital functional food, nutritional supplements which is currently underutilized. Phytochemical screening showed that it is phytochemically very rich source of polyphenols, flavonoids, tannins, alkaloids and saponins. These polyphenols have been proven for having disease fighting power. Good antioxidant activities were found in *B. lanzan* seeds which show its ability in reducing oxidative stress. GC/MS profile shows, abundance of fatty acids, polyphenols, phytosterols, sitosterol, and stigmasterol in seed extract, these phytosterols are receiving so much attention because of their clinical and nutritional significances (Khatoon *et al.*, 2015). In our study also we have found that the leaves extract of chrironji contains high phenols and flavonoids along with other phytochemicals which suggest their possible role in antipesticidal activity.

L. aspera is an annual herb found throughout India as a weed in cultivated fields, wastelands and roadsides. The flowers are used along with honey to treat cough and cold in children. The leaves are applied to the bites of serpents, poisonous insects and scorpion sting. *L. aspera* leaves are also used as insecticides and mosquito repellent in rural area (Reddy *et al.*, 1993). Phytochemical and pharmacological investigations on this plant revealed the presence of various chemical constituents like terpenes, sterols, glycosides, lignans, flavonoids and long-chain fatty compounds which are responsible for biological activities such as antimicrobial, antioxidant, anti-inflammatory, anticancer, antidiabetic, larvicidal, antivenom and phytotoxic activity.

Flavonoids, (a large group of naturally occurring plant polyphenolic compounds including flavones, flavonols, isoflavones, flavonones and chalcones), also known as nature's tender drugs, possess numerous biological and/or pharmacological activities. Recent reports of antiviral, antifungal, antioxidant, anti-inflammatory, antiallergenic, antithrombic, anticarcinogenic, hepatoprotective, and cytotoxic activities of flavonoids have generated interest in studies of flavonoid-containing plants. Flavonoids have demonstrated to exert beneficial effects on some diseases involving lipid peroxidation. The capability to interact with protein phosphorylation and the antioxidant, iron chelating, and free radical scavenging activity may account for the wide pharmacological profile of flavonoids (Saija, *et al.*, 1995; Van Acker, *et al.*, 1996; Kim *et al.*2004; Moon *et al.*, 2006; Veitch, 2007; Jiang *et al.*, 2008).

Phenols are very important plant constituents. There exist a highly positive relationship between total phenols and antioxidant activity of many plant species, because of the scavenging ability of their hydroxyl groups. The phenolic contents of the extract can also scavenge hydrogen peroxide by donating electrons (Yen *et al.*, 1993) and thereby neutralizing it to water (Osawa, 1994). It was also reported that phenolic compounds are effective hydrogen donors, making them very good antioxidants. The antioxidant activity of phenolic compounds is mainly due to their redox properties, which can play an important role in adsorbing and neutralizing super oxide anion (O_2^-), hydroxyl radical or peroxy radicals, quenching singlet and triplet oxygen or decomposing peroxides (Osawa, 1994). Polyphenolic contents appear to function as good electron and hydrogen atom donors, and therefore, be able to terminate radical chain reaction involved in lipid peroxidation by converting free radicals and reactive oxygen species to more stable products. Results of phytochemical screening showed the presence of heavy amount of flavonoids and phenols in the leaves extracts of all the three plants. As evident from several literatures it has been shown that the phenolic and flavonoids contribute in the defense system of plants by making them pest resistance.

The biopesticides formulated will be helpful in carrying out the farming in a better way using organic pesticides instead of inorganic pesticides. All the three plants extracts used in our study exhibited pesticidal effect individually as well as in combination, with reduction of larval population within 72 h and adults from 24 h to 48 h. The late pesticidal response against larva in comparison to the adult might be due to the fact that the larva being small in size which help them to hide in the folds of the leaves and thus escape the extract action during spray. Thus, pesticidal activities of the leaves extracts of all the three plants might be attributed to the presence of flavonoid and phenolic contents and can provide better eco-friendly solution to the problems of pest infestation like red spider mites not only in brinjal but also in other commercially important plants.

References

Binage, A.B., Suryawanshi, D.S., Munde, A.T., Mane, P.D., Salunke, S.G. and Kedar, P.B. (2004). Studies on efficacy of some botanicals against major pests of cabbage. Journal of Soils and Crops, 14: 163-165.

Bolland, H.R., Gutierrez, J. and Flechtmann, C.H.W. (1998). World catalogue of the spider mite family (Acari: Tetranychidae). Koninklijke Brill NV, Leiden, the Netherlands.

Boudreaux, H.B. (1956). Revision of the twospotted spider mite (*Acarina, Tetranychidae*) complex, *Tetranychus telarius* (Linnaeus). Annals of the Entomological Society of America, 49: 43-48.

Brandenburg, R.L. and Kennedy, G.G. (1987). Ecological and agricultural considerations in the management of two spotted spider mite (*Tetranychus urticae* Koch). Agricultural Zoology Reviews, 2: 185-236.

Chhillar, B.S., Gulati, R. and Bhatnagar, P. (2007). Agricultural acarology. Daya Publishing House, New Delhi, 355.

Haq, A. (2006). Efficacy of different neem products against sucking complex on okra *Abelmoschus esculentus*. M.Sc. Thesis (Entomology) submitted to Sindh Agriculture University Tandojam. 17-25.

Helle, W., Sabelis, M.W. (1985). Spider mites: Their biology, natural enemies, and control. Vol. A. Amsterdam: Elsevier Publishers. 1–405.

Herbert, H.J. (1981). Biology, life tables, and innate capacity for increase of the twospotted spider mite, *Tetranychus urticae* Koch (Acarina: Tetranychidae). The Canadian Entomologist, 113: 371-378.

Jiang, H., Zhan, W.Q., Liu, X. and Jiang, S.X. (2008). Antioxidant activities of extracts and flavonoid compounds from Oxytropis falcate Bunge. Natural Product Research, 22: 1650-1656.

Khatoon, N., Rajinder, K., Gupta K. and Tyagi Y.K. (2015). Nutraceutical potential and phytochemical screening of *Buchanania lanzan*, an underutilized exotic Indian nut and its use as a source of functional food. Journal of Pharmacognosy and Phytochemistry, 4: 87-94.

Kim, H.P., Son, K.H., Chang, H.W. and Kang, S.S. (2004). Anti-inflammatory Plant Flavonoids and Cellular Action Mechanisms. Journal of Phamacological. Sciencea, 96: 229-245.

Kumar S (2015). Biopesticide: an environment friendly pest management strategy. Journal of Biofertilizer and Biopesticides, 6: 127

Kumar, M. and Singh, A.K. (2001). Bioefficacy of insecticides against the cotton jassid, *Amrasca biguttula* (Ishida), on okra. Pest Management and Economic Zoology, 9: 55-58.

Lowery, D.T. and Isman, M.B. (1994). Insect growth regulating effects of neem extract and azadirachtin on aphids. Entomologia Experimentalis et Applicata, 72: 77-84.

Mazid, S., Kalida, J.C. and Rajkhowa, R.C. (2011). A review on the use of biopesticides in insect pest management. International Journal of Science and Advanced Technology, 7: 169-178.

Moon, Y.J., Wang, X. and Morris, M.E. (2006). Dietary flavonoids: Effects on xenobiotic and carcinogen metabolism. Toxicology, 20: 187-210.

Osawa, T. (1994). Novel natural antioxidants for utilization in food and biological systems; Post harvest biochemistry of plant food materials in the tropics. Japan Scientific Societies Press, Tokyo, Japan, 241- 251.

Reddy, K.M., Viswanathan, S., Thirugnanasabmantham, D., Santa, R. and Lalitha, K. (1993). Analgesic activity of *L. Aspera.* Fitoterapia, 64: 151-154.

Saija, A., Scalese, M., Lanza, M., Marzullo, D. and Bonina. (1995). Flavonoids as antioxidant agents: importance of their interaction with biomembranes. Free Radical Biology and Medicine. 19: 481-486.

Singh, A.K. and Kumar, M. (2003). Efficacy and economics of neem based products against cotton jassid, *Amrasca biguttulla* Ishida in okra. Crop Research (Hisar) 26: 271-274.

Tiwari, B.K. and K.J. Srivastava. (2004). Studies on bio-efficacy of some plant extracts against pathogens of onion. News Letter National Horticultural Research and Development. 24(1): 6-10.

Van Acker, S. A. B. E., van den Berg, D.J., Tromp, M. N. J. L., Griffioen, D. H., van Bennekom, W. P., van der Vijgh, W. J. F. (1996). Structural aspects of antioxidant activity of flavonoids. Free Radical Biology and Medicine, 20: 331–342.

Veitch, N.C. (2007). Isoflavonoids of the Leguminosae. Natural Product Report, 24: 417-464.

Vinson, J.A. Yong, H. Xuchui, S. and Zubik, L. (1998). Phenol antioxidant quantity and quality in foods: vegetables. Journal of Agricultural and Food Chemistry, 46: 3630-3634.

Yen, G.C., Duh, P.D and Tsai C.L. (1993). The relationship between antioxidant activity and maturity of peanut hulls. Journal of Agricultural and Food Chemistry, 41: 67-70.

Microbial Biotechnology

26

Therapeutic Potentials of Wild Mushrooms with a Note on Mushrooms of Simlipal Biosphere Reserve, Odisha

M.K. Panda[1], *K. Tayung*[2] *and H.N. Thatoi*[1]*

[1]*Department of Biotechnology, North Orissa University,*
Baripada – 757 003, Odisha
[2]*Department of Botany, Guwahati University,*
Guwahati – 781 014, Assam
**E-mail: hn_thatoi@rediffmail.com*

ABSTRACT

Wild mushrooms have long been associated with humankind as a delicious foods and ethnomedicinal uses. In recent years pharmacological research confirms large parts of traditional knowledge regarding the medicinal effects of mushrooms for their anticancer, antifungal, antibacterial, antioxidant, anti-diabetic, anti-inflammatory and antiviral properties. Mushroom based natural products with different health benefits are available in the market. However the mechanisms of the various health benefits of mushrooms to humans still require intensive investigation. In this present paper useful information on the ethnomedicinal uses of wild mushrooms and their therapeutic potentials have been discussed with a note on the medicinal mushrooms of Simlipal Biosphere Reserve, in Mayurbhanj district of Odisha.

Keywords: *Bioactivities, Ethnomedicinal uses, Simlipal Biosphere reserve, Wild mushrooms.*

Introduction

Mushroom is a macro fungus with a distinctive fruiting body, which can be either epigeous or hypogeous. It is large enough to be seen with naked eye and to be picked by hand (Chang and Miles, 1992). According to this definition, mushrooms can be either Basidiomycetes or Ascomycetes, aerial or under-ground (Chang, 1993). Large numbers of mushrooms are found in forests because of the presence of rotting wood and organic matters in soil. Mushrooms collected from forests are regarded as wild mushrooms and have been widely utilized as human foods for centuries because of taste and flavor as well as some medicinal and tonic attributes (Manzi *et al.*, 2001). Mushrooms are low in calories and fat but rich in proteins and dietary fibers therefore are suitable for vegetarians and vegans (Manzi *et al.*, 1999). The protein from mushrooms contains all the nine essential amino acids required by humans. Apart from protein, mushrooms are good source of the nutrients like phosphorus, iron and vitamins, including thiamine, riboflavin, ascorbic acid, ergo sterol, and niacin (Barros *et al.*, 2008). Though wild mushrooms have a great potential for the production of useful bioactive metabolites, harvesting of mushrooms from wild is a very challenging activity. Wild mushrooms are the important sources of food for tribal communities who mostly live in the forest and depend on natural resources for their livelihood. In recent times, depletion of forest areas has adversely affected the availability of wild edible mushrooms. Therefore, it is very important to document the diversity and distribution of wild mushrooms for the welfare of mankind as a rich natural source of bioactive compounds. However, the awareness of wild mushrooms as a healthy food and as an important source of biological active substances with medicinal value has been acknowledged over the years (Cheung *et al.*, 2003).

The number of mushroom species on the earth is estimated about 1,40 000, suggesting that only 10 per cent are known and the proportion of useful mushrooms among the undiscovered and unexamined mushrooms will be only 5 per cent, which implies 7000 yet undiscovered species will be of possible benefit to mankind (Hawksworth, 2001). Among the 14,000 known species, 2000 are safe and about 600 have significant pharmacological properties (Wasser *et al.*, 2000). Most of the traditional knowledge about medicinal properties comes from the Far East (China, Japan, Korea, and Russia). However, in India there are references in Charaka Samhita (3000 ± 500 BC) about the use of mushrooms as food and medicine from the ancient times. The total recorded mushrooms in India are approximately 850 species (Deshmukh, 2004). Many scientific studies, including a number of clinical trials, have been conducted over the past few decades that support the findings of traditional medicine and provide evidence for the therapeutic properties of wild mushrooms in humans as well as animals. Therefore, wild mushrooms are not only fruits of the forest used merely for human consumption, but are increasingly recognized as a cultivated natural source of pharmacologically active compounds with medicinal properties.

Ethnomedicinal Uses

Wild mushrooms have been used in folk medicine for thousands of years. Mushrooms such as *Astreatus hygrometricus* and *Russula delica* are used to prevent malnutrition as well as nutritional disorders where as *Russula albinogra* and *Lentinus tuber-regium* are known to cure cold and cough problems (Sachan *et al.*, 2013). *Volvariella volvacea* is used to lowering the high blood pressure. The species such as *Pleurotus ostreatus* and *Pleurotus* sp. are known to cure asthma and help in high blood pressure (Sachan *et al.*, 2013). In Nigeria, Puff balls (*Lycoperdon pusilum* and *Calvatia gigantea*) are used to cure sores, abrasion or bruises, deep cut, haemorrhages and urinary infections (Buswell and Chang, 1993). Mushrooms belong to the species *Ganoderma* sp. like *G. lucidum* and *G. applanatum* are used for treatment of asthama and wound healing. Similarly, *Termitomyces eurrhizus* and *Termitomyces* sp. are used to treat rheumatism, diarrhea and high blood pressure (Sachan *et al.*, 2013). *Termitomyces* sp. collected from forest areas of Gorakhpur are used by ethnic tribes for malnutrition, weakness and other nutritional disorders (Srivastava *et al.*, 2011). Powdered fruit bodies of *Daldinia concentrica* and *Pisolithus arhizus* were mixed with coconut (*Cocos nucifera*) oil and applied to the skin for getting relief from burning, itching and healing minor skin infections. Several other mushrooms such as fruit body of *Termitomyces clypeatus* were pasted and applied to the skin for the treatment of pox, whereas powdered fruitbody of *Cordyceps sinensis* is consumed mixing up with tea or in the form of soup as aphrodisiac, invigorative *etc.* (Dutta and Acharya, 2014).

Six different species of mushrooms *Auricularia auricula-judae*, *Daldinia concentric*, *V. volvacea*, *Termitomyces robustus* and *Termitomyces* sp. were identified in Anambra State, Nigeria (Okigbo and Nwatu, 2015) for their ethno-medicinal values. These mushrooms are used to cure infertility in both men and women, anaemia, stomach upset, weakness of the body and to control high blood pressure (Table 26.1). Adhikari *et al.* (2005) have reported eight wound curing mushroom species. *Grifola frondosa* and *Ramaria botrytis* were used to get relief from muscles pain and *Laetiporus sulphurous* was used to cure cancer (Adhikari *et al.*, 2005). Although mushrooms have potential to cure many diseases, yet their ethnomedicinal uses have been extremely reported for cold, cough, malnutrition, weakness and other nutritional disorders *etc.*

Table 26.1: Ethnomedicinal Uses of Wild Mushrooms

Mushrooms	*Ethnomedicinal Uses*	*References*
Amanita muscaria, Antrodia camphorate, Calvatia cyathiformis, C. craniiformis, C. excipuliformis, C. gigantean, C.utriformis Cordyceps militaris, Inonotus obliquus, Laetiporus sulphurous, Termitomyces clypeatus	Inflammation and cancer related complication	Biziulevicius and Vaitkuviene (2007): Wu *et al.* (1997), Coetzee and Wyk, (2009): Rao *et al.* (2010): Lemieszek *et al.* (2011): Adhikari *et al.* (2005): Maiti *et al.* (2008): Dutta and Acharya, (2014): Mondal *et al.* (2016)
Auricularia auricular-judae	Infertility in both men and women anaemia and weakness of the body	Okigbo and Nwatu, (2015)
Calvatia gigantean, Lycoperdon pusilum	Sores, abrasion or bruises, deep cut, haemorrhages and urinary infections	Buswell and Chang, (1993)

Mushrooms	Ethnomedicinal Uses	References
Cordyceps sinensis	Aphrodisiac, invigorative, revitaliser, anti-aging	Dutta and Acharya, (2014)
Coriolus hirsutus, Pycnoporus cinnabarinus, Schizophyllum commune	Wounds healing	Adhikari *et al.* (2005): Dutta and Acharya, (2014)
Daldinia concentrica	Pneumonia, wounds, stops bleeding constipation, and itchiness.	Adhikari *et al.* (2005): Jha *et al.* (2011)
	Relief from burning, itching and healing minor skin infections	Dutta and Acharya, (2014)
Ganoderma applanatum, Ganoderma lucidum, Ganoderma sp., *Pleurotus ostreatus, Pleurotus* sp.	Asthama, lowering high blood pressure	Sachan *et al.* (2013)
Geastrum sp.	Reduction of staunch bleeding and reduce swelling	Panda *et al.* (2015)
Grifola frondosa	Relive for muscle pains and to maintain, protect or restore immune health and anti diabetic.	Adhikari *et al.* (2005): Jha *et al.* (2011)
Lentinus tuber-regium, Russula albinogra	Cold and cough	Sachan *et al.* (2013)
Lycoperdon pyriforme	Relief from stomach pain, wounds healing, hallucinogenic	Adhikari *et al.* (2005): Jha *et al.* (2011)
Pisolithus arhizus	Relief from burning, itching and healing minor skin infections	Dutta and Acharya, (2014)
Pleurotus tuberregium	Headache, cold, fever, stomach ache and constipation	Akpaja *et al.* (2003)
Ramaria botrytis	Relief from muscles pain	Adhikari *et al.* (2005)
Termitomyces sp., Volvariella volvacea, Termitomyces robustus, Termitomyces eurrhizus, Russula delica, Astreatus hygrometricus	Malnutrition, weakness and other nutritional disorder rheumatism, diarrhea, anaemia, lowering high blood pressure	Srivastava *et al.* (2011): Sachan *et al.*,(2013): Okigbo and Nwatu, (2015): Sachan *et al.* (2013): Okigbo and Nwatu, (2015): Panda *et al.* (2015)

Pharmacological Potentials

Mushrooms represent a major and untapped source of potent new pharmaceutical products. A wide range of activities including antidiabetic, antimicrobial, anticancer, antioxidant, anti-inflammatory and antiviral are reported in mushrooms. In developing countries like India mushroom progress is a boon in the field of food, medicine, and in generating employment. Many pharmaceutical substances with potent and unique properties were recently extracted from mushrooms and make their way all around the world. A brief discussion on biological activities of mushrooms to cure various human diseases is given below.

Antidiabetic

Diabetes mellitus is a life-threatening chronic metabolic disease caused by lack of insulin and/or insulin dysfunction, characterized by high levels of glucose in the blood (hyperglycemia). Millions worldwide suffer from diabetes and its complications (WHO, 2011). Presently, there are many chemical and biochemical synthetic drugs that are used in treating diabetes and are effective in controlling hyperglycemia. However, as they may have harmful side-effects and fail to significantly alter the course of diabetic complications: wild mushrooms emerge as an alternative natural source of bioactive compounds having potential hypoglycemic and anti-diabetic properties. Bioactive metabolites including polysaccharides, proteins, dietary fibres, and many other biomolecules isolated from medicinal mushrooms and their cultured mycelia have been shown to be successful in diabetes treatment (Dilani *et al.*, 2012).

Wild mushrooms such as *Agaricus bisporus, A. subrufescens, Cordyceps sinensis, Coprinus comatus, Ganoderma lucidum, Inonotus obliquus, Phellinus linteus, Pleurotus* spp, *Poria cocos* and *Sparassis crispa* have been reported to have hypoglycemic effects (reduction of blood glucose levels) and anti-hyperglycemic effects (Cha *et al.*, 2006: Yang *et al.*, 2008: Lee *et al.*, 2010). Mushrooms are known to contain compounds which help in proper functioning of the liver (Wani *et al.*, 2010), pancreas and other endocrinal glands, thereby promoting formation of insulin and related hormones which ensure healthy metabolic functioning (Chen *et al.*, 2012). Consumption of mushrooms markedly decreases the lipid levels including total cholesterol, total triglyceride, and low-density lipoproteins: and increases the level of high-density lipoproteins (Lee *et al.*, 2012).

Antimicrobial

Although a number of synthetic antimicrobial agents have been developed to kill pathogenic microorganisms effectively but many of them have adverse side effects. Therefore, novel antimicrobial agents from different biological sources are continuously sought (Yamac and Bilgili, 2006). This situation provided the impetus to the search for new antimicrobial substances from various sources including wild mushrooms.

The petroleum ether, chloroform, acetone and aqueous extracts of mushroom *Osmoporus odoratus* has been studied for the antibacterial activity against *Staphylococcus aureus, Escherichia coli, Streptococcus pyogenes, Bacillus subtilis,* and *Pseudomonas aeruginosa* and amongst different solvents the aqueous extract alone showed antibacterial activity against the tested pathogens (Sivakumar *et al.*, 2006). The antimicrobial activity of various mushroom species like *Lycoperdon perlatum, Cantharellu scibarius, Ramaria formosa, Maramiuso reades Clavaria vermiculris,* and *P. pulmunarius* against pathogens is dependent on the concentration of bioactive components (Ramesh *et al.*, 2010). Quershi *et al.* (2010) has studied the antimicrobial activity of various solvent extracts (40µg/ml) of *Ganoderma lucidum,* in which acetone extract exhibited maximum antibacterial activity. Ganomycins, a terpenoid hydroquinones, produced by the European mushroom *Ganoderma pfeifferi* inhibits the growth of methicillin resistant *S. aureus* (Mothana *et al.*, 2000). Terpenes isolated from *G. pfeifferi,* showed moderate activity against *E. coli, Proteus mirabilis* and *Serratiama rcescens* (Mothana *et al.*, 2000). Therefore, the compounds from natural resources, such as wild mushrooms, could be isolated and used as antimicrobial drugs. However, the toxicity of the mushroom extracts and individual compounds need to be evaluated for suitability of its application.

Anticancer

Cancer is one of the leading causes of death worldwide now days. The anti-cancer drugs currently available in the market are not target specific and pose several side-effects. These, highlights the urgent need for novel effective and less-toxic anticancer therapeutic agents. In this context, some medicinal mushrooms with validated anti-cancer properties are of immense interest. The wild mushrooms effective against cancer belong to the genus Phellinus, Pleurotus, Agaricus, Ganoderma, Clitocybe, Antrodia, Trametes, Cordyceps, Xerocomus, Calvatia, Schizophyllum, Flammulina, Suillus, Inonotus, Inocybe, Funlia, Lactarius, Albatrellus, Russula and Fomes (Patel and Goyal, 2012).

Biologically active substances of wild mushrooms with anticancer property comprise polysaccharides, polysaccharide-protein complexes, dietary fiber, certain types of proteins, terpenoids, steroids, phenols, *etc*. The polysaccharide, β-glucan is the most versatile metabolite due to its broad spectrum biological activity. (Chen and Seviour, 2007). Phenolic compounds (phenolic acids and flavoid) from mushrooms provide health benefits associated with reduced risk of chronic diseases as well as possibility of cancer. Betulinan A from *Lenzites betulina* shows cytotoxicity against human cancer cells H460, SF268 and MCF7. Caffeic acid phenethyl ester (CAPE) isolated from *Lentinula edodes* inhibited NF-êB binding to DNA and thus suppress aromatase activity (Mattila *et al.*, 2001). This compound is also available from *Phellinus linteus* which also inhibit NF-êB binding to DNA and prohibit cancer development (Nakamura *et al.*, 2003). Ganoderic acid T, most abundant triterpenic acid found in *G. lucidum* is believed to be responsible for anti-cancer effect both under *in vitro* and *in vivo* conditions (Tang *et al.*, 2006). Ganoderiol F (GA-F) is a tetracyclic triterpene found in *Ganoderma* species (Chang *et al.*, 2006), which has shown cytotoxicity *in vitro* against Lewis lung carcinoma, Meth-A, Sarcoma-180, and T-47D cell lines (Gao *et al.*, 2010). Leaianafulvene, a sesquiterpenoid fulvene obtained from *Mycena leaiana* have shown cytostatic properties (Harttig *et al.*, 1990). Zhankuic acid A, C, maleic and succinic acid derivatives from Taiwanofungus camphorates (*Antrodia camphorata*) induce apoptosis and prevent cancer (Chen *et al.*, 2010). Grifolin, a secondary phenolic metabolite from *Albatrellus confluens*, causes deprivation of kinases and G1 arrest which prevent growth of cancer cells (Luo *et al.*, 2011). In addition it can upregulate the death-associated protein kinase1 DAPK1 via p53 in nasopharyngeal carcinoma cells (Luo *et al.*, 2011). 3β-hydroxy-lanosta-8, 24-dien-21-al, inotodiol have shown to inhibit proliferation of cancer cells through caspase-3 dependent apoptosis (Nomura *et al.*, 2008).

Antioxidant

The free radicals are the main cause of oxidative stress by which serious health problems such as neurological disorders, cardiologic diseases, cirrhosis, diabetes, leukemia and necrosis could occur (Akgul *et al.*, 2017). Antioxidants play an active role in inhibition of the oxidative stress in living organisms. Due to the toxic and mutagenic effects of many synthetic drugs, natural products are preferred now days for a healthy life and a better physical defense system (Stajic *et al.*, 2013). Wild mushrooms are the good choices because of the presence of antioxidants which help to reduce the oxidative damage caused by environmental and inherent factors in living organisms.

A number of studies have been reported on antioxidant capability of wild mushroom species (Ferreira *et al.*, 2009). A number of compounds from mushrooms related to antioxidant activity have been identified, including polyphenols (phenolic acids, phenylpropanoids, lignin, melanins, tannins, and flavonoids), β-caroten, and ascorbic acid (Barros *et al.*, 2009). Ethyl acetate and n-hexane fractions of a wild strain of *Cantharellus cibarius* have been reported to possess antioxidant activity which has high amounts of total phenols and flavonoids (Sharareh *et al.*, 2016). *Ganoderma* (*Reishi*) has been called the king of herbal medicines and the phytoconstituents efficiently scavenged the free radicals (Rathee *et al.*, 2012). Wild mushrooms like *Ganoderma lucidum, Phellinus rimosus, Pleurotus florida* and *Pleurotus pulmonary* occurring in South India are valuable sources of antioxidant phenolic compounds (Ajith *et al.*, 2007). Pharmaceutically studies on antioxidant activity of wild mushrooms are still very few. Thus, intensive studies are required for the discovery of new antioxidant agents from wild mushrooms.

Anti-inflammatory

Inflammation is a complex set of interactions among soluble factors and cells that can arise in any tissue in response to trauma, infections, or postischaemic, toxic, or autoimmune injury (Nathan, 2002). It is usually associated with the pathogenesis of diseases such as diabetes, arthritis, obesity, metabolic syndrome, cancer and several cardiovascular diseases (Bellik *et al.*, 2012). Natural products are good resources for development of therapeutic compounds with anti-inûammatory potential and without or lower toxic effects (Yuan *et al.*, 2006). Bioactive compound such as Pleuran, isolated from fruiting bodies of oyster mushrooms has been reported to possess anti-inflammatory activity (Nosalova *et al.*, 2001). Ajith and Janardhanan (2001) found that the methanolic extract of cracked cap mushroom *Phellinus rimosus* significantly inhibited acute and chronic inflammations in mice.

Antiviral

Water soluble lignins isolated from *Inonotus obliquus*, commonly known as 'Chaga', inhibited HIV protease with an IC_{50} value of 2.5 µg ml^{-1} (Ichimura *et al.*, 1998). Anti-HIV activities were reported for mycelial culture medium of *L. edodes* (Tochikura *et al.*, 1988). Sulfated lentinan from *L. edodes* completely prevented HIV-induced cytopathic effect (Yoshida *et al.*, 1988). The protein-bound polysaccharides PSK and PSP from *Trametes versicolor* were also found to have an antiviral effect on HIV and cytomegalovirus *in vitro* (Tochikura *et al.*, 1987). Besides immunostimulation, other effects of the polysaccharide protein complexes contribute to the antiviral activity, *e.g.* inhibition of binding of HIV-1 gp120 to immobilized CD4 receptor and of reverse transcriptase activity of viruses (Colins, 1997). Several triterpenes from *Ganoderma lucidum i.e.* ganoderiol F, ganodermanontriol and ganoderic acid B are active as antiviral agents against human immunodeficiency virus type 1 (HIV-1). *In vitro* antiviral activity against influenza viruses type A and B was demonstrated for mycelial extracts of *Kuehneromyces mutabilis*, *Inonotus hispidus* and ergosterol peroxide, present in several mushrooms (Mentel *et al.*, 1994: Lindequist *et al.*, 1989: Awadh *et al.*, 2003).

Therapeutic Potentials of Mushrooms from Simlipal

Simlipal Biosphere Reserve (SBR) in Mayurbhanj district of Odisha, eastern India (20°17′ to 22°34′N and 85°40′ to 87°10′E) covering an area of 5569 Km^2 is one of the tropical forest ecosystems, which forms one of the mega biodiversity zones of the country with a rich diversity of flora and fauna including mushrooms. The landscape of Simlipal comprises numerous rolling hills. The vegetation comprises of tropical semi evergreen forest, tropical moist deciduous forest, dry deciduous hill forest, high level Sal forest and grass land. Although the climate of this region is most favorable for mushroom growth and development but the study on mushroom is lacking. Survey was conducted by the authors to document the diversity of wild mushrooms from this region through implementation of major research projects from DST, Govt. of Odisha and DST, Govt. of India. Apart from mushroom diversity therapeutic potentials of the wild mushrooms and molecular investigations leading to their proper identification were also conducted. Information on indigenous knowledge regarding the use of wild edible mushrooms for food as well as medicine purpose by few ethnic tribes of SBR was collected comprehensively through structured questionnaires (Sachan *et al.*, 2013). In a study conducted by the authors, 14 wild mushroom species were reported to have food and ethnomedicinal values. In another report by Panda *et al.* (2015), identified 24 different fleshy mushroom species out of which 14 mushroom species have been documented to have ethnomedicinal uses from the northern region of Odisha including SBR.

The medicinal potential of 2 wild edible mushroom species *Lentinus sajor-caju* and *Lentinus torulosus* was studied for their antioxidant and antibacterial activities (Sachan *et al.*, 2013). The solvent extracts (ethanol, methanol, and water) of the mushrooms exhibited strong antioxidant properties (ABTS, DPPH, H_2O_2, and metal chelating activities) with scavenging activity up to 70.54 per cent along with phenol, flavonoid, and total antioxidant capacity. Both the mushrooms showed moderate antibacterial activity (11.0-18.33 mm inhibition zones) against *Streptococcus aureus* and *Vibrio cholera*. Another three wild edible mushrooms *viz.*, *Russula vesca, R. delica* and *T. eurrhizus* of SBR were analyzed for their nutritional antioxidant and antibacterial potential by Sachan *et al.* (2014). Their findings showed that these three mushrooms are rich sources of antioxidants (ABTS, DPPH, H_2O_2 and metal chelating activities) in three different solvent extracts ethanol, methanol and aqueous with scavenging potential up to 89 per cent at concentration 100 mg/ml. Total phenol content was found between 21.92–41.99 mg catechol/g extract and flavonoid 2.53–7.52 mg quercetin/g extract. The studied mushrooms possess moderate antibacterial properties with zones of inhibition ranging from 13 to 30 mm against six human pathogenic bacteria which are comparable with Amphoxyllin standard. Being a source of nutrients and molecules with medicinal potential, the studied mushrooms can be used in human diet as nutraceuticals/functional foods for maintaining and promoting health, longevity and life quality. Medicinal mushrooms

Figure 26.1: Some of the Wild Medicinal Mushrooms of Simlipal (A) *Ganoderma applanatum* (B) *Ganoderma tsugae* (C) *Volvorella volvacea* (D) *Astraeus hygrometricus* (E) *Geastrum* sp. (F) *Lentinus sajor-caju* (G) *Russula delica* (H) *Russula vesca* (I) *Termitomyces eurrhizus*.

viz. Ganoderma applanatum, Ganoderma lipsiense, Ganoderma chalceum and *Ganoderma tsugae* are important forest products of SBR, Odisha were evaluated for the chemical constituents, antioxidant and antibacterial potential (Sachan *et al.*, 2017). The chemical constituents were determined in terms of macronutrients, micronutrients and phenolic contents. The macronutrient content of these mushrooms revealed high amounts of protein (17.10–20.50/100 g), carbohydrate (41.80–59.16/100 g) and low fats (1.2–3.0/100 g) and possessed good quantities micronutrients (vitamins, carotenoids). Mushrooms showed strong antioxidant properties due to the presence of high phenol (27.40–220.41 mg catechol/g extract) and flavonoid (3.42–171.44 mg quercetin/g extract) content. All these mushrooms possess moderate antibacterial properties with zone of inhibition ranging from 11.76 to 18.83 mm against five human pathogenic bacteria. Based on their rich bioactive compounds (vitamins and phenolics), antioxidant and antibacterial activity, they might be exploited by pharmaceutical and cosmetic industries.

Panda *et al.* (2017) have successfully developed the mycelia from the macro-fungal spore of *Geastrum* sp., a rarely found as wild edible mushroom species from Simlipal and established the antimicrobial activity of the fungal metabolites which was reported earlier for its ethnomedicinal uses as wound healing by tribal peoples of this region (Panda *et al.*, 2013). Antimicrobial activity was evaluated by the authors using three liquid media (*viz.* PDB, MEB, CDB)) against six bacterial and three fungal clinically significant human pathogens at different incubation periods *viz.* 7 days, 14 days, 21 days. They have demonstrated that mushroom possesses considerable antimicrobial activity of highest zone of inhibition (30 mm) against the fungal

Figure 26.2: Collection and Investigation of Wild Mushrooms from Simlipal for their Use as Food/Ethnomedicine.

pathogen *Trychophyton mentagrophytes* - MTCC 8476 and out of six test bacterial pathogens higher inhibitory activity was observed against two gram-negative bacteria *i.e. Shigella flexneri*- MTCC 1457 (20 mm), *Klebsiella pneumonia* - MTCC 3384 (22 mm) and one gram-positive bacteria *i.e. Staphylococcus aureus* - MTCC 737 (22 mm). This indicated that metabolites are broad spectrum in nature and it could be exploited for therapeutic applications. However, a more indepth study is required for isolation and identification of active principle responsible for such bioactivity.

Conclusion

This paper summarized the therapeutic potential of wild mushrooms by highlighting their ethnomedicinal uses and pharmacological potential from mushrooms. The therapeutic potential and ethnomedicinal knowledge of wild mushrooms from Simlipal discussed in the present paper might provide useful information for discovery of new therapeutic agents to combat growing health problems of the world. As mentioned, the mushroom species presently identified and utilized only account for a small percentage of total amounts, which indicates more species yet to be discovered, verified, and cultivated which will help in further exploitation and boosting of the development of the pharmaceutical industry by interdisciplinary studies combined with genomics, proteomics, metabolomics, and systems pharmacology.

References

Adhikari, M.K., Devkota, S. and Tiwari, R.D. (2005). Ethnomycolgical knowledge on uses of wild mushrooms in western and central Nepal. Our Nature, 3: 13-19.

Ajith, T.A. and Janardhanan, K.K. (2001). Antioxidant and antiinfammatory activities of methanol extract of *Phellinus rimosus*. Indian Journal of Experimental Biology, 39: 1166-1169.

Ajith, T.A., Usha, S. and Nivitha, V. (2007) Ascorbic acid and α-tocopherol protect anticancer drug cisplatin induced nephrotoxicity in mice: a comparative study. Clinica Chimica Acta, 375: 82-86.

Akgul, H., Sevindik, M., Coban, C., Alli, H. and Selamoglu, Z. (2017). New approaches in traditional and complementary alternative medicine practices: *Auricularia auricula* and *Trametes versicolor*. Journal of Traditional Medicine and Clinical Naturopathy, 6: 4.

Akpaja, E.O., Isikhuemhen, O.S. and Okhuoya J.A. (2003). Ethnomycology and usage of edible and medicinal mushrooms among the Igbo people of Nigeria. International Journal of Medicinal Mushroom, 5: 13: 313–319.

Awadh, A.A.N., Mothana, R.A.A., Lesnau, A., Pilgrim, H. and Lindequist, U. (2003). Antiviral activity of extracts and compounds from *Inonotus hispidus*. Fitotherapia, 74: 483–485.

Barros, L., Cruz, T., Baptista, P., Estevinho, L.M and Ferreira, I.C. (2008). Wild and commercial mushrooms as source of nutrients and nutraceuticals. Food Chemistry and Toxicology 46: 2742-2747.

Barros, L., Dueñas, M. and Ferreira I.C. (2009). Phenolic acids determination by HPLC-DAD-ESI/MS in sixteen different Portuguese wild mushrooms species. Food Chemistry and Toxicology, 47: 1076-1079.

Bellik, Y., Boukraâ, L., Alzahrani, H. A., Bakhotmah, B. A., Abdellah, F., Hammoudi, S. M., and Iguer-Ouada, M. (2012). Molecular mechanism underlying anti-inflammatory and anti-allergic activities of phytochemicals: an update. Molecules, 18: 322–353.

Biziulevicius, G.A., Vaitkuviene, A. (2007). vantage of the experience in ethnomedicinal use of mushrooms: anti-inflammatory and related pharmacological activities of fly agaric (*Amanita muscaria*) ethanolic extract deserve a modern evaluation. Medical Hypotheses, 69: 4: 946-947.

Buswell, J.A. and Chang, S.T. (1993). Edible Mushrooms. Attributes and Applications. In: Chang S.T., Buswell J.A. and Miles, P.G (eds). Genetics and Breeding of Edible Mushrooms. Gordon and Breach, Philadelphia. pp. 297-394.

Cha, J.Y., Jun, B.S., Kim, J.W., Park, S.H., Lee, C.H. and Cho Y.S. (2006). Hypoglycemic effects of fermented Chaga mushroom (*Inonotus obliquus*) in the diabetic Otsuka Long-Evans Tokushima Fatty (OLETF) rat. Food Science Biotechnology, 15: 739–745.

Chang, S.T. (1993). Mushroom biology: the impact on mushroom production and mushroom products. In: Chang ST, Buswell JA, Chiu SW (eds) Mushroom biology and mushroom products. The Chinese University Press, Hong-Kong, pp 3–20.

Chang, S.T. and Miles, P.G. (1992). Mushrooms biology - a new discipline. Mycologist, 6: 64–5.

Chang, U.M.L., Chyi H. L., Liang H., Cheng P. K., Lou S.L. and Shwu Bin. (2006). Ganoderiol F, a ganoderma triterpene, induces senescence in hepatoma HepG2 cells. Life sciences, 79: 1129-39.

Chen, H., Lu, X., Qu, Z., Wang, Z. and Zhang, L. (2010) Glycosidase inhibitory activity and antioxidant properties of a polysaccharide from mushroom *Inonotus obliquus*. Journal of Food Biochemistry, 34: 178–219.

Chen, J., Mao, D., Yong, Y., Li, J., Wei, H. and Lu, L. (2012). Hepatoprotective and hypolipidemic effects of water-soluble polysaccharidic extract of *Pleurotus eryngii*. Food Chemistry, 130: 687–694.

Chen, J. and Seviour, R. (2007). Medicinal importance of fungal beta-(1–>3), (1–>6)-glucans. Mycological Research, 111(6): 635-652.

Cheung, L.M., Cheung, P.C. and Ooi, V.E. (2003). Antioxidant activity and total phenolics of edible mushroom extracts. Food Chemistry, 81: 249–255.

Coetzee, J.C. and Wyk, A.E.V. (2009) The genus Calvatia ('Gasteromycetes', Lycoperdaceae): A review of its ethnomycology and biotechnological potential. African Journal of Biotechnology, 8(22): 6007-6015.

Colins, R.A. and Ng T.B. (1997). Polysaccharopeptide from Coriolus versicolor has potential for use against human immunodeficiency virus type 1 infection. Life Sciences, 60: 383–387.

Deshmukh, S.K. (2004) Biodiversity of tropical basidiomycetes as sources of novel secondary metabolites. In: Jain PC (ed). Microbiology and Biotechnology for Sustainable Development. CBS Publishers and Distributors, New Delhi, India. pp 121-140.

Dilani, D.D., Rapior, S. H. and Kevin, B. A. (2012). Medicinal mushrooms in prevention and control of diabetes mellitus. Fungal Diversity, 56: 1-29.

Dutta, A.K. and Acharya, K. (2014) Traditional and ethno-medicinal knowledge of Mushrooms in West Bengal, India. Asian Journal of Pharmaceutical and Clinical Research, 7: 4.

Ferreira, I.C.F.R., Barros L. and Abreu, R.M.V. (2009). Antioxidants in wild mushrooms. Curreent Medicinal Chemistry, 16: 1543-1560.

Gao, X., Deeb, D., Hao, J., Liu, Y., Arbab, A.S., Dulchavsky, S.A., Gautam, S.C. (2010). Synthetic triterpenoids inhibit growth, induce apoptosis and suppress pro-survival Akt, mTOR and NF-êB signaling proteins in colorectal cancer cells. Anticancer Research, 30(3): 785–792.

Harttig, U., Anke, T., Scherer, A. and Steglich, W. (1990). Leaianafulvene a sesquiterpenoid fulvene derivative from cultures of *Mycena leaiana*. Phytochemistry, 29(12): 3942-3944.

Hawksworth, D.L. (2001). Mushrooms: The extent of the unexplored potential. International Journal of Medicinal Mushrooms, 3: 333–340

Ichimura, T., Watanabe, O. and Maruyama, S. (1998). Inhibition of HIV-1 protease by water-soluble lignin-like substance from an edible mushroom, *Fuscoporia obliqua*. Bioscience Biotechnology Biochemistry, 62: 575–7.

Jha, K.S., Kumar V. and Tripathi N.N. (2011). Survey of ethnomedicinal macrofungi of Nagarjuna and Phulchowki areas of Kathmandu Valley, Nepal. International Journal of Pharmaceutical Science Review Research, 11: 147-151.

Lee, K.H., Morris, N.S.L., Yang, X., Huang, R., Zhou, T., Wu, S., Shi, Q. and Itokawa, H. (2012). Recent progress of research on medicinal mushrooms, foods, and other herbal products used in traditional Chinese medicine. Journal of Traditional and Complemental Medicine, 2(2): 84–95.

Lee, Y.S., Kang, I.J., Won, M.H., Lee, J.Y., Kim, J.K. and Lim, S.S. (2010). Inhibition of protein tyrosine phosphatase 1beta by hispidin derivatives isolated from the fruiting body of *Phellinus linteus*. Natural Product Communications, 5(12): 1927–1930.

Lemieszek, M.K., Langner, E., Kaczor, J., Kandefer S.M., Sanecka. B., Mazurkiewicz, W. and Rzeski, W. (2011). Anticancer effects of fraction isolated from fruiting bodies of Chaga medicinal mushroom, *Inonotus obliquus* (Pers.: Fr.) Pilát (Aphyllophoromycetideae): *In vitro* studies. International Journal of Medicinal Mushrooms, 13(2): 131-43.

Lindequist, U., Lesnau, A., Teuscher, E. and Pilgrim, H. (1989). Untersuchungen zur antiviralen Wirksamkeit von Ergosterolperoxid. Pharmazie, 44: 579–580.

Luo, X.J., Li, W., Yang, L.F., Yu, X.F., Xiao, L.B., Tang M., Dong X., Deng Q.P., Bode A.M., Liu J.K. and Cao, Y. (2011). DAPK1 mediates the G1 phase arrest in human nasopharyngeal carcinoma cells induced by grifolin, a potential antitumor natural product. European Journal of Pharmacology, 670: 427–434.

Maiti, S., Bhutia, S.K., Mallick, S.K., Kumar, A., Khadgi, N and Maiti, T.K. (2008). Antiproliferative and immunostimulatory protein fraction from edible mushrooms. Environmental Toxicology and Pharmacology, 26: 187–191.

Manzi, P.A., Agguzzi, A. and Pizzoferrato, L. (2001). Nutritional mushrooms widely consumed in Italy. Journal of Food Chemistry, 73: 321-325.

Manzi, P., Gambelli, L., Marconi, S., Vivanti, V. and Pizzoferrato, L. (1999). Nutrients in edible mushrooms: an inter-species comparative study. Food Chemistry, 65: 477–482.

Mattila, P., Könkö, K., Eurola, M., Pihlava, J.M., Astola, J., Vahteristo, L., Hietaniemi, V., Kumpulainen, J., Valtonen, M. and Piironen, V. (2001). Contents of vitamins, mineral elements, and some phenolic compounds in cultivated mushrooms. Journal of Agriculture Food Chemistry, 49(5): 2343-2348.

Mentel, R., Meinsen, D., Pilgrim, H., Herrmann, B. and Lindequist, U. (1994) *In vitro* antiviral effect of extracts of *Kuehneromyces mutabilis* on influenza virus. Pharmazie, 49: 859–860.

Mondal, S., Chakraborty, I., Rout, D and Islam S.S. (2006). Isolation and structural elucidation of a water-soluble polysaccharide (PS-I) of a wild edible mushroom, *Termitomyces striatus*. Carbohydrate Research, 341: 7: 878-886.

Mothana, R.A.A., Jansen, R., Jülich, W.D. and Lindequist, U. (2000). Ganomycin A and B, new antimicrobial farnesyl hydroquinones from the basidiomycete *Ganoderma pfeifferi*. Journal of Natural Product, 63: 416–418.

Nakamura, M., Kono, Y. and Takatsuki, A. (2003). Mepanipyrim, a novel inhibitor of pharmacologically induced Golgi dispersion. Bioscience, Biotechnology, and Biochemistry, 67: 139–150.

Nathan, C. (2002). Points of control in inflammation. Nature, 420: 846-852.

Nomura, M., Takahashi, T., Uesugi, A., Tanaka, R. and Kobayashi, S. (2008). Inotodiol, a lanostane triterpenoid, from *Inonotus obliquus* inhibits cell proliferation through caspase-3-dependent apoptosis. Anticancer Research, 28: 2691-2696.

Nosál'ová, V., Bobek, P., Cerná, S., Galbavy, S. and Stvrtina, S. (2001) Effects of pleuran (beta-glucan isolated from *Pleurotus ostreatus*) on experimental colitis in rats. Physiological research / Academia Scientiarum Bohemoslovaca, 50: 575-581.

Okigbo, R.N. and Nwatu, C.M. (2015). Ethnostudy and usage of edible and medicinal mushrooms in some parts of Anambra state, Nigeria. Natural Resources, 6: 79-89.

Panda, M.K. and Tayung, K. (2015). Documentation and ethnomedicinal knowledge on wild edible mushrooms among ethnic tribes of northern Odisha, India. Asian Journal of Pharmaceutical and Clinical Research, 8: 139-143.

Panda, M.K., Thatoi, H.N. and Tayung, K. (2017). Antimicrobial potentials of spore culture of geastrum sp., a rare wild edible mushroom of Simlipal biosphere reserve, odisha, india against some significant human pathogens. International Journal of Advance Research, 5: 9.

Patel, S. and Goyal, A. (2012). Recent developments in mushrooms as anti-cancer therapeutics: A review. Biotech, 2: 1–15.

Quereshi, S., Pandey, A.K. and Sandhu, S.S. (2010). Evaluation of antibacterial activity of different *Ganoderma lucidum* extracts. People's Journal of Scientific Research, 3: 9-13.

Ramesh, C. and Pattar, M.G. (2010). Antimicrobial properties, antioxidant activity and bioactive compounds from six wild edible mushrooms of Western Ghats of Karnataka, India. Pharmacognosy Research, 2: 107-112.

Rao, Y.K., Fang, S.H., Wu, W.S. and Tzeng, Y.M. (2010). Constituents isolated from Cordyceps militaris suppress enhanced inflammatory mediator's production and human cancer cell proliferation. Journal Ethnopharmacology, 131: 363-367.

Rathee, S., Rathee, D., Kumar, V. and Rathee, P. (2012). Mushrooms as therapeutic agents. Brazilian Journal of Pharmacognosy, 22: 459-474.

Sachan, S.K.S., Patra J.K., Tayung, K., Sarangi, K. and Thatoi H.N. (2014). Evaluation of nutritional and nutraceutical potentials of three wild edible mushrooms from Simlipal Biosphere Reserve, Odisha, India. Journal of Consumer Protection and Food Safety, 9: 111-120.

Sachan, S.K.S., Patra, J.K. and Thatoi, H.N. (2013). Indigenous knowledge of ethnic tribes for utilization of wild mushrooms as food and medicine in Simlipal biosphere reserve, Odisha, India. Journal of Agriculture Technology, 9: 335-348.

Sachan, S.K.S., Patra, J.K. and Thatoi, H.N. (2013). Nutritional and bioactive potential of two wild edible mushrooms (*Lentinus sajor-caju* and *Lentinus torulosus*) from Simlipal Biosphere Reserve, India. Food Science Biotechnology, 22: 137-145.

Sachan, S.K. and Patra, J.K. (2017). Chemical constituents, antioxidative and antibacterial properties of medicinal mushrooms collected from Similipal Biosphere Reserve, Odisha, India. Proceedings of National Academic of Science, India, Sect. B Biological Science, 87: 559.

Sharareh, R., Sara, S., Taher, N.S. and Amin, M. (2016). Antioxidant potential and other medicinal properties of edible mushrooms naturally grown in Iran. Biomedical Research India, 27: 1.

Sivakumar, R., Vetrichelvan, T., Rajendran, N.N., Indira D.M., Sundaramoorthi, K., Shankar, A.S.K. and Shanmugam, S. (2006). Antibacterial activity of mushroom *Osmoporus odoratus*. Indian Journal of Pharmaceutical Science, 68: 523-524.

Srivastava, B., Dwivedi, A.K. and Pandey, V.N. (2011). Ethnobotanical survey, distribution and utilization of Termitomyces species in Gorakhpur forest division. Plant Sciences Feed, 1: 28-33.

Stajic, M., Vukojevic, J., Kneževiæ, A., Duletic, S., and Milovanovic, I. (2013). Antioxidant protective effects of mushroom metabolites. Current Topics in Medicinal Chemistry, 13.

Tochikura, T.S., Nakashima, H., Hirose, K. and Yamamoto, N. (1987). A biological response modifier, PSK, inhibits human immunodeficiency virus infection *in vitro*. Biochemistry and Biophysics Research Communication, 148: 726–733.

Tochikura, T.S., Nakashima, H., Ohash, i Y. and Yamamoto N. (1988). Inhibition (*in vitro*) of replication and of the cytopathic effect of human immunodeficiency virus by an extract of the culture medium of *Lentinus edodes*mycelia. Medical Microbiology and Immunology, 177: 35–44.

Wani, B.A., Bodha, R.H. and Wani, A.H. (2010). Nutritional and medicinal importance of mushrooms. Journal of Medicinal Plants Research, 4: 2598–2604.

Wasser, S.P. (2000). Dietary supplements from medicinal mushrooms: Diversity of types and variety of regulations. International Journal of Medicinal Mushrooms, 2: 20-21.

WHO. (2011). World Health Organization. Diabetes program. http: //www.who.int/mediacentre/factsheets/fs312/en/

Wu, S.H., Ryvarden, L. and Chang, T.T. (1997). *Antrodia camphorata* ('niu-chang-chih'), new combination of a medicinal fungus in Taiwan. Botanical Bulletin of Academia Sinica. 38: 4: 273–275.

Yamac, M and Bilgili, F. (2006) Antimicrobial activities of fruit bodies and/or mycelia cultures of some mushroom isolates. Pharmaceutical Biology, 44(9): 660-667.

Yoshida, O., Nakashima, H., Yoshida, T., Kaneko, Y., Yamamoto, I. and Matsuzaki, K. (1988). Sulfation of the immunomodulating polysaccharide lentinan: a novel strategy for antivirals to human immunodeficiency virus (HIV). Biochemistry Pharmacology, 37: 2887–2891.

Yuan, G., Wahlqvist, M.L., He, G. and Yang, M. (2006) Natural products and antiinflammatory activity. Asia Pacific Journal of Clinical Nutrition, 15: 143-152.

27

Termitomyces Mushrooms: A Natural Resource of Biomolecules with Enormous Bioactive Potentials for Biotechnological Exploitation

*K. Singha[1], D.K. Sahoo[1], A. Banerjee[1,2] and P.K. Das Mohapatra[3]**

[1]Department of Microbiology, Vidyasagar University,
[2]Department of Biotechnology, Oriental Institute of Science and Technology, Midnapore – 721 102, West Bengal
[3]Department of Microbiology, Raiganj University, Raiganj – 733 134, Uttar Dinajpur, West Bengal
**E-mail: pkdmvu@gmail.com*

ABSTRACT

In present century, food and medicine industry have shown an increased interest in the development of next generation therapeutics and nutraceuticals. Currently many researches are going on for the discovery of bioactive compounds and their potential applications from natural resources. Termitomyces is a genus of edible mushroom collected from the wild and commonly consumed in Africa and Asia. Termitomyces mushrooms grow as symbionts in the termite nests, produce various enzymes to help termites for the digestion of lignocellulosic substrates from plant materials. Many species of Termitomyces are used by different ethnic tribes with profound ethnomedicinal knowledge. Major bioactive components of Termitomyces mushrooms include polysaccharides, phenolics, fatty acids and lignocellulolytic enzymes have potential uses as antioxidants, immunomodulators, antitumors, and antimicrobials. The present study depicts major bioactive compounds from Termitomyces species as well as their medicinal and immunotherapeutic properties which may emphasise their conservation, cultivation and further biotechnological exploitation.

Keywords: *Antioxidant, Lignocellulolytic, Phenolics, Polysaccharides, Teremitomyces.*

Introduction

Mushrooms have been used for many centuries in Asian and other traditional cuisine and also as medicines due to their culinary and medicinal properties respectively. But their properties remained unknown to the wide scientific community for a long time. Edible mushrooms are excellent sources of protein, have low-fat content, free of cholesterol and also a good source of minerals, serves as low-calorie diets and of great importance to vegetarians. They are potential protein sources especially in developing countries where animal protein is scarce and expensive (Fink and Hoppenhaus, 1958; Suzuki and Oshima, 1976; Gruen and Wong, 1982). In central southern Africa, China, India, Wild Edible Fungi (WEF) are a significant source of nutrition for indigenous people who live in rural and tribal belts. WEF are being collected for food and for money in more than 80 countries (Mattila*et al.*, 2000; Rai, 1994).Wild and cultivated mushrooms contain a huge diversity of biomolecules with nutritional (Kalaè, 2009) and medicinal properties (Borchers *et al.*, 2004; Lindequist *et al.*, 2005). Due to these properties, they have been known as functional foods, and as a source for the development of medicines and nutraceuticals. Mushroom fruiting bodies and mycelia contain a variety of bioactive metabolites with immunomodulatory, cardiovascular, liver-protective, antifibrotic, anti-inflammatory, antidiabetic, antiviral, antioxidant, antitumor, and antimicrobial properties (Wasser *et al.*, 1999; Ishikawa *et al.*, 2001; Poucheret *et al.*, 2006; Zhang *et al.*, 2007). Rural people believe mushrooms have medicinal values and can serve as blood tonic, reduce obesity and lower blood pressure in hypertensive patients. Their pharmacological and therapeutic applications in promoting human health have been known for thousands of years (Öztürk *et al.*, 2011; Xu *et al.*, 2011). Mushrooms could be a source of natural antimicrobials which can be low molecular weight (LMW) and high molecular weight (HMW) compunds respectively. LMW compounds are mostly secondary metabolites such as sesquiterpenes and other terpenes, steroids, cinnamic acid and benzoic acid derivatives, and quinolines. HMW compounds mainly include peptides and proteins along with long chain polysaccharides. It is estimated that there are about 140000 species of mushrooms on earth, and of these only 22000 are known and only a small percentage (5 per cent) has been investigated. Therefore, there is immense scope to explore new bioactive compounds from mushrooms and their potential applications (Lindequist *et al.*, 2005).

In this scenario, the mushroom *Termitomyces*, a genus of macrofungi belonging to the family Lyophyllaceae, class Basidiomycetes, was distinguished and discovered by Roger Heim in 1942 for a group of agarics mutualistically associated with termites in Africa and Asia (Aanen *et al.*, 2009; Mueller *et al.*, 2005). The genus *Termitomyces* was first classified by Berkley, (1847) and thereafter many researchers (Heim 1977; Zang 1981) have studied the diversity and ecology of *Termitomyces.* Approximately 20 species of *Termitomyces* are good in flavour and texture of fruiting bodies. These mushrooms occur during rainy seasons, regularly collected both for domestic consumption and for sale in local markets (Ghorai *et al.*, 2009; Sangvichien and Taylor-Hawksworth 2001). *Termitomyces*is very slow-grower in culture medium and also it is difficult to isolate from the natural habitat where many other fungus propagules are present. In areas where staple foods with low nutrition values are consumed and result in poor health, *Termitomyces* mushrooms become an important source of nutrients as well as microelements including vitamins and minerals (Nakalembe *et al.*, 2015; Sargunam *et al.*, 2012). It was estimated that the protein content of approximately 100 g fruiting body of *Termitomyces* species varied from 33- 45g on dry weight basis (Parent and Thoen, 1977). According to Chinese Academy of Sciences (1978), *Termitomyces* sp. contains 7.3 per cent of dry matter (Table 27.1) and 92.7 per cent of water.

Table 27.1: Various Components of Dry Matter Present in Fruiting Body of *Termitomyces* Species

Components	*Concentration*
Crude fat	3.40 per cent
Soluble sugar	4.57 per cent
Ash	7.17 per cent
Hydrolysis of sugar	9.59 per cent
Crude fiber	13.91 per cent
Pure protein	21 per cent
Crude protein	36.94 per cent
Soluble non-nitrogen quality materials	38.42 per cent

Several species of *Termitomyces* are commonly used as ethnomedicine by indigenous communities for health promotion and treatment of illnesses. In India, *T. heimii* and *T. microcarpus* are used in treatment for fever, cold, and fungal infections (Venkatachalapathi and Paulsamy 2016); *T heimii* is used as blood tonics during wound healing and blood coagulation (Chandrawati *et al.*, 2014); and *T. eurrhizus* is for curing of rheumatic disorder and diarrhoea, and for lowering high blood pressure (Sachan *et al.*, 2013).The *Termitomyces* fungus in the comb decomposes 167 to 341 kg of organic matter a year (Kamat, 1971).

In this context due to enormous nutritional and medicinal values of *Termitomyces* mushrooms, the present chapter highlighted on the major bioactive compounds from *Termitomyces* species along with their corresponding biological potentials.

Major Bioactive Compounds from *Termitomyces* Mushroom and their Potential Bioactivity

Polysaccharides

Various types of polysaccharides have been isolated from *Termitomyces* species for structure analysis, with their implications as immunomodulators and antioxidants. A water-soluble heteropolysaccharide was isolated from fruiting bodies of *T. striatus* composed of D-glucose, D-galactose, D-mannose, and L-fucose exhibited significant immune response (Mondal *et al.*, 2006). One water-soluble and one water-insoluble β-glucan were isolated from a hot water extract from fruiting bodies of *T. robustus* showed immunomodulatory effects (Bhanja *et al.*, 2012). A water-soluble β-glucan was isolated from *T. heimii*, exhibited potent antioxidant activities (Manna *et al.*, 2015). A water-soluble heteroglycan, isolated from fruiting bodies of *T. clypeatus*, consisted of D-glucose, D-galactose, D-mannose, and L-fucose showed antioxidant properties (Pattanayak *et al.*, 2015). Polysaccharides extracted from *T. albuminosus* has been reported to reduce atrophy of both spleen and thymus of immune suppressed mice (Wang 2013). Many polysaccharides from various mushrooms also exhibit antitumor properties (Zhang *et al.*, 2007).

Immunomodulatory Properties

A positive immunomodulatory effect was found in splenocytes by heteropolysaccharide present in fruiting bodies of *T. striatus* (Mondal *et al.*, 2006). Hot water extract from fruiting bodies of *T. robustus* contain two types of β-glucans, one is linear and water-soluble, another is branched and water-insoluble (Bhanja *et al.*, 2012). These β-glucans help in Immunomodulation by significantly activating splenocytes, macrophages and thymocytes. In another study, the people of Tanzania consumed edible mushrooms as *T. titanicus, T. letestui, T. eurrhizus, and T. aurantiacus* in different gastrointestinal problem and the *T. microcarpus* during long illness to boost their immune system (Tibuhwa 2012). In 2013 Wang reported that, the atropy of spleen and thymus of immunosuppressed mice was reduced by the polysaccharide of *T. albuminosus.*

Antioxidant Activity

Various types of polysaccharides have been isolated from *Termitomyces* have been used as antioxidants (Manna *et al.*, 2015, Hsieh and Ju, 2018). It was reported that the β-glucan of *T. heimii* have significant antioxidant property. They also exert protective function against nicotine toxicity (Manna *et al.*, 2015). A similar antioxidant property was proven by a water-soluble heteroglycan isolated from fruiting bodies of *T. clypeatus* (Pattanayak *et al.*, 2015). Exopolysaccharides (EPS) and their native depolymerized forms (DEPS) were found in fermentation medium of *T. albuminosus*, and it was found that the DEPS have superior antioxidant activity compared to that of EPS (Hsieh and Ju, 2018).

Detoxification

Two important enzymes named alcohol dehydrogenase ADH2 and aldehyde dehydrogenase ALDH2 have important role in alcohol detoxification. Zhao *et al.* (2015) reported that the polysaccharides from *T. albuminosus* exert protective function on an alcohol-induced injury of mouse livers through increasing the production of these enzymes.

Treatment of Atherosclerosis

Polysaccharides extracted from *T. albuminosus* exhibited hepatoprotective effects against hyperlipidemia (Zhao *et al.*, 2016; Zhao *et al.*, 2017), which is a risk factor of atherosclerosis and cardiovascular diseases due to high levels of total cholesterol, triglyceride, and low density lipoprotein cholesterol and a low level of high density lipoprotein cholesterol (Nelson 2013).

Treatment of Hemorrhoids

In traditional Chinese medicine, *T. albuminosus* is used to treat hemorrhoids. Phytochemical assays revealed the presence of polysaccharides, saponins, cerebrosides, coumarin and ergosterol in this fungus (Qi *et al.*, 2001) have potent analgesic and anti-inflammatory effects (Lu, 2008; Hsieh and Ju, 2018) for hemorrhoids treatment.

Treatment of Leishmaniasis

The causative agents of leishmaniasis is *Leishmania* sp., a parasitic protozoans are usually transmitted between vertebrate hosts by the bite of blood sucking female sand flies. Between the year 2004 to 2008, an estimated 200,000–400,000 cases and 20,000– 40,000 deaths was reported by visceral leishmaniasis (VL) globally (Alvar *et al.*, 2012). In another report, VL leads to substantial health problems or death for up to 400,000 people per year. A water-soluble polysaccharide fraction from *T. eurrhizus* was efficiently inhibit the replication of intracellular amastigotes of *Leishmania donovani* (Mallick *et al.*, 2014).

Phenolics

In recent years, there are an increasing number of reports on phenolic compounds in *Termitomyces* mushrooms species. Phenolic acids including cinnamic and benzoic acid derivatives have been pointed out as the most common. The phenolic compound caffeic acid showed *in vitro* inhibitory effect on the proliferation of cancer cell (Prasad *et al.*, 2011) whereas some phenolic compounds present in few edible mushroom also exhibited an inhibitory effect on inflammation *in vitro* (Moro *et al.*, 2012). Phenolic compounds extracted from fruit bodies from 23 species edible mushrooms using water and methanol as solvent (Puttaraju *et al.*, 2006). It was found that among all these species, *T. heimii, T. mammiformis* showed highest antioxidant activity in water extract compared to its methanolic extract due to presence of high amount of phenolic compounds (tannic acid and gallic acid, gentisic acid, and protocatechuic acid, vanillic acid, syringic acid, pcoumaric acid, caffeic acid, ferulic acid, and cinnamic acid) in water extract. In 2009, Ferreria *et al.*, also indicated that antioxidant found in wild mushroom is predominantly phenolic compounds. Cheah and Halliwell (2012) described that the ergothioneine have numerous antioxidant along with cytoprotective functions. Similar study was also reported by Abdullah *et al.* (2012), among 17 medicinal mushrooms, water extract of *T. Heimii* showed significant antioxidant activity (IC_{50} = 26.839±0.189 mg/ml). The major component of methanolic extracts from fruiting bodies of *T. heimii* and *T. microcarpus* (Johnsy and Kaviyarasan 2014), and fruiting bodies of *T. clypeatus, T. microcarpus,* and *T. letestui* (Woldegiorgis *et al.*, 2014) were phenolic compounds and showed antioxidant activity along with scavenging and chelating properties. The phenolic content of *T. Microcarpus* was found 7.258±0.323 µg/mg which is significantly higher, have powerful chain breaking antioxidant and also be used as food supplement to fight against stress related disorders (Mitra *et al.*, 2016).

Table 27.2: Antioxidant Activity of different Phenolics and Flavonoides from *Termitomyces* Species

Bioactive Compounds	*Mushroom*	*Reference*
Phenolics		
Protocatechuic acid	*Termitomyces sp.*	Ferreira *et al.* (2009)
Gallic acid	*Termitomyces sp., T. heimii, T. microcarpus*	Ferreira *et al.* (2009), Mitra *et al.* (2015), Mitra *et al.* (2016)
Vanillic acid	*T. heimii, T. microcarpus*	Mitra *et al.* (2015), Mitra *et al.* (2016)
Pyrogallol	*T. heimii, T. microcarpus*	Mitra *et al.* (2015), Mitra *et al.* (2016)
Syringic acid		
Salicylic acid	*T. heimii, T. microcarpus*	Mitra *et al.* (2015), Mitra *et al.* (2016)
Cinnamic acid	*T. heimii, T. microcarpus*	Mitra *et al.* (2015), Mitra *et al.* (2016)
p-coumaricacid	*T. heimii, T. microcarpus*	Mitra *et al.* (2015), Mitra *et al.* (2016)
Chlorogenic acid	*T. heimii, T. microcarpus*	Mitra *et al.* (2015), Mitra *et al.* (2016)
Caffeic acid		
Ferulic acid	*T. heimii, T. microcarpus*	Mitra *et al.* (2015), Mitra *et al.* (2016)
Flavonoides		
Myricetin	*T. heimii, T. microcarpus*	Mitra *et al.* (2015), Mitra *et al.* (2016)
Kaempferol	*T. heimii*	Mitra *et al.* (2015)
Quercetin	*T. heimii, T. microcarpus*	Mitra *et al.* (2015), Mitra *et al.* (2016)
Kaempferol	*T. heimii, T. microcarpus*	Mitra *et al.* (2015), Mitra *et al.* (2016)

Cerebrosides

Eight novel cerebrosides named termitomycesphins (A-H) were isolated from the ethanolic extract of *T. Albuminosus* (Qi *et al.*, 2000, 2001, Qu *et al.*, 2012). Termitomycesphins A, C, E, and G, containing a C16 hydroxy fatty acid, exhibit higher neuritogenic activities than those of B, D, F, and H, which contain a C18 hydroxy fatty acid. The maximum activity of neuronal differentiation can reach 30 per cent by termitomycesphin E at a concentration of 14 μM with a standardized NGF as a positive control. Cerebroside 7, a major cerebroside (0.042 per cent) with a C16 hydroxy fatty acid, obtained from the same mushroom was inactive against PC12 cells. The application of di- and tetrahydroxylation into the inactive cerebroside 7 resulted in the enhancement of its neuritogenic activity (Qi *et al.*, 2001). These termitomycesphins were shown to induce neuronal differentiation in rat PC12 cells, a clonal rat pheochromocytoma cell line.

Fatty Acids

Five termitomycamides (A to E), the fatty acid amides of *T. titanicus* were isolated and analyzed for structural elucidation, and it was found that termitomycamides B and E showed protective activity against the cell death due to stress of endoplasmic reticulum (ER) (Choi *et al.*, 2010). So, it may also say that, protective activity was achieved against various diseases like ischemia/reperfusion injury, heart disease, hypoxia, and neurodegeneration that are involved in ER-mediated cell death pathway (Xu *et al.*, 2005).

Enzymes

Softening and Leavening of Bread

Various enzymes are used in bakery for softening, leavening for increasing palatability of the product and also for inhibiting staling important for increasing the shelf life of bakery products. A novel enzymatic formulation from *Termitomyces clypeatus* used for improvement of leavening of bakery products (Sengupta *et al.*, 1999a). The enzymes hemicellulase, amylases and cellulases concentrated from culture media were added to flour along with other routine ingredients (salt, sugar, yeast, additives *etc.*) resulting much softer and bigger bread due to better leavening.

Clarification of Non-citrus Fruit Juice

A mixed enzyme preparation containing pectinase, cellulase, hemicellulase, arabinase and xylanase from *Termitomyces clypeatus* was employed for clarification of non-citrus apple juice. It was found that the mixture of pectinase and xylanase has potential activity in clarification of non-citrus fruit juice (Senguptaet al.1999b). Pectinase from *T. clypeatus* was also used in the extraction, clarification, filtration, and depectinization of fruit juices and wines (Kashyap *et al.*, 2000).

Bioremediation

A novel alkaline protease (30 k Da) was isolated from the wild edible mushroom *Termitomyces albuminosus* (Zheng *et al.*, 2011). An extracellular alkaline protease (AkP) from *T. clypeatus* has significant multifunctional bioremediation potency. AkP could efficiently dehair the goat skin, detached bird feather vanes from its shaft and reduce BOD, COD and pH of tannery waste effluent. As an ecofriendly alternative, the enzyme showed significant promise for bioremediation and industrial applications through time-saving bioprocesses. *Termitomyces* fungus has almost all the available enzyme genes needed to decompose the large, complex molecules of plant cells. *Termitomyces* mushrooms are rich sources of cellulose (Ahmed *et al.*, 1989). *Termitomyces clypeatus* cultivated in agro-residues of mustard produced lignocellulolytic enzymes used for substrate saccharification (Pal *et al.*, 2013).

Anticancer Activity

An AkP from *T. clypeatus* MTCC-5091 showed biochemical property and cytotoxic potency. It could cleave cell surface proteoglycans, exhibiting antiproliferative activity against growth cells of a human liver cancer cell line through the induction of apoptosis (Majumder *et al.*, 2016).

Ergostane

Five ergostanes were extracted from fresh fruiting bodies of *T. microcarpus*, and among them dimethylincisterol and 5α, 8α- epidioxyergosta-6, 22-dien-3β-ol have cytotoxic activities and also significantly active against most tumor cell lines (Njue *et al.*, 2018). Betulinic acid (BA) from *T. microcarpus* exhibits antitumor property related to itsability to trigger mitochondrial membrane permeabilization, which is a central event in the apoptotic process, of cancer cells while non-neoplastic cells and normal tissue remain relatively resistant to BA (Njue *et al.*, 2018).

Antimicrobial Activity

Methanol extracts prepared from fruiting bodies *Termitomyces robustus* could inhibit growths of three fungi, *Aspergillus niger, A.flavus,* and *Candida albicans* as well as three bacteria, *Bacillus cereus, Escherichia coli,* and *Klebsiella pneumonia*; those from *T. Microcarpus* showed bactericidal effects against *Proteus vulgaris* (Jonathan *et al.*, 2008). The acetone extract from fruiting bodies of *T. heimii* exhibited noticeable antibacterial potential against *Staphylococcus aureus* and *Klebsiella pneumonia* among ten investigated human pathogenic bacteria (Singha *et al.*, 2017).

Cultivation and Biotechnological Exploitation

Mushrooms are one of the most popular, non- conventional food accepted globally and the increased consumer demand over the years has led to the quantum jump in their cultivation. The cultivation of *Termitomyces* for the fruiting body production is unavailable in laboratory scale. A *Termitomyces* mushroom is cultivated by the black-winged subterranean termite *Odontotermes formosanus* Shirakiat fungus gardens (Taiwan). The mycelia grown from underground with fecal balls excreted by termite workers. Mutualistic relationship between *O. Formosanus* and its *Termitomyces* symbiont was demonstrated by rearing of *O. formosanus* in laboratory (Hsieh *et al.*, 2017). Previously associations between African macrotermitine termites and *Termitomyces* mushrooms have been established by artifcially inoculating incipient termite colonies with *Termitomyces* spores (Sands 1960). In the absence of suitable cultivation method for fructification of *Termitomyces,* the submerged mycelial culture development of the *Termitomyces* species could be the best alternative to achieve their broad spectrum potentials (National Food Industry Product Quality Testing Guangzhou- China, Zhao Chengyu, 1988). The identification of different proteins is essential that have potential roles in the development of *T. heimii* mushroom could provide useful information regarding fruiting body formation. Altered protein expression can facilitate changes in metabolism and may also directly involved in the fruiting body development of *T. heimii*. Various proteomic tools like liquid chromatography-based separation and label-free quantitative proteomics can be used to improve the cultivation methods of *T. Heimii* by studying each developmental stage of the fungus.

Conclusion

Studies on *Termitomyces* mushrooms revealed their enormous nutritional and medicinal values established them as functional foods. Interestingly these mushrooms are purely natural resources only available from the wild and popularly consumed by peoples of Africa and Asia. *Termitomyces* species effectively showed antioxidative, anti-inflammatory, antitumor, antimicrobial, bioremedial activities could be a natural resource of bioactive molecule within dietary supplements. The cultivation of *Termitomyces* for the fruiting body production is unavailable currently.

In this context a sustainable biotechnological approach is needed for their large-scale mycelia production. Exploitation of further bioactive compounds from *Termitomyces* mushrooms and their potential uses could solve our increasing health problems.

References

Aanen, D.K., de Fine Licht, H.H., Debets, A.J., Kerstes, N.A., Hoekstra, R.F. and Boomsma, J.J. (2009) High symbiont relatedness stabilizes mutualistic cooperation in fungus-growing termites. Science (Wash D C), 326: 1103–1106.

Abdullah, N., Ismail, S.M., Aminudin, N., Shuib, A.S. and Lau, B.F. (2012) Evaluation of selected culinary-medicinal mushrooms for antioxidant and ACE inhibitory activities. Evid Based Complement Alternate Medicine 464238: 1–12.

Ahmed, N. Abo-Khatwa. (1989) *Termitomyces*: A new source of potent cellulases. JKAU Science, pp. 51-59.

Alvar, J., Velez, I.D., Bern, C., Herrero, M., Desjeux, P., Cano, J., Jannin, J., den Boer, M. and Team, W.H.O.L.C. (2012) Leishmaniasis worldwide and global estimates of its incidence. PLoS One. 7; e35671.

Bhanja, S.K., Nandan, C.K., Mandal, S., Bhunia, B., Maiti, T.K., Mondal, S. and Islam, S.S. (2012) Isolation and characterization of the immunostimulating β-glucans of an edible mushroom *Termitomyces robustus* var. Carbohydrate Research, 357: 83–89.

Borchers, A., Keen, C.L. and Gershwin, M.E. (2004) Mushrooms, tumors, and immunity: an update. Experimental Biology and Medicine, 229: 393–406.

Chandrawati, P.S., Narendra, K. and Tripathi, N.N. (2014) Macrofungal wealth of Kusumhi forest of Gorakhpur, UP, India. American International Journal of Research In Formal, Applied and Natural Sciences, 5: 71–75.

Cheah, I. K. and Halliwell, B. (2012) Ergothioneine; antioxidant potential, physiological function and role in disease. Biochimicaet Biophysica Acta (BBA). Molecular Basis of Disease, 1822: 784–793.

Choi, J.H., Maeda, K., Nagai, K., Harada, E., Kawade, M., Hirai, H. and Kawagishi, H. (2010) Termitomycamides A to E, fatty acid amides isolated from the mushroom *Termitomyces titanicus*, suppress endoplasmic reticulum stress. Organic Letters, 12: 5012–5015.

Ferreira, I.C., Barros, L. and Abreu, R.M.V. (2009) Antioxidants in wild mushrooms. Current Medical Chemistry, 16; 1543–1560.

Ghorai, S., Banik, S.P., Verma, D., Chowdhury, S., Mukherjee, S. and Khowala, S. (2009) Fungal biotechnology in food and feed processing. Food Research International, 42 : 577–587.

Gruen, V.E.C. and Wong, H.X. (1982). Immunodulatory and antitumour activities of a polysaccharide-peptide complex from a mycelial culture of *Trichoderma* sp. Sciences, 57: 269-281.

Halliwell, B. (2012) Free radicals and antioxidants: updating a personal view. Nutritional Reviews, 70(5): 257-265.

Heim, R. (1941) Etudes descriptive set experiment alessur les agarics termitophilesd'Afriquetropicale. Mem de l'Acad Science, 64: 25-29.

Heim, R. (1942) Nouvellesétudes descriptivessur les agarics termitophilesd' Afriquetropicale. Archives du Muséum National d'HistoireNaturelle (in French), 18(6): 107–166.

Heim, R. (1977) Termites et Champignons. Boubee Ed., Paris, pp. 207.

Heim, R. (1977). Termites and mushrooms: Termitophilic fungi of Black Africa and Southern Asia. Boubee Ed., Paris, pp. 1-207.

Hsieh, H.M., Chung, M.C., Chen, P.Y., Hsu, F.M., Liao, W.W., Sung, A.N., Lin, C.R., Wang, C.J.R., Kao, Y.H., Fang, M.J., Lai, C.Y., Huang, C.C., Chou, J.C., Chou, W.N., Chang, B.C.H. and Ju, Y.M. (2017) A termite symbiotic mushroom maximizing sexual activity at growing tips of vegetative hyphae. Botanical Studies, 58(39): 1–14.

Hsieh, H.M. and Ju, Y.M. (2018) Medicinal components in *Termitomyces* mushrooms. Applied Microbiology and Biotechnology, 102: 4987–4994.

Ishikawa, N.K., Kasuya, M.C.M. and Vanetti, M.C.D. (2001) Antibacterial activity of Lentinulaedodes grown in liquid medium. Brazilian Journal of Microbiology. 32L 206– 210.

Johnsy, G. and Kaviyarasan, V. (2014) Evaluation of antioxidant activities and determination of bioactive compounds in two wild edible ermitomyces (*T. microcarpus* and *T. heimii*). World Journal of Dairy Food Science, 9: 10–19.

Jonathan, S.G., Kigigha, L.T. and Ohimain, E. (2008) Evaluation of the inhibitory potentials of eight higher Nigerian fungi against pathogenic microorganisms. African Journal of Biomedical Research, 11: 197–202.

Kalaè, P. (2009) Chemical composition and nutritional value of European species of wild growing mushrooms: A review. Food Chemistry, 113: 9–16.

Kamat, M.N. (1971) Fungi of Maharashtra, Mahatma Phule Krishi Vidyapeeth, Rahuri (M.S.) Bulletin, 1, pp. 124.

Kashyap, D. R., Chandra, S., Kaul, A. and Tewari, R. (2000) Production, purification and characterization of pectinase from a *Bacillus* sp. DT7. World Journal of Microbiology and Biotechnology, 16: 277–282.

Lindequist, U., Niedermeyer, T.H.J. and Jülich, W.D. (2005) The pharmacological potential of mushrooms. eCAM. 2: 285–299.

Lu, Y.Y., Ao, Z.H., Lu, Z.M., Xu, H.Y., Zhang, X.M., Dou, W.F. and Xu, Z.H. (2008) Analgesic and anti-inflammatory effects of the dry matter of culture broth of *Termitomyces albuminosus* and its extracts. Journal of Ethnopharmacology, 120: 432–436.

Majumder, R., Banik, S.P. and Khowala, S. (2016) AkP from mushroom *Termitomyces clypeatus* is a proteoglycan specific protease with a poptotic effect on HepG2. International Journal of Biological Macromolecules, 91: 198–207.

Mallick, S., Dutta, A., Dey, S., Ghosh, J., Mukherjee, D., Sultana, S.S., Mandal, S., Paloi, S., Khatua, S., Acharya, K. and Pal, C. (2014) Selective inhibition of *Leishmania donovani* by active extracts of wild mushrooms used by the tribal population of India: an *in vitro* exploration for new leads against parasitic protozoans. Experimental Parasitology, 138: 9–17.

Manna, D.K., Nandi, A.K., Pattanayak, M., Maity, P., Tripathy, S., Mandal, A.K., Roy, S., Tripathy, S.S., Gupta, N. and Islam, S.S. (2015) A water soluble β-glucan of an edible mushroom *Termitomyces heimii*: Structural and biological investigation. Carbohydrate Polymer, 134: 375–384.

Mattila, P., Suonpää, K. and Piironen, V. (2000) Functional properties of edible mushrooms. Nutrition, 16(7-8): 694-66.

Mitra, P., Mandal, N.C. and Acharya, K. (2015) Polyphenolic extract of *Termitomyces heimii*: antioxidant activity and phytochemical constituents. Journal of Consumer Protection and Food Safety.

Mitra, P., Mandal, N.C. and Acharya, K. (2016) Mycochemicals and antioxidant activity of polyphenol-rich fraction of *Termitomyces microcarpus*. International Food Research Journal, 23(6): 2384-2389.

Mondal, S., Chakraborty, I., Rout, D. and Islam, S.S. (2006) Isolation and structural elucidation of a water-soluble polysaccharide (PS-I) of a wild edible mushroom, *Termitomyces striatus*. Carbohydrate Research, 341: 878–886.

Moro, C., Palacios, I., Lozano, M., D'Arrigo, M., Guillamón, E., Villares, A., Martínez, J.A. and García-Lafuente, A. (2012) Anti-inflammatory activity of methanolic extracts from edible mushrooms in LPS activated RAW 264.7 macrophages. Food Chemistry, 130: 350–355.

Mueller, U.G., Gerardo, N.M., Aanen, D.K., Six, D.L. and Schultz, T.R. (2005) The evolution of agriculture in insects. Annual Reviev of Ecology, Evolution and Systematics, 36L 563–595.

Nakalembe, I., Kabasa, J.D. and Olila, D. (2015) Comparative nutrient composition of selected wild edible mushrooms from two agro-ecological zones, Uganda. Spring. 4: 433–447.

Nelson, R.H.. (2013) Hyperlipidemia as a risk factor for cardiovascular disease. Primary Care, 40(1): 195-211.

Njue, A.W., Omolo, J.O., Cheplogoi, P.K., Langat, M.K. and Mulholland, D.A. (2018) Cytotoxic ergostane derivatives from the edible mushroom *Termitomyces microcarpus* (Lyophyllaceae). Biochemica System Ecology, 76: 12–14.

Öztürk, M., Duru, M.E., Kivrak, S., Mercan-Doðan, N., Türkoglu, A. and Özler, M.A. (2011) *In vitro* antioxidant, anticholinesterase and antimicrobial activity studies on three Agaricus species with fatty acid compositions and iron contents: a comparative study on the three most edible mushrooms. Food Chemistry and Toxicology, 49: 1353–1360.

Pal, S., Banik, S.P. and Khowala, S. (2013) Mustard stalk and straw: A new source for production of Lignocellulolytic enzymes by the fungus *Termitomyces clypeatus* and as a substrate for saccharification. Industrial Crop and Products, 41: 283-288.

Parent, G. and Thoen, D. (1977) Food value of edible mushrooms from Upper-Shaba region. Economic Botany, 31(4): 436-445.

Pattanayak, M., Samanta, S., Maity, P., Sen, I.K, Nandi, A.K., Manna, D.K., Mitra, P., Acharya, K. and Islam, S.S. (2015) Heteroglycan of an edible mushroom *Termitomyces clypeatus*: Structure elucidation and antioxidant properties. Carbohydrate Research, 413: 30–36.

Poucheret, P., Fons, F. and Rapior, S. (2006) Biological and pharmacological activity of higher fungi: 20-Year retrospective analysis. Mycologie, 27: 311–333.

Prasad, N.R., Karthikeyan, A., Karthikeyan, S. and Reddy, B.V. (2011) Inhibitory effect of caffeic acid on cancer cell proliferation by oxidative mechanism in human HT-1080 fibrosarcoma cell line. Molecular Cellular Biochemistry, 349: 11–19.

Puttaraju, N.G., Venkateshaiah, S.U., Dharmesh, S.M., Urs, S.M.N. and Somasundaram, R. (2006) Antioxidant activity of indigenous edible mushrooms. Journal of Agricultural Food Chemistry, 54: 9764–9772.

Qi, J.H., Ojika, M. and Sakagami, Y. (2000) Termitomycesphins A-D, novel neuritogeniccerebrosides from the edible Chinese mushroom *Termitomyces albuminosus*. Tetrahedron, 56: 5835–5841.

Qi, J.H., Ojika, M. and Sakagami, Y. (2001) Neuritogeniccerebrosides from anedible Chinese mushroom. Part 2: structures of two additional termitomycesphins and activity enhancement of an inactive cerebroside by hydroxylation. Bioorganic Medicinal Chemistry, 9: 2171–2177.

Qu, Y., Sun, K., Gao, L., Sakagami, Y., Kawagishi, H., Ojika, M. and Qi, J. (2012) Termitomyces phins G and H, additional cerebrosides from the edible Chinese mushroom *Termitomyces albuminosus*. Bioscience, Biotechnology and Biochemistry, 76: 791–793.

Rai, R.D. (1994) Nutritional and medicinal values of mushrooms. In: Advances in Horticulture. (Chadha, K.L., Sharma, S.R. eds.), Malhotra Publishing House, New Delhi, pp. 537-551.

Sachan, S.K.S., Patra, J.K. and Thatoi, H.N. (2013) Indigenous knowledge of ethnic tribe for utilization of wild mushrooms as food and medicine in similipal biosphere reserve, Odisha, India. Journal of Agricultural Technology, 9: 335–348.

Sands, W. A. (1960) The initiation of fungus comb construction in laboratory colonies of *Ancistrotermes guineensis* (Silvestri). Insect Society, 7: 251–259.

Sangvichien, E. and Taylor-Hawksworth, P.A. (2001) *Termitomyces* mushrooms: A tropical delicacy. Mycologist,15: 31–33.

Sargunam, S.D., Johnsy, G., Samuel, A.S. and Kaviyarasan, V. (2012) Mushrooms in the food culture of the Kaani tribe of Kanyakumari district. Indian Journal of Traditional Knowledge, 11: 150–153.

Sengupta, S., Ghosh, A. K., Naskar, A. K., Sengupta, D., and Jana, M. L. (1999b) A process for the preparation of an enzyme composition containing a mixture of pectinase and xylanase useful for clarification of non-citrus fruit juice. Indian Patent NF/686/99.

Sengupta, S., Sengupta, D., Naskar, A. K. and Jana, M. L. (1999a). A process for the preparation of a novel enzymatic formulation useful for improved leavening of bakery products. Indian Patent NF/403/99.

Singha, K., Pati, B.R., Mondal, K.C. and Das Mohapatra, P.K. (2017) Study of nutritional and antibacterial potential of some wild edible mushrooms from Gurguripal Ecoforest, West Bengal, India. Indian Journal of Biotechnology, 16: 222–227.

Suzuki, S. and Oshima, S. (1976) Influence of Shii-te-ke (Lentinusedodes) on human serum cholesterol. Mushroom Science, 1: 463-467.

Tibuhwa, D.D. (2012) Folk taxonomy and use of mushrooms in communities around Ngorongoro and Serengeti National Park, Tanzania. Journal of Ethnobiology and Ethnomedicine, 8: 36–44.

Wang, S-L. (2013) The antagonism of *Termitomyces albuminosus* polysaccharide to cyclophosphamide on immune organ injury in mice. Progressive Veterinary Medicine, 34: 56–59.

Wasser, S.P. and Weis, A.L. (1999) Medicinal properties of substances occurring in higher Basidiomycetes mushrooms: Current perspectives (Review). International Journal of Medicinal Mushrooms, 1: 31–62.

Woldegiorgis, A.Z., Abate, D., Haki, G.D. and Ziegler, G.R. (2014) Antioxidant property of edible mushrooms collected from Ethiopia. Food Chemistry 157; 30–36.

Xu, J. (2005) Fundamentals of fungal molecular population genetic analyses. In Evolutionary Genetics of Fungi (Xu J, ed). Horizon Scientific Press, England. pp. 87–116.

Xu, X., Yan, H., Chen, J. and Zhang, X. (2011) Bioactive proteins from mushrooms. Biotechnol Adv., 29: 667–674.

Zang, M. (1981) *Sinotermitomyces*, a new genus of Amanitaceae from Yunnan, China. Mycotaxon, 13: 171–174.

Zhang, M., Cui, S.W., Cheung, P.C.K. and Wang, Q. (2007) Antitumor polysaccharides from mushrooms: A review on their isolation process, structural characteristics and antitumor activity. Trends in Food Science and Technology, 18: 4–19.

Zhao, H., Li, S., Zhang, J., Che, G., Zhou, M., Liu, M., Zhang, C., Xu, N., Lin, L., Liu, Y. and Jia, L. (2016) Theantihyperlipidemic activities of enzymatic and acidic intracellular polysaccharides by *Termitomyces albuminosus*. Carbohydrate Polymer, 151: 1227–1234.

Zhao, H.J., Li, J., Zhang, J.J., Wang, X.X., Liu, M., Zhang, C. and Jia, L. (2017) Hepatoprotective and *in vitro* antioxidant effects of native depolymerisedexopolysaccharides derived from *Termitomyces albuminosus*. Science Reporter 7(3910), 1–13.

Zhao, Y.X., Tao, M.X., Cheng, G.Y., Xing, J. and Lu, W.J. (2015) Effects of polysaccharides from *Termitomyces albuminosus* on pathological liver ultrastructure and gene expression in mice with alcoholic liver injury. Food Science, 36: 195–199.

Zheng, S., Wang, H. and Zhang, G. (2011) A Novel alkaline protease from wild edible mushroom *Termitomyces albuminosus*. Acta Biochimica Polonica, 58: 269-273.

28

Preparation of Giant Unilamellar Vesicles (GUVs) from Bacterial Polar Lipid Extract: Developing a Prokaryotic Model Membrane System

*H.G. Behuria, N.S. Pal, R.N. Munda and S.K. Sahu**

Department of Biotechnology, North Orissa University, Baripada – 757 003, Odisha

****E-mail: drsantoshnou@gmail.com***

ABSTRACT

Giant unilamellar vesicles (GUVs) are large fluid filled vesicles in which a lipid bilayer encloses an aqueous interior that act as a pure model membrane system resembling plasma membrane. These GUVs vary from 10µ to 300 µ in diameter which is much larger compared to prokaryotic (1-2µ) or eukaryotic cells (10-20µ), that can be easily manipulated, imaged and could be studied in isolation from other membrane-mediated activities. In bacteria, GUVs have been an attractive model membrane system for studies involving structure and function of membrane proteins, drug resistance, permeability metabolites and mechanism of membrane-mediated anti-microbial activities. However, preparation of GUVs involves expensive materials and cumbersome methodologies, hence difficult to prepare in most average furnished laboratories. In the present work, we have prepared GUVs from bacterial polar lipid extract by electroformation method using copper electodes and a locally prepared transformer. The GUVs prepared varied in size varied from 50 to150 µm depending upon the composition of the lipid extract. The GUVs analyzed after staining with 10, N-nonyl acridine orange (NAO) that specifically bound to cardiolipin (CL) and phosphatidyl ethanolamine (PE). To confirm the integrity of GUVs, we trapped two different fluorescent dyes Sulforhodamine B and 5(6)-Carboxyfluorescein and visualized using Texas Red/Cy3/Y2E/C filter (λ_{ex}= 565 nm and λ_{em}= 586 nm) FITC λ_{ex} = 515 nm and λ_{em} = 550 nm and B-2A filter (λ_{ex}= 495 nm and λ_{em}= 517 nm) respectively. The results showed that our *method of GUV preparation is easy, inexpensive, reproducible and maintains of properties of bacterial membrane are ideal for investigating membrane-mediated events such as peptide-mediated membrane lysis and studies on membrane proteins.*

Keywords: *Carboxy fluorescene, Electroformation, Giant unilamellar vesicles, NAO, Phospholipids, Sulforhodamine B.*

Introduction

Membrane constitutes an integral component of cell that serves as a boundary between the living cytosol and non-living extracellular medium. Apart from its function as a determinant of cellular boundary, it's other important functions include transport of nutrients, being a protective barrier from the invading pathogens, host-parasite interaction, virulence, cell-cell communication (*e.g.* horizontal gene transfer), biochemical activities (*e.g.* electron transport chain), bio-film formation and resistance to a multiple anti-microbial drugs (Dimova *et al.*, 2006: Walde *et al.*, 2010: Weso³owsk *et al.*, 2009: Jalmar *et al.*, 2010). The inner bacterial membrane (IBM) is a dynamic entity that actively participates in a multitude of cellular activities including release of microvesicles and quorum sensing, permeability to different drugs, ATP synthesis and cell division. Recent investigations show that membrane-derived vesicles have several biotechnological applications in microbial biotechnology. To list a few, membrane-derived vesicles are useful in development of vaccines, drug delivery and activation of enzymes (Knol *et al.*, 1996: Leite *et al.*, 2015).

However, bacteria are microscopic (1-2 µ) and hence, are difficult to be investigated in detailed under commonly available microscopic techniques. Further, bacteria possess cell wall around the membrane that sometimes needs to be removed before investigation membrane-associated events. Bacterial membranes are studded with multiple proteins that might interfere with investigation of membrane-mediated activity. Pathogenic bacteria are difficult to be investigated under most common laboratories that require a specific biosafety norms to be followed. One alternative technique to investigate membrane-associated activities is to purify the membrane from bacteria. However, this isolation process requires expensive instruments such as ultracentrifuge, hence is difficult to be performed in many common laboratories.

Model membranes (*e.g.*, liposomes) prepared from lipids derived bacterial membranes have become one of the most attractive tool among the membrane biologists for investigation of membrane-associated events in bacteria. Thus, giant unilamellar vesicles (GUVs) are a very useful tool for studying membrane properties. In addition, the possibility to directly visualize certain interactions in a small but optically resolvable volume (about ten picolitres) encapsulated by the vesicle membrane has provoked the development of various ideas from interdisciplinary fields. To mention a few recent ones, giant vesicles have been considered as (i) a possible construct of microreactors (Noireaux and Libchaber. 2004: Fischer *et al.*, 2002: Michel *et al.*, 2004) (ii) a means to study micro-compartmentalization and cellular crowding (Helfrich *et al.*, 2002: Long *et al.*, 2005) (iii) microscopic vessels for protein crystallization (Yamashita *et al.*, 2002). In addition the recent finding that certain amphiphilic diblock copolymers can form giant vesicles, or polymersomes (Discher *et al.*, 1999) has reinforced work related to development of drug delivery tools. The main application of GUVs, though, remains the characterization of the mechanical and rheological properties of single- or multi-component membranes as well as investigation of bilayer interactions. Although liposomes vary in their size and lamelarity depending upon the method used in their formation, small unilamellar vesicles (SUVs) and large unilamellar vesicles (LUVs) have become two highly useful model membrane systems for investigating membrane-associated events (Figure 28.1).

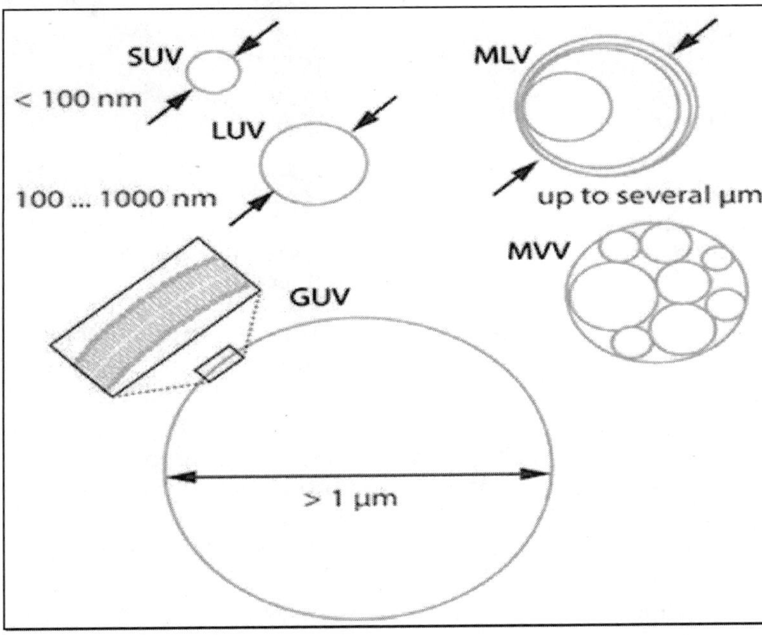

Figure 28.1: Different Types of Model Membranes Prepared from Lipids Derived from Cells and their Size. SUV: small unilamellar vesicle, LUV: large unilamellar vesicles, MLV: multilamellar vesicle, MVV: multivesicular vesicle, GUV: giant unilamellar vesicle

SUVs are also termed as liposomal nanoparticles and are useful for drug delivery and activation of membrane-bound enzymes, whereas, LUVs are useful for investigation of proteins bound to membranes and transport of molecules across the membrane. However, the LUVs are below 1 µ in diameter that needs powerful microscopic techniques to be observed. GUVs are spherical vesicles made from lipids extracted from biological membranes (*e.g.* bacterial plasma membrane) that are 10-300 µ in diameter (Figure 28.2). These vesicles are similar in composition to the bacterial inner membrane and can replace bacterium itself to investigate their membrane-associated events.

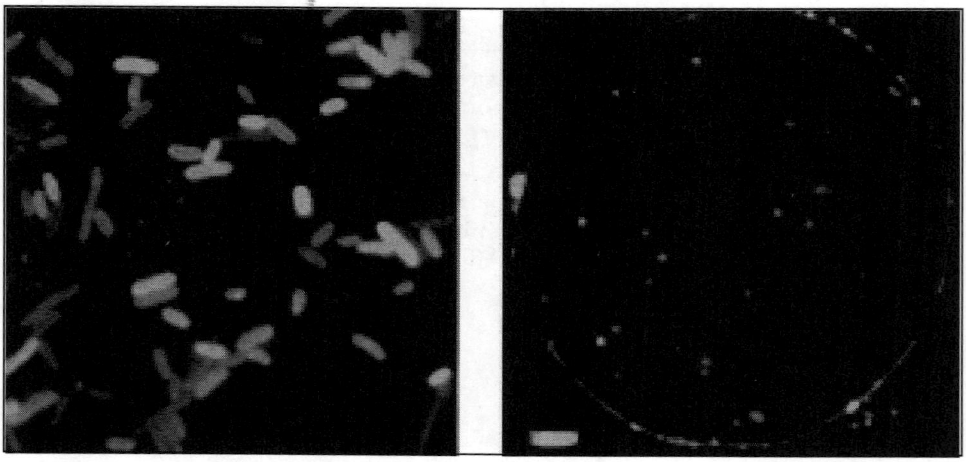

Figure 28.2: Comparison between the Size of Bacteria and Size of GUVs. Bacteria (*e.g. E.coli*) are 1-2μ in length to be visualized in minute detailed (left). GUV prepared in our laboratory stained with 10-N, Nonyl-Acridine Orange (NAO) (inset: a single bacterium for comparision of size). Scale: 1 μ. The images were taken using a Nikkon inverted fluorescence microscope (Japan) using 100 X objective.

GUVs are an attractive model system to investigate membrane-mediated activities as (i) They are pure, (ii) much larger (iii) stable and (iv) can enclose water soluble content. Further, they are non-toxic, biocompatible, biodegradable and non-immunogenic (Nagahama *et al.*, 2006: Weso³owska *et al.*, 2009). Currently available methods for preparation of GUVs from bacterial membrane lipids include gentle swelling processes such as (i) hydration and (ii) electroformation. Electroformation method is the most favored method as it is rapid and produces uniform sized unilamellar vesicles (Pereno *et al.*, 2017). This method involves hydration of dried lipid layers deposited on platinum electrodes, steel electrodes or titanium oxide (TiO_2) coated slides in low salt buffer. However, the biggest drawback in this method is that it is very expensive to be afforded by most laboratories (Moscho *et al.*, 1996). Since then, many protocols have been developed to electroform the vesicles with varying lipid composition of hydration (MacDonald *et al.*, 1991: Costa *et al.*, 2014: Lira *et al.*, 2014: Pereno *et al.*, 2017). However most of these existing methods of GUV preparation are (i) expensive, (ii) time consuming and (iii) require sophisticated instrumentation. Hence, easy, inexpensive and quicker methods are required to be developed for preparation of GUVs in common laboratories with basic instrumentation facilities. In the present work, we have used copper electrodes and a locally-made transformer as AC source to prepare GUVs from the polar lipid extract of *E. coli*.

Materials and Methods

Bacteria

E. coli (DH5α) maintained at Department of Biotechnology, North Orissa University, Baripada, Odisha was used in this study.

Growth of *E. coli* (DH5α)

E. coli (DH5α) was grown in LB medium by inoculating 100 ml broth with 1 ml seed culture grown for 12 h at 25 °C and 200 rpm. The cells were grown for 16 hrs at 25 °C and 200 rpm on a table top shaker. Growth of bacteria was monitored by measuring OD at 600 nm at 2 h intervals up to 16 hrs (early saturation phase).

Collection and Re-suspension of Cells

Cells were collected by centrifugation at 5000 × g for 7 min at 25 °C and re-suspended at 10 mg/ml in re-suspension buffer (50 mM Tris-HCl, pH 7.5, 100 mM NaCl, 5 mM BHT, 1 mM PMSF) and used immediately for further experiments.

Cell Lysis and Protein Estimation

Cells were lysed in resuspension buffer (50 mM Tris-HCl, pH 7.5, 100 mM NaCl, 5 mM BHT, 1 mM PMSF) by sequentially incubating with lysozyme (1 mg/ml) and triton X-100 (1 per cent w/v) at 37 °C for 30 min. The protein content of the whole cell lysate was estimated as per Lowry *et al.* (1951) with slight modification.

Extraction of Total Lipid from *E. coli* (DH5α)

Total lipid from *E. coli* (DH5α) was extracted using aqueous two phase method as per Bligh and Dyer (1959) with slight modification. Briefly, total lipid from 0.5 mg cells in 0.5 ml 20 mM Tris-HCl, (pH 8.0) was extracted by aqueous two phase made from methanol, chloroform and water (2: 1:1 by volume). The bottom layer was collected and dried in oven below 45 °C. The dried samples were dissolved at approximately 1 µmol/ml phospholipid (PL) in $CHCl_3$ and stored at -20 °C for further analysis.

Quantification of Phospholipid in Total Lipid Extract

PL content in total lipid extract was quantified by phosphate assay (Fiske and Subarrow 1925). Briefly, 100 µl of total lipid extract in chloroform was completely dried and hydrolyzed by heating in 70 per cent perchloric acid at 160 °C. PO_4 thus released was quantified by method of PO_4 estimation using ascorbic acid and ammonium molybdate reagents. Absorbance of the sample were taken at 797 nm.

Quantification of Lipid Peroxidation

Peroxidation of lipids in total lipid extract was quantified by measuring diene conjugation (Howlett and Avery 1997). Briefly, dried lipid extract containing 1 µmol of PL was dissolved in 3 ml cyclohexane. Absorbance of the samples was scanned from 200 nm to 400 nm. Two absorbance peaks were observed at 230 nm (peak1) ($A_{230\,nm}$) and 274 nm ($A_{274\,nm}$) (peak 2) respectively. The ratio $A_{230\,nm}/A_{274\,nm}$ gives the relative amount of conjugated-dienes formed in the lipid.

Two Dimensional Thin Layer Chromatography (2D-TLC)

2D-TLC of total lipid extract from *E. coli* (DH5α) was performed using methods of Schneiter and Daum (2006). Briefly, lipid extract containing 500 nmol PL in 50 µl $CHCl_3$ was applied on a 20 cm × 20 cm × 0.0002 cm silica gel GF 254 TLC plate. The sample was sequentially developed in solvent ($CHCl_3$: CH_3O : H_2O 65 : 25 : 4

v/v) and ($CHCl_3$: CH_3OH : CH_3COOH : H_2O 80 : 12 : 15 : 4 v/v) sequentially and the spots were detected using iodine vapour. The spots were identified using PL standards developed in the same condition.

Quantification of Phospholipids from Spots on TLC Plates

Known weight of silica from the spots identified on TLC plates was scrapped into 12 mm × 125 mm assay tubes. PL adsorbed to silica powder was hydrolyzed to release phosphate by heating with Perchloric acid at 160 °C for 2 hrs. Phosphate thus released was quantified by the method described by Rouser *et al.* (1970).

Preparation of GUVs

Polar lipids from total lipid extract containing 2 mg PL was separated from non-polar (*e.g.* neutral and storage) lipids using a 12 × 150 mm silica gel column containing silica beads of 60-120 µ mesh $CHCl_3$:CH_3OH 1:1 (v/v) was used as mobile phase (Crider *et al.*, 1964) at a flow rate of 1 ml/min. GUVs were prepared from following the procedure of Pereno *et al.* (2016) with modification. Briefly, 1 mg polar lipid (corresponding to 1300 nmol PL) was dried completely and dissolved in 100 µl chloroform was coated on a thin copper electrode (5 cm x 0.5 cm x 1 mm) and dried at 45 °C for 1 hr in a hot air oven. The dried lipid film was hydrated by placing inside a small (1.2 cm x 5 cm) round bottomed glass bottle containing 3 ml low salt buffer (10 mM KCl and 100 mM Sucrose). The GUVs were formed by placing the electrodes in 0.5 cm apart and passing an alternating current of 10 Hz and 30 V for 3 hrs at 25 °C.

Staining and Imaging of GUVs

Formation of GUVs was analyzed using an inverted fluorescent microscope (Nickon, Model-100, Japan) with 40 objective. The vesicles were stained with 0.01 mM NAO(Sigma, USA) following the procedure of Fernandez (Fernandez *et al.*, 2003). Briefly, the vesicles were stained in DMSO, containing 0.01mM NAO. Images were taken both under bright and fluorescent field. Fluorescence images were taken using green filter (excitation 440-480 nm and emission 510 - 560 nm). Size of GUVs were measured using NIS-Elements Basic Research software. To confirm the integrity of GUVs, we trapped two different fluorescent dyes inside the vesicles (Estes and Mayer. 2005). Sulforhodamine B (SRB) (Sigma) and 5(6)-Carboxyfluorescein (CF) (Sigma, USA) were trapped inside the GUVs by electroforming the vesicles on LSB containing 50 mM SRB and 100 mM CF respectively. Vesicles that trapped SRB and CF were visualized using Texas Red/Cy3/Y2E/C filter (λ_{ex}= 565 nm and λ_{em}= 586 nm) and FITC and B-2A filter (λ_{ex}= 495 nm and λ_{em}= 517 nm) respectively.

Statistical Analysis

The statistical significance between the treatments was evaluated by students "t" test for the biochemical studies and one-way ANOVA test was used to compare the results whenever more than two experimental groups were compared. All the data are expressed as Mean ± SEM (standard error of the mean).

Results

Separation of Polar Lipids from Neutral Lipids

E. coli ((DH5α)) used as a model system for extraction of lipids was found to contain three different phospholipids as identified by two dimensional thin layer chromatography and quantitative estimation of phosphate. These PLs are phosphatidyl ethanolamine (PE), phosphatidyl glycerol (PG) and cardiolipin (CL). The phospholipids were identified by comparing with standard PLs (Figure 28.3A).

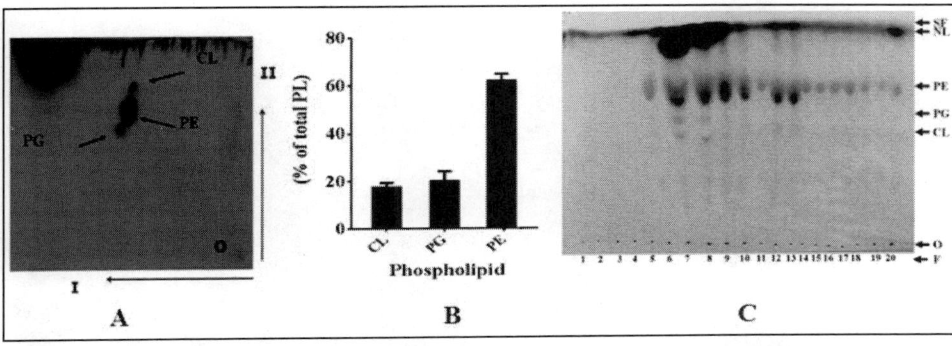

Figure 28.3: Extraction and Estimation of Phospholipids from Bacteria: (A) 2-D TLC of total bacterial lipid extract showing phosphatidyl ethnolamine (PE), phosphatidyl glycerol (PG) and cardiolipin (CL). O designates the origin where the lipid was loaded. I and II denote the directions of solvents for first and second dimension respectively. (B) Percentage of each phospholipid component in the total lipid extract. (C) TLC showing fractionation of total lipid extract to separate polar lipids from neutral lipids. The numbers at the bottom are the number of fractions (F), O represent the point of application of sample, SF: solvent front, NL: neutral lipids.

PE was the predominant PL in *E.coli* (DH5α) that constituted up to 70 per cent of total PLs. PG and CL constituted up to 20 per cent and 10 per cent of total PLs respectively. Each of the three PLs was quantified by scrapping the spots from TLC plates and quantifying the phosphate content in them. Each of PE and PG contain single PO_4 group. However, CL contain 2 PO_4 group(Figure 28.3B). However, a large quantity of neutral PLs was present in the total extract as visible in the upper left corner of the TLC plate (Figure 28.3A). Neutral lipids are mostly non-membrane forming lipids, that interfere in formation of membranes. Hence, neutral lipids were eliminated from the polar lipid extract by passing through a 12 × 150 mm silica gel column containing silica beads of 60-120 µ mesh size using chloroform: methanol 1:1 (v/v) as mobile phase. Total lipid extract containing 2 mg PLs were fractionated to collect 20 fractions each containing 1 ml solvent. Solvent was evaporated at < 45°C in a hot air oven, dissolved in 50 µl chloroform and analyzed on a 1-dimensional (1D TLC) (Figure 28.3C). Fractions 8-20 that were devoid of neutral lipids were pooled together to obtain ~ 1 mg total PLs that was used to make GUVs.

Preparation of GUVs

Polar lipid extract from *E. coli* (DH5α) containing 1mg PL was coated on a thin copper electrode (5 cm × 0.5 cm × 1 mm) and dried at 45 °C for 1 hr in a hot air oven. The dried lipid film was hydrated by placing inside a small (1.2cm x 5cm) round bottomed glass bottle containing 3 ml low salt buffer (10 mM KCl and 100 mM Sucrose). GUVs were formed by placing the electrodes 0.5 cm apart and passing an alternating current of 10 Hz and 30 V for 2-3 hrs at 25 °C. A cloudy mass was shown to be developed at positively charged electrode that was analyzed under microscope using 40 X objectives for the presence of GUVs.

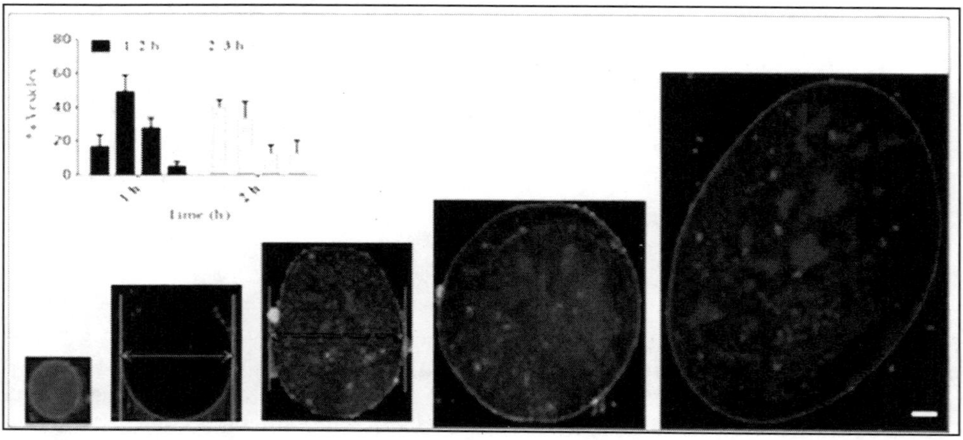

Figure 28.4: GUVs Observed under Fluoresncence (Nikkon, Japan) Microscope using 40X Objective. Vesicles with diameter ranging from 20-150 µm were obtained within 2-3 hrs. The vesicles were stained with 1 µM NAO for 1 hr at room temperature in LSB and were observed using green filter. The inset shows time-dependent size distribution of vesicles. The columns from left to right represent vesicles of size 20-50 µ, 51-100 µ, 101-150 µ and 151-200 µ respectively. Data were analyzed in graph-pad prism (version 6). Significance of variation between different data groups were tested by student's t-test. Scale: 10 µ.

Size of the vesicles ranged from 25 to 150 µ. However, 50-100 µ sized vesicles were of predominant size constituting 40-50 per cent of total vesicles (Figure 28.4). However, incubation for longer time led to fusion of vesicles to mega sized structures. Shape of the vesicles also changed from rounded shape to oval and irregular shape. These GUVs were checked for their integrity after trapping water soluble fluorescent dyes 5, 6-carboxy fluorescein and sulforhodamine B (data not shown). These results show that our method successfully prepared the GUVs that could be used as ideal prokaryotic model membrane system for analysis of membrane-related events.

Discussion

Currently the routine electroformation method used for preparation of GUVs has used expensive equipments such as electroformation chambers and titanium oxide (TiO_2) coated slides that are too expensive to be used by common laboratories. The average cost of preparation of GUVs using above methods is ~ 1000 per preparation (Pereno *et al.*, 2017). In the present procedure, we have used copper electrodes that reduced the cost of GUV preparation by a hundred fold. Further, it is easy, convenient and does not need expensive instrumentation. We have confirmed the formation GUVs by analyzing the GUVs in a phase contrast microscope using 40 x objectives. Further, we trapped two fluorescent dyes 5, 6-carboxy fluorescein and sulforhodamine-B inside the GUVs to determine their integrity (data not shown). Our results show that the fluorescent dyes get concentrated inside the GUVs within 2-3 h and remain at an elevated concentration during the experiments. We repeated the GUV preparation with polar lipids extracted from five different bacteria (unpublished data) and could reproduce the preparation of GUVs. However, the abundancy of GUVs differed with alteration of lipid composition as observed by many other workers (MacDonald *et al.*, 1991: Costa *et al.*, 2014: Lira *et al.*, 2014). Elevated anionic lipid (PG and CL) content produced more number of GUVs compared to low content of these lipids. However, further experiments are required to confirm these findings. As lipid peroxidation during drying of samples is an important determinant of lipid peroxidation, we compared the effect of lipid peroxidation on GUVs (unpublished data). We observed < 15 per cent peroxidized lipids in the samples as compared to freshly extracted samples. Many workers have shown that increased peroxidation in fatty acyl chains of PLs results in formation of irregular shaped vesicles (Schnitzer *et al.*, 2007; Heuvingh and Bonneau. 2009). As lipid peroxidation is known to alter the shapes and integrity of vesicles we determined the physico-chemical characteristics of GUVs prepared from lipids with elevated lipid peroxidation (data not shown). Our results show that the shape of the vesicles exhibits more distortion in samples with enhanced lipid peroxidation. Presence of ethanol and temperature variation is important factors that determine the shape of GUVs. These factors alter the mobility and packing of lipids in liposomes, thereby, change the shape, integrity and permeability (Angelova *et al.*, 1999). One of the most undesirable changes encountered by the lipids was increased peroxidation. However, careful handling of lipids and use of anti-oxidants such as butylated hydroxyl toluene (BHT) and vitamin E reduced this effect and led to formation of smooth and uniform sized vesicles that were not leaky to the water soluble dyes trapped inside them.

Conclusion

In conclusion, we have developed an easy, cheap and convenient protocol to produce giant unilamellar vesicles from bacterial polar lipid extract. The average size of vesicles ranged from 50 to 100 μm depending upon the lipid composition, voltage and time of swelling. We have confirmed the procedure with PLs extracted from multiple bacteria. By using these GUVs as a model membrane system, one can investigate the efficacy of membrane-targeted drugs, efficiency of lipid translocators and lipid-protein interaction in bacterial membrane. Our method of GUV preparation will enable many laboratories with basic facilities to investigate membrane-associated events in bacteria such as mechanism of action for AMPs, drug resistance and function of membrane proteins.

References

Angelova, M.I., Mutafchieva, R., Dimova, R. and Tenchov, B. (1999). Shape transformations of giant unilamellar vesicles induced by ethanol and temperature variations. Colloids and Surfaces A: Physicochemical and Engineering Aspects, 149: 201-205.

Bligh, E.G. and Dyer, W.J. (1959). A rapid method for total lipid extraction and purification. Canadian Journal of Biochemistry and Physiology, 37: 911-917.

Costa, B.A., Sanches, L., Gomide, A.B., Bizerra, F., Mas, C.D., Oliveira, E.B., Perez, K.R., Itri, R., Oguiura, N. and Hayashi, M.A.F. (2014). Interaction of the rattlesnake toxin crotamine with model membranes. Journal of Physical Chemistry, 118: 5471-5479.

Dimova, R., Aranda, S., Bezlyepkina, N., Nikolov, V., Riske, K.A. and Lipowsky, R. (2006). A practical guide to giant vesicles. Probing the membrane nanoregime via optical microscopy. Journal of Physics-Condensed Matter, 18 (28): S1151-S1176.

Discher, B., Won, Y.Y., Ege, D., Lee, J., Bates, F., Discher, D. and Hammer, D. (1999). Polymersomes: tough vesicles made from diblock copolymers. Science, 284: 1143-1146.

Estes, D.J. and Mayer, M. (2005). Giant liposomes in physiological buffer using electroformation in a flow chamber. Biochimica et Biophysica Acta, 1712: 152-160.

Fischer, A., Franco, A. and Oberholzer, T. (2002). Giant vesicles as microreactors for enzymatic mRNA synthesis. European Journal of Chemical Biology, 3: 409-417.

Fiske, C. and Subarrow, Y. (1925). The colorimetric determination of phosphorous. The Journal of Biological Chemistry, 66: 375-400.

Helfrich, M.R., Mangeney-Slavin, L.K., Long, M.S., Djoko, K.Y. and Keating, C.D. (2002). Aqueous phase separation in giant vesicles. Journal of American Chemical Society, 124: 13374-13385.

Heuvingh, J. and Bonneau, S. (2009). Asymmetric oxidation of giant vesicles triggers curvature-associated shape transition and permeabilization. Biophysical Journal, 97: 2904-2912.

Howlett, N.G. and Avery, S.V. (1997). Induction of lipid peroxidation during heavy metal stress in *Saccharomyces cerevisiae* and influence of plasma membrane fatty acid unsaturation. Applied and Environmental Microbiology, 63: 2971-2976.

Inmaculada, M., Fernandez, G., Ceccarelli, D. and Muscatello, U. (2014). Use of the fuorescent dye 10-N-nonyl acridine orange in quantitative and location assays of cardiolipin: a study on different experimental models. Analytical Biochemistry, 328: 174-180.

Jalmar, O., Garcýa-Saez, A.J., Berland, L., Gonzalvez, F. and Petit, P.X. (2010). Giant unilamellar vesicles (GUVs) as a new tool for analysis of caspase-8/ Bid-FL complex binding to cardiolipin and its functional activity. Cell Death and Disease, 1: e103.

Knol, J., Veenhoff, L., Liang, W., Henderson, P.J.F., Leblanci, G. and Poolman, B. (1996). Unidirectional reconstitution into detergent-destabilized liposomes of the purified lactose transport system of *Streptococcus thermophilus*. Journal of Biological Chemistry, 271: 15358–15366

Leite, N.B., Roberts, A., Palma, M.S., Connell, S.D., Neto, J.R. and Beales, P.A. (2015). PE and PS lipids synergistically enhance membrane poration by a peptide with anticancer properties. Biophysical Journal, 109: 936–947.

Lira, R.B., Dimova, R. and Riske, K.A. (2014). Giant unilamellar vesicles formed by hybrid films of agarose and lipids display altered mechanical properties. Biophysical Journal. 107: 1609-1619.

Long, M.S., Jones, C.D., Helfrich, M.R., Mangeney-Slavin, L.K. and Keating, C.D. (2005). Dynamic microcompartmentation in synthetic cells. Progress in Colloid and Polymer Science, 102: 5920-5925.

Lowry, O.H., Rosebrough, N.J., Farr, A.L. and Randall, R.J. (1951). Protein measurement with the Folin phenol reagent. Journal of Biological Chemistry, 193: 265-275.

MacDonald, R.C., MacDonald, R.I., Menco, B.P.M., Takeshita, K., Subbarao, N.K. and Hu, L. (1991). Small-volume extrusion apparatus for preparation of large unilamellar vesicles. Biochimica et Biophysica Acta, 1061: 297-303.

Michel, M., Winterhalter, M., Darbois, L., Hemmerle, J., Voegel, J.C., Schaaf, P. and Ball, V. (2004). Giant liposomemicroreactors for controlled production of calcium phosphate crystals. Langmuir, 20: 6127-6133.

Moscho, A., Daniel, O.O., Chiu, T., Modi, B.P. and Zare, R.N. (1996). Rapid preparation of giant unilamellar vesicles. Proceedings of the National Academy of Sciences of the United States of America, 93: 11443-11447.

Nagahama, M., Hara, H., Fernandez-Miyakawa, M., Itohayashi, Y. and Sakurai, J. (2006). Oligomerization of clostridium perfringens å toxin is dependent upon membrane fluidity in liposomes. Biochemistry, 45: 296-302.

Noireaux, V. and Libchaber, A. (2004). A vesicle bioreactor as a step toward an artificial cell assembly. Proceedings of the National Academy of Sciences of the United States of America, 101: 17669-17674.

Pereno, V., Carugo, D., Bau, L., Sezgin, E., De la Serna, J.B., Eggeling, C. and Stride, E. (2017). Electroformation of giant unilamellar vesicles on stainless steel, electrodes. ACS Omega, 2: 994-1002.

Quincey, C., Petara L., Illsberry, E.H., Cathyy, E.N. and Bradford, R. (1964). Separation of lipids by Silica Gel G column chromatography. Journal of Lipid Research, 5: 479-481.

Rouser, G., Fkeischer, S. and Yamamoto, A. (1970). Two dimensional then layer chromatographic separation of polar lipids and determination of phospholipids by phosphorus analysis of spots. Lipids, 5(5): 494-496.

Schneiter, R. and Daum, G. (2006). Analysis of yeast lipids. Methods in Molecular Biology, 313: 75-84.

Schnitzer, E., Pinchuk, I. and Lichtenberg, D. (2017). Peroxidation of liposomal lipids. European Biophysics Journal, 36: 499-515.

Skipski, V.P., Peterson, R.F. and Barclay, M. (1964). Quantitative analysis of phospholipids by thin-layer chromatography. The Biochemical Journal, 90: 374-378.

Walde, P., Cosentino, K., Engel, H. and Stano, P. (2010). Giant vesicles: preparations and applications. European Journal of Chemical Biology, 11(7): 848-865.

Weso³owska, O., Michalak, K., Maniewska, J. and Hendrich, A.B. (2009). Giant unilamellar vesicles: a perfect tool to visualize phase separation and lipid rafts in model systems. Acta Biochemica Polonica, 56(1): 33-39.

Weso³owska, O., Michalak, K., Maniewska, J. and Hendrich, A.B. (2009). Giant unilamellar vesicles: a perfect tool to visualize phase separation and lipid rafts in model systems. Polska Akademia Nauk Komitet Biochemiczny, 56(1): 33-39.

Yamashita, Y., Oka, M., Tanaka, T. and Yamazaki, M. (2002). A new method for the preparation of giant liposomes in high salt concentrations and growth of protein microcrystals in them. Biochimstry Biophysics Acta, 1561: 129-134.

29

Structural Diversity of Heterotrophic and Aerobic Denitrifying Bacteria in Domestic Wastewater

S.K. Padhi[1]*, *S. Tripathy*[2], *S. Mohanty*[3] and *N.K. Maiti*[3]

[1]*College of Food, Agriculture and Natural Resource Sciences, University of Minnesota, Saint Paul, Minnesota, USA*
[2]*Department of Biotechnology, Banki Autonomous College, Banki – 754 008, Cuttack, Odisha*
[3]*Department of Fish Health Management, ICAR-CIFA, Kausalyaganga, Bhubaneswar – 751 002, Odisha*
**E-mail: soumesh_skp@yahoo.co.in*

ABSTRACT

Anthropogenic activities increase rates of nitrogen (N) input to the environment and removal of this N is controlled by several factors, including nitrification and denitrification. In this work, 36 heterotrophic nitrifying and aerobic denitrifying bacteria were isolated from domestic wastewater and biochemically characterized. 16S rDNA and DNA gyrase genes were targeted to study the phylogenetic relationship between these isolates. Bacteria from Enterobacteriaceae, Pseudomonadaceae and Bacillales were predominating population found to be responsible of nitrogen removal process. This study demonstrated the diversity of bacterial population present in domestic wastewater which could possibly explored as efficient candidates in nitrogen bioremediation.

Keywords: *Heterotrophic nitrification, Aerobic denitrification, 16S rDNA, DNA gyrase.*

Introduction

Biological nitrogen removal is a critical biochemical process in modern wastewater pollution control facilities. Elimination of active nitrogen species such as ammonia, nitrate and nitrite from wastewater is very essential in order to prevent eutrophication of receiving water. Microbial involvement in these processes is still under investigation due to highly variable physiological and biochemical features. The conventional biological process was basically based on anaerobic ammonium oxidation (Anammox) (Strous *et al.*, 1997), complete autotrophic nitrogen removal over nitrite (Canon) (Sliekers *et al.*, 2002), oxygen-limited autotrophic nitrification-denitrification (Oland) (Kuai and Verstraete, 1998) and short-cut nitrification and denitrification (Peng and Zhu, 2006). Recently individual heterotrophic bacteria are under focus due to their ability to remove nitrogen through simultaneous nitrification and denitrification aerobically (Chen and Ni, 2011; Duan *et al.*, 2015). Furthermore, organic substrate utilization, oxygen tolerance properties and higher rate of nitrogen removal as compared to autotroph makes heterotrophs a potential candidate in water bioremediation (Third *et al.*, 2005).

In the present study, heterotrophic bacterial strains capable of doing nitrification and aerobic denitrification were isolated from domestic wastewater sediments. The identification of isolated strains was determined through PCR product sequencing of 16s ribosomal DNA and DNA gyrase. The biochemical characteristic like protease, lipase, amylase and cellulose activities of each strain was carried out. This study puts limelight on the possible prevalence of different heterotrophic bacterial population present in domestic wastewater sediments and each of the strains could be used as a potential candidate for application in wastewater bioremediation to promote nitrogen removal.

Materials and Methods

Isolation of Heterotrophic Bacteria

Isolation of nitrifying heterotroph, was made in an enrichment medium containing 0.2 per cent acetamide as carbon source in an inorganic medium containing 8.2 g of KH_2PO_4, 1.6 g of NaOH, 0.5 g of $MgSO_4.7H_2O$, 0.5 g of KCl, and 0.5 mg each of $CaSO_42H_2O$, $CuSO_45H_2O$, $FeCl_3.6H_2O$, and 0.5 mg of $ZnSO_4H_2O$ per litre of distilled water. The medium was adjusted to pH 7.0 and autoclaved for 15 min at 121 °C. Wastewater and sediments were collected from domestic sewerage area of Bhubaneswar, Odisha, India (20°13′31.8″ N; 85°50′32.6″ E). Five ml of domestic sewage water was inoculated into 200 ml of medium incubated on rotary shaker (120 × rpm) at 30 °C for 1 week. Resultant bacterial suspension was streaked on enrichment agar media. Separate colonies were picked up and individually tested for nitrogen removal activity in basal medium.

The ingredients of basal medium in 1000 ml distilled water (pH 7.0) were as follows: 0.1 g $MgSO_47H_2O$, 7.9 g $Na_2HPO_412H_2O$, 1.5 g KH_2PO_4 supplied with 1 per cent (v/v) trace elements solution (pH 6.0), which consisted of 50.00 g EDTA, 22.00 g $ZnSO_4$, 5.54 g $CaCl_2$, 5.06 g $MnCl_2$, 4.99 g $FeSO_4$, 1.10 g $(NH_4)_6Mo_7O_2$, 1.57 g $CuSO_4$ and 1.61 g $CoCl_2$ (per litre) (Shi *et al.*, 2013). The amount of nitrogen and carbon sources added to the basal medium is described in Table 29.1. The media were autoclaved for 15 min at 121°C. The nitrogenous sources were sterilized separately and added as needed. Experiments were carried out in triplicate in 500 ml of Erlenmeyer flask containing 100 ml of basal medium and incubated on rotary shaker (120 × rpm) at 30°C. The tributyrin agar plate, modified skim milk agar, carboxymethyl cellulase (CMC) agar and starch agar plates were use to determine lipase, protease, cellulose and amylase, respectively. The chemicals were purchased from Sigma (USA) and Hi-media (India).

Table 29.1: Basal Media Composition

Processes	*Nitrogen Source*		*Carbon Source*	
	NH_4Cl (1 mg/ml of NH^+_4-N)	*$NaNO_3$ (1 mg/ml of NO^-_3-N)*	*$NaNO_2$ (1 mg/ml of NO^-_2-N)*	*Glucose (mg)*
Nitrification media	10 ml	-	-	250
Denitrification media (DM-1)	-	10ml	-	250
Denitrification media (DM-2)	-	-	10 ml	250

Estimation of Heterotrophic Nitrification and Aerobic Denitrification Ability

One-milliliter (1 per cent inoculum) active culture of selected bacterial strains grown in LB overnight was inoculated into 100 mL of basal medium (Table 29.1) in a 500-mL Erlenmeyer flask. The flasks were incubated aerobically with constant shaking of 120 rpm at 30°C. The appropriate medium without inoculation was used as the control. The samples were taken from flask at 12 h and 24 h interval to determine the optical density at 600 nm (OD600), concentration of ammonium nitrogen (NH_4^+-N), nitrate nitrogen NO_3^--N, and nitrite nitrogenNO_2^--N.

Analytical Methods

Samples were withdrawn at 3 h intervals for estimating the changes in biomass, ammonia, nitrate and nitrite concentration as per standard methods (APHAAWWAWPCF, 2005). Briefly, the growth of the isolates was measured by spectrophotometer at a wavelength of 600 nm. Ammonia was determined by the method based on Berthelot reaction between ammonia, chlorine and phenolic compounds to form indophenol dyes. Nitrite was determined by N- (1-naphthyl)-ethylene di amine photometry method. Nitrate was measured by hydrochloric acid photometry method. Hydroxylamine was estimated by indirect spectrophotometry (Frear and Burrell, 1955).

Amplification of Conserved Genes 16s rDNA and gyrB

The genes encoding 16S rRNA were amplified from the extracted genomic DNA by polymerase chain reaction (PCR) using forward primer 27f (AGAGTTTGATCCTGGCTCAG) and reverse primer 1492r (TACGGTTACCTTGTTACGACTT), described by Stanley *et al.* (1995). The PCR was carried out as follows: 94 °C for 5 min: 35 cycles at 94 °C for 0.5 min: 49 °C for 0.5 min: 72 °C for 1 min, and final extension at 72°C for 9 min. DNA gyrase subunit B gene of strain CF-S36 was amplified with degenerate PCR primers UP-1 (GAAGTCATCATGACCGTTCTGCA(T/C)GC(N)GG(N)GG(N)AA(A/G)TT(T/C)

Figures 29.1a and b: Phylogenetic Tree Based on Comparison of Partial 16S rRNA Gene (a) and gyrB (b) Sequences Using MEGA 5.0. The phylogenetic relationships between isolated heterotrophic nitrifier and aerobic denitrifier was generated using neighbour joining method with a Bootstrap values of 1000 replications.

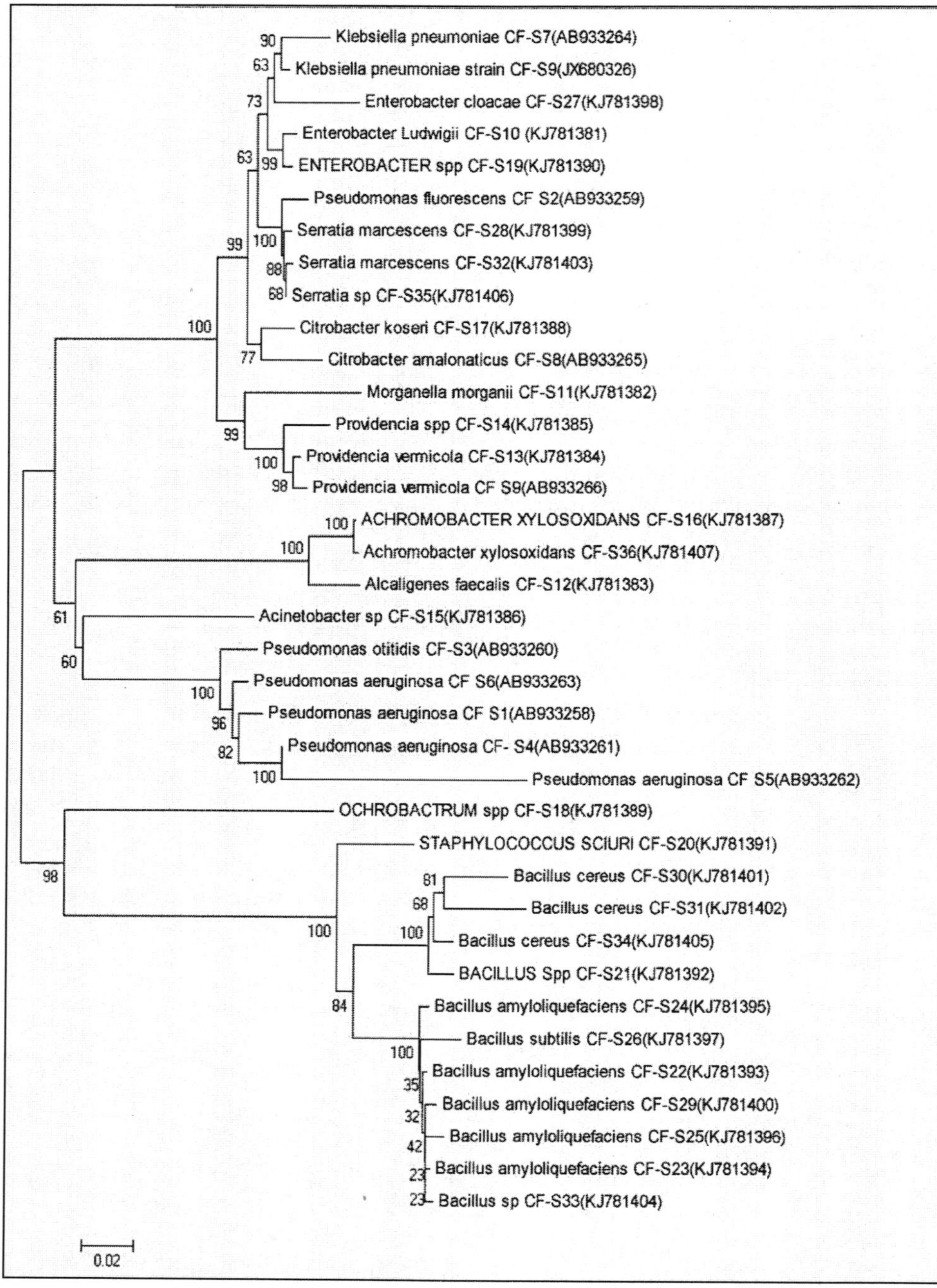

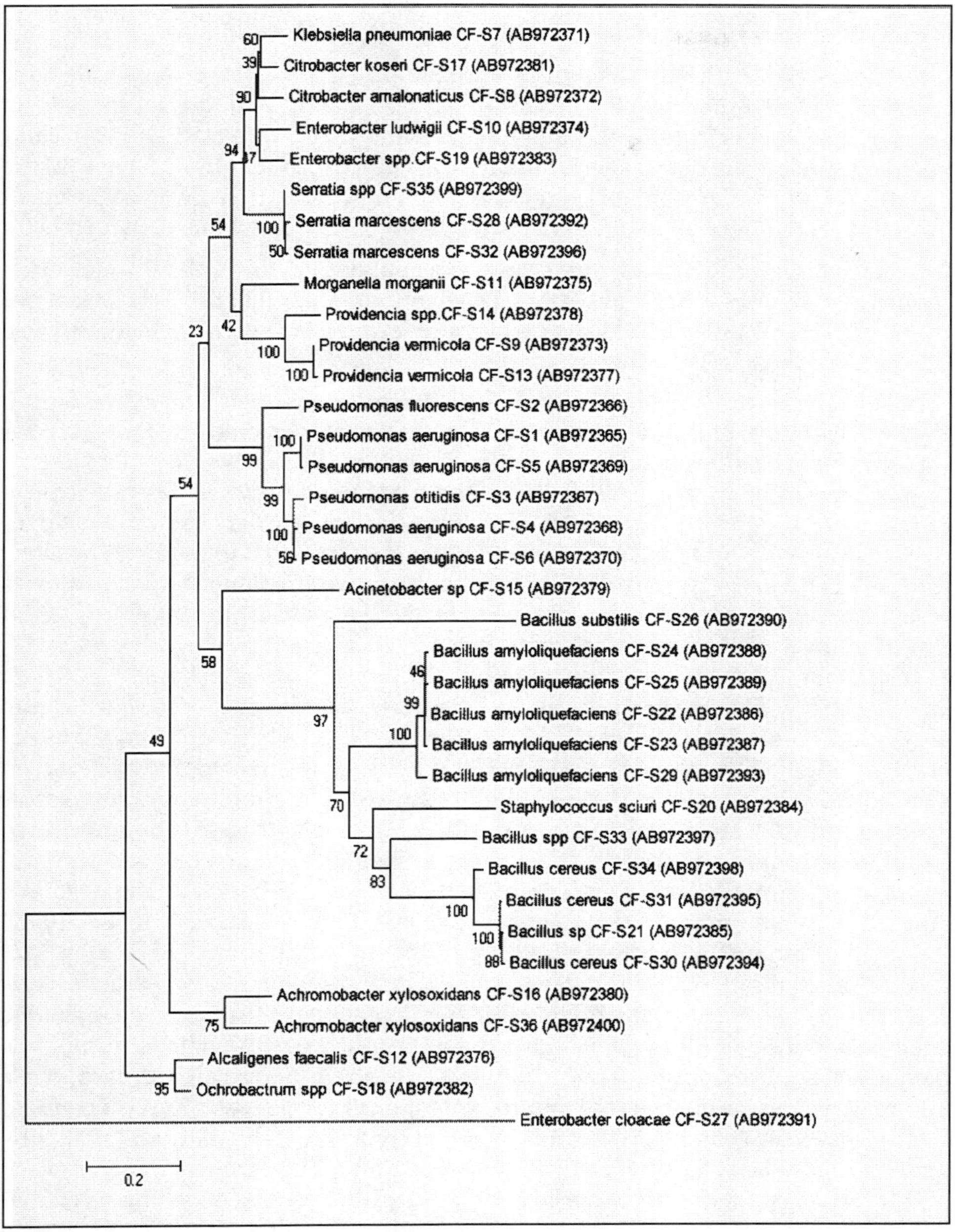
Klebsiella pneumoniae CF-S7 (AB972371)
Citrobacter koseri CF-S17 (AB972381)
Citrobacter amalonaticus CF-S8 (AB972372)
Enterobacter ludwigii CF-S10 (AB972374)
Enterobacter spp.CF-S19 (AB972383)
Serratia spp CF-S35 (AB972399)
Serratia marcescens CF-S28 (AB972392)
Serratia marcescens CF-S32 (AB972396)
Morganella morganii CF-S11 (AB972375)
Providencia spp.CF-S14 (AB972378)
Providencia vermicola CF-S9 (AB972373)
Providencia vermicola CF-S13 (AB972377)
Pseudomonas fluorescens CF-S2 (AB972366)
Pseudomonas aeruginosa CF-S1 (AB972365)
Pseudomonas aeruginosa CF-S5 (AB972369)
Pseudomonas otitidis CF-S3 (AB972367)
Pseudomonas aeruginosa CF-S4 (AB972368)
Pseudomonas aeruginosa CF-S6 (AB972370)
Acinetobacter sp CF-S15 (AB972379)
Bacillus substilis CF-S26 (AB972390)
Bacillus amyloliquefaciens CF-S24 (AB972388)
Bacillus amyloliquefaciens CF-S25 (AB972389)
Bacillus amyloliquefaciens CF-S22 (AB972386)
Bacillus amyloliquefaciens CF-S23 (AB972387)
Bacillus amyloliquefaciens CF-S29 (AB972393)
Staphylococcus sciuri CF-S20 (AB972384)
Bacillus spp CF-S33 (AB972397)
Bacillus cereus CF-S34 (AB972398)
Bacillus cereus CF-S31 (AB972395)
Bacillus sp CF-S21 (AB972385)
Bacillus cereus CF-S30 (AB972394)
Achromobacter xylosoxidans CF-S16 (AB972380)
Achromobacter xylosoxidans CF-S36 (AB972400)
Alcaligenes faecalis CF-S12 (AB972376)
Ochrobactrum spp CF-S18 (AB972382)
Enterobacter cloacae CF-S27 (AB972391)
0.2

GA) and UP-2r (AGCAGGGTACGGATGTGCGAGCC(A/G)TC(N)AC(A/G)TC(N) GC(A/G)TC(N)GTCAT) as described by Yamamoto and Harayama (1995).

Results and Discussion

Thirty-six purified isolates from enrichment agar medium were tested for nitrifying and denitrifying capabilities in basal media by periodically measuring ammonia, nitrite and nitrate. By Sanger sequencing, PCR product of 16S rDNA and gyrB sequences were obtained. Homology was compared with the sequences of the isolates with those in GenBank database. The phylogenetic tree was constructed based on partial 16S rDNA and gyrB sequences of the isolates (Figures 29.1a and b).

Nucleotide sequence of the isolates was submitted to GenBank nucleotide sequence databases (NCBI) and DNA Databank of Japan (DDBJ). All the isolates were tested for cellulase, lipase, protease, amylase, aerobic growth, heterotrophic nitrification and aerobic denitrification (Table 29.2).

The general perception throughout the wastewater industry was that, *Pseudomonas* spp. was the predominant heterotrophic bacteria responsible for nitrogen removal during activated sludge treatment (Drysdale *et al.*, 1999). It was also presumed that this genus of microorganisms were responsible for complete reduction of nitrates, via nitrites, to nitrous oxides and nitrogen gas. However, in confirmation with findings by Glass *et al.* (1997), the results of this study showed the presence of different types of culturable heterotroph responsible for nitrification as well as aerobic denitrification process present in wastewater sediments. The present results showed that different species with different genera such as *P. aeruginosa, P. fluorescens, P. stutzeri, Alcaligenes faecalis, Achromobacter xylosoxidans, Klebsiella pneumoniae, Providencia* spp., *Bacillus* spp., responsible for heterotrophic nitrification as well as aerobic denitrification. Wagner *et al.* (1994) demonstrated that the use of enriched media actually selects against gram-positive heterotrophic bacteria, on the contrary the present experiment demonstrated that although gram-negative heterotrophic bacteria are predominant in wastewater sediment, gram-positive heterotrophic bacteria also playing a greater role. In this study, certain bacterial strains were found tofail in utilizingeither nitrate or nitrite for nitrogen source and substantiate findings by Rheinheimer (1985), Robertson and Kuenen (1992). Removal of ammonia, nitrite and nitrate nitrogen by heterotrophic nitrification-aerobic denitrification strain is an economical and effective method (Khardenavis *et al.*, 2007). This study gives a brief idea about diversity of culturable heterotrophic bacteria and their nitrogen utilization capabilities. A variety of molecular methods have been applied to investigate nitrifying and denitrifying communities in the natural environments or wastewater treatment facilities. These culture-dependent approaches aim at characterizing the microbial community diversity and structure, quantifying the abundance of nitrifiers and denitrifiers. The phylogenetic analysis of highly conserved region of bacterial genome *i.e.*, 16S rDNA and gyrB showed that DNA sequences of 16S rDNA were separated with more clad. Thus 16S rDNA was more preferred molecular marker to study the phylogenetic relationship between isolated bacteria than gyrB.

Table 29.2: Characterizations of Isolated Bacteria Isolated from Domestic Wastewater of Bhubaneswar

Isolates	*Protease*	*Lipase*	*Amylase*	*Cellulase*	*Accession No.*		*Heterotrophic Nitrification and Denitrification*	
					16S	*gryB*	*Nitrification*	*Denitrification*
Pseudomonas aeruginosa CF-S1	+	+	-	+	AB933258	AB972365	+	+
Pseudomonas fluorescens CF-S2	+	+	+		AB933259	AB972366	+	-
Pseudomonas otitidis CF-S3	+	+	-	+	AB933260	AB972367	+	-
Pseudomonas aeruginosa CF-S4	+	+	-	+	AB933261	AB972368	+	+
Pseudomonas aeruginosa CF-S5	+	+	-	+	AB933262	AB972369	+	+
Pseudomonas aeruginosa CF-S6	+	+	-	+	AB933263	AB972370	+	+
Klebsiella pneumoniae CF-S7	+	+	+	+	AB933264	AB972371	+	+
Citrobactera malonaticus CF-S8	-	-	+	+	AB933265	AB972372	+	+
Providencia vermicola CF-S9	-	-	+	+	AB933266	AB972373	+	+
Enterobacter ludwigii CF-S10	-	-	+	+	KJ781381	AB972374	+	+
Morganellamorganii CF-S11	-	-	-	-	KJ781382	AB972375	+	+
Alcaligenes faecalis CF-S12	-	-	-	-	KJ781383	AB972376	+	+
Providencia vermicola CF-S13	-	-	+	+	KJ781384	AB972377	+	+
Providencia spp. CF-S14	-	-	+	+	KJ781385	AB972378	+	+
Acinetobacter sp. *CF-S15*	-	-	+		KJ781386	AB972379	+	+
Achromobacter xylosoxidans CF-S16	+	+	-	-	KJ781387	AB972380	+	+
Citrobacter koseri CF-S17	-	-	+	+	KJ781388	AB972381	+	+
Ochrobactrum spp. CF-S18	-	-	+	-	KJ781389	AB972382	+	+
Enterobacter spp. CF-S19	-	-	+	+	KJ781390	AB972383	+	+
Staphylococcus sciuri CF-S20	-	-	+	+	KJ781391	AB972384	+	-
Bacillus spp. CF-S21	+	+	+	+	KJ781392	AB972385	+	-
Bacillus amyloliquefaciens CF-S22	+	+	-	-	KJ781393	AB972386	+	-
Bacillus amyloliquefaciens CF-S23	+	+	-	-	KJ781394	AB972387	+	-

Isolates	*Protease*	*Lipase*	*Amylase*	*Cellulase*	*Accession No.*		*Heterotrophic Nitrification and Denitrification*	
					16S	*gryB*	*Nitrification*	*Denitrification*
Bacillus amyloliquefaciensCF-S24	+	+	-	-	KJ781395	AB972388	+	-
Bacillus amyloliquefaciens CF-S25	+	+	-	-	KJ781396	AB972389	+	+
Bacillus subtilis CF-S26	+	+	-	+	KJ781397	AB972390	+	+
Enterobacter cloacae CF-S27	+		+	+	KJ781398	AB972391	+	+
Serratia marcescens CF-S28	+	+	+	+	KJ781399	AB972392	+	+
Bacillus amyloliquefaciens CF-S29	+	+		-	KJ781400	AB972393	+	+
Bacillus cereus CF-S30	-	-	+	-	KJ781401	AB972394	+	+
Bacillus cereus CF-S31	-	-	+	-	KJ781402	AB972395	+	+
Serratia marcescens CF-S32	+	-	+	-	KJ781403	AB972396	+	+
Bacillus sp. CF-S33	+	-	+	+	KJ781404	AB972397	+	+
Bacillus cereusCF-S34	+	-	+	+	KJ781405	AB972398	+	+
Serratia sp. CF-S35		-	+		KJ781406	AB972399	+	+
Achromobacter xylosoxidans CF-S36	+	-	+	+	KJ781407	AB972400	+	+

References

APHAAWWAWPCF. (2005). Standard Methods for the Examination of Water and Wastewater, twenty-first ed. Washington D.C.

Chen, Q. and Ni, J. (2011). Heterotrophic nitrification–aerobic denitrification by novel isolated bacteria. Jouranl of Indurstial. Microbiology Biotchnology, 38: 1305-1310.

Duan, J., Fang, H., Su, B., Chen, J., and Lin, J. (2014). Characterization of a Halophilic Heterotrophic Nitrification-Aerobic Denitrification Bacterium and its Application on Treatment of Saline Wastewater. Bioresource Technology, 179: 421-428

Drysdale, G.D., Kasan, H.C. and Bux, F. (1999). Denitrification by heterotrophic bacteria during activated sludge treatment. Water sa-pretoria, 25: 357-362.

Frear, D.S. and Burrell, R.C. (1955). Spectrophotometric methods for determining hydroxylamine reductase activity in higher plants. Analytical Chemistry, 27: 1664-1665.

Glass, C., Silverstein, J., and Oh, J. (1997). Inhibition of denitrification in activated sludge by nitrite. Water Environment Research, 69: 1086-1093.

Khardenavis, A. A., Kapley, A. and Purohit, K.J. (2007). Simultaneous nitrification and denitrification by diverse *Diaphorobacter* sp. Applied Microbiology and Biotechnology, 77: 403-409.

Kuai L.and Verstraete W. (1998). Ammonium removal by the oxygen limited autotrophic nitrifcation–denitrifcation (oland) system. Applied and Environmental Microbiology, 64: 4500–4506

Peng Y. and Zhu G. (2006). Biological nitrogen removal with nitrification and denitrifcation via nitrite pathway. Applied and Environmental Microbiology, l73: 15–26

Rheinheimer, G. (1985). Aquatic Microbiology, (3rd edn.). VEB Gustav Fischer Verlag: Jena.

Robertson, L.A. and Kuenen, G. (1992). Nitrogen removal from water and waste. Society for General Microbiology. Symposium, 48: 227-267.

Shi Z., Zhang Y., Zhou J., Chen M. and Wang X. (2013). Biological removal of nitrate and ammonium under aerobic atmosphere by *Paracoccus versutus* LYM. Bioresource Technology 148: 144-148.

Sliekers A. O., Derwort N., Strous M., KuenenJ. G. and Jetten M.S.M. (2002). Completely autotrophic nitrogen removal over nitrite in one single reactor. Water Research, 36: 2475–2482

Stanley, J., Baquar, N. and Burnens, A. (1995). Molecular subtyping scheme for Salmonella panama. Journal of Clinical Microbiology, 33: 1206-1211.

Strous, M., Van Gerven, E., Zheng, P., Kuenen, J.G. and Jetten M.S.M. (1997). Ammonium removal from concentrated waste streams with the anaerobic ammonium oxidation (ANAMMOX) process indifferent reaction configurations. Water Research, 31: 1955–1962

Third, K.A., Gibbs, B., Newland, M., and Cord-Ruwisch, R. (2005). Long-term aeration management for improved N-removal via SND in a sequencing batch reactor. Water Research, 39: 3523–3530.

Wagner, M., Erhart, R., Manz, W., Amann, R., Lemmer, H., Wedi, D. and Schleifer, K. H. (1994). Development of an rRNA-targeted oligonucleotide probe specific for the genus Acinetobacter and its application for in situ monitoring in activated sludge. Applied and Environmental Microbiology, 60 (3): 792- 800.

Yamamoto, S. and Harayama, S. (1995). PCR amplification and direct sequencing of gyrB genes with universal primers and their application to the detection and taxonomic analysis of *Pseudomonas putida* strains. Applied and Environmental Microbiology, 61(3): 1104-1109.

30

Impact of Heavy Metal on Growth and Amelioration of Antioxidant Defense Mechanism of *Terribacillus goriensis* Strain MNSMP5 with Improved Production of Acid Protease

M.M. Panda[1], S.K. Sahoo[2] and S. Sahoo[2]*

[1]Department of Pharmaceutical Sciences,
[2]Department of Biotechnology,
Utkal University, Bhubaneswar – 751 004, Odisha
****E-mail: sabujbiotech@rediffmail.com***

ABSTRACT

The present investigation was undertaken with an objective to screen, isolate, and identify the microorganism having acid protease production potential. Further the impact of heavy metal induced stress on growth response of the isolate Terribacillus goriensis strain MNSMP5 from manganese open ore mines, Keonjhar and amelioration of antioxidant defence mechanism was intensively studied. The bacterial strains were isolated on the basis of hydrolysis zone on skimmed milk agar and was identified to be T. goriensis strain MNSMP5 Gen Bank Accession Number: KY111915.1, by detailed morphological followed by 16S rDNA based molecular method. Acid protease production by T. goriensis strain MNSMP5 employing submerged fermentation using casein as a substrate was found to be 4.70 U/ml. Antibiogram pattern revealed the isolate to be susceptible to gentamycin, ciprofloxcin, tetracycline and resistant to amoxycillin, bacitracin, cefixime, ampicillin and chloramphenicol. Effect of metals on growth pattern of strain was estimated. Cu^{++} significantly decrease the growth rate of bacteria followed by of Cd^{++} and Hg^{++}. Effect of hydrogen peroxide on the strain treated with Cd^{++} and Cu^{++} into the medium enhanced the activity but the activity decreased in medium containing

Hg^{++}. Cellular Glutathione (GSH) content on various oxidative stress conditions was also determined wherein Cd^{++} treated strain exhibited enhanced production of GSH than Hg^{++} treatment and minimal activity was observed by the treatment with Cu^{++} when compared to Hg^{++}. The strain was found to be catalase negative when exposed to metal induced stress.

Keywords: *Acid protease, Terribacillus goriensis, Metal induced stress, Antioxidant.*

Introduction

Proteases are the group of enzymes whose main function is to break down proteins to small peptides or free amino acid (Bhatnagar *et al.*, 2010). Proteases are obtained from a wide diversity of sources such as plants, animals and microorganisms but due to the low production of proteases from above sources, they are mainly produced by microorganisms (West and Westport *et al.*, 1983; Raimbaulta *et al.*, 2013; Fazilat *et al.*, 2016). The microbial proteases account about 60 per cent of the worldwide enzyme commercialization (Gupta *et al.*, 2002). They are highly thermostable with high specific activities and better physical and chemical characteristics which seem to be good for future biotechnological applications (Rao *et al.*1998; Temiz *et al.*, 2008). Proteases are classified into various groups (Dash *et al.*, 2003) but our aim is to focus upon acid protease. Acid proteases are without doubt the most interesting group of proteases. Acid proteases are endopeptidases, having molecular masses in the range of 30 to 45 kDa. The optimum activities are observed at a wide pH range between 2.0 to 7.0 and temperature range (30 °C – 70 °C) (Castillo *et al.*, 2004; Sethi *et al.*, 2015). It is one of most significant enzymes produced by the biotechnology industry (Kumar *et al.*, 2012) and also contribute several chemical and biochemical reactions that take place in the food, beverage (Pant *et al.*, 2014), pharmaceutical, cosmetics and other Industries, (Uyar and Baysal, 2004, Okumura *et al.*, 1997; Mohr *et al.*, 1986). They also have clinical applications as digestive aids in the treatment of gastrointestinal disorders (Rakte *et al.*, 2014) and perform significant role related to nutrition. The present investigation was undertaken to Screen, isolate and identify the microorganism with improved acid protease production with high purity. Also to findout the impact of heavy metal induced stress on growth response of the isolate *T. goriensis* strain and amelioration of its antioxidant defence mechanism

Materials and Methods

Sample Collection

Different types of soil samples like cashew factory soil, soil from manganese open ore mines, Keonjhar, Tarabolo hot water, Odisha, slaughtering house soil from different parts of Odisha were collected in sterile containers and bacteria were isolated under laboratory condition

Screening of the Acid Protease Producing Bacterial Isolate

The single colonies were screened for protease production by using (2 per cent) skim milk agar medium plates (Alnahdi *et al.*, 2012) with pH 3-5. The plates were incubated at 37 °C for 24 h. Proteolytic activities of bacteria were detected on the basis of appearance of clear zones around the bacterial colonies. The colony showing highest zone of inhibition at low pH was selected for further study. The colony having highest hydrolysis zone was selected and streaked on nutrient agar plate to obtained isolated colony followed by preserved at 4 °C on nutrient agar slant.

Identification of the Bacterial Isolate

The bacterial isolate sample showing highest enzyme activity was further subjected to different morphological and taxonomical identification. Gram's staining method was used for identification of the strain morphologically. For taxonomical identification 16s rDNA gene sequencing was done. DNA was isolated from the culture. Its quality was evaluated on 1.0 per cent agarose gel, a single band of high-molecular weight DNA has been observed. Fragment of 16s rDNA gene was amplified by 27F and 1492R primers. The PCR amplicon was purified to remove contaminants. Forward and reverse DNA sequencing reaction of PCR amplicon was carried out with forward primer and reverse primers using BDT v3.1 Cycle sequencing kit on ABI 3730xl Genetic Analyzer. Consensus sequence of 16S rDNA gene was generated from forward and reverse sequence data using aligner software. The 16S rDNA gene sequence was used to carry out BLAST with the database of NCBI genbank database. Based on maximum identity score first ten sequences were selected and aligned using multiple alignment software program Clustal W. Distance matrix was generated using RDP database and the phylogenetic tree was constructed using MEGA 4. (Dopazo *et al.*, 1994; Rzhetsky *et al.*, 1992).

Effect of Antibiotics

The antibiogram pattern of the bacterial isolate was tested for sensitivity against different antibiotic discs (Himedia, Mumbai) *viz.*, ciprofloxacin (10 mcg/disc), ampicillin (10 mcg/disc), cefixime (5mcg/disc), gentamicin (10 mcg/disc) tetracycline (30 mcg/disc), bacitracin (8 unit/disc), chloramphenicol (10 mcg/disc), amoxycillin (30 mcg/disc) in sterilized nutrient agar plate by disc diffusion method.

Enzymatic Assay

Acid proteases activity was determined according by the method of (Tsuchida *et al.*, 1986; Mukhtar *et al.*, 2009; Kumura *et al.*, 2011) by using casein as a substrate. Mixture of 500µl of 1 per cent casein in 50 mM phosphate buffer (pH 7.0) and 200µl crude enzyme extract were incubated in a water bath at 40 °C for 20 min. After incubation the enzyme reaction was terminated by the addition of 1ml of 10 per cent TCA and kept at room temperature for 15min. Then the reaction mixture was centrifuged at 10,000 × rpm for 5 min to separate unreacted casein. The supernatant was mixed with 2.5 ml of 0.4 M Na_2CO_3 and finally 1ml of 3-fold of diluted Follin Ciocalteus phenol reagent was added. The resulting solution was incubated at room temperature in the dark for 30 min till blue colour development. After incubation absorbance was measured at 660 nm against a reaction blank using tyrosine as standard. The amount of protease produced was estimated and expressed in microgram of tyrosine released under standard assay conditions.

Effect of Heavy Metals on Growth

To examine the effect of different metals like $Pb(NO_3)$, $CdCl_2$, CuCl, As_2O_3, HgCl and FeCl on the test sample, a loop full of freshly grown bacterial culture was inoculated into 50ml of nutrient broth supplemented with 1mM of individual metals of concentration for 24 h. Both experimental (with metal) and control sets (without metal) were incubated at 37 °C for 6 days. Bacterial growth was determined by measuring the absorbance at 540 nm on 2nd, 4th, and 6th days of post inoculation.

Catalase Activity

Catalase sample was prepared by adding 5 µl of absolute ethanol to 50 µl of sample, and incubated for 30 min in ice bath. To this 56µl of 10 per cent Triton X-100 was added. The enzymatic reaction started by adding 0.1ml sample to 2.9 ml of 12 mM H_2O_2.The absorbance was recorded at 240 nm immediately at 15 secs interval till 2 min and activity was expressed as pkat per mg protein (1pkat=10^{-12}katal). One katal (kat) is defined as the amount of enzyme that transforms one mole of substrate per second.

Estimation H_2O_2 Content

H_2O_2 content was determined spectrophotometrically according to the method of Pick and Keisari (1981) using horse radish peroxidase (HRP) and phenol red. The assay mixture containing 50 mM phosphate buffer, (pH 7.4), 0.1 ml of phenol red solution and 50 ml of HRP was incubated for 5min at room temperature. Finally, 0.1 ml sample (1mg protein) was added to it, followed by immediate addition of 50 ml of NaOH to stop the reaction. The absorbance was read at 610nm and the H_2O_2 content was expressed as n mol per mg protein.

Estimation of Intracellular GSH

Determination of total sulphydryl groups, protein-bound sulphydryl groups, and free sulphydryl groups in biological samples using DTNB was done by Ellman's Method. GSH concentration can be determined by measuring absorbance at 412 nm. The crude homogenate sample was used for this assay. Reaction wascarried out in triplicate. 50 µl of sample was added with 50µl of NADPH and 50 µl of DTNB to the respective wells followed by addition of 100 µl of assay buffer to the wells and incubated 2-3min.after incubation absorbance was taken at 412 nm.

Results

Screening of the Acid Protease Producing Bacteria

Protease activity was observed from zone of hydrolysis given in Figure 30.1. Out of five samples two samples showed good result at pH 4.0 and 5.0 and were designated as Mn_4 (Manganese pH 5.0), Mn_4 (Manganese pH 4.0) among these Mn_4 sample was selected and named as MNSMP5 and further biochemical assays was performed by this strain.

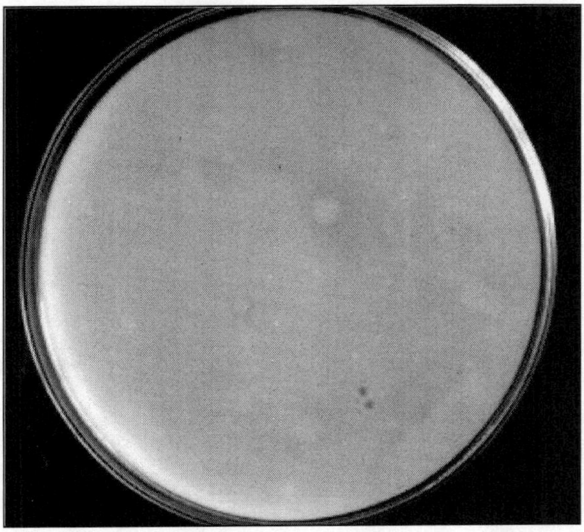

Figure 30.1: Plate Showing Hydrolysis Zone by the Isolate.

Identification of the Bacterial Isolate Sample

From the Gram staining photo of isolated microorganism it was concluded that, the microorganism was Gram negative in nature having size ranging from 0.55-3.12 µm.

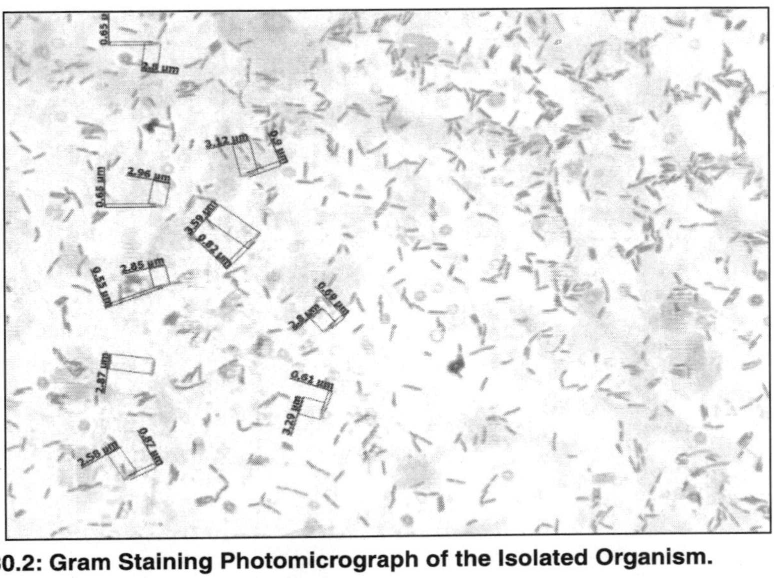

Figure 30.2: Gram Staining Photomicrograph of the Isolated Organism.

Taxonomical Identification

Sample labeled as MNSMP5 was found to be *Terribacillus goriensis* based on nucleotide homology and phylogenetic analysis.

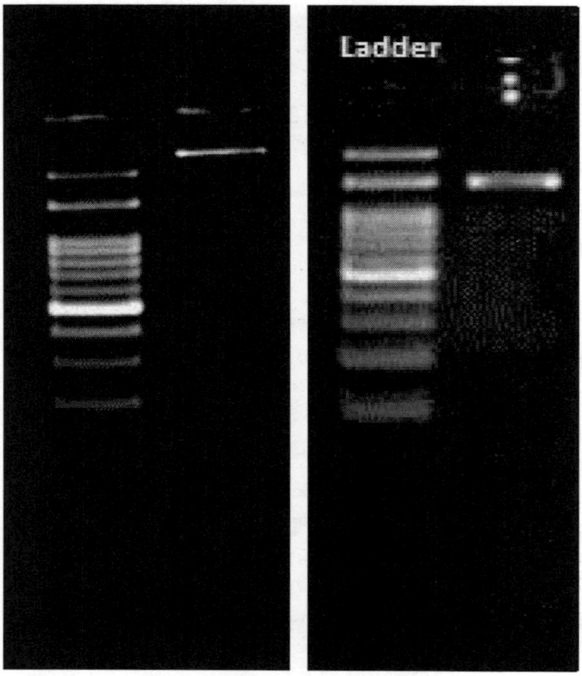

Figure 30.3: Gel Image of gDNA16S PCR Amplicon.

Effect of Antibiotics

Antibiogram pattern revealed the isolate to be susceptible to gentamycin, ciprofloxcin, tetracycline and resistant to amoxycillin, bacitracin, cefixime, ampicillin and chloramphenicol.

Table 30.1: Antibiogram Pattern of the Isolate

Antibiotics	*Zone of Inhibition (mm) Interpretation*
Bacitracin	Resistant
Amoxycillin	Resistant
Tetracycline	Sensitive
Gentamycin	Sensitive
Chloramphenicol	Resistant
Ciprofloxacin	Sensitive
Cefixime	Resistant
Ampicillin	Resistant

Enzymatic Assay

Enzyme acid protease produced by strain *T. goriensis* (MNSMP5) was found to be 4.70 U/ml

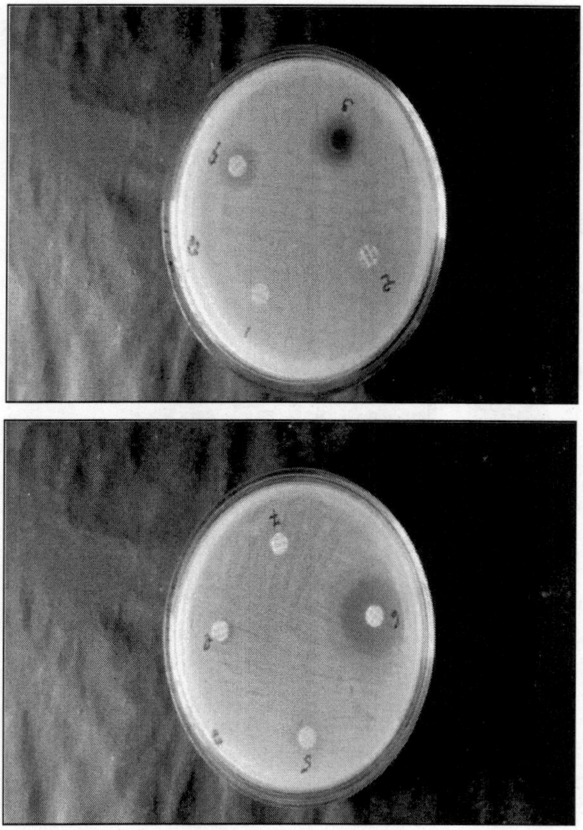

Figure 30.4: Zone of Inhibition by using Various Antibiotics.

Effect of Heavy Metals on Growth of *T. goriensis*

The effect of metal ion on sample MNSMP5 is depicted in Figure 30.5. Results clearly indicated that addition of these metals ion to the culture medium reduced the growth of sample strain MNSMP5. Copper, mercury and cadmium treated sample was further taken for determination of antioxidant assays (Table 30.3).

Table 30.2: Effect of Heavy Metals on Growth of *T. goriensis*

Days	*Reading (O.D) at 540nm*						
	Control	*Arsenic*	*Lead*	*Cadmium*	*Copper*	*Mercury*	*Iron*
0	0.02	0.03	0.16	0.04	0.06	0.02	0.14
Day 1	1.17	0.41	0.50	0.10	0.34	0.02	0.44
Day 2	1.38	0.53	0.58	0.36	0.60	0.02	0.60
Day 3	1.50	0.54	0.47	0.48	0.55	0.02	0.57
Day 4	1.49	0.42	0.48	0.58	0.43	0.03	0.81
Day 5	1.52	0.27	0.32	0.49	0.37	0.03	0.62
Day 6	1.55	0.25	0.29	0.30	0.32	0.03	0.56

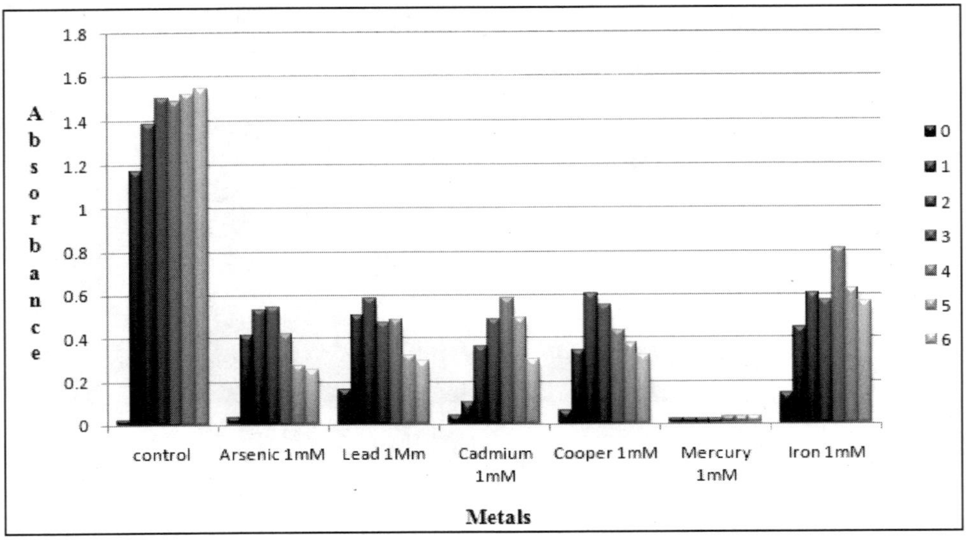

Figure 30.5: Effect of Various Heavy Metals (1mM) on Growth of *T. goriensis* upto 6 Days Post Inoculation

Catalase Activity (pkat/mg protein)

The isolate *T. goriensis* was found to be catalase negative when exposed to metal induced stress.

Estimation of H_2O_2 Content

It has been also concluded that optimum hydrogen peroxide activity was seen high in Cd treated *T. goriensis* and minimal activity was observed in Cu treated followed by Hg at day 2. But the activity was decreased at day 4. Hydrogen peroxide content on Cd was more than Hg and Cu.

Estimation of Intracellular GSH

Cu enhanced production of GSH than Cd, Hg treated strain in day 2 but at day 4 production of GSH was more in Cd followed by Hg and Cu treated strain.

Table 30.3: Effect of different Metals on Catalase Activity (pkat/mg protein), Cellular GSH Activity (μmol g^{-1} cells), H_2O_2 Activity (nmol mg^{-1}) of *T. goriensis* strain

Sample	*Catalase Activity (pkat/mg protein)*	*H_2O_2 Present (μmole/ml)*	*GSH (nmole/ml)*
D2Mn4Hg	-ve	3.299	5.27
D2Mn4Cd	-ve	135.53	4.82
D2Mn4Cu	-ve	12.99	5.32
D4Mn4Hg	-ve	4.21	5.73
D4Mn4Cd	-ve	4.77	6.00
D4Mn4Cu	-ve	0.8629	5.42

Discussion

The enzyme production by using different media was reported in different literature employing different sources for acid protease production. Skimmed milk agar media reported by Narendra *et al.* in 2012 for production of acid protease by bacterial source was used with slight modification. As a result of our studies colonies having highest hydrolysis was zone in skimmed milk agar shown in Figure 30.1 was selected and used for fermentation. After completion of fermentation, filtrate was estimated for acid protease activity as per method mentioned by Tsuchida *et al.* (1986) tyrosine was used as a standard found that the strain showed good proteolytic activity *i.e.* 4.70 U/ml. Morphological identification showed that the strain was gram positive rod shaped bacillus shown in Figure 30.2 and 16S rDNA results showed that the strain was *Terribacillus gorienswas.* Antibiogram assay was an important attribute for selection acid protease producing strain we found that the strain was susceptible to gentamycin, ciprofloxcin, tetracycline and reswastant to amoxycillin, bacitracin, cefixime, ampicillin and chloramphenicol.

Effect of various metals on growth pattern of strain MNSMP5 was corroborated as reported by Behera *et al.* (2014) and shown in Figure 30.5 which clearly indicated that effect of metals in the medium significantly decrease growth rate of the bacteria This shows that strain *T. goriens* was sensitive to metal ion. Effect of hydrogen peroxide, GSH, Catalase was depicted in Table 30.3. From the table it was clearly observed that hydrogen peroxide activity was seen high in Cd treated strain and minimal activity was observed in Hg treated bacteria. Again from the Table 30.3, it was observed effect of cellular GSH content on various oxidative stress conditions. GSH acts against ROS including hydrogen peroxide. From the above graph it was clearly understood that effect of Cu enhanced production of GSH than Cd treated strain. Minimal activity was observed by the treatment with Hg. So it was confirmed that strain *T. goriens* produces GSH against ROS or H_2O_2. But catalse activity against various oxidative stress of the strain has shown negative response proved that strain *T. goriensis* was catalase negative.

Conclusion

A promising isolate from manganese ore mines Keonjhar identified as *T. goriensis* was found to be a potent producer of extracellular acid protease. The extracellular acid protease was found to be optimally active at pH 4.0.The effect of heavy metals (Cd, Cu, Hg) on the growth and antioxidant enzymes of the strain. It has been also concluded that optimum hydrogen peroxide activity was seen high in Cd treated strain and minimal activity was observed in Hg treated bacteria. It has been also concluded that effect of Cu enhanced production of GSH than Cd, Hg treated strain in day 2 but at day 4 production of GSH is more in Cd followed by Hg and Cu treated strain. The isolated strain has shown negative response proved that strain *T. goriensis* MNSMP5 was catalase negative.

References

Alnahdi, H S. (2012). Isolation and screening of extracellular proteases produced by new isolated *Bacillus* sp. Journal of Applied Pharmaceutical Science, 2(9): 071-074.

Behera, M., Dandapat, J. and Rath, C.C. (2014). Effect of heavy metals on growth response and antioxidant defense protection in Bacillus cereus. Journal of Basic Microbiology, 54(11): 1201-1209.

Bhatnagar, D., Joseph, I and R, Paul, (2010). Amylase and acid protease production by solid state fermentation using *Aspergillus niger* from mangrove swamp. Indian Journal of Fisheries, 57(1): 45-51.

Castillo, F.J. and Aguilar, R. (2004). Characterization of acidic proteolytic enzymes from Monterey sardine (*Sardinops sagax caerulea*) viscera. Food Chemistry, 85: 343–350.

Dash, C. Kulkarni, A. Dunn, B. and. Rao, M. (2003). Aspartic peptidase inhibitors: implications in drug development.Critical Reviews in Biochemistry and Molecular Biology **38**: 89-119.

Dopazo, J. (1994). Estimating errors and confidence intervals for branch lengths in phylogenetic trees by a bootstrap approach. Journal of Molecular Evolution, 38: 300-304.

Fazilat, A. (2016). Production, isolation, purification and partial characterization of an extracellular acid protease from *Aspergillus niger*. International Journal of Advanced Research in Biological Sciences, 3(3): 32-38.

Gupta, R. Beg, Q.K. and Lorenz, P. (2002). Bacterial alkaline proteases: molecular approaches and industrial application. Applied Microbiology and Biotechnology, 59: 15-32.

Kumar, V., Sahai, V., and Bisaria, V.S. (2012). Production of amylase and chlamydospores by Piriformospora indica, a root endophytic fungus. Biocatalysis and Agricultural Biotechnology, http: / / dx.doi.org / 10.1016 / j.bcab.

Kumura, H. Ishido, T. and Shimazaki, K. (2011). Production and partial purification of proteases from *Aspergillus oryzae* grown in a medium based on whey protein as an exclusive nitrogen source. Journal of Dairy Science, 94: 657–667.

kumura, H.O. Watanabe, R. Kotoura, Y. Nakane, Y. Tangiku, O. (1997). Effects of a proteolytic enzyme preparations used concomintantly with an antibiotic Osteroarticular infections. The Japanese Journal of Antibiotics, 30(3): 223-7.

Mohr, V. and Vincent, J. (1986). Proteases of Antarctic Krill- a new system for effective enzymatic debriment of necrotic ulcerations. Cellular Molecular Life Science, 42(4): 403-404.

Mukhtar, H. and Haq, I.U.l. (2009). Production of Acid Protease by *Aspergillus niger* using Solid StateFermentation. Pakistan Journal of Zoology, 41(4): 253-260.

Narendra, D. (2012). Isolation and characterization of protease producing bacterial from soil and estimation of protease by spectrophotometer. International Journal of Science and Technology, 1(1): 1-7.

Panta, G. Prakasha, A. Pavania, J.V.P., Bera, S. Devirama, G.V.N.S. Kumara, A., Panchpurib, M.and Prasunaa, R.G. (2015). Production, optimization and partial purification of protease from *Bacillus subtilis*. Journal of Taibah University for Science, 9: 50–55.

Raimbaulta, A.K. Fodila, Y.Z. Solerb, A. and Carvalhoa, M. (2013). A novel aspartic acid protease gene from pineapple fruit (*Ananas comosus*): Cloning, characterization and relation to postharvest chilling stress resistanceJournal of Plant Physiology, 170: 1536-1540.

Rakte, A.S. and Nanjwade, B.K. (2014). Proteolytic Enzymes Delivery Systems: A Review. International Journal for Pharmaceutical Research Scholars, 3: I-2.

Rao, M.B., Aparna, M., Tanksale, Ghatge, M.S. And Deshpande, V.V. (1998). Molecular and biotechnological aspects of microbial proteases. Microbiology molecular Biology Review, 62: 597-635.

Reed, G. and Westport, C.T. (1983). Prescott and Dunn's industrial microbiology, 4th Ed. AVI Publishing Company, Inc., Westport, Connecticut, pp. 684-691.

Rzhetsky, A. and Nei, M. (1992). A simple method for estimating and testing minimum evolution trees Molecular Biology and Evolution, 9: 945-967.

Sethi, B.K., Nanda, Jana, A., Dasmohapatra, P.K. and Sahoo, S.L. (2015). Thermostable acidic protease production in *Aspergillus terreus* NCFT 4269.10 using chickling vetch peels. Journal of Taibah University for Science.

Sumantha A.C., Larroche, A., Pandey. (2006). Microbiology and industrial biotechnology of food-grade proteases: a perspective. Food Technolgy Biotechnolgy, 44: 211–20.76.

Temiz, H., Okumus, E., Aykut, U., Dervisoglu, M. and Yazici, F. (2008). Partial purification of pepsin from turkey proventriculus. Wld. Journal of Microbiological Biotechnology, 24: 1851–1855.

Tsuchida, O., Yamagata, Y. and Ishizuka, J. (1986). An Alkaline proteinase an alkalophilic Bacillus sp. Current Microbiology, 14: 7-12.

Uyar, F., and Baysal, Z. (2004). Production and optimization of process parameters for alkaline protease production by a newly isolated Bacillus sp. under solid-state fermentation. Process Biochemistry, 39: 1893–1898.

31

In vitro and *In vivo* Characterization of Bacteria Isolated from different Sources for Possible Probiotic Use in Aquaculture Practices

*M. Behera, L. Nayak, Lopamudra and S. K. Nayak**

Department of Biotechnology,
North Orissa University, Baripada – 757 003, Odisha
**E-mail: sukantanayak@rediffmail.com*

ABSTRACT

Aquaculture is emerging as one of the most viable, dynamic and fastest growing enterprises which not only keeps pace with the surging need for animal protein but also provides nutritional and food security to humans. However, the success of modern aquaculture requires good management practices but with the advancement of intensive aquaculture practices led to outbreaks of diseases and are being increasingly recognized as one of the major constraints to the aquaculture production in many countries and India is no exception. Use of probiotics in aquaculture practices as non-antibiotic agent has been accelerated over the year. To acquire suitable probiotic strain, In vitro characterisation of bacteria is the most essential steps to be emphasized. The present study aimed to screen the bacterial isolates from different sources like gastrointestinal (GI) tract of different fishes, soil and water to develop a suitable probiotic strain. In vitro characterisation like antimicrobial activity study, antibiotic sensitivity assay, haemolytic activity, proteolytic activity, presence of plasmid DNA of the isolates were studied. A total 96 bacteria were Screened, out of which 1 bacterial isolates from fish gut namely C3 exhibiting antibacterial activity against Aeromonas hydrophila. Like that11 bacterial isolates from soil namely B3, B6, B13, B16, B17, D1, D2, D3, E1, E3, SB1exhibited antibacterial activity against Aeromonas hydrophila and Edwardsiella tarda. Based on haemolytic nature out of 12 bacterial isolates C3, B16 and B17 found to be haemolysed human RBCs, whereas D2 haemolysed fish and chicken

RBCs. Like that D3 haemolysed fish RBCs and SB1 haemolysed human, fish, chicken and goat RBCs. Other isolates namely B3, B6, B13, D1, E1 and E3 were found to be non-haemolytic to any of the tested erythrocytes. Among all the isolates E1 and E3 found to be resistance to a huge numbers of antibiotics. Isolates E1 and E3 doesn't bear plasmid DNA and also these have no proteolytic activity.

Keywords: *Antibiotics, Aquaculture, Probiotics, Fish.*

Introduction

Aquaculture is currently the fastest growing food production sector in the world that contributes one-third of world fisheries production (Lowther, 2005). It provides food security, poverty alleviation as well as income generation among people. Over the year aquaculture has undergone sea changes from extensive to intensive practices. Such changes lead to higher production beyond any doubt but simultaneously brought a plethora of problems like diseases which is a major problem in fish growth and its expansion. To control diseases antibiotics are commonly used but their excessive use not only effects the environment and fish health but also developed drug resistance pattern among pathogens which lead to their restricted use. Hence, the use of probiotics as non-chemotherapeutic method has been adapted for treatment of diseases in aquaculture which proved to be beneficial to the industry in the long run.

Probiotics are usually natural microflora, which after administration colonize in the intestine and subsequently reduce the proliferation of pathogens and promote the optimum utilization of feed (Mukherjee and Nayak, 2004). Probiotics may prevent bacterial diseases through a variety of mechanisms, which includes creation of a hostile environment for pathogens by the production of inhibitory compounds, by competing for essential nutrients and adhesion sites or by modulating the immune responses (Balcazar *et al.*, 2006; Merrifield *et al.*, 2010). Currently several probiotics are commercially available for aquaculture use.

Although probiotics promised as an alternative to chemicals and antibiotics for aquaculture practices (Gatesoupe, 1999; Vaseeharan and Ramasamy, 2003; Li *et al.*, 2006; Nayak and Mukherjee, 2011), several reports indicate deleterious effects like growth reducing agent,non-stimulation of immune responses and no significant protection by various probiotics in fish (Murthy and Naik, 2002; Abraham *et al.*, 2007; Shelby *et al.*, 2007). Hence the selection of probiotics is usually based on *in vitro* antagonism and growth in intestinal mucus (Verschuereet *et al.*, 2000; Irianto and Austin, 2002; Vine *et al.*, 2004; Nayak and Mukherjee, 2011). Several factors including non-pathogenic, non-haemolytic ability of a probiotic strain are often overlooked (Panigrahi and Azad, 2007; Nayak and Mukherjee, 2011).

Therefore, probiotics should be carefully selected for species-specific advantages (Ghosh *et al.*, 2007; Hai *et al.*, 2007; Nayak and Mukherjee, 2011).Information on characteristics like antibiotic resistance, haemolytic nature, pathogenic potential, Proteolytic activity and growth characteristics under different conditions will help to select a safe as well as effective probiotic strain. Hence invitro studies are the most important challenge during selection of a suitable and beneficial probiotic strain in order to perform *in vivo* challenges. In this study we tried to develop a suitable probiotic strain from different sources like gastrointestinal (GI) tract of different fishes, soil and water to evaluated the isolates on various invitro studies including antimicrobial activity assay, antibiotic sensitivity assay, haemolytic nature, proteolytic effect, plasmid DNA presence, pathogenicity assay and growth range study before choosing a candidate probiotic species best suited for aquaculture practices.

Materials and Methods

Isolation of Gastro Intestinal Bacteria

The bacteria were isolated from the gastro intestinal tract of different fish species like rohu (*Labeo rohita*), catla (*Catla catla*), mrigala (*Cirrhinus mrigala*), kou (*Anabus testudinis*) live fish belonging to different species were purchased from nearby farms and villages with no history of probiotics and/or any bacterial adjuvant use.1gm of triturated gut sample was serially diluted in 9ml of sterile distilled water. Approximately 0.1 ml of diluted sample was plated into Nutrient Agar by spread plate technique. All plates were incubated at 37 °C for overnight and after incubation selected colonies were randomly picked and subcultured into NA plates. Such sub culturing was done for 3 times to get pure colonies. Finally pure colonies were maintained in slants for further use.

Isolation of Bacteria from other Sources

Bacteria were also isolated from different soil and water. Briefly 1 gm. sample were serially diluted in 9 ml. of sterile distilled water. Approximately 100µl of the solution was plated into Nutrient Agar by spread plate technique. All plates were incubated at 37 °C for overnight and after incubation selected colonies were randomly picked and subcultured into Nutrient Agar plates. Such sub culturing was done for 3 times to get pure colonies. Finally pure colonies were maintained in slants for further use.

Pathogens

Bacterial pathogens like *Aeromonas hydrophila* (Ah1v, Ahv, Ahr), *Bacillus subtilis*, *Escherichia coli*, *Staphylococcus aureus* and *Edwardsiella tarda* species maintained at Dept. Biotechnology, North Orissa University, Takatpur, Baripada were used in the study.

Antibacterial Activity of Isolated Bacteria

The antibacterial activities of the bacteria isolated from different sources were carried out by both well diffusion and disc diffusion method against five different pathogens. First the isolated bacteria were grown in the nutrient broth for 12 h at 37 °C and wells were made by the help of sterile puncture. Then nutrient agar plates were separately swabbed with pathogenic bacteria. Finally the wells were filled with overnight grown bacterial isolates and the plates were then incubated at 37°C for overnight. After incubation the pates were observed for zone of inhibition around the well. Bacteria exhibiting inhibitory zone size were selected for further study.

Like that in disc diffusion method sterilized discs (Wattman No.1) were soaked with 20ml of overnight grown bacterial culture. Then discs were firmly placed over the plates pre-swabbed with bacterial isolates. The plates were then incubated at 37 °C for overnight to observe inhibitory zone.

Different *In vitro* Screening Parameters

The isolated bacteria with antimicrobial activity against pathogens were further processed to *in vitro* growth characteristics.

Haemolytic Activity Study

The haemolytic nature of bacteria with antibacterial activity against different pathogens were tested in blood agar plates containing erythrocytes suspension of different species like human, fish (*L. rohita*), goat and chicken. The blood agar plates were spot inoculated with overnight grown bacterial isolates and then incubated overnight at 37 °C. The zone of clearance around the colony was considered haemolytic or positive while no change in colony around the inoculated region indicated negative reaction or non-haemolytic nature of isolates.

Antibiotic Resistance Study

The antibiotic resistance study of isolated bacteria exhibiting antibacterial activity against different antibiotics was performed as per standard microbiological method (Baurer *et al.*, 1966). Briefly, the overnight grown individual bacterial isolates were separately lawn cultured on nutrient agar plates and then antibiotic discs were firmly placed into the seeded nutrient agar plates. Finally plates were incubated at 37 °C for 24 h and after incubation, the zones of inhibition were measured and interpreted as sensitive, resistant and inter mediate sensitive as per the chart provided by Hi-media (India).

Proteolytic Activity Study

Bacterial isolates with antibacterial activity were further processed to study their proteolytic activity. The proteolytic activity study was carried out by spot inoculation of overnight grown bacterial isolates into 1 per cent skim milk agar plates and then incubated at 37°C for 24 hrs. After incubation the zone of clearance was observed around the colony indicated negative reaction/clear zone and positive reaction/non proteolytic nature.

Detection of Plasmid DNA

The presences of plasmid in bacterial isolates were studied as per the following procedure. Briefly 5 ml Luria Bertani (LB media) containing antibiotics were inoculated with individual bacterial isolates and then tubes were incubated overnight at 37 °C for overnight. After incubation cultures were centrifuged at 3500 × rpm for 20 min at 4 °C. Then the supernatant was discarded. Bacterial pellet were resuspended in 150 µl ice-cold lysis buffer I (50 mM glucose, 10 mM EDTA, 25 mM Tris, pH 8.0) and stored for 5 minutes at room temperature.

Then 300µl freshly prepared lysis buffer II (0.2N NaOH, 1 per cent SDS) was mixed by inversion for 5 minutes. Then 225µl ice-cold 3M KAC (pH 6.0) were added and mixed gently. Tubes were stored on ice for 5 minutes to precipitate the chromosomal bacterial DNA. The mixture was centrifuged at 13000 × rpm for 15 min at 4 °C and supernatant was transferred to a fresh tube. The protein contamination was removed by vortexing with 400µl phenol adding with 400µl chloroform followed by centrifugation for 2 minutes at 13000 × rpm. Then aqueous layer was gently removed and mixed with 600µl isopropanol at room temperature and then mixture was vortexed mixed. Then the tubes were allowed to stand for 2-3 minutes. Tubes were then centrifuged for 10 minutes at 13000 × rpm and 4°C. Supernatant was discarded and the eppendrop was allowed to dry in an inverted position. Then purified by adding 1ml of 70 per cent ethanol to the pellet and recovered the DNA by centrifugation. Finally the pellet was kept at room temperature and was added 10µl of TE buffer (pH 8.0). 5µl of RNAase A 20 µg/ml was added to avoid RNA contamination and to get plasmid DNA in the solution. Agarose gel electrophoresis was run to confirm the presence of plasmid DNA.

Growth Range Study

The bacteria with antibacterial activity against different pathogens were grown at different salt concentration of (0.5gm, 1 gm and 5gm), temperature (22 °C, 30 °C 37 °C.) and pH (5.5, 7.0 and 8.5) and to find out their growth range along with optimum growth requirements.

In vivo Characteristics Study

L. rohita fingerlings of average body weight 50-60gm were taken for *in vivo* characteristics study. The fingerlings were fed with commercial diet @ 3 per cent body weight and acclimatized 15 days before start of the experiment. Four groups in duplicate containing 5 fishes per group were made. Three groups were intraperitoneally injected with 0.5ml of bacterial suspension ranging from 10^8-10^{10} CFU/ml. Another group injected with 0.5ml of sterile PBS (pH 7.2) as control. The mortality in each group was recorded up to 15 days post injection.

Results

During the present study, 96 bacteria were isolated from different sources like fish, soil and water. Out of which, 32 were isolated from gastro-intestinal tract of fishes, 49 from soil and 15 from different water samples (Table 31.1). The bacterial isolates were further processed for the study of antimicrobial activity against various Pathogens such as different strains of *A. hydrophila* (Ah1v, Ahv, Ahr), *B. subtilis, E. coli, S. aureus* and *E. tarda*. Table 31.2 shows the antimicrobial activity of bacterial isolates against these pathogens. The bacterial isolates were further processed for various parameters studies such as haemolytic activity against erythrocytes of different species, antibiotics resistant, proteolytic activity and presence of plasmid DNA. Table 31.3 shows the haemolytic activity of bacterial isolates against erythrocytes of different species such as human, fish, goat, chicken. Table 31.4 shows the antibiotics resistant capacity, proteolytic activity and presence of plasmid DNA of bacterial isolates.

Table 31.1: The Number of Bacterial Isolates along with their Codes Isolated from different Sources

Sl.No.	*Sources*	*Type/Area*	*Total Isolates*	*Isolates with Codes*
1	Fish	*L.rohita*	9	R1-R9
		C.mrigala	17	M1-M17
		C.catla	3	C1-C3
		A.testudinis	3	K1-K3
2	Soil	Rice field, Baripada	19	B1-B19
		Rice field, Cuttack	13	D1-D13
		Groundnut field, Balasore	9	E1-E9
		Pastures/Local field, Baripada	8	SB1-SB8
3	Water	Pond, Baripada	15	W1-W15

Antimicrobial Activity of Bacterial Isolates

Table 31.2 shows the antibacterial activity of bacterial isolates against 3 strains of *A. hydrophila* (Ah1v, Ahv, and Ahr), *E. coli, S. aureus, B. subtilis* and *E. tarda*. Out of 96 bacterial isolates, 1 isolates from fish gut namely C3 exhibited antibacterial activity against 2 strains of *A. hydrophila* (Ah1v and Ahv), *S. aureus* and B. *subtilis*. Like that 11 bacterial isolates from soil namely B3, B6, B13, B16, B17, D1, D2, D3, E1, E3 and SB1 exhibited antibacterial activity against 3strains of *A. hydrophila* (Ah1v, Ahv, and Ahr), *E. coli, B. subtilis* and *E. tarda*.

Table 31.2: Antimicrobial Activity along with Inhibitory Zone (in cm) of Bacterial Isolates from different Sources against Various Pathogens

Sl.No.	*Bacteria*	*Pathogens*						
		A. hydrophila			*E. coli*	*S. aureous*	*B.subtilis*	*E.tarda*
		Ah1v	*Ahv*	*Ahr*				
1	C3	+(1.2cm)	+(1.1cm)	-	-	-	-	-
2	B3	+(1.0cm)	+(1.5cm)	-	ND	ND	ND	+(0.6cm)
3	B6	-	+(0.4cm)	-	ND	ND	ND	-
4	B13	+(0.8cm)	+(1.6cm)	-	ND	ND	ND	+(1.2cm)
5	B16	+(0.6cm)	+(0.8cm)	-	ND	ND	ND	+(0.8cm)
6	B17	+(1.0cm)	-	-	ND	ND	ND	+(0.8cm)
7	D1	-	+(1.6cm)	-	ND	ND	ND	+(1.2cm)
8	D2	-	-	-	ND	ND	ND	+(0.6cm)
9	D3	+(1.6cm)	-	-	ND	ND	ND	+(1.0cm)
10	E1	+(1.6cm)	+(1.2cm)	+ (0.8cm)	ND	ND	ND	+(1.0cm)
11	E3	+(1.2cm)	+(1.0cm)	-	ND	ND	ND	+(1.0cm)
12	SB1	+(1cm)	+(1.2cm)	+(0.9cm)	+(3.5cm)	-	+(2.8cm)	-

'+':- Zone of inhibition; '-':- No zone, ND- Not Done

Haemolytic Activity Study of Isolates

Haemolytic activity of bacterial isolates against RBC of different species is shown in Table 31.3. Among bacterial isolates like B3, B6, B13, D1, E1 and E3 were found to be non-haemolytic. While bacterial isolates C3, B16 and B17 were found to be haemolytic to human RBC. D2 was haemolytic to fish and chicken RBCs and SB1 was haemolytic to human, fish, goat and chicken RBCs.

Table 31.3: Bacterial Isolates with Antibacterial Activity Showing Haemolytic Activity against Erythrocytes of different Species

Sl.No.	*Bacteria*	*RBCS*			
		Human	*Fish*	*Goat*	*Chicken*
1	C3	+	-	-	-
2	B3	-	-	-	-
3	B6	-	-	-	-
4	B13	-	-	-	-
5	B16	+	-	-	-
6	B17	+	-	-	-
7	D1	-	-	-	-
8	D2	-	+	-	+
9	D3	-	+	-	-
10	E1	-	-	-	-
11	E3	-	-	-	-
12	SB1	+	+	+	+

+ve: Haemolytic, -ve: Non-haemolytic.

Antibiotics Resistant, Proteolytic Activity and Detection of Plasmid DNA Study

Antibiotics resistant, proteolytic activity and presence of plasmid DNA were shown in Table 31.4. Bacterial isolates D1, E1 and E3 Found to be resistant to most of the antibiotics. Whereas other isolates were found to be resistant to lesser antibiotics. Like that in proteolytic activity study, bacterial isolates B3,B17,D1,D2,D3,E1 and E3 found to be non-proteolytic in nature. In plasmid DNA detection study, similarly, all the bacterial isolates except E1 and E3 bear were found to possess plasmid DNA.

Table 31.4: Bacterial Isolates Showing Resistant to Antibiotics, Proteolytic Activity and Presence of Plasmid DNA

Sl.No.	*Bacteria*	*Resistant to Antibiotics*	*Proteolytic Activity*	*Presence of Plasmid DNA*
1	C3	Pc, A, Cn, Ca	+ve	+ve
2	B3	Cs, Mr, Cpm	-ve	+ve
3	B6	Cs, Mr, Cpm, Ca, Cep	+ve	+ve
4	B13	Pc, A, Cn, Ca	+ve	+ve

Sl.No.	Bacteria	Resistant to Antibiotics	Proteolytic Activity	Presence of Plasmid DNA
5	B16	Cs, Mr, Cpm, Ca	+ve	+ve
6	B17	C, Naf, T, S, Pc, C	-ve	+ve
7	D1	Cs, Mr, Cpm, Ca, Cep, Pc, C, Ci, Cn, A	-ve	+ve
8	D2	Ca, Cep, Pc, C	-ve	+ve
9	D3	Cep, Pc, C, Ci, Cn, A	-ve	+ve
10	E1	Cs, Mr, Cpm, Ca, Cep, Pc, C, Ci, Cn, A	-ve	-ve
11	E3	Cs, Mr, Cpm, Ca, Cep, Pc, Ci, Cn, A	-ve	-ve
12	SB1	C, Naf, T, S, Sf, Pc, C	+ve	+ve

+ve: Proteolytic activity/Plasmid present: Non-proteolytic activity/plasmid Absence; A - Ampicillin; C - Chloramphenicol; CI - Ceftizoxime; CN - Cephalexin; CS - Colistin; CA - Ceftriaxone; CEP - Cefpodoxime; Naf - Nafcillin; CPM - Cefepime; G - Gentamicin; MR - Meropenem; PG-Penicillin-G.

Effect of different Parameters on the Growth of Bacterial Isolates

Bacterial isolates showing antibacterial activity were found to grow at different salt concentration, pH and temperature (Table 31.5). The optimum growth among different isolates B13, D1 and E3 were found at in 0.5gm salt concentration. Other bacterial isolates B6 and E1 grew optimum at 1gm salt concentration. In 5gm salt concentrations no optimum growth of bacteria found. The optimum growth of bacterial isolates B6, B13, E1 and E3, was recorded at pH 8.5.The highest OD value of 0.943, 1.221 and 1.281 was recorded for B6, E1 and E2 respectively at temperature 37°C. Similarly the optimum growth for bacterial isolates D1 was 1.353 at pH 8.5 was recorded.

Table 31.5: Bacterial Growth at different Salt Concentration

Sl.No.	Bacteria No.	Salt Concentrations		
		0.5 gm	1 gm	5 gm
1	B06	0.597	0.722	0.619
2	B13	0.690	0.598	0.514
3	D01	1.065	0.924	0.463
4	E01	1.188	1.146	0.509
5	E03	0.899	0.861	0.553

In vivo Characterisation of Isolated Bacteria showing Antibacterial Activity

Pathogenicity Study

The pathogenicity study indicated that none of the bacterial isolates *i.e.*, E1 and E3 were virulent to *L. rohita*. No mortality could be recorded after 15 days. Post

Table 31.6: Bacterial Growth at different pH and Temperature

Sl.No.	*Bacteria No.*	*Temperature 22°C*			*Temperature 30°C*			*Temperature 37°C*		
		pH 5.5	*pH 7*	*pH 8.5*	*pH 5.5*	*pH 7*	*pH 8.5*	*pH 5.5*	*pH 7*	*pH 8.5*
1	B6	0.489	0.863	0.763	0.753	0.652	0.699	0.830	0.798	0.943
2	B13	0.181	0.325	0.515	0.103	0.352	0.420	0.075	0.194	0.332
3	D1	0.955	1.353	1.001	1.093	0.959	1.128	1.129	1.268	1.174
4	E1	1.057	1.174	1.100	1.211	1.174	1.168	1.085	1.072	1.221
5	E3	1.196	1.085	1.071	1.085	0.997	0.939	1.186	1.180	1.281

challenge in any of the intraperitonealy injected bacterial isolates up to 10^{10} CFU/ml. No mortality could also be recorded in PBS (pH 7.2) injected group as well.

Discussion

The bacteria were isolated from gastrointestinal tract of different fishes, soil and water. Fishes were collected from local market, soil and water samples were collected from different areas of Odisha. 32 bacteria were isolated from gastrointestinal tract of different fishes, 49 bacteria were isolated from soil and 15 bacteria from water samples. A total of 96 isolated bacteria were screening for antibacterial activity via well and/or disc diffusion method. Isolated bacteria showed antibacterial activity were processed for identification based on morphological characters and biochemical tests, *In vitro* characterization (haemolytic test, proteolytic test, antibiotic sensitivity test, growth parameters), plasmid isolation as well as *in vivo* characterization to get a good probiotic bacteria.

Probiotics help in generating specific microbiota like Bifidobacter, Lactobacillus species in the host, (Houdijk *et al.*, 1998; Torrecillas *et al.*, 2007; Costalos *et al.*, 2008). Selection of probiotic is very critical because failure can lead to the selection of in appropriate microorganisms, (Gomez-Gil, 1998). In aquaculture the current trend of disease management is to replace the antibiotics with eco-friendly probiotics (Gatesoupe, 2005). However, commercially available probiotics are often relatively ineffective because many are from non-fish sources, which are believed to be unable to survive and/or remain viable at optimum concentrations in the fish intestine (Moriarty, 1996; Ghosh *et al.*, 2007).

In most cases, much emphasis is given to the production of antibacterial compounds and invitro inhibitory activities (Verschuere *et al.*, 2000). The use of Bacillus species in aquaculture practices has been reported previously (Rengpipat *et al.*, 2000; Ghosh *et al.*, 2007). Moreover the advantages of *B. subtilis* over other potential probiotic bacteria are considerable. This bacterium is non-haemolytic, non-antibiotic resistant, non-pathogenic and has broad spectrum activity. This species is also known to form spores, which allows for easy preparation and storage as well as incorporation into artificial diets (Gatesoupe,1993; Casula and Cutting, 2002; Moriarty, 2003; Hong *et al.*, 2005). Here in this study, we have succeeded in isolating a suitable probiotic species from the natural resources, which can be useful for aquaculture practices.

Conclusion

Bacteria isolated from different sources were screened for antibacterial activity. Out of 96 isolates 12 bacterial isolates showed antibacterial activity against various pathogens. The bacteria showing antibacterial activity were processed for further *in vitro* and *in vivo* characterization studies. From this study bacterial isolates SB1 showed antibacterial activity against 5 pathogens, but found to be haemolytic in nature. Bacterial isolates E1 showed antibacterial activity against 4 various pathogens and found non-haemolytic, non proteolytic, resistant to antibiotics and grew best at salt concentration 0.5gm. Bacterial isolate E3 showed antibacterial activity against three pathogens, non-haemolytic, non proteolytic, resistant to antibiotics and was not found to be plasmid bearing. Both E1 and E3 were not pathogenic to rohu (*L. rohita*) fingerlings. Hence bacterial isolates E1 and E3 possessed all the desired characters of good probiotic bacteria for fish.

References

Abraham, T.J., Babu, C.H.S., Mondal S. and Banerjee, T. (2007). Effects of dietary supplementation of commercial human probiotic and antibiotic on the growth rate and content of intestinal microflora in ornamental fishes. Bangladesh Journal of Fisheries Research, 11: 57-63.

Balcazar, J.L., de Blas, I., Ruiz-Zarzuela, I., Cunningham, D., Vendrell, D. and Muzquiz, J.L. (2006). The role of probiotics in aquaculture. Veterinary Microbiology, 114: 173–186.

Gatesoupe, F.J. (1999). The use of probiotics in aquaculture. Aquaculture, 180: 147-165.

Ghosh, S., Sinha, A. and Sahu, C. (2007). Isolation of putative probionts from the intestines of Indian major carps. The Israeli Journal of Aquaculture -Bamidgeh 59: 127-132.

Gildberg, A. and Mikkelsen, H. (1998). Effect of supplementing the diet to Atlantic cod (*Gadus morhua*) fry with lactic acid bacteria and immuno-stimulating peptides during a challenge trial with *Vibrio anguillarum*. Aquacultur, 167: 103-113.

Hai, N.V., Fotedar, R. and Buller, N. (2007). Selection of probiotics by various inhibition test methods for use in the culture of western king prawns, *Penaeus latisulcatus* (Kishinouye). Aquaculture, 272: 231-239.

Irianto, A. and Austin, B. (2002). Probiotics in aquaculture. Journal of Fish Diseases, 25: 633-642.

Li, J., Tan, B., Mai, K., Ai, Q., Zhang, W., Xu, W., Liufu, Z. and Ma, H. (2006). Comparative study between probiotic bacterium Arthrobacter XE-7 and chloramphenicol on protection of *Penaeus chinensis* post-larvae from pathogenic Vibrios. Aquaculture, 253: 140-147.

Lowther, A. (2005). Highlights from the FAO database on aquaculture production statistics–fishery information. FAO Aquaculture Newsletter, 33: 22–24.

Merrifield, D.L., Bradley, G., Baker, R.T.M. and Davies, S.J. (2010b). Probiotic applications for rainbow trout (*Oncorhynchus mykiss* Walbaum) II. Effects on growth performance, feed utilization, intestinal microbiota and related health criteria post antibiotic treatment. Aquaculture Nutrition, 16: 496–503.

Mukherjee, S.C. and Nayak, S.K. (2004). Probiotics: an overview. Science and Culture, 1-2: 46-48.

Murthy, H.S. and Naik, A.T.R. (2002). Growth performance of *Cirrhinus mrigala* in relation to a natural probiotic feed supplement, Biovet-YC. Fishing Chimes, 22: 54-56.

Nayak, S. and Mukherjee, S.C. (2011). Screening of gastrointestinal bacteria of Indian major carps for a candidate probiotic species for aquaculture practices. Aquaculture Research, 42: 1034-1041.

Panigrahi, A. and Azad, I.S. (2007). Microbial intervention for better fish health in aquaculture: the Indian scenario. Fish Physiology and Biochemistry, 33: 429-440.

Shelby, R.A., Lim, C.E., Aksoy M. and Klesius, P.H. (2007). Effects of probiotic bacteria as dietary supplements on growth and disease resistance in young channel catfish, *Ictalurus punctatus* (Rafinesque). Journal of Applied Aquaculture, 19: 81-91.

Vaseeharan B. and Ramasamy P. (2003). Control of pathogenic Vibrio spp. by Bacillus subtilis BT23, a possible probiotic treatment for black tiger shrimp *Penaeus monodon.* Letters in Applied Microbiology, 36: 83-87.

Verschuere, L., Rombaut, G., Sorgeloos, P. and Verstraete, W. (2000). Probiotic bacteria as biological control agents in aquaculture. Microbiology and Molecular Biological Reviews, 64: 655-671.

Vine, N.G., Leukes, W.D. and Kaiser, H. (2004). *In vitro* growth characteristics of five candidate aquaculture probiotics and two fish pathogens grown in fish intestinal mucus. FEMS Microbiological Letters, 231: 145-152.

Vaseeharan, B. and Ramasamy, P. (2003). Control of pathogenic Vibrio spp. by *Bacillus subtilis* BT23, a possible probiotic treatment for black tiger shrimp *Penaeus monodon*. Letters in Applied Microbiology, 36: 83-87.

32

Bioremediaton of Selenite using Salt Tolerant Bacteria from Mangrove Soils of Bhitarkanika, Odisha

R.R. Mishra[1]* and H.N. Thatoi[2]

[1]Department of Biotechnology, MITS School of Biotechnology, Bhubaneswar – 751 024, Odisha
[2]Department of Biotechnology, North Orissa University, Baripada – 757 003, Odisha
****E-mail: rashmiranjan93@gmail.com***

ABSTRACT

Mangrove soil of Bhitarkanika harbours a large number of halotolerant and halophilic bacteria. In the present study, 20 halotolerant bacterial strains (BSB1-BSB20) were isolated from soil samples collected from mangroves of Bhitarkanika, Odisha. All the isolates were screened for their salt tolerance in an increasing concentration of NaCl. In this study, two bacterial strains (BSB6 and BSB12) were found to tolerate 12 per cent NaCl (w/v) while all other bacterial strains could able to tolerate 10 per cent NaCl (w/v). With a view to identify these two highly salt tolerant bacterial strains both phenotypic and molecular characterization were carried out. Both the strains (BSB6 and BSB12) were Gram positive, endospore forming motile rods. Based on biochemical data, both the strains were found to the genus Bacillus. On 16S rRNA gene sequencing, both the strains exhibited more than 95 per cent similarity with Bacillus megaterim. A qualitative screening towards Se(IV) tolerance and reduction was carried out by both the strains which exhibited greater selenite resistance. Both the strains effectively reduced Se(IV) and complete reduction of selenite (up to 0.25 mM) was achieved within 40 hrs.

***Keywords**: Bacillus megaterium, Halotolerant, Mangrove soil, Selenium reduction.*

Introduction

Selenium is of considerable environmental importance as it is essential at low concentrations but toxic at high concentrations for animals and humans. Selenium occurs in different oxidation states as reduced form (selenide, Se^{2-}), least mobile elemental selenium (Se^{0}) and water soluble selenite (SeO_3^{2-})/selenate (SeO_4^{2-}) oxyanions. Despite its essentiality, selenium is a potential toxic element to natural ecosystems due to its bioaccumulation potential (Tan *et al.*, 2016). Bioaccumulation of selenium as an agricultural drainage and effluents is the main cause of selenium contamination (Yanai *et al.*, 2015). Moreover, waste discharge from industries like tanneries, glass production industry, plastic industry, paint and pigment industry, oil refineries, and power stations are sources of water-soluble selenium contamination to our aquatic life form (Nissim *et al.*, 2015). Over the last few decades, various technologies have been developed for the treatment of selenium-containing wastewaters. However, biological selenium reduction has emerged as the leading technology for removing selenium from wastewaters. Studies have shown that Se undergoes oxidation and reduction reactions mediated by microorganisms. In recent years, several reports have described the bacterial reduction of selenate and selenite to the less soluble and less toxic form, elemental selenium (Losi and Frankenberger, 1997) suggesting that this capability be used as a remediation strategy for the treatment of selenium-contaminated waste water. Among the different microorganisms, bacteria isolated from marine/saline environment are considered to be better candidate for bioremediation of toxic metals including selenium. Halophilic and halotolerant microorganisms are suitable candidates for bioremediation as they have exceptional properties, that is high concentrations of anions and cations are necessary for the growth of these microorganisms (Ventosa *et al.*, 1998), hence they are not only naturally tolerant to some elements that are toxic to other microorganisms, but also they have a requirement for these elements. In this context mangroves being saline environments occurring between land and sea harbour both halophilic and halotolerant bacteria which can be isolated and employed for efficient bioremediation of toxic metals such as selenite.

Bhitarkanika, in the state of Odisha is one of richest mangrove forests of India. This mangrove forest is known for plant and animal diversity. This mangrove forest is suitable habitat for saline microorganisms which might show immense biotechnological potentials. However, only a few studies have been conducted for assessment of microbial diversity so far from this part of the forest (Mishra *et al.*, 2009). Therefore, in the present investigation, attempt has been made to isolate and characterize the salt tolerant microorganisms from Bhitarkanika with an aim to explore their potential for bioremediation of selenium, a toxic heavy metal commonly found as contaminant of many aquatic environment.

Materials and Methods

Study Site: Mangroves of Bhitarkanika

The Bhitarkanika mangrove forest lies between 20^{0} 30′ N to 20^{0} 50′N latitude and 86^{0} 30′ E to 87^{0} 6′ E longitude, located in the confluence of the rivers Brahmani and Baitarani in the Kendrapara district of Odisha is the study site in the present experiment. Mangroves of Bhitarkanika, between the Dhamara mouth to Barunei mouth on the east coast has been notified as Bhitarkanika sanctuary (672 sq km.) in 1975. It is the largest mangrove forest of Odisha. The core area of the sanctuary is notified as National park in Septemmber, 1998. Bhitarkanika sanctuary area has been designated as a Ramsar site during 2002.

Isolation of Salt Tolerant Bacteria

Soil samples were collected from the Bhitarkanika mangrove forest and brought to the laboratory in sealed sterile polythene packets. The pH and conductivity of the soil samples were measured and were found in the ranges 5.9–7.5 and 13.0–17.3 mS cm^{-1}, respectively. The bacteria were isolated using nutrient agar (NA) medium containing 4–12 per cent NaCl (w/v) at pH 7.0 following the dilution plate technique method. Morphologically distinct, bacterial colonies were picked and the pure cultures were maintained in NA slants and stored in refrigerator at 4 ^{0}C for further experiment.

Screening of Bacteria for Salt Tolerance

Twenty halotolerant bacterial isolates (BSB1-20) were screened in an increase concentration of NaCl and sea salt. NA plates with varied concentration of NaCl (5. 7, 12, 13) per cent were prepared. Then the bacterial strains were spot inoculated in the surface of the nutrient agar plates. After spot inoculation, the NA plates were incubated at 37 ^{0}C for overnight and growth of the bacterial colonies was recorded.

Screening of Selenium Tolerance

A qualitative screening of selenite tolerance was carried out to find out the efficient selenite tolerant bacterial isolates. For this purpose, NA plates supplemented with 5 mM of sodium selenite at pH 7.5 were spot inoculated with isolates and incubated for 48 hrs at 37 ^{0}C. The intensity of orange–red/red colour around the bacterial colonies was taken as a qualitative efficiency for the reduction of selenite to elemental selenium.

Phenotypic Characterization

The characters of the organisms were studied following standard microbiological methods. Morphology, vegetative cell and spore characters were observed under a phase contrast microscope (100 × objective) from 12 hrs old cultures grown on rotary shaker at 100 × rpm, 30 ± 1 °C. The physiological and biochemical characters *viz.* indole production oxidase, catalase, urease hydrolysis, acid from glucose, mannitol, arabinose, xylose, citrate, and propionate utilization and tyrosine hydrolysis were studied. Assay of casein, gelatin, starch hydrolysis *etc.* were also checked.

Molecular Characterization (16S rRNA gene sequencing)

Molecular identification of bacteria was done using 16S rRNA gene sequencing. Partial 16S rRNA gene sequencing was done by ABI 3130xl analyzer based on Sanger's dideoxy termination method at Institute of Microbial Technology, Chandigarh, India. The sequences obtained were analyzed using BLAST at ncbi.nlm.

nih.gov and the phylogenetic tree was constructed after multiple sequence alignment using cluster algorithm (Chumakov and Yushmanov, 1998). The phylogenetic tree is built on the matrix of pair distances between sequences. In the boot strap a multiple alignment is resampled 100 times.

Reduction of Selenite

The assay for selenium reduction was carried out in 100 mL NB in 250 mL Erlenmeyer flasks which were incubated at 37 ^{0}C in a BOD incubator. Bacteria were cultured in the NB medium (pH 7.5), at 37 ^{0}C for 8 hrs. 10 ml culture was taken and centrifuged at 8000 × g for 10 min at 4 ^{0}C and the pellet was resuspended in 2 mL of Tris–HCl buffer (pH, 8.0). The cell suspension (100 µL) was inoculated into the test medium (100 mL) containing 0.25mM of SeO_3^{2-}. Control experiments without selenite were performed simultaneously. At regular intervals, 2 mL culture was withdrawn from each flask and centrifuged at 10,000 × g for 10 min and the remaining selenite content in the supernatant was analyzed. Selenium contents both in supernatant and pellet fractions were determined. The concentration of SeO_3^{2-} in the supernatant fraction was estimated spectrophotometrically as selenite equivalent following Afsar *et al.* (1989) using 2-mercapto ethanol (0.05 M) as the reducing/complexing agent and/or Kessi *et al.* (1999). For the estimation of total selenium reduced, the pellets associated with selenium particles were centrifuged, washed with deionized water and subjected to an acid digestion for 5 min with 10 mL of HNO3 and 0.5 mL of H_2SO_4 followed by reduction of selenite with 6 M HCl at 100^{0}C for 30 min following Ike *et al.* (2000). The selenite content in the acid digested solution was estimated as described above.

SEM and EDX Analysis of the Reduced Product

The bacterial cells associated with selenium particles, after 48 hrs incubation in presence of selenite (0.25 mM) at optimum pH (7.5) and temperature (37^{0}C), were filtered through polycarbonate micropore filters (0.22 lm) and washed with Tris–HCl buffer (pH 8.0) three times and fixed with 3 per cent glutaraldehyde in 0.1 M phosphate buffer (pH 7.4) for 60 min. The suspension was centrifuged at 10,000 × g for 10 min at 4.0 ^{0}C and the pellet was washed with Tris–HCl buffer followed by deionized water three times. The samples were dehydrated with 70 per cent ethanol, mounted on an aluminium stub, coated with gold and examined under JEOL (840A, Japan) SEM at 200 kV. Analysis of elemental selenium was also carried out simultaneously by energy dispersive X-ray spectroscopy (EDX).

Results

Isolation and Study of Morphological Characteristics of Halotolerant Bacteria

In the present study, 20 bacteria were isolated from Bhitarkanika mangrove soil using NA supplemented with 7 per cent NaCl. The isolates were designated as Bhitarkanika soil bacteria (BSB1-BSB20). The morphological characteristics of these 20 halotolerant bacterial isolated are given in Table 1. Majority of the isolates had irregular colonies with undulate margins, but some isolates had filamentous (BSB9,

BSB10, BSB11, BSB 18) or entire (BSB1, BSB12, BSB20) margins. Colony of most of the isolates had flat, excluding BSB2 and BSB17 (raised colonies). Colony colours of the isolates were white (BSB6, 7, 8, 10, 13, 14, 15, 16), yellow (BSB1) and brown (BSB2, BSB3, BSB4, BSB9, BSB11, BSB12, BSB17, and BSB20). The bacterial isolates showed variation in size and they ranged from 0.1-4.0 mm. The colony characters of the isolates (BSB1-BSB7) were irregular, filamentous, punctiform or circular, yellowish-brown, brown, creamy white or pale brown. Range of colony size was 1.0-1.4 mm on nutrient agar, gummy and sticky consistency. Seven out of twenty isolates (BSB1-5, 6, 12) were Gram (+)ve and the rest thirteen isolates (BSB7-11, 13-20) were Gram (-)ve (Table 32.1).

Table 32.1: Morphological Characteristics of 20 Halotolerant Bacterial Strains Isolated from Mangroves of Bhitarkanika, Odisha

Bacterial Isolates No.	*Form*	*Colour*	*Elevation*	*Margin*	*Size (mm)*	*Consistency*	*Gram Stain*
BSB 1	Circular	Yellow	Flat	Entire	0.1-1.5	Sticky	+
BSB 2	Irregular	Brown	Raised	Undulate	0.1-1.8	Gummy	+
BSB 3	Irregular	Brown	Convex	Undulate	0.2-1.2	Sticky	+
BSB 4	Irregular	Brown	Convex	Undulate	0.3-1.2	Sticky	+
BSB 5	Circular	White	Convex	Undulate	0.2-2.0	Gummy	+
BSB 6	Irregular	White	Flat	Undulate	0.1-1.5	Gummy	+
BSB 7	Punctiform	White	Convex	Undulate	0.1-1.5	Gummy	-
BSB 8	Irregular	White	Flat	Undulate	0.4-1.5	Sticky	-
BSB 9	Irregular	Brown	Flat	Filamentous	0.4-1.5	Sticky	-
BSB 10	Irregular	White	Convex	Filamentous	0.1-1.3	Gummy	-
BSB 11	Irregular	Brown	Flat	Filamentous	0.1-1.5	Gummy	-
BSB 12	Circular	Brown	Convex	Entire	0.2-1.8	Sticky	+
BSB 13	Circular	White	Flat	Undulate	0.3-1.3	Gummy	-
BSB 14	Circular	White	Flat	Undulate	0.2-1.2	Sticky	-
BSB 15	Punctiform	White	Convex	Lobate	0.3-1.2	Sticky	-
BSB 16	Irregular	White	Flat	Undulate	0.4-2.0	Sticky	-
BSB 17	Irregular	Brown	Raised	Undulate	1.4-2.3	Sticky	-
BSB 18	Filamentous	White	Flat	Filamentous	1.5-2.5	Gummy	-
BSB 19	Irregular	White	Flat	Undulate	2.5-4.0	Gummy	-
BSB 20	Punctiform	Brown	Convex	Entire	1.0-2.0	Gummy	-

Screening for Salt Tolerance

Seven Gram positive bacterial strains (BSB1-BSB5, BSB6 and BSB12) were subjected to screening test for salt tolerance in nutrient media supplemented with NaCl and sea salt and growth under different pH and temperature as given in Table 32.2. All isolates could grow in medium containing 10 per cent NaCl (w/v) and 10 per cent sea salt (w/v). Two isolates (BSB6, BSB12) could able to tolerate up to 12 per cent NaCl on solid medium but tolerating only up to 10 per cent sea salt (Table

32.2). Generally growth was better in the medium supplemented with NaCl and the optimum concentration of NaCl for growth of the isolates varied. Optimum NaCl concentration for growth of majority of the strains was 7 per cent NaCl (Table 3202). Better growth of the isolates was observed at pH 7.0 and at temperature of 35⁰C (Table 32.2). As both the strains could able to tolerate 12 per cent NaCl, they were considered as moderately halotolerant bacteria.

Table 32.2: Screening of Salt Tolerance of Seven Gram Positive Bacterial Strains

Growth Medium	*BSB 1*	*BSB 2*	*BSB 3*	*BSB 4*	*BSB 5*	*BSB 6*	*BSB 12*
NA	+	+	+	+	+	+	+
NA+ NaCl (per cent)							
5	+	+	+	+	+	+	+
7	+	+	+	+	+	+	+
10	+	+	+	+	+	+	+
12	-	-	-	-	+	-	+
13	-	-	-	-	-	-	-
NA + Sea salt (per cent)							
10	+	+	+	+	+	+	+
15	-	-	-	-	-	-	-
NB + Sea salt (per cent)							
10	+	+	+	+	+	+	+
15	-	-	-	-	-	-	-
Anaerobic growth in NB	-	-	-	+	-	-	-
Growth at Temperature							
30 ⁰C	+	+	+	+	+	+	+
35 ⁰C	+	+	+	+	+	+	+
Growth at pH							
pH 7	+	+	+	+	+	+	+
pH 5	+	+	-	-	-	+	-

Screening for Selenite Tolerance

Two halotolerant bacterial strains (BSB6 and BSB12) isolated from mangrove soil of Bhitarkanika were also screened for their heavy metal tolerance potential against different concentration of selenite (400, 600, 800, and 1000) mM. Both the strains could able to tolerate (1000 mM) of selenite in nutrient broth. Due to high tolerance towards Na_2SeO_3, these two strains were further evaluated for their selenite reduction potential.

Biochemical Identification

Two bacterial strains, BSB6 and BSB12, showing resistance and potential for high Se(IV) were characterized by different biochemical tests. The strains were found to be Gram positive, endospore forming motile rods. The organisms did not hydrolize cholesterol, tween 80, chitin and pectin but tributyrin hydrolysis was positive. However, the fermentation of carbon components by the organisms was not identical and most of them were non-fermentative. The isolates were negative for urease and positive for starch and gelatin hydrolysis and they utilized D-glucose, L-arabinose, D-fructose and sucrose but not D-mannitol and D-xylose and citrate. They were positive for catalase, nitrate reduction and aerobic acid formation of glucose but negative for indole production. Based on morphological and biochemical data both the strains were identified as the members of genus *Bacillus*.

16S rRNA Gene Sequencing

Further species level identification of the two strains was done by 16S rDNA sequencing. The phylogenetic analyses, based on the maximum-parsimony algorithm, revealed that both the strains (BSB6 and BSB12) branched with *B. megaterium* with 100 per cent and 97 per cent identity, respectively. The 16S rDNA homology of the two isolates showed that these two organisms may be different strains of *B. megaterium* with some differences in their genome. The strains, BSB6 and BSB12, were deposited in the Microbial Type Culture Collection Centre: Chandigarh (India) with MTCC numbers MTCC-9204 and MTCC-9205 respectively. The 16S rDNA sequences of the strains were submitted to Gene Bank with accession numbers FJ853653 and FJ973576.

Reduction of Selenium

The representative time-courses of selenite reduction by two *B. megaterium* strains BSB6 and BSB12 strains at neutral pH are shown in Figure 32.1. There was practically no change of selenite concentration in the control during the same period. Under the static condition both the strains were found effective for reduction of Se^{4+} to Se^0 with a lag-time of approximately 10-12 hrs. The appearance of orange-red colour in the culture media was indicative of formation of amorphous allotropic form of elemental selenium. It was observed that 0.25 mM of selenite was almost completely reduced within 30-40 hrs. After complete reduction, the reduced product

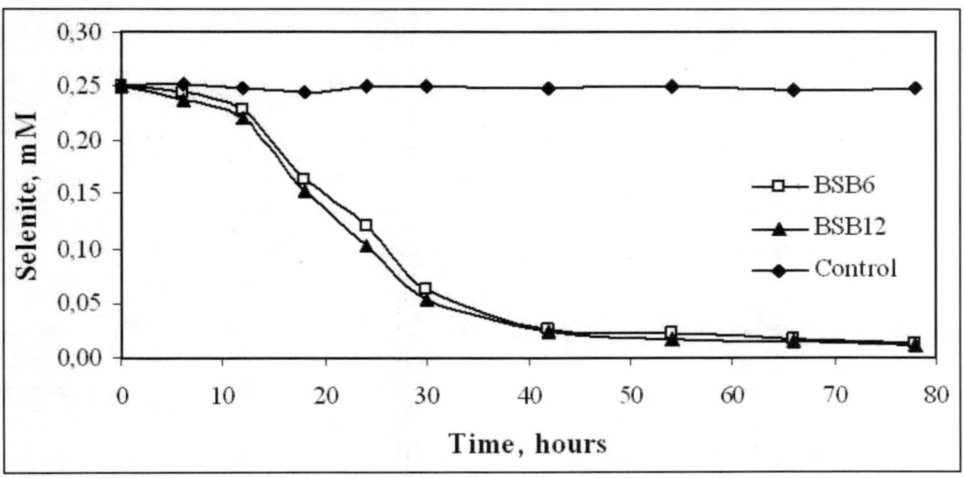

Figure 32.1: Selenite Reduction by *B. megaterim* at Neutral pH.

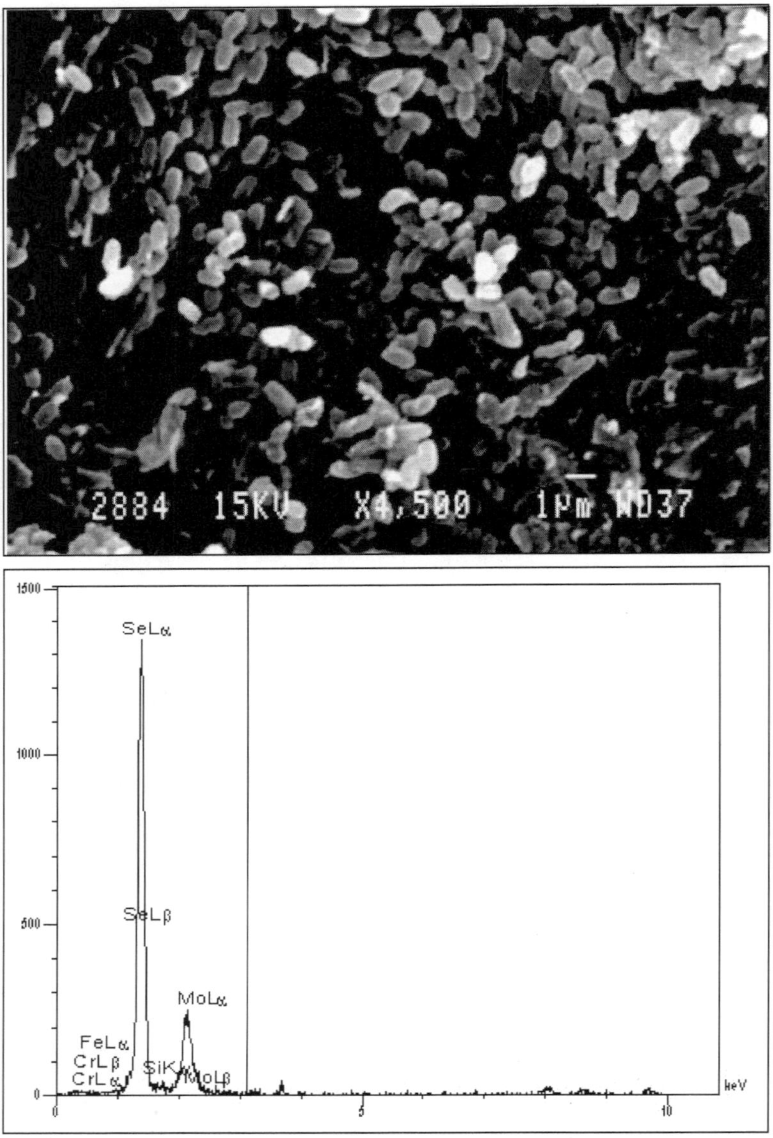

Figure 32.2: SEM and EDX Photograph of *B. megaterium* Treated with Selenite.

was characterized using SEM and EDX. SEM with energy dispersive X-ray revealed the formation of nano size spherical selenium particles in and around the bacterial cells. These bacterial strains with selenite reduction potential may be exploited further for bioremediation of Se (IV) at relatively high salt concentrations and green synthesis of selenium nanoparticles.

Discussion

Mangroves are typically tropical coastal ecosystems found in the inter-tidal zones of river deltas and back water areas. They represent highly dynamic and fragile ecosystems yet they are the most productive and biologically diversified habitats of various life forms including plants, animals and microorganisms. These ecosystems are characterized by periodic tidal flooding which makes the environmental factors such as salinity and nutrient availability highly variable resulting in unique and specific characteristics. Mangrove soils at the interface of sea and land are saline in nature. Hence bacteria occurring in mangrove soil are natural salt tolerant. The mangrove ecosystem is dominated by both halotolerant and halophilic bacteria (Zaharan, 1997). However, the salt tolerance of bacteria in mangrove soil varies from species to species. In the present study, 20 halotolerant bacteria were isolated. Seven out of 20 bacteria were Gram-positive. Among the different groups of bacteria, the Gram-positive bacteria are well represented in saline habitat and the members of the genus *Bacillus* and Micrococcus are found to be dominating among them (Zaharan *et al.,* 1992). The microorganisms those are surviving in high saline condition, possesses special adaptation strategies that make them interesting not only for fundamental research but also towards exploration of various biotechnological potentials (Margesin and Schinner, 2001). It is well-known that, microbes play important role in bioremediation of heavy metals such as Senenite which has attempted in the present study.

The Gram (+)ve halotolerant bacteria initially isolated from mangrove soils using 7 per cent NaCl but later it has been found that they could able to tolerate relatively high concentration of NaCl (10-12 per cent). The high salt tolerant character of the organisms might be due to the constantly growing in saline coastal soils so they might be adopted high salt tolerant character in course of time. Growth of the organisms without NaCl or salt also proved that they were halotolerant, and tolerance of 7-12 per cent NaCl, grouped them as moderately halotolerant but not halophilic (Vreeland, 1987: Vreeland, 1984). Based on screening test two strains (BSB6 and BSB12) exhibited tolerance up to 12 per cent NaCl and considered to be halotolerant. The phenotypic and molecular identification revel that the two strains were *Bacillus megaterium.*

Heavy metal tolerance of bacterial strains is useful trait which could be exploited biotechnologically for bioremediation heavy metal from environment. Keeping this in view, initial screening of both the well characterized moderately halotolerant strains of *B. megaterium* were made for selenium tolerance. From the study, it was found that the bacteria could tolerate a very high concentration of selenite (having MIC values of 1.0 M). The values are comparable with those reported for bacteria isolated from selenium contaminated soils (Ghosh *et al.,* 2008: Desouza *et al.,* 2001). Resistance of halotolerant strains towards selenite has also been reported elsewhere (Rathgeber *et al.,* 2002). Moreover, the members of the genus *Bacillus* are usually known for their resistance to various metals and metalloids (Stolz *et al.,* 2006). A preliminary study showed that both the bacterial strains were very much effective for reduction of Se^{4+} to Se^0 in the concentration less than 1 mM. The reduction of Se^{4+} to Se^0 at neutral to alkaline pH has been reported earlier (Lortie *et al.,* 1992) and also resulted in similar observation in the present investigation made by the authors. A recent study reported the resistance of *B. megaterium* TKW3, isolated from heavily metal polluted marine sediment, toward various oxyanions including selenite (Cheung and Gu, 2005). It may be noted that several phylogenetically bacteria like *Thauera, Sulfurospirillum, Bacillus, Ralstonia, Desulfotomaculum, Desulfovibrio,* and *Shewanella* can effectively reduce SeO_4^{2-} (Se^{6+}) or SeO_3^{2-} (Se^{4+}) to Se^0 (Klonowska *et al.,* 2005). Thus the present study indicated that the bacteria isolated from soils without selenium contamination can also tolerate salt concentration and have the ability to reduce a relatively high concentration of Se^{4+} to Se^0. Furthermore, the study also resulted that reduced selenite particles are deposited on the surface of the bacterial cells. The SEM micrograph with EDX analyses of precipitates along with the associated bacteria also confirmed the formation of Se^0 particles in the nanometric range. Since *B. megaterium* is capable of reducing selenite in presence of salt, it could be useful for bioremediation of industrial effluents containing high amount of salt.

Conclusion

From the above study, it has been concluded that, mangroves of Bhitarkanika harbours a diverse groups of halotolerant bacteria. Gram-positive halotolerant bacteria were suitable for bioremediation purpose since they exhibit tolerance towards heavy metals. In the present study two moderately salt tolerant *Bacillus megaterium* strains isolated from mangrove soil of Bhitarkanika, showed selenite tolerance. Bioremediation experiment with these two *B. megaterim* species revealed that both the strains could able to reduce toxic selenite to red elemental selenium with a relatively small period of time. Thus, these strains could be useful for bioremediation of selenite in various industrial effluents containing high salt.

References

Afsar, H., Apak, R. and Izzet, T. (1989). Reduction of selenium (IV) and complexation of Se(0) with mercaptoethanol. Analyst, 114: 1315–1318.

Cheung, K.H. and GU, J.D. (2003). Reduction of chromate (CrO_4^{2}) by an enrichment consortium and an isolate of marine sulfate-reducing bacteria. Chemosphere, 52: 1523–1529.

Chumakov, K.M. and Yushmanov, S.Y.U. (1998). Maximum topological similarity principle in molecular taxonomy. Molekularnaya Genetika, Mikrobiologijai Virusologija 3, 3-9.

Desouza, M.P., Amini, A., Dojka, M.A., Pickering, I.J., Dawson, S.C., Pace, N.R. and Terry, N. (2001). Identification and characterization of bacteria in a selenium-contaminated hypersaline evaporation pond. Applied Environment Microbiology, 53: 185-188.

Ghosh, A., Mohod, K.M., Paknikar, M. and Jain, R.K. (2008). Isolation and characterization of selenite- and selenate-tolerant microorganisms from selenium contaminated sites. World Journal of Microbiology Biotechnology 24: 1607–1611.

Imhoff, J.F. (1986). Osmoregulation and compatible solutes in eubacteria. FEMS Microbiology Review, 39: 57-66.

Ike, M., Takahashi, K., Fujita, T., Kasiwa, M. and Fujita, M. (2000). Selenate reduction by bacteria isolated from aquatic environment free from selenium contamination. Water Research, 34: 3019–3025.

Kessi, J., Ramuz, M., Wehrli, E., Spycher, M. and Bachofen, R. (1999). Reduction of selenite and detoxification of elemental selenium by the phototrophic bacterium *Rhodospirillum rubrum*. Applied Environment Microbiology, 65: 4734–4740.

Klonowska, A., Heulin, T. and Vermeglio, A. (2005). Selenite and tellurite reduction by *Shewanella oneidensis*. Applied Environment Microbiology, 71: 5607-5609.

Lortie, L., Gould, W.D., Rajan, S., McCready, R.G.L. and Cheng K.J. (1992). Reduction of selenate and selenite to elemental selenium by a *Pseudomonas stutzeri* isolate. Applied Environment Microbiology, 58: 4042-4044.

Losi, M.E. and Frankenberger, W.T. (1997). Reduction of selenium oxyanions by *Enterobacter cloacae* SLD1a-1: isolation and growth of the bacterium and its expulsion of selenium particles. Applied Environment Microbiology, 63: 3079–3084.

Margesin, R. and Schinner, F. (2001). Potential of halotolerant and halophilic microorganisms for biotechnology. Extremophiles, 5: 73-83.

Mishra, R.R., Dangar, T.K., Rath, B. and Thatoi, H.N. (2009). Characterization and evaluation of stress and heavy metal tolerance of some predominant Gram negative halotolerant bacteria from mangrove soils of Bhitarkanika, Odisha. African Journal of Biotechnology, 8(10): 2224-2231.

Nissim, W.G., Hasbroucq, S., Kadri, H., *et al.* (2015). Potential of selected Canadian plant species for phytoextraction of trace elements from selenium-rich soil contaminated by industrial activity. International Journal of Phytoremediation, 17: 745–752.

Rathgeber, C., Yurkova, N., Stackebrandt, E., Beatty, J.T and Yurkov, V. (2002). Isolation of tellurite and selenite resistant bacteria from hydrothermal vents of the Juan de Fuca Ridge in the Pacific Ocean. Applied Environment Microbiology, 68: 4613-4622.

Stolz, J.f., Basu, P., Santini, J.M and Oremland, R.S. (2006). Arsenic and selenium in microbial metabolism. Annual Review of Microbiology, 60: 107-130.

Tan, L.C., Nanchavaian, Y.V., Van Hullebusch, E.D. and Lens, P.N.L. (2016). Selenium: Environmental significance, pollution and biological treatment technologies. Biotechnology Advances, 34(5): 886-907.

Ventosa, A., Nieto, J.J. and Oren, A. (1998) Biology of moderately halophilic aerobic bacteria. Microbiology Molecular Biology Review, 62: 504-544.

Vreeland, R.H. (1984). *Halomonas* in Bergey's Manual of Systematic Bacteriology: Krieg, NR and Holt J.G. Eds. 1: 343.

Yanai, J., Mizuhara, S., Yamada, H. (2015). Soluble selenium content of agricultural soils in Japan and its determining factors with reference to soil type, land use and region. Soil Science and Plant Nutrition, 61: 312–318.

Zaharan, H.H., Moharram, A.M. and Moharammad, H.A. (1992). Some ecological and physiological studies on bacteria isolated from salt affected soils of Egypt. Journal of Basic Microbial Technology, 32: 405-413.

Zaharan, H.H. (1997). Diversity adaptation and activity of the bacterial flora in saline environments. Biology of Fertile Soils, 25: 211-223.

33

Enumeration of Soil Microbiota from Paddy Field and its Relation to Soil Microbial Biomass Carbon

M. Das[1], *P. Bhattacharyya*[2] *and S.K. Nayak*[1]*

[1]*Department of Biotechnology, North Orissa University, Baripada – 757 003, Odisha*

[2]*Crop Production Division, National Rice Research Institute, Cuttack – 753 006, Odisha*

**E-mail: sukantanayak@rediffmail.com*

ABSTRACT

An experiment was conducted to study the soil microbiota in paddy field with relation to their microbial biomass carbon during the rice cropping system. During the rice cropping system the microbial biomass carbon (MBC) was significantly high (280.04 µg/g of soil) in panicle initiation stage and the organic carbon (OC) (0.47 per cent) also in panicle initiation stage. Heterotrophic bacterial population was significantly high under panicle initiation stage of rice crop. The fungal population was not significant among the different stages of rice crop. The results suggested that microbial biomass carbon dynamics in rice soil are largely controlled by organic substances released from rice roots and its associates with microbiota of the soil. As microbiota increases in the soil the microbial biomass carbon also increases. Microbial biomass is a measure of the weight of microorganisms in soil, which mostly consists of bacteria, fungi and other microbes called archaea. Measures of microbial biomass usually measure either the weight of carbon or nitrogen in soil microorganisms. Microbial biomass is a useful indicator of soil quality. Soil microorganisms are involved in several processes that influence soil quality and microbial biomass changes rapidly in response to changes in soil properties.

***Keywords**: Cropping system, Heterotrophic population, Microbial biomass carbon, Organic carbon, Paddy field.*

Introduction

Soil is a complex ecosystem where living organisms play a key role in the maintenance of its properties. Soil biota comprises a huge diversity of organisms belonging to different taxonomic and physiologic groups, which interact at different levels within the community (Lopes *et al.*, 2010). Soil microorganisms constitute a source and sink for nutrients and are involved in numerous activities, such as transformation of C, N, P and S, degradation of xenobiotic organic compounds, formation of soil physical structure and enhancement of plants' nutrient uptake. Soil microorganisms, such as bacteria and fungi, actinobacteria help the soil decomposition and nutrient cycling and may serve as indicators of land-use change and ecosystem health. Soil microbial community structure and activity depend to a large extent on the status of their soil habitat. Within this habitat, soil organisms are eating, respiring, competing, cooperating, and responding to changes in their immediate environment. Indeed, the majority of the microbial community may be dormant at any given time in most soils, ready to respond as conditions for a particular group become favorable (Balser *et al.*, 2010).

Rice paddy soil is considered a unique agro-ecosystem as it is kept flooded during the rice growth and is drained during the off crop season. The rice paddy field comprises three compartments where the soil is subjected to different physical and chemical conditions: the oxic surface, the anoxic bulk and the rhizosphere plus rhizoplane (Liesack *et al.*, 2000). Rhizosphere bacteria may improve the uptake of nutrients to plants and/or produce plant growth promoting compounds. They also protect plant root surfaces from colonization by pathogenic microbes through direct competitive effects and production of antimicrobial agents. These bacteria can indirectly or directly affect plant growth (Mantelin and Touraine, 2004).

Soil microbial biomass defined holistically as the living component of soil organic matter (*i.e.* organisms with a volume less than 5,000 μm^3, *e.g.*, bacteria, fungi, protozoa) is actively involved in biogeochemical processes that occur in soil micro niches of paddy soils. These processes include organic matter decomposition, microbial oxido-reduction, and cycling of N, C, and plant nutrients. The nature and extent of these biogeochemical processes cannot be approximated without understanding the involvement of microorganisms in these processes (Inubushi and Acquaye, 2004). Keeping all this in mind the objective is formulated to study the relation between soil microbiota and soil microbial biomass carbon in rice cropping system.

Materials and Methods

Soil Sampling

Rhizospheric soil samples (0-15 cm depth) were collected by an auger from paddy field. The fresh soil samples were kept in refrigerator at 4 °C for enzymes analysis. The important growth stage was including active tillering (30 Day After Transplanting), maximum tillering (50 Day After Transplanting), panicle Initiation (80 Day After Transplanting) and harvesting (120 Day After Transplanting) under various treatments.

Soil Organic Carbon and Available Nitrogen

Soil organic carbon was estimated by wet digestion method by (Walkley and Black, 1934). Available Nitrogen was estimated by micro Kjeldahl methods using KL plus analyzer was given by Jackson (1973).

Enumeration of Microorganisms from Soil

Total number of population was estimated by using standard protocol. Standard procedure of serial dilution methods was using by Page *et al.* (1982). To find out the total number of aerobic heterotrophic population, anaerobic population, fungus and Actinobacterial population microbiological media are used like Nutrient hiveg agar, anaerobic hiveg agar, Rose bengal chloramphenicol agar and Actinomycetes isolation agar (Hi Media, Mumbai) was used for isolation of different microbes. After plating, the plates were incubated at 30±2 ^{0}C for heterptrophs and 28±2 ^{0}C for fungi and actinomycetes. The colony forming unit per gram of soil obtained from the valid plate count was calculated by using the following formula:

$$\text{CFU/g of Soil} = \frac{\text{No. of colonies (with replication)}}{\text{Quantity taken} \times \text{Dilution factor}}$$

Microbial Biomass Carbon (MBC)

MBC was estimated by chloroform-fumigation extraction methods. Soils are fumigated with chloroform vapour inducing the lysis of microbial cell membranes and subsequent release of microbial cell constituents which can be extracted and quantified. Soil MBC was determined by following the standard protocol of (Vance *et al.*, 1987). MBC was extracted from both fumigated and unfumigated samples with 0.5 M K_2SO_4 and digested with 0.325N potassium dichromate ($K_2Cr_2O_7$) and conc. H_2SO_4 and digestion for 30 mints over a hot plate. MBC was calculated by fumigated samples value minus from that measured unfumigated samples and dividing it by a Kc value of 0.25. The values of MBC were represented in µg g-1 dry soil. Organic carbon from dry soil was estimated by the digestion method by using 1N ($K_2Cr_2O_7$) and 20ml of conc. H_2SO_4 for 30min, after that titration against freshly prepared 0.5N ferrous ammoniun sulphate $(NH_4)_2SO_4$ using 1ml of diphenylamine indicator.

Results

During the rice cropping system, the MBC showed significantly higher in panicle Initiation stage of rice (P < 0.05) (Table 33.1). The MBC exhibited rapid response and high sensitivity to management practices and environmental changes (Emmerling *et al.*, 2002). The OC showed no significant among the crop growth stage of rice. The MBC/OC ratio reflects the mineralization potential of the soil organic matter (Garcia-Orenes *et al.*, 2010).

Soil Microbial Community Structure

Rhizospheric microbial population study was done by using standard plate count methods. The field soil contains different types of microorganisms like aerobic heterotrophs, anaerobes, fungi and actinobacterial population. Fungi and

Table 33.1: Microbial Biomass Carbon (µg/g of soil), Organic Carbon (per cent) and Available N (kg/h) under Various Crop Growth Stage

Crop Growth Stages	*MBC (µg/g of soil)*	*OC (per cent)*	*Available N (Kg/h)*
AT	110.08	0.32	118.30
MT	270.18	0.45*	132.08
PI	280.04*	0.47	98.75
H	116.82*	0.43	105.00

AT: active tillering; MT: maximum tillering; PI: panicle initiation; H: harvesting, * Significant level at $p < 0.05$.

actinobacteria are the important groups of decomposer. As microorganisms in soil regulate the dynamics of organic matter decomposition and plant nutrient availability, they play a key role in the responses of ecosystem to global climate change. During the rice cropping system the heterotrophic population was significantly different among the crop growth stages and it show highest (7.50 log CFU) under panicle initiation stage of the rice at ($p < 0.05$). Heterotrophic population was also significantly more among the population. Anaerobic bacterial population was found highest (6.25 log CFU) under panicle initiation stage of the rice crop. Anaerobic bacterial population was significantly different among the treatments and it was found in the order of PI > MT > AT > H. Fungal population was found less compare to other microbial population and also found highest in PI stage of rice. Actinobacterial population was not shown any significant different among the stages (Table 33.2). Population was found highest under panicle initiation stage of the crop growth stage.

Table 33.2: Microbial Population Under different Growth Stages under Rice Ecosystem.

Different Microbial Population		*Different Growth stages*		
	AT	*MT*	*PI*	*H*
Heterotrophicbacterial population	7.50^{d}	7.40^{d}	7.60^{b}	7.02^{b}
Anerobic bacterial population	6.09^{c}	6.06^{c}	6.25c	6.05^{c}
Fungal population	5.11^{b}	5.28^{b}	4.52^{a}	5.50^{a}
Actinobacterial population	4.54^{a}	4.92^{a}	5.94^{a}	4.77^{a}

AT: active tillering; MT: maximum tillering; PI: panicle initiation; H: harvesting, * Similar superscripts indicates no significant difference ($p < 0.05$) among various microbial population.

Discussion

Soil MBC regulates soil organic matter decomposition and nutrient cycling, and thus plays a key role in maintaining function and sustainability of terrestrial ecosystems. MBC has been included in current soil monitoring concepts due to its rapid response and high sensitivity to management practices and environmental changes (Bhattacharyya *et al.*, 2013). Organic Carbon did not very significantly ($p < 0.05$) among the treatment. Available N (kg/h) also showed significant variation among the growth stages of rice crop. The microbiological activity of a soil directly influences the ecosystem stability and fertility and it is widely accepted that a good level of microbiological activity is essential for maintaining soil quality (Dick *et al.*, 1993). Fungi are regarded as the main decomposers of recalcitrant litter with high C to N ratios. Fungi produce organic-binding agents and entangle soil particles, thereby stimulating aggregate formation and providing physical protection of SOM (Tisdall, 1994). Moreover, fungi have been shown to deposit litter C within these aggregates (Frey *et al.*, 2018). Soil organic matter is the substrate for many soil enzymes and protects them through the formation of enzyme complexes with clay and humus (Tabatabai, 1994).

Conclusion

Microbial biomass carbon is a useful indicator of soil quality and its dynamics in rice soil are largely controlled by organic substances released from rice roots and soil associated microbiota. Microbial biomass carbon usually measures either the weight of carbon or nitrogen in soil microorganisms. Thus, the findings of the present study elucidate this mechanism that helps in maintaining the stability of soil in paddy field.

References

Balser, T.C., Wixon, D., Moritz, L. K., and Lipps, L. (2010). The microbiology of natural soil. In: Wixon, D. R. and Tilson E.L. (Eds) Soil Microbiology and Sustainable Crop Production, pp. 27-58.

Bhattacharyya, P., Nayak, A.K., Mahanty, S., Tripathi, R., Sahid, M., Kumar, A., Raja, R., Panda, B.B., Roy, K.S., Neogi, S., Dash, P.K., Shukla, A.K. and Roa, K.S. (2013). Greenhouse gas emission in relation to labile soil C, N pools and functional microbial diversity as influenced by 39 years long-term fertilizer management in tropical rice. Soil and Tillage Research, 129: 93–105.

Dick, W.A. and Tabatabai, M.A. (1993). Significance and potential uses of soil enzymes, In: B. Metting (ed.), Soil Microbial Ecology, Marcel Dekker, New York. Pp. 95-127.

Emmerling, C., Schloter, M., Hartmann, A. and Kandeler, E. (2002). Functional diversity of soils organisms-a review of recent research activities in Germany. Journal of Plant Nutrition and Soil Science, 165: 408-420.

Frey, M.M., Stewartb, R.W., Mcconne, J.R., and Balesd, R.C. (2018). Atmospheric hydroperoxides in West Antarctica: links to stratospheric ozone and atmospheric oxidation capacity. Journal of Geophysical Research: Atmospheres, 110, D23301, doi: 10.1029/2005JD006110.

Garcia-Orenes, F., Guerrero, C., Roldan, A., Mataix-Solera, J., Cerda, A., Campoy, M., Zornoza, R., Barcenas, G. and Caravaca, F. (2010). Soil microbial biomass and activity under different agricultural management systems in a semiarid Mediterranean agro-ecosystem. Soil and Tillage Research, 109: 110-115.

Inubushi, K. and Acquaye, S. (2004). Role of microbial biomass in biogeochemical processes in paddy soil environments. Soil Science and Plant Nutrition, 50(6): 793-805.

Jackson, M.L. (1973). Soil Chemical Analysis. Pentice Hall of India Pvt. Ltd., New Delhi.

Liesack, W., Schnell, S., and Revsbech, N. P. (2000). Microbiology of flooded rice paddies. FEMS Microbiology Review, 24: 625-645

Lopes, A.R., Faria, C., Prieto-fernández, Á., Trasar-cepeda, C., Manaia, C. M., and Nunes, O. C. (2010). Soil biology and biochemistry comparative study of the microbial diversity of bulk paddy soil of two rice fields subjected to organic and conventional farming. Soil Biology and Biochemistry, 43 (1): 1-11.

Mantelin, S. and Touraine, B. (2004). Plant growth-promoting bacteria and nitrate availability: Impacts on root development and nitrate uptake. Journal of Experimental Botany, 55: 27-34.

Tabatabai, M.A. (1994). Soil enzymes. In: Weaver, R.W., Angle, J.S., Bottomley, P.S. (Eds.), Methods of Soil Analysis. Part 2. Microbiological and Biochemical Properties. SSSA, Madison, WI, 775–833.

Tisdall, J. M. (1994). Possible role of soil microorganisms in aggregation in soils. Plant and Soil, 159(1): 115-121.

Walkley A.J, and Black, C.A. (1934). An estimation of the Degtjareff method for determining soil organic matter and a proposed modification of the chromic acid titration method. Soil Science, 37: 29–38.

34

Effect of Natural Plant Oils on *Candida albicans* Clinical Isolates from Endophthalmitis Cases

G.C. Patra[1], P.S. Sahu[2], P. Panda[3] and S. Sahoo[4]*

[1]District Public Health Laboratory, Government Medical College, Balangir – 767 001, Odisha
[2]School of Biotechnology, KIIT University, Bhubaneswar – 751 024, Odisha
[3]Department of Microbiology, PRM Medical College, Baripada – 757 001, Odisha
[4]Department of Botany, Utkal University, Vanivihar, Bhubaneswar – 751 002, Odisha
**E-mail: guru123in@gmail.com*

ABSTRACT

The natural oils were extracted from three plants such as Kusum (Schleichera trijuga), Karanja (Pongamia pinnata) and Mahula (Madhuca longifolia) of 'Similipal Biosphere Reserve' forest and were evaluated against Candida albicans clinical isolates obtained from endophthalmitis cases. The minimum inhibitory concentration (MIC) value of the Schleichera trijuga, and Pongamia pinnata oils at a concentration of 333.33µl/ml was observed to be inhibitory against the standard strain of C. albicans. The pure 'Mahula Oil' (1000µl/ml) was used for the observation of susceptibility against the C. albicals. The minimum detectable action time varied between 8 to 12 hrs. The 'Kusum Oil' was 81.25 per cent sensitive to C. albicans at its MIC level where as 'Karanja Oil' was 76.56 per cent sensitive and 'Mahula Oil' did not show any sensitivity against C. albicans. The standard antifungal, Fluconazole showed 59.375 per cent sensitivity against C. albicans. In this study the 'Kusum' and 'Karanja' oils were found to be effective against C. albicans clinical isolates that expanded the medicinal utility possibly as an alternative source of anti-candidal agent.

***Keywords**: Candida albicans, Endophthalmitis, Karanja oil, Kusum oil, Mahula oil.*

Introduction

Empiric knowledge about plants with medicinal properties has been accumulated through centuries, always following human evolution through the time (Ferro, 2006). Traditional healers and elders use several plants to treat different types of infections including those caused by fungal organisms. Many of the plant materials used in traditional medicine are readily available in rural areas at relatively cheaper than modern medicine (Mann *et al.*, 2008). Extracts (oils) of three plants *viz.*, *Schleichera trijuga*, *Pongamia pinnata* and *Madhuca longifolia* were found to have antimicrobial activities and thus evaluated in this present study to assess the anti-mycotic effect of these oils on *C. albicans*, a pathogenic fungi isolated from endophthalmitis cases. It has been reported that *C. albicans* is the most common cause of endogenous endophthalmitis resulting from hematogenous disseminations following trauma or eye surgery (Schiedler *et al.*, 2004). An earlier study from our hospital also authenticates *C. albicans* to be the most common pathogen isolated in endogenous fungal endophthalmitis cases (Patra *et al.*, 2012a). Endogenous fungal endophthalmitis, including both chorioretinitis and vitritis needs an early and appropriate therapy. Because the eye is a protected compartment, penetration of systemically administered anti-fungal agents is highly variable. Advances in the development of new anti-fungal, might offer some new hope of improving the future of anti-fungal therapy.

The estimated availability of *S. trijuda* seed is about 25, 000 oil potential per tones per annum. Seeds contain 40.3 per cent of yellowish brown colored oil (Mallela *et al.*, 2011; Chanida *et al.*, 2008). Oleic acid was previously well-known as major fatty acid in 'Kusum Oil' (CSIR, 1972). Though *S. trijuga* is well known before as the source of Indian macassar oil, very limited information is available on its medicinal properties. The oil which is traditionally used for the cure of itch, acne, burns, other skin troubles, rheumatism (external massage), hair dressing and promoting hair growth (Mallela *et al.*, 2011; CSIR, 1972; Maharashtra State Gazetteers, 1953). Powdered seeds of *S. trijuga* are applied to wounds and ulcers of cattle to remove maggots (Iwasa, 1997). The bark is astringent and used against skin imflammations and ulcers, while an infusion is taken against malaria. In traditional system of medicines, such as ayurveda and Unani, the *P. pinnata* plant is used for anti-inflammatory (Srinivasan *et al.*, 2001), anti-plasmodial, anti-nonciceptive, anti-hyperglycaemics, anti-lipidoxidative, anti-diarrhoeal, anti-ulcer, anti-hyperammonic, CNS depressant activity (Mahli, *et al.*, 1989) and antioxidant. The extracts of *P. pinnata also* used on "keratitis" and "eye ailments" for the ocular diseases (Chopada *et al.*, 2008). Leaves are active against *Candida albicans* and other microorganisms (Chandrashekar and Prasanna, 2010).

Different parts of the *P. pinnata* are traditionally used for the itch, herpes, skin disease, ulcers, leaves, anthelmintic, digestive, laxative, inflammation, piles, diabetes, urinary discharges, leucomia, worm, wound and so many others diseases (Chopada *et al.*, 2008). Roots are used for cleaning gums, teeth, and ulcers. Bark is used internally for bleeding piles. Juices from the plant as well as oil are anticeptic. The leaf juice is used for cold, coughs, diarrhea, dyspepsia, flatulence, gonorrhea and leprosy. Today the seed and oil are used as a liniment for rheumatism (Chopada *et al.*, 2008), anti malaria, skin disease, rheumatic and leprous sores (Bobbarala *et al.*, 2009). *M. longifolia* flower is edible and good item for tribals. These are use to make syrup and medicinal properties (Ministry of tribal affairs). *M. longifolia* oil has emollient properties and is used in skin diseases, rheumatism, headache, hair fixation, habitual constipation, piles and haemorrhoides and as a laxative (Ministry of tribal affairs; Kaushik, *et al.*, 2010). However there is a lack of published reports on identification of any specific compound from these plants and their role in specific microorganism. In the present study oils extracted from seeds of *S. trijuga, P. pinnata* and *M. longifolia* which are the unexplored plants from a local natural reserve forest are tested in order to explore a possible indigenous source of antifungal material.

Materials and Methods

Plants and Preparation of the Plant Extracts

The Kusum Plant and Preparation of the Plant Extract (Kusum Oil)

The *S. trijuga* plant and its fruits were identified as per the morphological descriptions (CSIR, 1972; Maharashtra State Gazetteers 1953; Iwasa, 1997). The ripen fruits of 'Kusum' were collected from forests of 'Similipal Biosphere Reserve' (Odisha) during fruiting season (*i.e.*, June and July). Natural appearance of Kusum fruit and seeds are shown in Figures 34.1 and 34.4. 'Kusum Oil' was extracted as per the method of Patra *et al.* (2012). Briefly, the outer covering sheath and fleshy layer were removed by crushing manually for harvesting the seeds which were then dried under natural sun light. Hard coat of the seeds were removed and ground. The ground seeds were steamed for about 30-45 min until the oil fraction separated out and the oil was extracted by pressing the ground seeds with the help of 'Wooden Ghana' (locally used tool that is used for extracting oil from seeds).

The Karanja Plant and Preparation of the Plant Extract (Karanja Oil)

P. pinnata is a tropical, mixed deciduous and semi evergreen plant. The fruits of *P. pinnata* were collected from the local forest areas of 'Similipal Biosphere Reserve' forest (Odisha) during the month of December and identified as described by Pandey *et al.* (2011). The natural appearance of the plant, flowers, pods and seeds are shown in Figures 34.2 and 34.5. Pods were separated by hammering the fruit and seeds were dried under natural sun light. Oil was extracted as described earlier by Patra *et al.* (2010). Briefly, sun-dried seeds were ground and grounded seeds were steamed for about 30-45 minutes. Oil was extracted by pressing with wooded Ghana.

The Mahula Plant and Preparation of the Plant Extract (Mahula Oil)

M. longifolia belongs to family Sapotaceae, is an Indian tropical tree found largely in central and north Indian planes and forest. Plant is adapted to arid environments, being a prominent tree in Odisha. *M. longifolia* is a large evergreen or semi-evergreen tree with dense spreading crown extensively cultivated in warm climates for its oil containing seeds (Alexander *et al.*, 2009). Plant was identified as per literature (Mundhra and Paria, 2009). The ripen fruits of Mahula were collected from 'Similipal Biosphere Reserve' forest (in Odisha) during July and August. Natural appearance of Mahula fruit and seeds are shown in Figures 34.3 and 34.6, respectively. The fleshy

Figure 34.1: Branch and Fruit of *Schleichera trijuga*.

Figure 34.2: Flowers of *Pongamia pinnata*.

Figure 34.3: Flowers *of Madhuca longifolia*.

Figure 34.4: Fruits and Seeds of *Schleichera trijuga*.

Figure 34.5: Fruits, Pods and Seeds of *Pongamia pinnata*.

Figure 34.6: Fruits and Seeds of Picture *Madhuca longifolia*.

Figures 34.1–34.6: Pictures Showing Plant Features (Leaves, flowers, fruits, pods, seeds) of *Schleichera trijuga, Pongamia pinnata* and *Madhuca longifolia*.

soften sheath was removed out and seeds were hammered to remove the hard coat. Then the half broken seeds were dried (about 48 to 72 hrs) under natural sun light for remove the water content. Sun dried seeds were ground and grounded seeds were steamed for about 30-45 minute and oil was extracted as described before.

Anti-fungal Assay by Disc Diffusion Technique

Oils were screened for their anti-fungal activity against isolated *C albicans* by disc diffusion method of Baur *et al.* (1966) on Sabouraud Dextrose Agar Sodium Taurocholate (SDAST) medium plate (Rath *et al.*, 2001). The sterile paper disces (5 mm diameter, Whatmans filter paper No.42) containing MIC levels of oils (*S. trijuga, P. pinnata, M. longifolia,*) were soaked and tested for antifungal activity in triplicate. Plates were incubated at 35±2 °C for 48 h. within 15 minutes after applied the discs. Zones of inhibition in mm were determined and comparison of zones of sensitivity with standard anti-fungals (Amphotericin-B, Clotrimazole, Fluconazole, Itraconazole, Nystatin). Fluconazole was taken for standard positive control in this present study as per CLSI guidelines (formerly known as NCCLS) for interpretation of the sensitive or resistance of 'Kusum Oil', 'Karanja Oil' and 'Mahula Oil'. A filter paper disc soaked with sterile distilled water was used as negative control. Zones of the oils were measured. The data of the parameters were statistically analyzed. Sensitive or resistant result was interpreted based on the diameter of the zone of inhibition after comparing with the standard anti-fungal disc as per NCCLS guidelines (2004).

Determination of Minimum Inhibitory Concentration (MIC)

The MIC of the 'Kusum Oil' was determined by tube dilution technique following the protocol described earlier (Janssen *et al.*, 1986). Briefly, Sabouraud Dextrose Broth Sodium Taurocholate (SDBST) medium was prepared by supplementing 0.75 per cent (w/v) Sodium taurocholate in to the Sabouraud dextrose broth (SDB). A series of ten sterile test tubes were taken and 2.0 ml of Kusum oil was added in to the first one. Rest were contained each 1.0 ml of sterile SDBST broth. Taking 1.0 ml of oil from 1st tube, two-fold serial dilution was made up to 9th tube and1.0 ml of suspension was discarded from the 9th tube. The tenth tube, which served as a negative control having no oil where as 1st one is positive control. Inoculated 1.0 ml of the diluted culture suspension (0.5-2.5 × 10^3 cfu/ml) of *C. albicans* into each of the tubes. The result was observed after 24 h at 35±2 °C. The minimum inhibitory concentration of the oil was determined as the least concentration inhibiting the growth of the *C. albicans* in the test tube. Similarly MIC of the 'Karanja Oil' and 'Mahula Oil' was determined and the effect was tested against *C. albicans.*

Determination of Action Time of the Oil

An *in-vitro* experiment was previously described (Patra *et al.*, 2012) with slightly modification to determine the time required to kill the *C. albicans* by 'Kusum Oil', 'Karanja Oil' and 'Mahula Oil'. A set of ten test tubes with each one ml of SDBST at MIC level of 'Oil' were inoculated with 50µl of freshly grown broth culture of *C. albicans* (0.5-2.5 × 10^3 cfu/ml) in to each tube and incubated at 35±2 °C. The incubated tubes were sub cultured on SDA plates at 0 minute, 30 minutes, 1h, 2h, 3h, 4h, 5h,

6h, 8h, 10h and after overnight incubation. The activity was observed by incubating the plates at 35±2 ^{0}C for 24 hrs. No growth on the streaking line was considered to be the time required by the oil to kill the *C. albicans*.

Results

Candida albicans strains of endophthalmitis were taken for the observation of the anti candidal activity of three oils (Figure 34.7) extracted from locally available plant sources (*S. trijuga, P. pinnata, M. longifolia*). The minimum inhibitory concentrations of 3 oils were determined by tube dilution technique. The concentration of 'Kusum Oil' and 'Karanja Oil' varied for each isolated strain of *C. albicans*. However, the MIC of the both oils at a concentration of 333.33µl/ml was observed to be inhibitory against the standard strain of *C. albicans*. The pure 'Mahula Oil' (1000µl/ml) was used for the observation of susceptibility against the *C. albicals*. The minimum detectable action time was found to be varying between 8 to 12 hrs for all the three oil extracts as observed in the present study. This preliminary study revealed 'Kusum Oil' to have inhibitory effect against *C. albicans* clinical isolates. The 'Kusum Oil' was 81.25 per cent sensitive to *C. albicans* at its MIC level where as 'Karanja Oil' was 76.56 per cent sensitive to *C. albicans* at its MIC level and 'Mahula Oil' did not show any sensitive against *C. albicans*. The standard antifungal, Fluconazole showed 59.375 per cent sensitivity against *C. albicans*. The susceptibility result was varied according to the different standard disc used in this study. Amphotericine-B

Table 34.1: Susceptibility of different Antifungals and 'Kusum Oil', Karanja Oil, Mahua Oil, against *C. albicans*

Antifungals	*Abbreviation*	*Concentration*	*Susceptibility Test Result (N=64)*	
			Sensitive	*Resistance*
Amphotericin-B	Ap	100 units/disc	52 (81.25 per cent)	12 (18.75 per cent)
Clotrimazole	Cc	10 mcg/disc	46 (71.875 per cent)	18 (28.125 per cent)
Fluconazole	Fu	25 mcg/disc	38 (59.375 per cent)	26 (40.625 per cent)
Itraconazole	It	10 mcg/disc	47 (73.44 per cent)	17 (26.56 per cent)
Nystatin	Ns	100 units/disc	44 (68.75 per cent)	20 (31.25 per cent)
KUSUM OIL *	KO	10 µl/disc	52 (81.25 per cent)	12 (18.75 per cent)
KARANJA OIL*	KrO	10 µl/disc	49 (76.56 per cent)	15 (23.44 per cent)
MAHULA OIL*	MO	10 µl/disc	-	64 (100 per cent)

N: total number of isolates. Sensitive or resistant result was interpreted based on the diameter of the zone of inhibition after comparing with the standard table as per CLSI guidelines 2004.

*The basis of interpretation of the result as 'sensitive' or 'resistant' of oil was taken according to the zone size of standard disc. Fluconazole (Fu). KO: Kusum Oil; KrO: Karanja Oil; MO: Mahula Oil.

showed 81.25 per cent sensitive against *C. albicans*, followed by Itraconazole (73.44 per cent), Clotrimazole (71.875 per cent), Nystatin (68.75 per cent) and Fluconazole (59.375 per cent) (Table 34.1).

Sensitivity or resistance was interpreted based on the diameter of the zone of inhibition following the CLSI guidelines. The inhibited zone sizes varied from 2 mm to 32 mm in diameter. The average zone of inhibition diameters in the susceptibility test to different antifungal agents as well as the 'Kusum, Karanja and Mahula Oils' are mentioned in Table 34.2. Antifungal sensitivity test on SDA culture plates are shown in Figures 34.8 and 34.9.

Table 34.2: Average Zone of Inhibition Diameters in the Susceptibility Test to different Antifungal Agents

Anti-Fungal Agent	*Number of Isolates (N)*		*Mean Average Diameter of ZOI*	*Standard Deviation (SD)*	*Significance (P value)*
Ap	Sensitive	52	21.32	4.56	0.0001
	Resistant	12	1.33	1.82	
Cc	Sensitive	46	19.56	3.73	0.0001
	Resistant	18	1.88	2.54	
Fu	Sensitive	38	21.36	3.74	0.0001
	Resistant	26	2.08	3.34	
It	Sensitive	47	21.25	4.09	0.0001
	Resistant	17	1.41	2.12	
Ns	Sensitive	44	20.95	4.17	0.0001
	Resistant	20	2.75	2.98	
KO	Sensitive	52	20.57	4.64	0.0001
	Resistant	12	3.41	4.14	
KrO	Sensitive	49	21.34	4.40	0.0001
	Resistant	15	2.93	3.84	
MO	Sensitive	-	-	-	-
	Resistant	64	1.03125	-	

ZOI: Zone of Inhibition; Ap: Amphotericin-B; Cc: Clotrimazole; Fu: Fluconazole; It: Itraconazole; Ns: Nystatin; KO: Kusum Oil; MO: Mahula Oil; KrO: Karanja Oil; (-): Not applicable; P values are based on the Unpaired *t* test results calculated using GraphPad QuickCals.

Discussion

Although the current anti-fungal therapies have been significantly improved, the outcome is still far from satisfactory, partly due to the limited number of classes of clinically available anti-fungals, the development of resistance to current anti-fungals, and the challenges of proper and early diagnosis (Zhai and Lin, 2011). *C. albicans* resistance to azoles has been reported (Niimi *et al.*, 2010). Because of the increasing of antimicrobial resistance of *Candida*, many attempts have been made to find out a new antimicrobial agent to prevent candidal endophthalmtis. Medicinal plants could be a source of alternative antimicrobial agents to treat

Figure 34.7: Picture Showing the Oils of *Schleichera trijuga* (Kusum), *Madhuca longifolia* (Mahula) and *Pongamia pinnata* (Karanja).

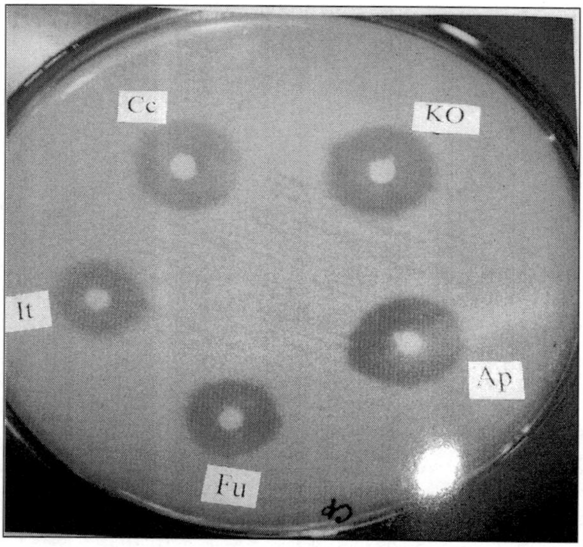

Figure 34.8: Picture Showing Zones of Inhibited by Amphotericin-B(Ap), Chlotrimazole (Cc), Fluconazole (Fu), Itraconazole (It) and Kusum Oil (KO).

infectious diseases. In an earlier study analyzing 1,662 vitreous fluid samples from fungal endophthalmitis cases, *C. albicans* was found to be the most common pathogen isolated in endogenous fungal endophthalmitis cases constituting 52.99 per cent (Patra *et al.*, 2012a). So the present study is justified to focus on the candidal endophthalmitis cases only where the effect of three oils ('Kusum, Karanja and Mahula Oil') are experimented *in vitro* to find its possible anti-*Candida* effect where these plants are known to be prevalent locally in this geographic region.

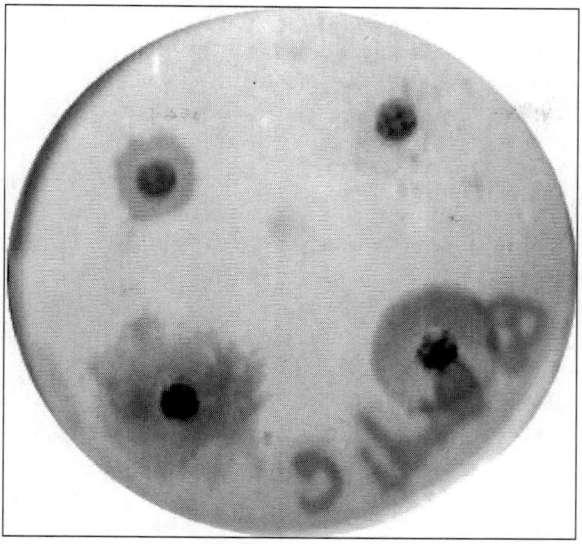

Figure 34.9: Picture Showing Zones Inhibited by Oils (*Schleichera trijuga, Madhuca longifolia, Pongamia pinnata*) and Fluconazole.

Due to limited no of anti-fungal drugs, *Candida* endophthalmitis are commonly treated empirically; hence, drugs are typically selected without regard to susceptibility (Lalita *et al.*, 2007). The present *in-vitro* study has shown that 'Kusum Oil' is highly effective against *C. albicans* when compared with the results from other medicinal plants such as *B. discolor, C. transvaalensis, P. capense, W. salutaris* (Samie *et al.*, 2010). In this study 'Kusum Oil' was effective against 81.25 per cent of the total isolates of *C albicans* clinical isolates. However, the effectiveness was found to be of different extent in case of different isolates of *C. albicans* when we consider diameter of the zone of inhibition as a parameter. Otherwise we can augment this finding when the concentration of the extract is considered. Therefore, more number of isolates were killed at 333.33 µl/ml of oil extract keeping the protocol same for each of the batch of experiment.

At the same MIC level the 'Karanja Oil' was 76.56 per cent effective against *C albicans* clinical isolates. The inhibited zone diameter was distributed between 15mm to 30mm, where as the leaf extract of karanja oil was effective against *Candida albicans* with zone sizes varies from 24 mm to 26 mm in diameter (Chandrashekar and Prasanna, 2010). Bark and seeds of *Pongamia pinnata* has also promising anti-mycotic activity (Bobbarala *et al.*, 2009). Hence, the plant can be used to discover bioactive natural products that may serve as leads for the development of new therapeutic needs. It is noted that this study would lead to the establishment of some compounds that could be used to formulate new and more potent anti-candidal drugs of natural origin. Undiluted raw 'Mahula Oil' was used for the observation of susceptibility against *C. albicans* clinical isolates. However, this oil didn't exhibit any activity against *C. albicans.* The result is in accordance with the earlier reports indicating that plant extract of *Madhuca longifolia* is not active against *C. albicans*

(Kaushik *et al.*, 2010). *Scleichera triguda* and *Pongamia pinnata* contain tannins as the major constituent; the antimicrobial activity of the extract may be due to presence of tannins (Chandrashekar and Prasanna, 2010; Bobbarala *et al.*, 2009). The tannins isolated from the medicinal plants possess remarkable toxic activity against bacteria and fungi and may assume pharmacological importance.

The different mechanisms proposed so far to explain tannin antimicrobial activity include inhibition of extracellular microbial enzymes, deprivation of the substrates required for microbial growth or direct action on microbial metabolism through inhibition of oxidative phosphorylation. A further mechanism involving iron deprivation is proposed (Augustin, 1991). 'Kusum oil' and 'Karanj oil' have novel substance that has been used by tribal people for their disease like, cold, itch, ulcer for both human and cattle. However, identifying specific fraction from the crude extract and further characterization is warranted which was beyond the scope of the present study.

References

Alexander, J., Auounsson, G.A., Benford, D., Cockburn, A., Cravedi, J., Furst, P., Galli, C., Granjean, P., Gzyl, J., Heinemeyer, G., Johansson, N., Mutti, A., Schlatter, J., Leeuwen, R.V., Peteghem, C.W. and Verger, P. (2009). Scientific opinion. Saponins in *Madhuca longifolia* L. as undesirable substances in animal feed. Scientific opinion of the panel and contaminants in the food chain. The EFSA Journal, 279: 1-36.

Augustin, S. (1991). Antimicrobial properties of tannins. Phytochemistry, 301(2): 3875-3883.

Baur, A.W., Kirby, W.M.M., Sherrisa, J.C. and Turk, M. (1966). Antibiotic susceptibility testing by standard disc diffusion method. American Journal of Clinical Pathology, 45: 493-496.

Bobbarala, V., Katikala, P.K., Naidu, K.C. and Penumajji, S. (2009). Antifungal activity of selected plant extracts against phytopathogenic fungi *Aspergillus niger* F2723. Indian Journal of Science Technology, 2(4): 87-90.

Chandrashekar, K.S. and Prasanna, K.S. (2010). Antimicrobial activity of *Pongamia pinnata* leaves. International Journal of Medicobiological Research, 11: 18-20.

Chanida, P. and Niran, V. (2008). Fatty acid constituents of *Schleichera oleosa* (Lour.) Oken. seed oil. Journal Health Research, 22(4): 203e212.

Chopade, V.V., Tankar, A.N., Pande1, V.V., Tekade1, A.R., Gowekar1, N.M., Bhandari1, S.R. and Khandake, S.N. (2008). *Pongamia pinnata*: Phytochemical constituents, Traditional uses and Pharmacological properties: A review, International Journal of Green Pharmacy, 2(2): 72-75.

CSIR-Council of Scientific and Industrial Research. (1972). The wealth of India: A dictionary of India raw materials and industrial products. New Delhi: Publications and Information Directorate.

Ferro, D. (2006). Fitoterapia: conceitos clinicos. Sao Paulo: Ateneu.

Iwasa, S. *Schleichera oleosa* (Lour.) Oken. In Faridah Hanum, I. and van der Maesen, L.J.G. (Eds.). (1997). Plant Resources of South-East Asia No. 11. Auxiliary Plants. Prosea Foundation, Bogor, Indonesia. 227-229.

Janssen, A.M., Chin, N.L., Scheffer, J.J.C. and Svendsen, A.B. (1986). Screening for antibacterial activity of soma essential oils by the agar overlay technique. Statistics and corelations. Pharm. Weekbl. Science, 8: 289-292.

Kaushik, P., Kaushik, D., Khokra, S.L., Sharma, C., Aneja, K.R., Khah, S., Mehra, R. and Singh, G. (2010). Evaluation of antioxidant and antimicrobial activity of *Madhuca indica*. Pharmacology online, 2: 1-8.

Lalita, P., Shapiro, B.L., Srinivasan, M., Prajna, N.V., Acharya, N.R., Fothergill, A.W. (2007). Antimicrobial susceptibility of *Fusarium, aspergillus*, and other filamentous fungi isolated from keratitis. Archives in Opthalmology, 125(6): 789-793.

Maharashtra State Gazetteers Department. (1953). Maharashtra State Gazetteers Series No-1: Botany Part-I, Medicinal Plants. On http: / /www.maharashtra.gov. in/pdf/gazeetter reprint/BOTANY/SECTION_I_MEDIC INAL_PLANTS.pdf.

Mahli, S.S., Basu, S.P., Sinha, K.P. and Banarjee, N.C. (1989). Pharmacological effects of Karanjin and Pongamol. Indian Journal of Animal Science. 59: 657-660.

Mallela, G., Ramu, N. and Bakkiya, R.S. (2011). Methyl ester production from *Schlichera oleosa*. International Journal of Pharmaceutical Science and Research. 2 (5): 1244-1250.

Mann, A., Banso, A. and Clifford, L.C. (2008). An antifungal property of crude plant extracts from *Anogeissus leiocarpus* and *Terminalia avicennioides*. Tanzania Journal of Health Research, 10 (1): 4-38.

Ministry of Tribal Affairs, Government of India (2013). Product profile, Mahuwa. Trifed.nic.in. 11-21.

Mundhra, A. and Paria, N.D. (2009). Epigeal cryptocotyly in *Madhuca indica* J.F Gmel. (Sapotaceae). International Journal of Botany, 5: 200-202.

NCCLS. (2004). Method of antifungal disc diffusion susceptibility testing by Yeast; approved Guidelines. NCCLS document M44-A, Vol 24. No-15.

Niimi, M., Firth, N.A. and Cannon, R.D. (2010). Antifungal drug resistance of oral fungi. Odontology, 98(1); 15-25.

Pandey, A., Khatri, P., Patel, R., Jakhetia, V., Sharma, S. (2011). Pharmacognostic and phytochemical evaluation of *Pongamia pinnata* Linn family fabaceae. International Journal of Pharmaceutical Research and Development, 3(2): 11-19.

Patra, G.C., Panda, P., Sahoo, S., Ranasingh, N. and Nanda, P.K. (2012a). Aetiology of Endophthalmitis in Odisha: Eastern India. Environment Ecology, 30 (1): 159-162.

Patra, G.C., Sahu, P.S., Panda, P., Sahoo, S., Mohapatra, S. and Bindhani, B.K. (2012b). Anti-Mycotic Effect of 'KUSUM OIL' Extract on *Candida albicans* clinical isolates from endophthalmitis cases. International Journal of Pharmaceutical and Biological Sciences, 3(2): 475-484.

Rath, C.C., Dash, S.K., Mishra, R.K. and Ramachandraias, O.S. (2001). A Comparative analysis of *in vitro* antifungal activity of pure and fractionally distilled turmeric (*Curcuma longa* L.) leaf essential oil. Indian Drugs. 39(1): 18-22.

Samie, A., Tambani, T., Harshfield, E., Green, E., Ramalivhana, J. N. and Bessong, P. O. (2010). Antifungal activities of selected venda medicinal plants against *Candida albicans*, *Candida krusei* and *Cryptococcus neoformans* isolated from South African AIDS African patients. Journal of Biotechnology, 9(20): 2965-2976.

Schiedler, V., Scott, I., Flynn, H.W., Janet, D.L., Benz, M.S., Miller, D. (2004). Culture-proven endogenous endophthalmitis clinical features and visual outcomes. Americal Journal of Ophthalmology, 137: 725-731.

Srinivasan, K., Muruganandan, S. and Lal, J. (2001). Evaluation of antiinflammatory activity of *Pongamia pinnata* leaves in rats. Journal of Ethnopharmacology. 78: 151-157.

Zhai, B. and Lin, X. (2011). Recent progress on antifungal drug development. Current Pharmaceutical Biotechnology, 12(8): 1255-1262.

35

Phytochemistry and Antibacterial Activity of *Spirogyra bichromatophora* (Randhawa) Transeau, Sourced from Barehipani Water Fall

B.S. Sipra*[1], *Emamee Sahu*[1], *Sukumar Bhakta*[2] *and Akshaya Kumar Bastia*[1]

[1]Department of Botany, North Orissa University, Baripada, Mayurbhanj – 757 003, Odisha
[2]Botanical Survey of India, Western Regional Centre, Pune – 411 001, Maharashtra
****E-mail: bastianou@gmail.com***

ABSTRACT

Spirogyra bichromatophora, a filamentous green alga, isolated from Barehipani waterfall of Similipal Biosphere Reserve, Odisha was studied for its phycochemistry and antibacterial activity. The biochemical composition revealed presence of adequate amount of primary metabolites like proteins, carbohydrates and lipids. Intracellular metabolites were extracted fractionally with different polar and non-polar solvents viz. petroleum ether, acetone, hexane, ethanol, methanol and aqueous. Crude extract of S. bichromatophora was screened for antimicrobial activity against three pathogenic bacteria (Escherichia coli, Pseudomanas aeruginosa and Staphyllococcus aureus) adopting agar cup diffusion method. The crude metabolites extracted with acetone showed considerable activity against all the test pathogens followed by ethanol and hexane extracts. Maximum zone of inhibition (20 mm) was recorded against S. aureus by the aqueous extract followed by acetone, which indicated the nature of the active metabolites that are mostly polar. MIC and MBC values were found to be maximum against

S. aureus by methanolic and ethanol crude extract respectively. Presence of secondary metabolites is probably the reason for their vigorous growth in the water bodies that indicates the good health of the water bodies in the region.

Keywords: *Spirogyra bichromatophora, Barehipani waterfall, Phycochemistry, Antibacterial activity, Pathogenic bacteria.*

Introduction

The use of algae in various fields has a long history. They are known for producing several biologically active compounds with antiviral, antibacterial, antifungal and anticancer activities. Secondary and primary metabolites produced by these organisms may be potential bioactive compounds of interest in pharmaceutical industry. There are a number of reports on the evaluation of antioxidant activity in microalgae belonging to the genera *Botryococcus* (Rao *et al.*, 2006), *Chlorella* (Wu *et al.*, 2005; Goh *et al.*, 2010), *Dunaliella* (Herrero *et al.*, 2006), *Nostoc* (Li *et al.*, 2007), *Phaeodactylum* (Guzman *et al.*, 2001), *Spirulina* (Miranda *et al.*, 1998; Jaime *et al.*, 2005; Mendiola *et al.*, 2007), *Haematococcus* (Cerón *et al.*, 2007) and *Chaetoceros* (Goh *et al.*, 2010). *Spirogyra* is one of the commonest green filamentous algae of order Zygnematales, named so because of helical or spiral arrangement of the chloroplasts. The organism is considered as a very good source of bioactive compounds and produces a great variety of secondary metabolites characterized by broad spectrum of biological activities. Earlier studies have reported that *Spirogyra weberi* have high content of lipids, carbohydrates, proteins and inorganic elemental composition. More recently, *Spirogyra* biomass has been reported to be an efficient energy source for biofuel (Hossain *et al.*, 2008; Eshaq *et al.*, 2010; John *et al.*, 2011). Very few reports are available on the phytoconstituents and antimicrobial activity of some *Spirogyra* sp. The present investigation aims to determine the macromolecular contents and antibacterial activity of *Spirogyra bichromatophora* in different organic solvent extracts.

Materials and Methods

Sample Collection

Spirogyra bichromatophora, a common green alga, is collected from Barehipani water fall of Similipal Biosphere Reserve (Figures 35.1–35.3). The Barehipani water fall is situated at 21.93N 86.38E on the Budhabalanga River flowing over the Meghasani mountain. Samples were collected using clean sampling bottles, forceps, polythene bags, brush, petridishes, scalpel *etc.* The collected samples were thoroughly washed with tap water, spreaded for shade drying. The dried samples were grounded to fine powder and stored for further study.

Macromolecular Estimation

The total carbohydrate content of the algal sample was estimated by Anthrone Reagent method using glucose as standard taken within the range 10-50 µg/ml. Total cellular protein was estimated following Folin-Phenol method (Lowry *et al.*,

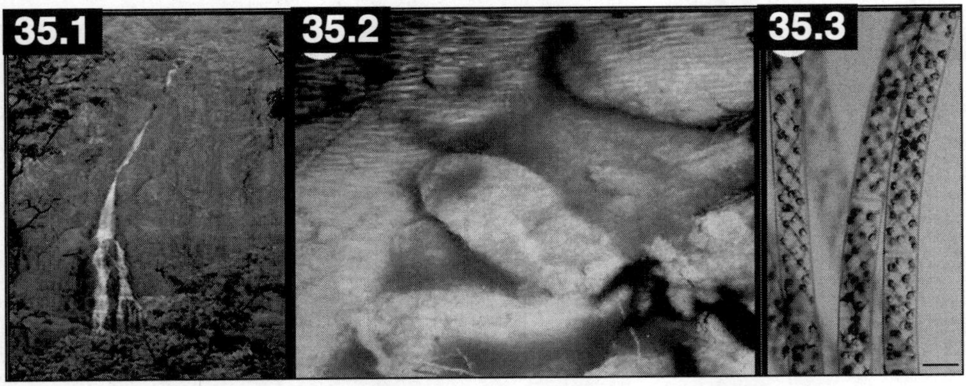

Figures 35.1–35.3: 35.1. Barehipani waterfall; 35.2. Occurrence of *Spirogyra* mat; 35.3. Microscopic Photograph of *S. bichromatophora*.

1951). Extraction and estimation of lipid was done by gravimetric method (Bligh and Dyer, 1959).

Detection of Phytosterols

To detect the phytosterols (qualitatively), Salkowski and Libermann Buchard test was adopted.

Detection of Fixed Oils and Fats

Small quantity of the extract biomass was pressed between filter papers. Oil stains were obtained which indicated the presence of fixed oils. To detect the presence of fat, a few drops of 0.5 N alcoholic Potassium hydroxide were added to the algal extract with few drops of phenolphthalein. The mixture was heated in a water bath for 1-2 hrs. This resulted in formation of soap.

Antimicrobial Study

The filaments were rinsed with tap water and were shade dried. The dried biomass was grinded in to powder form. Intracellular metabolites were extracted fractionally using different polar and non-polar solvents *i.e.* Petroleum ether, acetone, hexane, ethanol, methanol and aqueous. Crude solvent extracts of *S. bichromatophora* was screened for their antimicrobial activity against three pathogenic bacteria (*Escherichia coli*, *Pseudomonas aeruginosa* and *Staphyllococcus aureus*) adopting agar cup diffusion method. Moreover, the minimum inhibitory concentration (MIC) and minimum bactericidal concentration (MBC) were determined by 96 well microplates (Eloff, 1998).

Results

Systematic Enumeration

Division - Chlorophyta

Class - Conjugatophyceae

Order - Zygnematales

Family - Zygnemataceae

Genus - *Spirogyra*

Species - *bichromatophora*

Single species under the genus *Spirogyra* was collected which grows abundantly in freshwater habitat.The morphometric enumerations of the species are given below.

Spirogyra bichromatophora (Randhawa) Transeau

Randhawa, 1959, P. 328, Fig. 315

Vegetative cells 60-75 × 96-160µm,with plane end walls; 2 chloroplasts, making 4 to 6 turns, conjugation scalariform; large tubes formed by both gametangia; fertile cells cylindric or enlarged; zygospores ellipsoid, 54-60×80-90µm; median spore wall smooth, brown.

Macromolecular and Phycochemical Analysis

The total carbohydrate content of the algal sample was estimated and found to be 2.4 mg/gm of dried biomass. Protein was estimated by Folincioalteau's reagent. The protein content of 1 gm dried biomass of *Spirogyra* was estimated 3.2 mg/gm and the lipid to be 3.3 mg/gm respectively. The phycochemical analysis of the algal sample showed the presence phytosterols, fixed oils and fats of algae under study (Table 35.1).

Table 35.1: Presence of Bioactive Compounds in the *S. bichromatophora*

Algal Taxa	*Carbohydrate (mg/g)*	*Protein (mg/g)*	*Lipid (mg/g)*	*Phytosterol*	*Fixed Oil*	*Fat*
S. bichromatophora	2.4	3.2	3.3	+	+	+

'+' indicates the presence of metabolites.

Antibacterial Activity

The intracellular metabolites were extracted with different organic solvents as well as aqueous in a decreased gradient of polarity. The petroleum ether extract of *S. bichromatophora* showed moderate antibacterial activity against two pathogen *viz. P. aeruginosa,* and *E. coli* with zone of inhibition 10 mm in each. However, the acetone extract showed significant antibacterial activity against all the three pathogen *viz. P. aeruginosa, S. aureus* and *E. coli* with inhibition zone 10, 11 and15 mm respectively. The maximum inhibition zone is observed in aqueous extracts against two pathogenic organisms *viz.* 20 mm against *S. aureus* and 10 mm against *P. aeruginosa.* Hexane showed considerable inhibition zone of 11, 10 mm against *P. aeruginosa* and *S.aureus* respectively. The ethanol extract showed antibacterial activity against two pathogen *viz., P. aeruginosa, S. aureus* with inhibition zone 14 and 10 mm. Methanol showed inhibition zone of 10, 11 mm against *S.aureus, Escherichia coli* respectively. In contrast, the control antibiotic chloramphenicol showed zone of inhibition (15-25 mm) against the pathogens (Table 35.2).

Table 35.2: Antimicrobial Activity of Crude Extracts of *S. bichromatophora* against the Pathogenic Organisms

Solvent	*Zone of Inhibition (mm) against Pathogens*		
	E. coli	*P. aeruginosa*	*S. aureus*
Pet. ether	10	10	-
Acetone	11	10	15
Hexane	-	11	10
Ethanol	-	14	10
Methanol	11	-	10
Aqueous	-	10	20

MIC and MBC Test

MIC was determined by taking different solvent extracts like acetone, hexane, ethanol and methanol that showed good antibacterial activity. The acetone and hexane extracts have been tested against all the three pathogens whereas methanol and ethanol against *S. aureus* only. The MIC value of acetone extract was found to be 166 µg, 166 µg and 333 µg and MBC of 2.65 in each against *E. coli, P. aeruginosa* and *S. aureus* respectively. The MIC of hexane extract was 62.5 µg, 125 µg, 250 µg and MBC was 2.0 or ≥ 2.0 against all the pathogens. Ethanol showed MIC and MBC of 175 µg and ≥ 2.8 mg and in methanol, the MIC was 825 µg and MBC was ≥ 1.65 mg against the test pathogen *S. aureus* (Table 35.3).

Table 35.3: Determination of Minimum Inhibitory Concentration (MIC values in µg) and Minimum Bactericidal Concentration (MBC values in mg) of different Solvent Extracts against different Pathogens

Organisms	*Acetone*		*Hexane*		*Ethanol*		*Methanol*	
	MIC	*MBC*	*MIC*	*MBC*	*MIC*	*MBC*	*MIC*	*MBC*
E. coli	166	2.65	62.5	2.0	-	-	-	-
P. aeruginosa	166	2.65	125	≥ 2	-	-	-	-
S. aureus	333	2.65	250	2.0	175	≥ 2.8	825	≥ 1.65

Discussion

Algae have a significant attraction as natural source of bioactive molecules with a broad range of biological activities (Borowitzka, 1988; Skulberg, 2006). They can biosynthesize, metabolize, accumulate and secrete a great diversity of metabolites including pigments, phenolic compounds, sulphated compounds, phycosterols, polysaturated fatty acids, heat induced proteins and vitamins *etc.* The biomass and algae-derived compounds have a very wide range potential applications including medicines and pharmaceuticals which have positive effects on the health of man and animals (Pulz and Gross, 2004).

Present investigation revealed that the algal species *S. bichromatophora* is associated with structurally and biologically active primary or secondary metabolites. Biochemical studies revealed the presence of high concentration of macromolecules like carbohydrate, protein, lipid and phyco-constituents like phytosterols, fixed oils and fats. Antibacterial study revealed that *S. aureus* was significantly inhibited by all the solvent extracts (especially aqueous and acetone) except petroleum ether. *P. aeruginosa* showed resistant only to methanol extract with better sensitivity towards ethanol. *E. coli* was found sensitive to petroleum ether, acetone and methanol extracts. MIC and MBC analysis revealed the antibacterial property of the crude extracts even if in low concentrations against the test pathogens.

Conclusion

This study gives an indication that the alga *S. bichromatophora* is rich in primary as well as secondary metabolites which are bioactive in nature. The antibiotic activity may be due to bioactivity of different phycoconstituents. Presence of secondary metabolites as alkaloids and sterols suggest high rate of metabolism in them due to the presence of alternative pathways. Further investigation is required aiming at isolation and purification of active constituents responsible for antimicrobial activity.

Acknowledgement

The authors are grateful to the authorities North Orissa University, Baripada for providing laboratory facilities to carry out the research work.

References

Bligh, E.G. and Dyer, W. J. (1995) A rapid method for total lipid extraction and purification. Journal of Biochemistry and Physiology, 37: 911-917.

Borowitzka, M. A. (1988) Algal growth media and sources of algal cultures. In: Borowitzka, M. A., and Borowitzka L. J. (eds) Micro-algal Biotechnology. Cambridge University Press, Cambridge, pp 456-465.

Cerón, M. C., García-Malea, M. C., Rivas, J., Acien, F. G., Fernandez, J. M., Del Río, E., Guerrero, M. G. and Molina, E. (2007) Antioxidant activity of *Haematococcus pluvialis* cells grown in continuous culture as a function of their carotenoid and fatty acid content. Applied Microbiology and Biotechnology, 74: 1112–1119.

Eloff, J. N. (1998) A sensitive and quick microplate method to determine the minimum inhibitory concentration of the plate extracts for bacteria. Planta Medicine, 64: 711-714.

Eshaq, F. S., Ali, M. N. and Mohd, M. K. (2010) *Spirogyra* biomass a renewable source for biofuel (bioethanol) production. International Journal of Environmental Science Technology 2: 7045-7054.

Goh, S. H, Yusoff, F. M. and Loh, S. P. (2010) A comparison of the antioxidant properties and total phenolic content in a diatom, *Chaetoceros* sp. and a green microalga, *Nannochloropsis* sp. Journal of Agricultural Science, 2: 123–130.

Guzman, M., Galve-Roperh, I. and Sanchez, C. (2001). Ceramide: a new second messenger of cannabinoid action. Trends in Pharmacological Sciences, 22: 19–22.

Herrero, M., Jaime, L., Martín-Alvarez, P.J., Cifuentes, A. and Ibáñez, E. (2006) Optimization of the extraction of antioxidants from *Dunaliella salina* microalga by pressurized liquids. Journal of Agricultural Food Chemistry, 54: 5597–5603.

Hossain, A. B. M.S. and Salleh, A. (2008) Biodiesel fuel production from algae as renewable energy. American Journal of Biochemistry and Biotechnology, 4: 250-254.

Jaime, L., Mendiola, J., Herrero, M., Soler-Rivas, C., Santoyo, S., Senorans, F., Cifuentes, A. and Ibanez, E. (2005) Separation and characterization of antioxidants from *Spirulina platensis* microalga combining pressurized liquid extraction, TLC and HPLC-DAD. Journal of Sep Science, 28: 2111–2119.

John, R. P., Anisha, G. S., Nampoothiri, K. M. and Pandey, A. (2011) Micro and macroalgal biomass: a renewable source for bioethanol. Bioresources and Technology, 102: 186-193.

Li, H., Cheng, K., Wong, C., Fan, K., Chen, F. and Jiang, Y. (2007) Evaluation of antioxidant capacity and total phenolic content of different fractions of selected microalgae. Food Chemistry, 102: 771–776.

Lowry, O. H., Rosebrough, N. J., Farr, A. L. and Randall, R. J. (1951) Protein measurement with the folin phenol reagent. Journal of Biology and Chemistry, 193: 265 – 275.

Mendiola, J., Jaime, L., Santoyo, S., Reglero, G., Cifuentes, A. and Ibanez, E. (2007) Screening of functional compounds in supercritical fluid extracts from *Spirulina platensis*. Food Chemistry, 102: 1357–1367.

Miranda, M. S, Cintra, R. G., Barros, S. B. and Mancini, F.J. (1998) Antioxidant activity of the microalga Spirulina maxima. Brazilian Journal of Medical Biological Research, 31: 1075–1079.

Pulz, O., and Gross, W. (2004) Valuable products from biotechnology of microalgae. Applied Microbiology and Biotechnology, 65: 635-648.

Randhawa, M. S. (1959) Zygnemaceae. Indian Council of Agricultural Research, New Delhi.

Rao, A. R., Sarada, R., Baskaran, V. and Ravishankar, G. A. (2006) Antioxidant activity of *Botryococcus braunii* extract elucidated *in vitro* models. Journal Agricultural Food Chemistry, 54: 4593–4599.

Skulberg, O. M. (2006) Bioactive chemicals in microalgae. In: Richmond, A. (ed) Handbook of Microalgal Culture, Biotechnology and Applied Phycology, Blackwell, Oxford, pp 485-512.

Wu, L., Ho, J. A. A, Shieh, M. and Lu, I. (2005) Antioxidant and antiproliferative activities of *Spirulina* and *Chlorella* water extracts. Journal Agricultural Food Chemistry, 53: 4207–4212.

36

Eco-Friendly Bioenergy Options in India: Role of Micro-Algal Biofuels

D.C. Bhattacharya

Department of Microbiology, Vidyasagar University, Midnapore – 721 102, West Bengal
E-mail: debu720@gmail.com

ABSTRACT

In the modern world, industrial activities are increasing in leaps and bounds keeping pace with the rapid urbanisation. The energy consumption is also rising in the said context. For a eco-friendly green world, one cannot ignore the bioenergy options available to us. Being a sub-tropical country, India has the unique advantage of higher availability of PAR and the subsequent photo-harvesting of solar energy by the micro-organisms. Biofuels from micro-algal species are one of the myriad choices that have several advantages over other options. Basing on a spatio-temporal analysis driven by modern techno-economic developments in India, the study discusses the role of micro-algal biofuels and their feasibility for an holistic approach towards attaining energy security for India.

***Keywords**: Biofuels, Microalgae, Biodiesel, Bioethanol.*

Introduction

The developing countries like India has few challenges that have attracted more or less equal emphasis in terms of environmental quality and sustainable development goals (SDGs), in order to safeguard the urban life through the use of green and clean technologies in the long run. In the face of rapid urbanization followed by industrial growth, the main challenge being the sustainable energy security with an aim towards the next millennium. In rural areas, efforts have been made to consider eco-friendly alternatives such as photovoltaics, biogas and other contemporary bioenergy sources in addition to the existing technologies. The causes that accelerated the argument on behalf of the environmental necessity of clean technology initiatives were mostly anthropogenic like uncontrolled urbanization, burgeoning population, increasing vehicular pollution, municipal and industrial wastes. On the dialogue on sustainability, these conditions were reiterated almost everywhere in the world so as to convince the social and political regime on these technological imperatives for the coming years. Sustainable energy needs and their fulfillment has gained momentum in the recent years paving the way for a more comprehensive, time-bound yet eco-friendly energy consumption pattern for the consumers. The techno-economic feasibility has also been investigated for the possible alternatives that have long been a verbal topic among the stakeholders including the oil marketing companies for both rural and urban populations. Bioenergy has been defined as a renewable energy source derived from natural biological sources. As it is classified under renewable resources, the sustainability of these energy sources can be assured with time if pursued vigorously.

Common Sources of Bioenergy

The Indian sub-continent had the luxury of getting abundant sunshine and rain from time immemorial and agricultural activities had priorities in its population. Thus bioenergy feedstocks were commonly used cowdung, straws *etc.* Recent studies have shown that in addition to common agricultural crop residues, semi-arid crops like sweet sorghum and exotic crop like sugar beet can also be the required substrate to yield bioenergy in the form of bio-ethanol following suitable bioprocess technologies. Critical analysis have already proved that agricultural crop residues have several disadvantages in terms of water content, ligno-cellulosic degradation and the cost incurred to get the energy. Under these circumstances few experiments have shown the yield advantage in terms of biomass of microalgal strains, their ability to accumulate lipids, carbohydrates *etc.* The technical classification of biofuel substrates has led to the development of several terminologies, in which biofuel produced from microalgal substrates are commonly called 3rd generation biofuel. Earlier common substrates were either food grain crops (barley, soybean) or ligno-cellulosic wastes and the biofuels were commonly designated as first and second generation biofuels respectively.

Biofuels: The Micro-Algal Advantage

Being a subtropical country, India is perfect for the cultivation of photosynthesizing organisms which needs sunshine or more specifically PAR (photosynthetically active radiation). Cultivation of microalgae is mainly of two types - freshwater algal culture and marine algal culture. Again the method of culture also varies according to the availability of resources such ponds, seawater, freshwater resources *etc.* For example seawater is most suitable for marine micro-algal culture but nowadays race-way ponds are also designed to cultivate marine algae. Apart from that common low-cost tank culture is also popular for some specific algal strains. Technologically advanced countries have adopted photobioreactor (PBR) based culture as the growth rate can be significantly improved with PBR. Biochemically, micro-algae have the advantage of better photosynthetic efficiency and CCM (carbon concentrating mechanism), an unique phenomenon in algal strains. They have also been investigated to photosynthesize in the presence of flue gas as source of CO_2. Hence, mitigation of CO_2 pollution resulting from vehicular exhaust can even be investigated with these micro-algal biofuel production systems.

Microalgal lipid accumulation has been studied widely so as to find the answer for a suitable economically viable answer to the problem of biodiesel production from cheap source of lipid. Many algal species have shown promise both in tank culture, raceway ponds/PBR in order to be considered for industrial biodiesel production. Common freshwater/marine micro-algal strains that can be used for the biofuel production are mentioned in the Table 36.1.

Table 36.1: Commonly Utilized Micro-Algal Genera for Biofuel Production

Micro-algae	*Source*	*Used for*	*Reference*
Chlorella vulgaris, Chlorella sp KR -1	Freshwater	bioethanol	Kim *et al.*, 2014; Agwa *et al.*, 2017, Lee *et al.*, 2015
Scenedesmus obliquus, S. bijugatus	Freshwater	Bioethanol/ biodiesel	Miranda *et al.*, 2012; Ashok Kumar *et al.*, 2015
Nannochloopsis oculata	Marine	Biodiesel	Sukarni *et al.*, 2014
N.oleoabundans	Freshwater	Biodiesel	Gouveia *et al.*, 2009
Phaeodactylum tricornutum	Marine	Biodiesel	Branco-Vieira *et al.*, 2018;
Schizochytrium sp., *Nitzschia* sp., *Tetraselmis sueica*	Marine	Biodiesel	Chisti *et al.*, 2007

Table 36.2: Microalgal Lipid Content and Lipid Productivity of Promising Microalgal Species

Microalgal Species	*Lipid Content (per cent dry wt)*	*Lipid Productivity (mg/L/day)*
Chaetoceros calcitrans	14.6 - 16.4/39.8	17.6
Chlorella sorokiniana	19.0 – 22.0	44.7
Chlorococcum sp.	19.3	53.7
Dunaliella tertiolecta	16.7 - 71.0	-
Dunaliella salina	6.0 – 25.0	116.0
Nannochloris sp.	20.0 -56.0	60.9 -76.5
Nannochloropsis oculata	22.7 – 29.7	84.0 -142.0
Neochloris oleoabundans	29.0 – 65.0	90.0 -134.0
Phaeodactylum tricornutum	18.0 – 57.0	46.8 -49.2
Scenedesmus sp.	19.6 – 21.1	40.8 -53.9

Source: Chisti (2007), Mousdale (2008) and author's unpublished data (for *P. tricornutum*).

Feasibility of Micro-Algal Biodiesel Production

In most cases studies have been made to ascertain the feasibility by depending on the single criteria of lipid accumulation within the micro-algal cell/biomass. Though this approach is not completely flawless, other approaches are found to be more inconclusive to determine the acceptable strains/species of microalgae. Hence, Chisti (2007) and other researchers have suggested the range of lipid accumulation to be considered before venturing into the biodiesel production system from the microalgal substrate. Table 36.2 summarizes the lipid content and lipid productivity of the promising micro-algal species suitable for bio-diesel production.

The common Indian debate on the suitable biodiesel production system had given rise to several questions earlier. The most common among them was the suitability of jatropha as a substrate for biodiesel. Also there was a technological dilemma on the so called "rural development challenges" being linked with the cultivation of jatropha on waste lands/problem soils *etc.* But considering the time limit required for the growth of jatropha plant and the availability of seeds in the plant, it is almost a non-feasible choice for industry. Chisti (2007) has also proved the non-feasibility of jatropha as its oil productivity is too low in comparison to micro-algal lipid accumulation (Table 36.3). The economic viability studies till date has no clear cut data for the South-Asian countries but Department of Energy, USA (DOE) has given an overall cost in terms of USD per gallon of biodiesel produced as represented by Figure 36.1.

Table 36.3: Comparative Oil Yield of different Sources for Biodiesel Production

Crop	*Oil Yield (lit/ha)*
Corn	172
Soybean	446
Canola	1190
Jatropha	1892
Coconut	2689
Oil palm	5950
Micro algae[a]	136900
Micro algae[b]	58700

a = 70 per cent oil by weight, b = 30 per cent oil by weight.

Source: Chisti (2007).

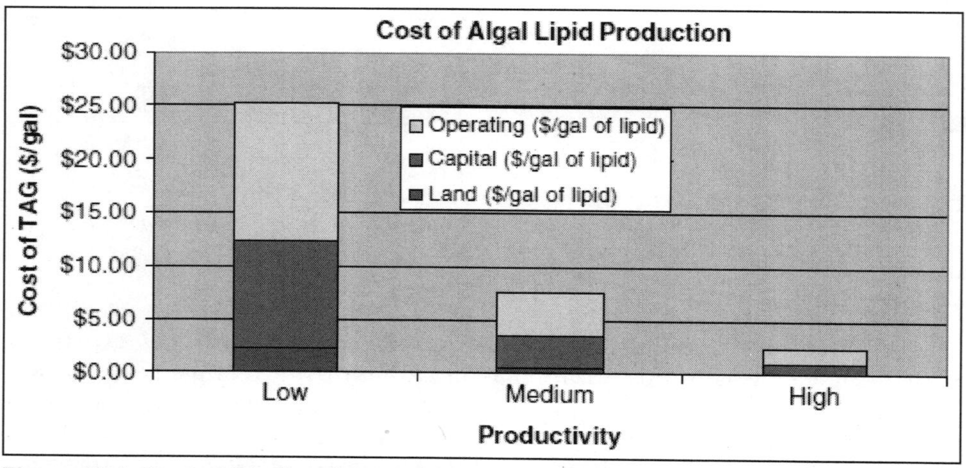

Figure 36.1: Cost of Biodiesel Production per Gallon (USD); *Source*: DOE, USA.

Though the technology for modern biofuel production systems and use has limited scope of pollution in terms of environmental accumulation of GHGs and vehicular exhaust, the biodiesel and bioethanol both are found to be better when compared to the emission of CO_2, carbon footprint and long term emissions for SO_2, NO *etc.*

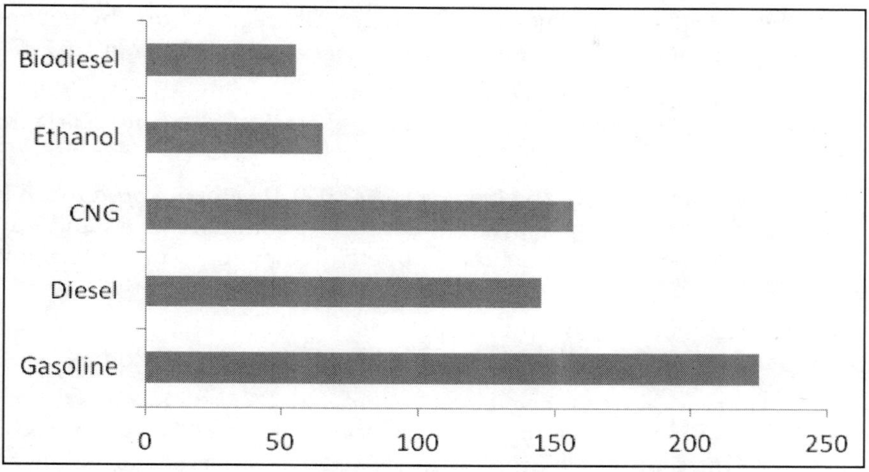

Figure 36.2: Comparative CO_2 Emission Data (g/km) for different Biofuels.
***Source*: Petroleum Planning and Analysis cell, Planning Commission (2003), Govt. of India.**

Thus the above discussions on the microalgal biofuels have already made the facts clear about the probable choices of substrates for future programmes on biofuels, there ease of use and economic advantage. The environmental effects will be a little more clear if the cost of production and the emission data are compared to per capita usage or usage over a particular period of time. Hence, it would not be proper for the skeptics and critics to predict the end of fossil fuels as the end of progress for the modern human devoid of fuel supplies. Biofuels have promising capability to meet the needs of tomorrow and at the same time these can be an eco-friendly choice if utilized with care.

References

Agwa, O.K., Nwosu I.G., and Abu, G.O. (2017). Bioethanol production from *Chlorella vulgaris* biomass cultivated with plantain (*M. paradisiaca*) peels extract. Advances in Bioscience and Biotechnology, 8: 478–490.

Ashok Kumar, V. Salam, Z. Tiwary, O.N., Chinnasamy, S., Mohammed, S. and Ani, F.N. (2015). An integrated approach for biodiesel and bioethanol production from *S.bijugatus* cultivated in a vertical tubular photobioreactor. Energy Conversion and Management, 101: 778–786.

Branco-Vieira, M. (2018). Potential of *P. tricornutum* for biodiesel production under natural condition in Chile. Energies, 154.

Chisti, Y. (2007). Biodiesel from microalgae. Biotechnology Advances, 25(3): 294- 306.

Gouveia, L., Marques, A.E., da Silva, T.L. and Reis A. (2009). *N. oleoabundans* utex 1185: a suitable renewable lipid source for biofuel production. Journal of Industrial Microbiology and Biotechnology, 36(6): 821- 826.

Kim, K.H., Choi, I.S., Kim, H.M., Wi, S.G. and Bae, H. (2014) Bioethanol production from the nutrient stress–induced microalga *Chlorella vulgaris* by enzymatic hydrolysis and immobilized fermentation. Bioresource Technology, 153: 47 – 54.

Lee, O.K., Oh, Y. and Lee, E.Y. (2015). Bioethanol production from carbohydrate enriched residual biomass obtained after lipid extraction of Chlorella sp.KR-1. Bioresource Technology, 196: 22- 27.

Miranda, J.R., Passarinho, P.C. and Gouveia, L. (2012). Bioethanol production from *Scenedesmus obliquus* sugars: the influence of photobioreactors and culture conditions on biomass production. Applied Microbiology and Biotechnology, 96(2): 555-564.

Mousdale, D.M. (2008). Biofuels-Biotechnology, Chemistry and Sustainable development. CRC Press, Taylor and Francis Group, Boca Raton, FL, USA.

Sukarni, S., Hamidi, N., Yanuhar, U. and Wardana, I.N.G. (2014). Potential and properties of marine microalgae *Nannochloropsis oculata* as biomass fuel feedstock. International Journal of Energy and Environmental Engineering. 5: 279- 290.

37

Antimicrobial Activity of Certain Marine Cyanobacteria from Talasari Coast, Odisha

*S. Dash, S. Parida and B. Rath**

Department of Biotechnology,
North Orissa University, Takatpur, Baripada – 757 003, Odisha
**E-mail: brath_2000@yahoo.com*

ABSTRACT

The present study was conducted to evaluate the antimicrobial activity of four species of cyanobacteria (Oscillatoria sancta, Oscillatoria geminata, Lyngbya birgei, Lyngbya aestuarii) isolated from Talasari, (Northeast coast of Odisha). The cyanobacterial extract was prepared using three solvents such as methanol, ethanol and hexane. These species were tested against four pathogenic bacteria (Bacillus subtilis MTCC-441, Escherichia coli MTCC-443, Staphylococcus aureus MTCC-96, Vibrio cholerae MTCC-3906) using agar well diffusion assay method. As per the findings, all the cyanobaterial isolates exhibited antimicrobial activity which varied according to different solvent extraction. Among the four species, ethanolic extract of O. geminata exhibited maximum inhibitory activity with an inhibitory zone size of 20 (±2) mm against B. subtilis and S. aureus followed by E. coli (18±1mm) and V. cholerae (15±1) mm) . The methanolic extract exhibited marginally less antimicrobial activity than ethanolic extract, where as hexane exhibited less activity compare to ethanol and methanol extract. The results showed that marine cyanobacterial species obtained from Odisha coast have immense potential of antibacterial compounds. Further research is required for the isolation and identification of these compounds.

***Keywords:** Antibacterial, Bioactive compounds, Cyanobacteria, Marine, Phytochemicals.*

Introduction

Cyanobacteria are an assemblage of Gram-Negative eubacteria. They are structurally diverse and widely distributed throughout the world and are later known as blue green algae. They are prokaryotic organisms capable of oxygenic photosynthesis (Ely *et al.*, 2003; Krishnaraju *et al.*, 2015). Cyanobacteria are very resistant to extreme environmental conditions because of their primary and secondary metabolites. They appeared to be a rich source of many useful products and are known to produce a number of bioactive compounds (Carmichael *et al.*, 2001; Codd *et al.*, 1997). Secondary metabolites are natural organic compounds which are diverse, having low molecular weight. These are not essential for normal growth and development of an organism but, they perform very potent biological activities like helping in defensive mechanisms against stress.

During the last 20 years, renewed interest has emerged to help in developing safer antimicrobial drugs from the natural sources. Cyanobacteria have been considered a rich source of secondary metabolites with potential biotechnological applications in the pharmacological field (Haripriya *et al.*, 2010; Alghazeer *et al.*, 2013; Lutzu *et al.*, 2016). Lately, production of bioactive compounds with commercial and medical applications has also increased interest in studying these organisms (Tan, 2007; Davies and Ryan, 2012). In fact, together with the production of potent toxins, cyanobacteria produce many substances that are interesting in terms of their antifungal, antibacterial, antiviral, anti-HIV (human immunodeficiency virus), anticoagulant, anti-inflammatory, antimalarial, antiprotozoal, antitumor and anticancerous activities (Dittmann *et al.*, 2007; Volk,2008). In recent years, scientists from different parts of the world have discovered various drugs for the treatment of such diseases. The chemical and biological diversity of the marine environment is immeasurable, making it an extraordinary resource for the discovery of new antibiotic drugs from cyanobacteria (Selvamaleeswaran *et al.*, 2010; Raghavendra *et al.*, 2006). Several work has been carried out on bioprospecting marine cyanobacteria across Indian coastal region but no such significant report is available from north east coast of India. Keeping this into consideration work has been initiated for collection, isolation and culture of different species of cynobacteria across Odisha coast and investigating their biotechnological potential with special emphasis on antimicrobial and antioxidant activities.

Materials and Methods

Sample Collection

The Cyanobacterial species (*Oscillatoria sancta, Oscillatoria geminata, Lyngbya birgei, Lyngbya aestuari*) were collected from the north-east coast of Odisha *i.e.*, Talasari of district Balasore (Figure 37.1). The samples were cleaned and necrotic parts were removed in the laboratory. Then the samples were rinsed with sterile water to remove any associated debris and cultured in ASN III medium. Axenic cultures were maintained under fluorescent light at a intensity of 2000-3000 lux with a photoperiod of 16hrs light and 8hrs dark and 25 ± 1^0C temperature. The cultures were hand shaken daily 2-3 times to provide uniform nutrient to the growing cells. Subsequently the organisms were dried for 24 – 48 h at room temperature and were crushed to obtain powered form.

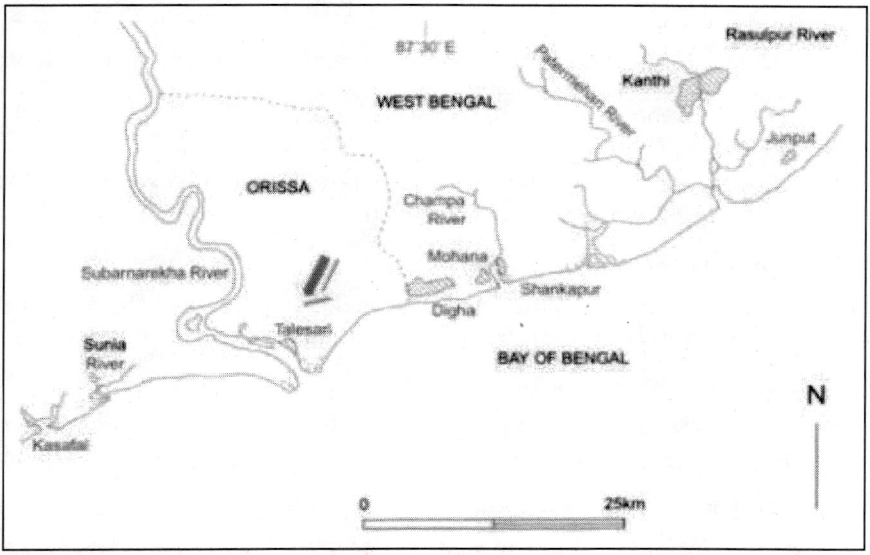

Figure 37.1: Map of Odisha Coast Showing Talasari Beach, Balasore District for Sample Collection.

Extract Preparation

The extraction of powdered Cyanobacterial samples was carried out using three solvents (methanol, ehanol and hexane). Aliquots of 0.5g of the powdered Cyanobacterial samples were added to 5 ml of solvents and kept the mixture in a shaker for 24 h. Then the mixture was poured in a petridish and was air dried initially for 24 h and then in the oven at 60⁰C for 2 h. The extracts so obtained was taken out and mixed with DMSO and then were assayed for the antimicrobial activity.

Antimicrobial Assay

Antimicrobial study was carried out by agar well diffusion method (Abedin and Taha 2008) against the four pathogenic bacteria such as *Bacillus subtilis* MTCC-441, *Escherichia coli* MTCC-443, *Staphylococcus aureus* MTCC-96 and *Vibrio cholerae* MTCC-3906. The respective bacterial culture was poured into the nutrient Mueller Hinton agar plates for uniform distribution of microorganisms. Using sterile cork borer, 8 mm wide well was made on each plate and filled with 100 µl cyanobacterial extract. The plates with bacteria were incubated at 37⁰C for 24 h. After incubation, the inhibition zones formed around the holes were measured and the results were recorded in mm (Attaie *et al.*, 1987).

Phytochemical Screening

Phytochemical analysis of the cyanobacterial extract was carried out using chemical method and tested for the presence of several phytochemicals like Phenolic compound, Flavonoids, Saponins, Alkaloids, Glycosides, Tannins, Steroids, Anthocyanin, Quinones and Resins (Dhanalakshmi *et al.*, 2013).

Test for Phenolic Compounds

In extract (1 ml) few drops of Hydrochloric acid was added. Yellowish brown color indicated the presence of Phenol compounds.

Test for Flavonoid (Ferric Chloride Test)

Few drops of Ferric chloride was added to the extracts and formation of brown precipitate indicates the presence of flavanoids.

Test for Saponins

The extract was taken in a tube and was shaken vigorously. The formation of stable foam was taken as an indication for the presence of saponins.

Test for Alkaloids (Wagner's Test)

Few drops of Wagner's reagent was added in the test tube containing the extract. The formation of reddish-brown precipitate indicates the presence of alkaloids.

Test for Glycoside

1 ml of concentrated sulph**uric acid** was added to the extract. The formation of reddish brown colour indicates the presence of glycosides.

Test for Tannins (Ferric Chloride Test)

Few drops of ferric chloride solution was added to the extract. The brackish precipitate showed the presence of Tannins.

Test for Steroids

To the extract few drops of acetic acid was added and then few drops of concentrated sulfuric acid was added to the mixture. Formation of reddish brown colour indicates the presence of steroids.

Test for Anthocyanin

A small amount of extract was treated with 2 ml of NaOH and observed. Blue green colour indicates the presence of anthocyanin.

Test for Quinones

To 1 ml of the extract alcoholic KOH solution was added separately. The change of colour from red to blue showed the presence of quinones.

Test for Resins

To 1 ml of the extract 2-3 ml of copper sulphate solution was added and the contents were mixed for 2 minute and then solution was allowed to separate. Green colour precipitate showed the presence of resin.

Results

In the present study, four species of cyanobacteria *viz., O. sancta, O. geminata, L. birgei,* and *L. aestuarii* isolated from the Talasari area situated at Northeast coast of Odisha were processed to find out the antibacterial potency of various solvent

Table 37.1: Phytochemical Screening of the Extract of Cyanobacteria from Talasari Coast, Odisha

Species	Solvent	Alkaloid	Glycoside	Tannin	Flavonoid	Steroid	Saponin	Resin	Quinones	Anthro-cyanin	Phenol
Oscillatoria sancta	Methanol	-	+	-	+	+	+	+	-	-	+
	Ethanol	+	+	-	+	+	+	+	+	+	+
	Hexane	+	-	+	+	+	+	+	+	+	+
Oscillatoria geminata	Methanol	+	+	-	+	+	+	+	-	-	+
	Ethanol	+	+	-	+	+	+	+	+	+	+
	Hexane	-	-	-	+	+	+	+	-	+	+
Lyngbya birgei	Methanol	+	+	-	+	+	+	+	-	-	+
	Ethanol	+	+	-	+	+	+	+	+	+	+
	Hexane	+	+	-	+	+	+	+	-	-	+
Lyngbya aestuarii	Methanol	+	-	+	-	-	+	+	-	-	+
	Ethanol	+	-	+	-	+	+	+	-	+	+
	Hexane	+	-	+	-	-	+	+	-	-	+

extracted compounds from these organisms. In cyanobacteria like *O. sancta, O. geminata* and *L. birgei,* various phytoconstituents like alkaloid, glycoside, steroid, saponin, flavonoid, resin, quinone, anthocyanin and phenol were found to be present. On the contrary, phytoconstituents like glycoside, flavonoid, quinone, were absent in all the extracts of *L. aestuarii* (Table 37.1). Whereas Tannin is absent in all the extract of three cyanobacteria except *L. aestuarii.*

In this study irrespective of cyanobacterial isolates, the methanolic, ethanolic and hexane extracts were found to show antibacterial activity against all test pathogens. However, the zone of inhibition varied among the cyanobacterial species, type of the solvent used and the test pathogens (Table 37.2).

Table 37.2: Antibacterial Activity of the Extract of Cyanobacteria from Talasari Coast, Odisha

Cyanobacterial Species	*Solvents*	*Inhibition Zone (Diameter size ±SD in mm)*			
		Bacillus subtilis	*Escherichia coli*	*Staphylococcus aureus*	*Vibrio cholerae*
Oscillatoria sancta	Methanol	16±1	14±1	17±2	12±1
	Ethanol	17±2	16±2	19±1	14±1
	Hexane	13±1	12±1	13±1	10±1
Oscillatoria geminata	Methanol	18±2	15±1	18±2	13±1
	Ethanol	20±2	18±2	20±2	15±2
	Hexane	13±1	11±1	14±1	11±1
Lyngbya birgei	Methanol	15±2	13±1	16±1	13±1
	Ethanol	17±2	15±2	18±2	15±2
	Hexane	14±1	12±1	11±1	10±1
Lyngbya aestuarii	Methanol	14±2	15±1	14±1	13±2
	Ethanol	15±2	18±2	17±2	10±1
	Hexane	10±1	12±1	10±1	10±1

The metabolic extract of *O. sancta* showed the antibacterial activity against three pathogens *viz.,* S. *aureus, B. subtilis* and *E. coli* with the inhibition zone size of 17mm,16mm and 14mm, respectively. On the other hand, the ethanolic extract inhibited all the four tested pathogens *i.e., S. aureus, B. subtilis, V. cholare* and *E. coli* with inhibition zone 16mm 19mm,17mm and 14mm respectively. The hexane extracts of *O. sancta* exhibited the antibacterial activity against two pathogens *i.e., S. aureus* and *B. subtilis* with inhibition zone of 13 mm each.

The methanolic extract of *O. geminata* exhibited the antibacterial activity against all the four pathogens *i.e. B. subtilis, E. coli, S. aureus* and *V. cholerae* with the zone of inhibition 18mm,15mm,18mm and 13mm, respectively. The ethanolic extracts of *O. geminata* exhibited the antibacterial activity against all the four pathogens *i.e. B. subtilis, E. coli, S. aureus* and V. *cholerae* with the zone of inhibition 20mm, 18mm, 20mm and 15mm respectively. The hexane extract of *O. geminata* inhibited S. *aureus* and *B. subtilis* only with a zone size of 14 mm and 13mm, respectively.

The methanolic extract of *L. birgei* show antibacterial activity against two pathogens *i.e., S. aureus, and B. subtilis* with the zone of inhibition 16mm and 15mm, respectively whereas the ethanolic extract of *L. birgei* inhibited all the test pathogens with a zone size of 18mm, 15mm, 17mm and 15mm against *S. aureus, E. coli, B. subtilis,* and *V. cholerae,* respectively. The hexane extract of *L. birgei* inhibited one pathogen *i.e. B. subtilis* while failed to inhibit rest of the pathogens. The methanolic extract of *L. aestuarii* exhibited the antibacterial activity against three pathogens *i.e. S. aureus, B. subtilis* and *E. coli* with the zone of inhibition 14mm,14mm and 15mm, respectively. The ethanolic extract of *L. aestuarii* exhibited the antibacterial activity against three pathogens *i.e. S. aureus,B. subtilis* and *E. coli* with the zone of inhibition 17mm, 15mm, and 18mm, respectively. The hexane extract exhibited less activity irrespective of pathogens.

Discussion

Microorganisms from marine environment are promising as proving as an excellent resources for the innovation of active secondary metabolites. Among them, marine cyanobacteria are important recourses for novel antibiotics. In the study, four marine cyanobacteria (*O. sancta, O. geminata, L. birgei* and *L. aestuarii*) isolated from Talasari coast, Balasore district, Odisha and their solvents (methanol, ethanol and hexane) extracts were tested against pathogens. The results revealed that the marine cyanobacteria are the excellent sources for antimicrobial property. All the extracts especially methanolic and ethanolic were found to inhibit pathogens whereas the hexane extract of four cyanobacterial species exhibited less antibacterial activity. The ethanolic extract of *O.geminata* exhibited higher antibacterial activity against *B.subtilis* and *S. aureus* followed by *E. coli.*

The result also proved that ethanol was the best solvent for extracting the antibacterial agents from *O.geminata* followed by methanol. Similar types of results were also demonstrated by earlier researchers (Tuney *et al.,* 2006, Moreau *et al.*1988). Furthermore, various solvent extracts of *Oscillatoria, Lyngbya* and *Phormidium* exhibited promising antimicrobial activity against bacterial pathogens (Ozdemir *et al.,* 2004; Sethubati and Prabu 2010).

Conclusion

The results of this present study on marine cyanobacteria using three different solvent extract against four different pathogens showed significant antimicrobial activity. However, further more in-depth research is required to find out the active ingredients responsible for such activity which in turn can be explored for their pharmaceutical applications.

References

Abedin, R.M.A. and Taha H.M. (2008) Antibacterial and antifungal activity of cyanobacteria and green microalgae. Global Journal of Biotechnology and Biochemistry, 3: 22-31.

Alghazeer, R., Whida, F., Abduelrhman, E., Gammoudi F. and Naili, M. (2013). *In vitro* antibacterial activity of alkaloid extracts from green, red and brown macroalgae from western coast of Libya. African Journal of Biotechnology, 12(51): 7086-7091.

Attaie, R. J., K.M. Whalen, Shahani and M.A. Arner (1987) Inhibition of growth of *S. aureus* during production of *cidophilus* yogurt. Journal of Food Protection, 50: 224- 228.

Davies, J. and Ryan, K.S. (2012). Introducing the parvome: bioactive compounds in the microbial world. ACS Chemical Biology, 7: 252-259.

Dhanalakshmi, M. and Angayarkanni, J. (2013). Phytochemistry and antibacterial activity of Chlorosarcinopsis species. International Journal of Scientific and Technology Research, 4 (10); 315-321.

Elsie, H.B. and DhanaRajan, M.S. (2010). Evaluation of antimicrobial activity and phytochemical screening of *Gelidium acerosa*. Journal of Pharmaceutical Sciences, 2(11): 704-707.

Ely, R., Supriya, T.and Naik, C.G. (2004). Antimicrobial activity of marine organisms collected off the coast of South East India. Journal of Experimental Marine Biology and Ecology, 309: 121-127.

Engene, N., Gunasekera, S., Gerwick, W.H. and Paul, V.J. (2013) Phylogenetic inferences reveal large extent of novel biodiversity in chemically rich tropical marine cyanobacteria. Applied and Environomental Microbiology, 79: 1882-1888.

García-Vilas, J. A., Martínez-Poveda, B., Quesada, A.R. and Medina, M. Á. (2015) Aeroplysinin-1, a sponge-derived multi-targeted bioactive marine drug. Marine Drugs, 14(1): https: //doi.org/10.3390/md14010001.

Haripriya, D., Selvan, N., Jeyakumar, N., Periasamy, R.S., Johnson, M. and Irudayaraj, V. (2010) The effect of extracts of *Selaginella involvens* and *Selaginella inaequalifolia* leaves on poultry pathogens. Asian Pacific Journal of Tropical Medicine, 3(9): 678-681.

Kannan, R.R.R., Arumugam, R.and Anantharaman, P. (2010). *In vitro* antioxidant activities of ethanol extract from *Enhalus acoroides* (L.F.) Royle. Asian Pacific Journal of Tropical Medicine, 3(11): 898-901.

Kinsalin, V.A., Prabhadevi, D., Duraipandiyan, V. and Ignacimuthu, S. (2014) Isolation, characterization and antimicrobial properties of marine bacteria and fungi from the surface of *Cyanobacterium Lyngbya* sp. International Journal of Pharmaceutical and Clinical Science, 4(1): 4-10.

Krishnaraju, A.V., Rao, T.V.N., Sundararaju, D., Vanisree, M., Tsay, H.S.and Subbaraju, G.V. (2005) Assessment of bioactivity of Indian medicinal plants using brine shrimp (*Artemia salina*) lethality assay. International Journal of Applied Sciences and Engineering, 2: 125-134.

Lutzu, G. A. and Parsaeimehr, A. (2016) Exploitation of algae as priceless source of potential commercial high-value compounds, In V. K. Gupta (Ed.), Daya Publishing House, pp 311.

Marhivanan, K., Ramamurthy, V., and Rajaram, R. (2010) Antimicrobial activity of *Oscillatoria princes* and *Lyngbya majuscule* against pathogenic microbes. International Journal of Current Research, 5: 97-101.

Moreau, J., D., Pesandoand, P.and Bernad, (1988). Seasonal variation in the production of antifungal substances by some Dictyotales (brown algae) from French Mediterranean coast. Hydrobiologyia, 162: 157-162.

Ozdemir, G., N. Karabay, M. Dolay and B. Pazarba (2010). Antibacterial activity of volatile extracts of *Spirulina platensis*. Phytotherapy Research, 18(9): 754-757.

Raghavendra, M.P, Satish, S.and Raveesha, K.A. (2006) Phytochemical analysis and antibacterial activity of *Oxalis corniculata*, a known medicinal plant. myScience, 1(1): 72-78.

Renu, A. (2010) Antibacterial activities of freshwater algae *Chlorella ellipsoidea.* International Journal of Basic and Applied Biology, 4(1 and 2): 22-26.

Sangeetha, S., Dhayanithi, N.B.and Sivakumar, N. (2014) Antibacterial activity of Sargassum longifolium and *Gracilaria corticata* from Gulf of Mannar against selected common shrimp pathogens. International Journal of Pharmaceutical and Biological Sciences, 5(2): 76-82.

Selvamaleeswaran, P., Wesely, E.G., Johnson, M., Velusamy, S.and Jeyakumar, N. (2010) The effect of leaves extracts of *Clitoria ternatea* Linn against the fish pathogens. Asian Pacific Journal of Tropical Medicine, 3(9): 723-726.

Sethubati, G V. B. and Prabu, V. A. (2010). Antibacterial activity of cyanobacterial species from Adirampattinam Coast, South East Coast of Palk Bay. Current Research Journal of Biological Sciences, 2(1): 24-26.

Singh, A.P.and Chaudhary, B.K. (2010) Preliminary phytochemical analysis and *in vitro* anti-bacterial activity screening of *Pithosphora oedogonia* (Mont.) Wittrock-a fresh water green alga forming mats in the water bodies. Journal of Algal Biomass and Utilisation, 1(2): 33-41.

Tuney, I., B.Carditci, Dunal and Sukatar, A. (2006).Antimicrobial activities of the extracts of marine algae from the coast of Ural (Izmir, Turkey).Turkish Journal of Biology, 30: 171-175.

38

Climate Change, Methane Emission from Rice Fields and Possible Mitigation Options

A.K. Rath[1]* *and S.R. Mishra*[2]

[1]*Department of Botany, Nayagarh Autonomous College, Nayagarh – 752 069, Odisha*

[2]*Department of Botany, Ramadevi Women's College, Bhubaneswar – 751 022, Odisha*

**E-mail: arunk_rath@yahoo.co.in*

ABSTRACT

Global warming is one of the current environmental issues and cause of climate change. Due to release of greenhouse gases in abundance, the equilibrium has been tilted and the global temperature has started increasing. There is no doubt that an overall warming trend on earth has been noticed in the recent years. Global temperature has been increasing at the rate of 0.3°C to 0.6°C over the last 130 years due to increasing concentration of greenhouse gases. The increasing concentrations of the trace gases such as carbon dioxide, methane and nitrous oxide in the earth's atmosphere have been a cause a great concern in the last decade for climate change, because of their potential influence on the atmospheric chemistry. The current evidence for global warming and prediction of significant climate change over the next century had prompted the world wide interest on green house gases. Worldwide researchers are currently focused on developing feasible technologies for mitigation of methane emission from flooded rice paddies. Intermittent flooding and drainage can significantly retard methane emissions. Nitrogenous fertilizer inhibited methane emission from flooded rice field (30-cm water depth.). Dual cropping of Azolla (equivalent to 30 kg N ha^{-1}) in conjunction with urea (30 kg N ha^{-1}) significantly retarded methane emission from the rice fields. Single super phosphate, ammonium thiosulphate, commercial formulation of Hexa Chloro Cyclohexane and carbofuran pesticides was found to inhibit methane emission from rice fields. The heavy metal Zn stimulated methane production in alluvial soil while Cd, Cu, Pb and Cr inhibited it.

Keywords: *Climate change, Methane emission, Methanogenic bacteria, Nitrogenous fertilizer, Sulphate-reducing bacteria.*

Introduction

The increasing concentrations of methane, a greenhouse gas implicated in global warming, in the atmosphere, is a topic of great concern. Flooded rice paddies are considered as a major anthropogenic source of atmospheric methane. Rice production is the greater agricultural priority in Asia. The demand for more rice production from dwindling land, water and other resources is increasing as the world's population is expected to reach 8.6 billion in 2025 (IRRI, 1993). The contribution of rice fields to the methane budget is about 5-15 per cent of the total emission from all sources and ranges between 25 and 60 Tg Yr^{-1} (Houghton *et al.*, 1996). The estimate of methane emission in U.S. from agricultural sources indicates increasing trend of methane emission (214.5 Tg Yr^{-1}) with declining trend of methane emission from rice fields (9.7 Tg Yr^{-1}) in 2007 (EIA, 2008). The projected increase in the intensity of rice cultivation would, however, lead to higher methane emission, as much as 20 per cent, over the next decade (Anastasi *et al.*, 1992). The concentration of methane in our Earth's atmosphere has tripled since pre-industrial times and continues to rapidly rise and the concentration of methane in the atmosphere has increased at an accelerating rate in the last decade. Worldwide research is currently focused on developing feasible farmer-friendly, eco-friendly and economically viable technologies for mitigation of methane emission from rice paddies.

Methane Budget of Indian Rice Fields

In India rice is grown under 44.5 m ha with varying water regimes. The US-EPA estimates, based on the data generated in developed countries (USA, Europe and Japan) and extrapolated to Indian rice fields, showed methane emission value of 37.8 Tg Yr^{-1} First efforts to obtain the actual estimate from Indian rice fields were made through the National Methane Campaign in 1991 at 18 locations covering different agro-climatic conditions. Based on the estimates using manual measurement system under National Methane Campaign in 1991 and at 4 centres under the ICAR AP-CESS Fund Coordinated Project with the coordinating centre at Central Rice Research Institute, Cuttack, the mean methane emission from irrigated and rainfed lowland systems in India is 4 to 6 Tg methane Yr^{-1} (Adhya *et al.*, 1994).

Possible Mitigation Options

Farmers adopt different agronomic practices in diverse rice growing areas. Various agronomic practices associated with rice growing, *viz.* land preparation, water management, direct seeding and transplanting, pest control and harvest affect methane emission. Researches worldwide have indicated that methane emission from rice fields can be reduced by combined strategies of water management, fertilizer management and use of specific inhibitors *etc.*

Water Management

The flood water regime is the single most important determinant for methane production and emission can be negligible in upland rice fields, due to lack of permanent flooding (Adhya *et al.*, 1994). On the contrary the irrigated rice fields can be a major source of methane emission due to continuous flooding (Adhya *et al.*, 1994; Yagi, 1997). Water management is an important strategy for controlling methane efflux from flooded rice paddies. Higher rate of water percolation in the paddy field soil, as in sandy soil or in soil not puddle can reduce total emission rates of methane during a rice season by 42-45 per cent (Yagi, 1997). Our results from pot experiments under greenhouse conditions planted to rice also confirm these findings (Mishra *et al.*, 1997) (Figure 38.1). The best water management method for mitigation of methane emission from irrigated rice involves a short-term drought period, created by drainage (Wassmann et *al.*, 1993) as the rice plants can withstand small period of water stress without any appreciable loss of grain yield. More frequent flooding and drying of the soil under rainfed conditions might affect the crop growth and rice yields, in addition to influencing methane emission. But, in rainfed lowland as in eastern and north- eastern India, drainage of water from rice fields is virtually impossible, due to higher water table. Our study indicates that subsurface application of urea supergranules was effective in reducing the methane flux over that of untreated control from a rainfed rice field (Table 38.1).

Table 38.1: Effect of Fertilizer Particles on Methane Efflux from Rice Field Plots with 30-cm Water Depth, Planted to Rice cv. Gayatri

Treatment	*Methane efflux in mg^{-2} h^{-1}**				
	Days after Transplanting				
	30	*70*	*100*	*125*	*140*
Control	8.26[a]	39.94[a]	62.80[ab]	102.86[a]	8.16[a]
Prilled Urea	5.70[a]	26.84[b]	71.57[a]	94.25[ab]	7.76[a]
Prilled Urea + Nimin	5.22[a]	27.13[b]	51.04[b]	77.20[a]	7.43[a]
Urea Super Granule	6.07[a]	30.72[a]	57.59[ab]	90.20[ab]	12.25[a]

*Mean of replicate observations; in a column, means followed by a common letter are not significantly different (P = 0.05) by Duncan's multiple range test.

Organic Amendment

Application of organic substrates, including green manure, often increases the methane flux from flooded rice paddy (Wassmann *et al.*, 1993). It is also reported that methane emission from green manure-amended field plots was higher than urea-amended plots (Rath *et al.*, 1999). However, the relationship between methane emission and organic input in rice fields is not clear. Application of rice straw to the flooded soil samples increased methane production significantly almost in proportion to the rate of rice straw application (Wang *et al.*, 1992). Use of composted organic materials is considered as one of the effective options for mitigating methane emission from rice fields. In a recent study, dual cropping of *Azolla* (equivalent to 30 kg N ha^{-1}) in conjunction with urea-N (30 kg N ha^{-1}) emitted less methane as compared to control (Figure 38.2). This study indicate that dual cropping of *Azolla* in conjunction with urea-N effected a negative influence on methane efflux while increasing the grain yield at per with that of urea-N application at 60 kg N ha^{-1}. This finding opens up the possibility of practicing *Azolla* dual cropping for sustaining higher yields and protecting the environment by minimizing the methane flux from flooded rice paddies.

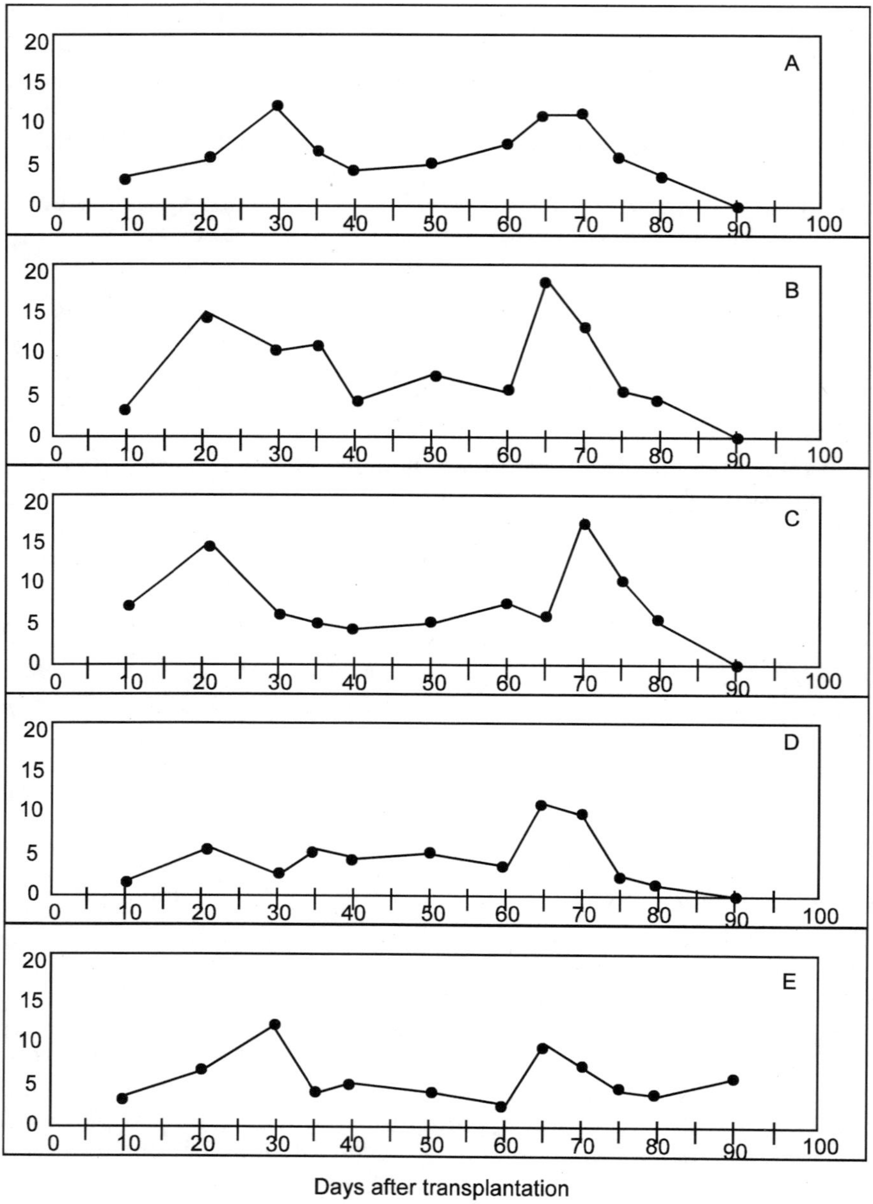

Figure 38.1: Cumulative Methane Emission from Pots Planted to Rice Plants (cv. Lalat) under different Water Regimes [A. continuously flooded B. continuously non-flooded, C. non-flooded (40 days) → flooded (40days) → non-flooded until maturity, D, Flooded (40 days) → non-flooded (40 days) → flooded until maturity, E. Flooded and non-flooded cycles alternate with each other at 20-days interval].

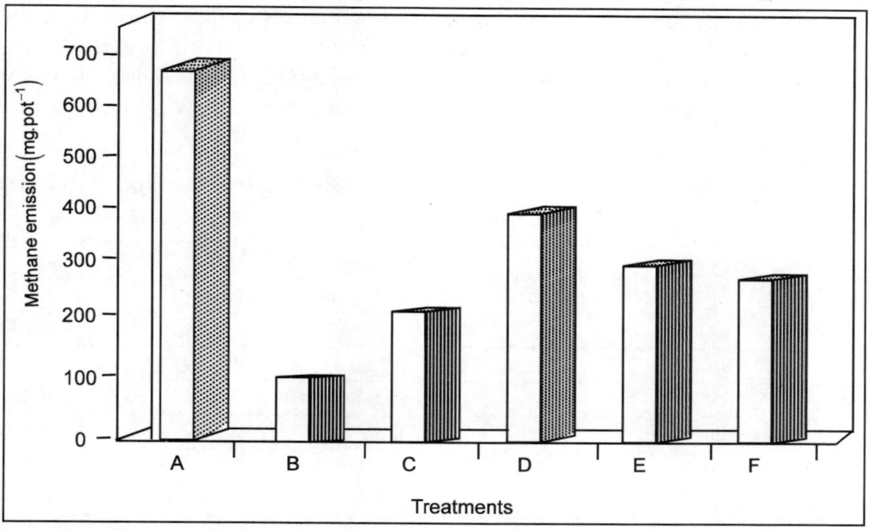

Figure 38.2: Effect of *Azolla* Urea Application on Methane Efflux from Flooded Alluvial Field Planted to Rice (cv. CR 749-20-2) [A. no N (control), B. Urea (60 kg N ha^{-1}), C. *Azolla* incorporated (30 kg N ha^{-1}) + Urea (30 kg N ha^{-1}), D. *Azolla* dual cropping (30 kg n ha^{-1}).

Fertilizer Management

The effects of different mineral fertilizers on methane flux from flooded rice paddies are not consistent and even contradictory (Minami, 1995). Urea and ammonium sulphate account for approximately 80 per cent and 6 per cent of the N fertilizer used in rice agriculture of Asia. Addition of ammonium releasing fertilizers to the flood water increase methane efflux from rice field by inhibiting methane oxidation in soil-flood water interface (Conrad and Rothfuss, 1991). Methane emission also depends on the mode of fertilizer application. Shallow application of urea however, resulted in a 50 per cent reduction in methane flux than in the fields with surface application. Our results suggest that application of prilled urea did not enhance methane emission significantly over that of untreated control in the flooded rice field (Rath *et al.*, 1999).

Unlike nitrogenous fertilizers, not much information is available on other mineral fertilizers including sulphur and phosphorus. Application of single super phosphate or Mussorie rock phosphate to flooded rice fields or to laboratory-incubated flooded soil samples effected a distinct inhibition on the production and emission of methane. Application of P as K_2HPO_4 increased methane production in P-deficit soil (Khuntuni), but not in P-normal soil (CRRI, alluvial), under flooded conditions. But addition of P as single super phosphate (SSP) or as rock phosphate (Mussorie rock phosphate) led to a distinct inhibition of methane production in laboratory studies (Table 38.2) and methane emission from rice fields. High sulphate content in SSP and rock phosphate was responsible for this inhibition (Adhya *et al.*, 1998). Application of P as SSP at 100 μg g^{-1} soil inhibited methane production distinctly in flooded alluvial rice soil, in the absence more than in the presence of rice straw (Rath *et al.*, 2005). Application of ammonium thiosulphate to laboratory incubated flooded soil (30 and 60 µg g^{-1} soil) and flooded rice fields (45.6 and 60 kg N ha^{-1}) effected a distinct inhibition of methane production and emission (Rath *et al.*, 2002).

Table 38.2: Phosphorus Sources and Methane Production in a Phosphorus Deficient (Khuntuni) Soil

Treatment	*Methane Efflux in* mg^{-2} h^{-1*}			
	Days after Transplanting			
	10	*20*	*30*	*40*
Control	159[a]	13.35[b]	335.40[bc]	2554.93[ab]
K_2HPO_4	155[a]	13.62[b]	994.76[a]	6425.98[a]
Single super phosphate	158[a]	6.55[b]	10.95[d]	4.30[c]
Mussorie rock phosphate	154[a]	48.21[a]	177.40[c]	1069.74[b]
Jordan rock phosphate	165[a]	50.82[a]	660.17[b]	6430.13[a]

* Mean of replicate observations; P was added at 100 µg g^{-1} soil; In a column followed by a common letter are not significantly different (P = 0.05) by Duncan's multiple range test.

Cropping System

Continuous submergence of rice fields is practiced in some tropical areas where water availability and favorable climate allow the cultivation of two or even three crops of rice in a year. Under these conditions, the soil remains constantly reduced leading to higher methane emission. On the contrary, growing a dry land crop can cause sufficient aeration of the soil and increase the redox potential (Neue, 1993), that in turn, may reduce methane emission. In a study on methane efflux from rice/ wheat agro-ecosystem, methane was shown to be consumed by soil during the dry land wheat crop and fallow period when soil were relatively dry and oxic, conditions that were favorable for methane oxidation (Singh *et al.*, 1997). Our study on rice based cropping system also indicates that methane emission is very low in other dry land crops except rice (Figure 38.3) (Adhya et *al.*, 1999). Thus crop diversification can be considered as feasible option to reduce total methane emission and increase production as well as income.

Chemical Inhibitors Including Pesticides and Heavy Metals

Metabolic inhibitors can also inhibit methane production and thereby reduce methane emissions from rice fields. Nitrification inhibitors, dicyandiamide and nimin are reported to inhibit methane in soils (Adhya *et al.*, 2000). Application of calcium carbide is an effective of mitigating methane emission from rice field without any adverse effect on rice grain yield (Banerjee *et al.*, 1990).

A few commonly used agricultural chemicals have also been shown to affect methane emission from rice soils. Application of commercial formulation of hexachlorocyclohexane (HCH) to flooded fields planted to rice significantly inhibited methane production (Satpathy *et al.*, 1997). Similarly, commercial formulation of

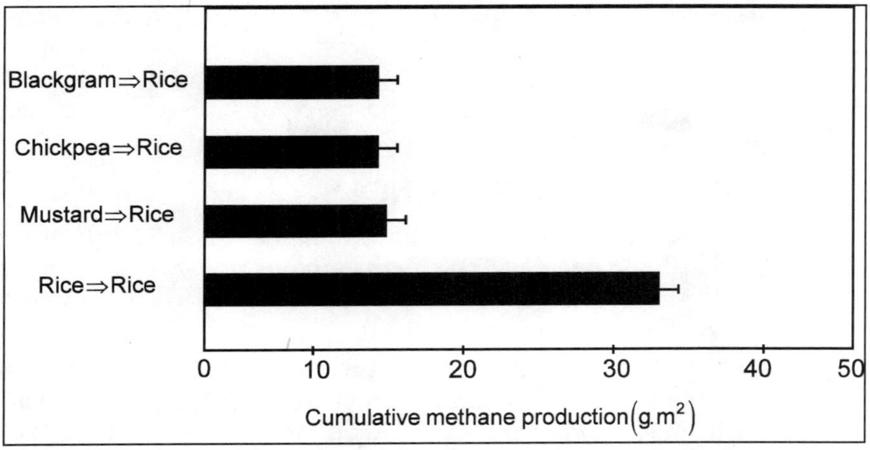

Figure 38.3: Cumulative Methane Emission Under different Crop Sequences (A. Blackgram → Rice, B. Chickpea → Rice, C Mustard → Rice, D Rice → Rice).

carbofuran, applied at 2 kg and 12 kg to a flooded field planted to rice, resulted in significant inhibition of methane emission (Kumaraswamy *et al.*, 1998). Heavy metals in soil affect the microbial populations and their activities. In a laboratory incubation study the effect of heavy metals on methane production in alluvial soil was investigated under flooded conditions. Cd, Cu, Pb and Cr inhibited methane production whereas Zn stimulated it (Table 38.3). The stimulatory effect of Zn and the inhibitory effect of Cd, Cu, Pb and Cr on methanogesis in alluvial soil were attributed to their respective stimulation or inhibition of methanogenic bacteria (Mishra *et al.*, 1999).

Table 38.3: Methane Production in a Flooded Alluvial Soil Amended with Heavy Metals

Treatment	*Methane Efflux in mg^{-2} h^{-1*}*			
	Days after Transplanting			
	10	*20*	*30*	*40*
Control	171.4[a]	447.6[ab]	5969.0[a]	22191.4[b]
Zn	67.7[a]	404.7[a]	7269.0[a]	33133.1[a]
Cu	66.9[a]	223.7[a]	5412.6[a]	15377.4[c]
Cr	39.3[a]	235.6[a]	469.1[b]	262.5[e]
Pb	111.1[a]	472.0[a]	3980.8[ab]	10543.6[d]
Cd	79.3[a]	463.7[a]	4040.5[ab]	9971.7[d]

*The values are means of five replicates; Heavy metals were amended at 20 μg g^{-1} soil; In a column, means followed by common letter are not significantly different (P=0.05) by Duncan's multiple range test.

Theoretically, it would be possible to reduce agricultural emissions of almost all the climatically relevant trace gases in a number of ways. However, the practicability of most of such measures appears doubtful in view of the new problems they may evoke and uncertainly as to their effectiveness. But, certain cultural practices with minor modifications can serve as viable options for mitigating methane emission from rice fields. Better understanding of the array of factors controlling methane emission and verifying the mitigation options at field sites throughout the world would be necessary to establish effective and practical strategies for mitigating methane emission from rice cultivation.

REFERENCES

Adhya, T.K., Mishra, S.R., Rath, A.K., Bharati K., Mohanty, S., Ramakrishnan,B., Rao, V.R and Sethunathan, N. (2000). Methane efflux from rice-based cropping system under humid tropical conditions of eastern India. Agricultural Ecosystem and Environment, 79: 85-80.

Adhya, T.K., Pattnaik, P. Satpathy, S.N., Kumaraswamy, S. and Sethunathan, N. (1998). Influence of phosphorus application on methane emission and production in flooded paddy soils. Soil Biology and Biochemistry, 30: 177- 181.

Adhya, T.K., Rath, A. K., Gupta P.K., Rao, V.R., Das S.N., Pairda, K.M., Parashar, D.C. and Senhunathan, N. (1994). Methane emission from flooded rice fields under irrigated conditions. Biological Fertile Soils, 18: 245-248,

Anastasi, C., Downding, M. and Simpson, V.J. (1992). Future CH_4 emission from rice production. Journal of Geophysics Research, 97: 7521-7525.

Banerjee, N.K., Mosier, A.R., Uppal, K.S and Goswami, N.N (1990). Use of encapsulated calcium carbide to reduce denitrification losses from urea-fertilized flooded soils. Mittle Deutsche Bodendkundl Gesellsch, 60: 245-248.

Bharati, K., Mohanty, S.R., Singh, D.P., Rao, V.R. and Adhya, T.K. (2000) Influence of incorporation or dual cropping of *Azolla* on methane emission from a flooded alluvial soil planted to rice in Eastern India. Agricultural Ecosystem and Environment, 79: 73-83.

Conrad, R. and Rothfuses, F. (1991). Methane oxidation in the soil surface layer of a flooded rice field and effect of ammonium. Biological Fertile Soils, 12: 28-32.

EIA, Emission of greenhouse gases. (2008). In U.S. 2007, DOE/EIA-0573 Report, Washington, DC.

Houghton, J.T., Meria, Filho, L.G., Calender, B.A., Harris N., Katterberg, A and Maskell, K. (1996). IPCC report on Climat Change: The Science on climate change: WGI contribution to the IPCC second assessment report on methane emission from rice cultivation. Cambridge University Press.

IRRI, (1993). Programme report for 1992. IRRI, Los Banos, Philippines.

Kumaraswamy, S., Rath, A.K., Satpathy, S.N., Ramakrishnan, B. Adhya, T.K. and Sethunathan, N. (1998). Spatial distribution of methane oxidizing activity in a flooded tropical rice soil. Biological Fertile Soils, 26: 362-366.

Minami, K. (1995). The effect of nitrogen fertilizer use and other practices on methane emission from flooded rice. Fertile Research, 40: 71-84.

Mishra, S.R., Bharati, K., Sethunathan, N. and Adhya, T.K (1999). Effects of heavy metals on methane production in tropical rice soils. Ecotoxicol. Environtal Safety, 44: 129-136.

Mishra, S.R., Rath, A.K., Adhya, T.K., V.R. and Sethunathan, N. (1997). Effect of continuous and alternate water regimes on methane efflux from rice under greenhouse conditions. Biological Fertile Soils, 24: 399-405

Neue, H.U. (1993). Methane emission from rice fields. BioScience, 43: 466-474.

Rath, A. K., Ramakrishnan and Sethunathan, N. (2002). Effect of application of ammonium thiosulphate on the production and emission of methane in a tropical rice soil. Agricultural Ecosystem and Environment, 90: 319-325.

Rath, A.K., Ramakrishnan and Sethunathan, N. (2005). Effects of rice straw and phosphorus application on production and emission of methane from tropical rice soil. Journal of Plant Nutrition and Soil Science, 248-254.

Rath, A.K., Swain, B., Panda, D., Ramakrishnan, B., Adhya T.K., Rao, V.R. and Sethunathan, N (1999). Influence of fertilizer management and water regime on methane emission from rice fields. Agricultural Ecosystem and Environment, 76: 99-107.

Satpathy, S.N., Rath, A.K., Mishra, S.R., Kumaraswamy, S., Ramkrishnan, B., Adhya. T.K and Sethunathan, N. (1997). Effect of hexachlorocyclohexane on methane production and emission from flooded rice soils. Chemosphere, 34: 2663-2671.

Singh, J.S., Singh, Smita, Raghubanshi,, Singh, A.S. Saranath, Kashyap, A. K. and Reddy D. V. S. (1997). Effect of soil nitrogen, carbon and moisture on methane uptake by dry tropical forest soils. Plant Soil 196: 115-121.

Wang, Z.P., Delaune, R.D., Lindau, C.W. and Patrick, Jr: W.H. (1992). Methane production from anaerobic soil amended with rice straw and nitrogen fertilizers. Fertile Research, 33: 115-121.

Wassman, R., Papen, H., Rennenberg, H. (1993). Methane emission from rice paddies and possible mitigation strategies. Chemosphere, 26: 201-217.

Index

A

16S rRNA Gene Sequencing 361
ABTS radical scavenging study 34
Acid protease 331-334, 340
Actinobacterial population 369
Aerobic denitrifying bacteria 321
Aerobic heterotrophs 369
Aeromonas hydrophila 343, 345
Agricultural vulnerability 3
Alkaline reagent test 44
Alkali pretreatment 110, 112
Allium cepa 240
Ammonium fibre explosion 111
Amplified fragment length polymorphism (AFLP) 12, 13, 14, 18, 19, 20
Anabus testudinis 345
Animal biotechnology 189
ANOVA 26, 27, 169, 251, 253, 255, 270-272, 314
Antheraea mylitta 167, 168, 177, 219-223
Antibacterial activity 47, 48, 293, 304, 345, 349, 350, 385, 386, 388, 404
Antibiotics resistant 349
Anticancer properties 214, 225, 229-234
Anti-diabetic 279, 283
Antidiabetic studies 65
Antifungal 56, 68, 84, 163, 274, 279, 373, 375, 377, 383, 386, 400, 405, 407
Anti-fungal assay 377
Antigen-presenting cells 79
Antimicrobial activity 43, 46, 78, 302, 348, 389, 399
Antioxidant activities 36
Antioxidant studies 34
Anti-proliferative 239
Aquaculture production 134, 343, 353
Aquatic ferns 181
Arthrospira platensis 257, 258
Asparagus recemosus 144
Avicennia officinalis 31, 32, 39, 65, 70, 74
Azadirachta indica 54, 56, 57, 59, 60, 155, 161

B

Bacillus megaterium 355, 363, 364
Bacterial isolate 332, 333, 335
Bacterial polar lipid extract 309
Bambara groundnut 18
Barehipani water fall 385
Bioactive compounds 399

Biochemical identification 360
Biodiesel 391, 393, 395, 396, 398
Bioethanol 105, 393, 395, 397, 398
Bioethanol production 105
Bioinformatics 85, 121, 131
Biopesticide 267, 275
Bioprospecting evaluation 51
Bioremediaton 355
Biotechnological exploitation 295, 302
Biotechnological tools 219
Biotechnology intervention 219, 221
Bridelia retusa 267, 269, 270, 273
Brinjal 267-269
Buchanania lanzan 145, 147, 267-275
Butea monosperma 54-60, 155, 161

C

Cancer cell lines 239, 242
Cancer rherapy 81, 236
Candida albicans 302, 373, 374, 378, 381, 383, 384
Carassius auratus 257, 258, 264, 265
Carbohydrate metabolism enzyme inhibition activity 38
Carboxy fluorescene 310
Careya arborea 23-27, 30
Catalase 249, 251, 253, 334, 338, 339
Catla catla 133, 135, 345
Cell morphology 242
Chemopreventive agents 236
Chemotherapy drugs 244
Chickpea 18, 415
Chloroform extract 232-234, 239, 243, 244
Cirrhinus mrigala 345, 353
Classification rrees 128
Climate change 3-9, 409
Climate resistant agriculture 3
Climate-stress condition 3, 6
Cnaphalocrosis medinalis 6
Colour enhancement 257-260
Column chromatographic separation 228
Combretum decandrum 144
Concentrated sulphuric acid test 44
Conservation assessment and management plan 147
Cow milk 162
Crop based technologies 3, 6
Crop calendar 6
Crop improvement 3, 6, 16, 219, 220, 222
Cropping system 367, 414
Crude plant extract 241
Cryptogamic flora 181
Curcuma longa 56-60, 82, 384
Cynodon dactylon 155, 161

D

Delignification 105, 106, 112, 113
Deoxyribonucleic acid 13
Disc diffusion technique 377
Diseases 151
Diversity index 141
DNA gyrase 321-323
DNA markers 13, 14, 18, 21
Domestic wastewater 321, 322
Dragendorff's test 43
Drug delivery system 80

E

Earthworms 249, 250
Eco-friendly bioenergy options 393
Economical issues 83
Edwardsiella tarda 345
Endophthalmitis cases 373
Environmental hazards 182
Environmental toxicity 181
Environment decontamination 181, 186

Enzymatic assay 333, 336
Escherichia coli 41, 43, 283, 302, 345, 385, 387, 388, 399, 401, 404
Ethnobotanical value 58
Ethnomedicinal uses 65, 68, 69, 151, 165, 279-281
Ethnomedicinal value 51
Ethno-medicines 151
Eudrilus eugeniae 249-255
Evaluation metrics 124
Extension growth 23, 26, 27
Extract treatment 228

F

Farm animals 6
Fatty acids 300
Feature subset selection (FSS) 126
Feed formulation 259
Ferric chloride test 43, 44, 402
Fertilizer management 413
Flavonoids 33, 36, 44, 46, 71, 229-231, 235, 243, 244, 274-276, 401-403
Food and drug administration 81
Food security 3
Freshwater prawn 133

G

Gastrointestinal (GI) tract 343, 345
Gelatin test 44
Gene expression 185
Genetic improvement 11
Genetic manipulation 219
Genomics 7, 13, 19, 122, 197, 222, 289
Giant unilamellar vesicles 309
Glycolytic enzymes 32
Greenhouse gas 4, 410
Growth rate 259
Gynecological disorders 55, 161

H

Hand-pollinated crosses 11
Heavy metals 181, 185, 187, 334, 337, 338, 414, 415
Hemorrhoids 299
Herbal drugs 151
Heterotrophic nitrification 321, 329
Hexane extracts 229, 231, 232
Hollarhena antydesenterica 144, 147
Horseshoe crabs 213
Human glioma cells 225, 232-234
Hybrid napier grass 112, 113
Hydrochloric acid 44, 108, 112, 323
Hydrogen peroxide activity 35
Hyper accumulation 183

I

Immune-tolerent 203
Immunohistochemistry 81
Immunomodulatory properties 206, 298
Indian major carps 133, 353, 354
Indian system of medicine 53
Indigenous peoples 151
Integrated nutrient management 7
International food research institute 5
Inter simple sequence repeats 15

K

Kalanchoe pinnata 54-60
Karanja oil 373, 375, 377-381
Kellar kiliani test 44
Kuldiha wildlife sanctuary 141, 142, 143, 145, 147
Kusum oil 373-381

L

Labeo rohita 84, 133, 135, 345
Leaf extracts 41, 46, 71, 72
Leishmaniasis 299, 303

Leucas aspera 267-273
Libermann-Buchard test 45
Lignin content 108, 110
Lignocellulolytic 295, 306
Lipid peroxidation 249, 251, 255, 313
Lysosomal enzyme 191

M

Maceration 32
Machine learning applications 121
Macromolecular estimation 386
Magnesium oxide 79
Mahula oil 373-381
Mangrove plant 31, 65
Mangrove soils 355
Marker assisted breeding 11
MaxEnt model outputs 172
Maximum entropy model 167
Mean germination time 23, 26
Medicinal flora 161
Medicinal plants diversity 141
Medicinal signaling cells 203
Mesenchymal stem cells 203
Metal induced stress 332
Metal nanoparticles 79
Methane emission 409, 412, 415
Methanogenic bacteria 409
Micro-algal biofuels 393
Microarrays 122
Microbial biomass carbon (MBC) 367-371
Microbial biotechnology 277
Micro-habitat conditions 23
Minimum inhibitory concentration 377, 389
Mitigation and adaptation strategies 3
Molecular characterization 357
Molecular markers 11-13, 16, 18
Molecular studies 221
Mungbean 17

N

Nanobarcodes 83
Nanotechnology 77-84
Nanovaccine 77
Natural carotenoids feed supplement 257
Natural plant oils 373
Nearest neighbour 129
Neem-coated urea 8
Network project on climate change 5
Nickel chloride 249-255
Nilaparvata lugens 6
Nitrogen bioremediation 321
Nitrogenous fertilizer 409
Nutrient agar 333, 345, 346, 357, 359
Nyctanthes arbor-tristis 41, 42, 45-49, 227-232

O

Organic amendment 411
Organic carbon 367, 369
Oryza sativa 85, 86, 92, 107

P

Paddy field 367
P. aeroginosa 47
Pathogenic bacteria 310, 386
Perivitelline fluid 213, 214
Pharmacological potentials 282
Phenolic compounds 402
Phenols 33, 36, 60, 71, 274
Phospholipids 310, 314, 315
Phyllanthus emblica 54-60
Physiological disorders 61
Physiological interventions 181
Phytochemical activity 41
Phytochemical analysis 33, 43, 54, 71, 242

Phytochemical assay 36, 45
Phytochemicals 42, 46, 51, 60, 66, 225, 230, 239, 267, 268, 399
Phytochemical screening 31, 35, 36, 71, 401, 403
Phytochemical studies 51
Phytochemistry 48, 62, 74, 75, 117, 119, 247, 275, 291, 382, 385, 406
Phytosociological aspect 141
Phytosociology 141, 142, 145
Plant extraction 228
Plasmid DNA 343-347, 349
Platelet derived growth factor 226
Polygenic traits 18
Polymerase chain reaction 13, 15
Polysaccharides 108, 109, 110, 116, 283, 284, 285, 295-299, 307, 308
Prokaryotic model membrane system 309
Protein detection 81
Proteolytic activity 346, 349
Proteomics 122, 197, 289, 302
Pteridophytic species 182
Pterocarpus marsupium 56-60, 146, 147
Pueraria tuberosa 161
Pulse breeding 18
Pulse improvement 11

Q

Qualitative phytochemical analysis 33
Quantitative trait loci 18, 85, 86, 221
Quinones 401, 402, 403

R

Rabi cropping season 269
Random amplified polymorphic DNA 12
Randomly amplified microsatellite polymorphism 16
Randomly-amplified polymorphic DNA marker 14
Reactive oxygen species 37, 79, 199, 274
Receiver operating characteristics 125, 170
Red spider mites 267-272
Reduced glutathione 249, 251, 254
Remediation 181
Remote sensing 167
Resource conservation technologies 7
Rhizospheric soil samples 368
Rice cultivation 410, 416
Rice fields 409, 410
RNA secondary structure 122

S

Salt tolerant bacteria 355, 357
Saponins 33, 36, 45, 71, 229, 235, 237, 382, 401, 402
Saraca asoca 54, 55, 58, 59, 60
Seed germination 23-29
Selenite tolerance 360
Selenium reduction 355
Sequence characterized amplified region 15
Sequence-related amplified polymorphism 16
Sequestering system 185
Shorea robusta 144, 146, 147, 156, 167, 177, 239, 241, 244, 246, 247
Silver oxide 79
Similipal biosphere reserve 23-25, 29, 40, 142, 143, 149-154, 158-161, 164, 167, 171, 293, 373, 375, 385, 386
Simple primer amplification reaction 16
Simple sequence repeat 13, 17
Simultaneous saccharification and co-fermentation (SSCF) process 115
Smilax macrophylla 156, 161
Socio-economic benefits 7
Soil conservation 8
Soil microbial biomass carbon 367

Soil microbiota 367
Soil organic carbon 369
Solanum melongena 267, 268
Soybean 16
Spirogyra bichromatophora 385, 386, 388
Spirulina 257-265
Staphylococcus aureus 41, 43, 46, 283, 289, 302, 345, 399, 401, 404
Statistical methods 251
Steroids 33, 36, 71, 75, 235, 243, 401, 402
Sulphate-reducing bacteria 409
Superoxide dismutase 249, 250, 253
Super-oxide dismutase scavenging activity 34
Supervised classification 121, 124
Sustainable development goals 393
Synovial fluid 204
Systems biology 122, 192

T

Tannic acid test 43
Target region amplification polymorphism 16
Taxonomical Identification 335
Terminalia arjuna 55, 57, 59, 60, 168
Termitomyces mushrooms 295
Terpenoids 33, 36, 71, 229, 231, 244
Terribacillus goriensis 331, 332, 335
Tetranychus urticae 267, 268, 269, 275
Text mining 122
The cancer genome atlas (TCGA) network 226
Therapeutic potentials , 279, 286
Thin layer chromatography 228, 241
Thiobarbituric acid 251, 256
Thyroid hormone 193, 197-202
Titanium dioxide 79
Tolerant bacteria 355, 357
Transfer of technology 5
Tribal community 151
Tropical tasar culture 219
Two dimensional thin layer chromatography 313

U

Ultrasonication 105, 111
Urdbean 12, 17
Urginea indica 144, 145, 147, 157

V

Vedic culture 51
Vitex negundo 54-60, 157, 162

W

Wagner's test 60, 402
Water management 410
Wild mushrooms 279, 281, 288
Wild tasar silkworm 167, 173

Z

Zero tillage 7

Figures 6.2a-h: List of Plants Available in and Around Udala Area Used for Traditional Medicines [a. *Kalanchoe pinnata* (Amarpoi); b. *Butea monosperma* (Palas); c. *Pterocarpus marsupium* (Piasal); d. *Azadirachta indica* (Neem); e. *Phyllanthus emblica* (Anla); f. *Vitex negundo* (Begunia); g. *Terminalia arjuna* (Arjuna); h. *Tridax procumbens* (Bisalyakarani)]. p. (57)

Figures 6.2i-k: List of Plants Available in and Around Udala Area Used for Traditional Medicines [i. *Saraca asoca* (Ashoka); j. *Rouvolfia serpentine* (Patalagaruda); k. *Curcuma longa* (Haladi)]. p. (58)

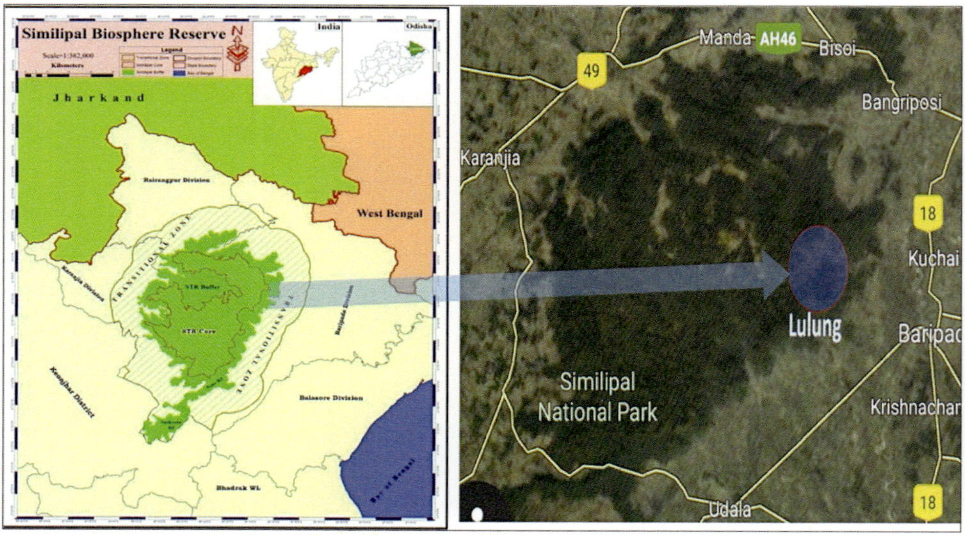

Figure 14.1: Map of Study Area. p. (153)

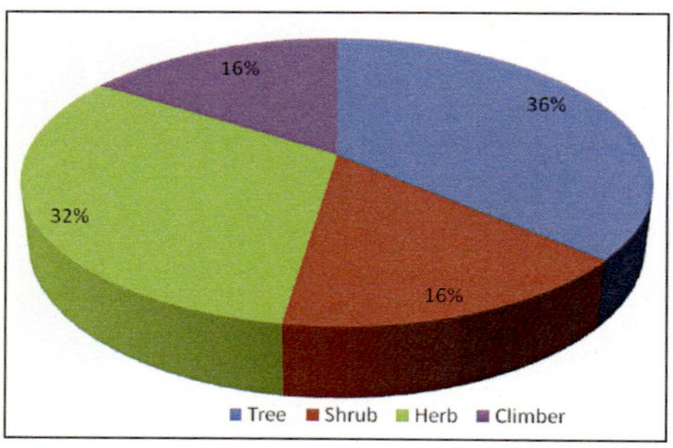

Figure 14.2: Life Form of Identified Plant Species at Lulung, Similipal Biosphere Reserve. p. (158)

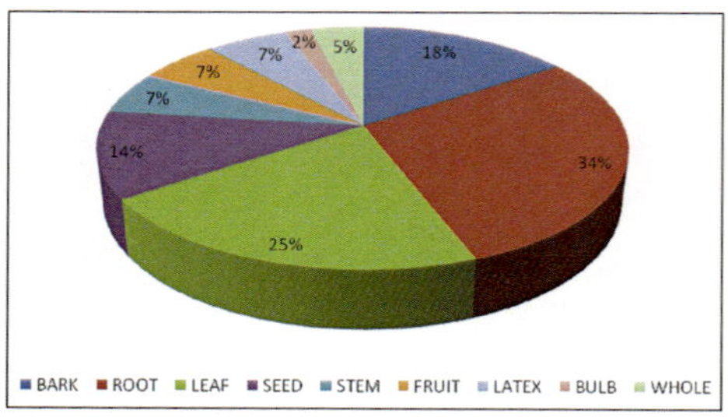

Figure 14.3: Plant Parts Used for Medicine Purpose by the Local Tribes of Lulung, Similipal Biosphere Reserve. p. (158)

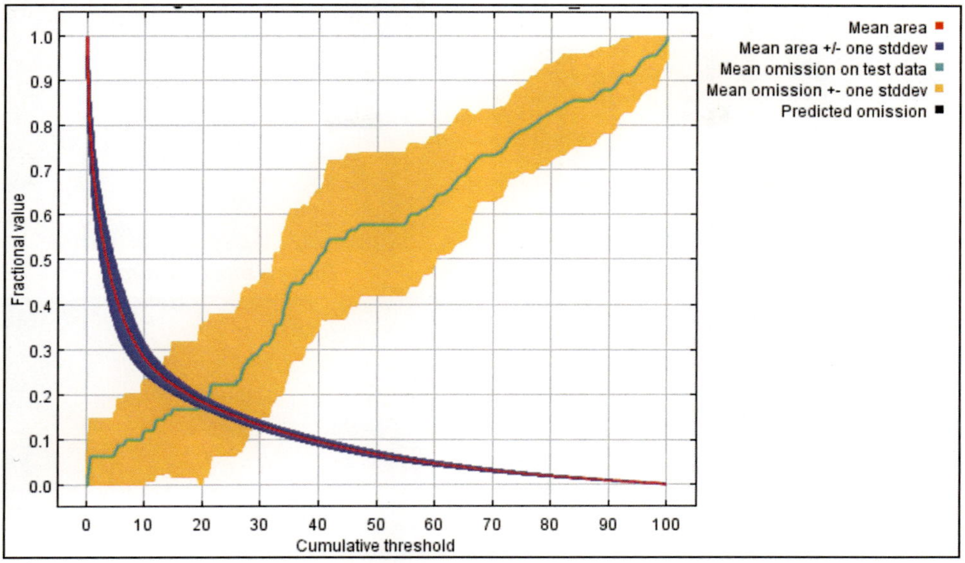

Figure 15.2: Average Omission and Predicted Area. p. (172)

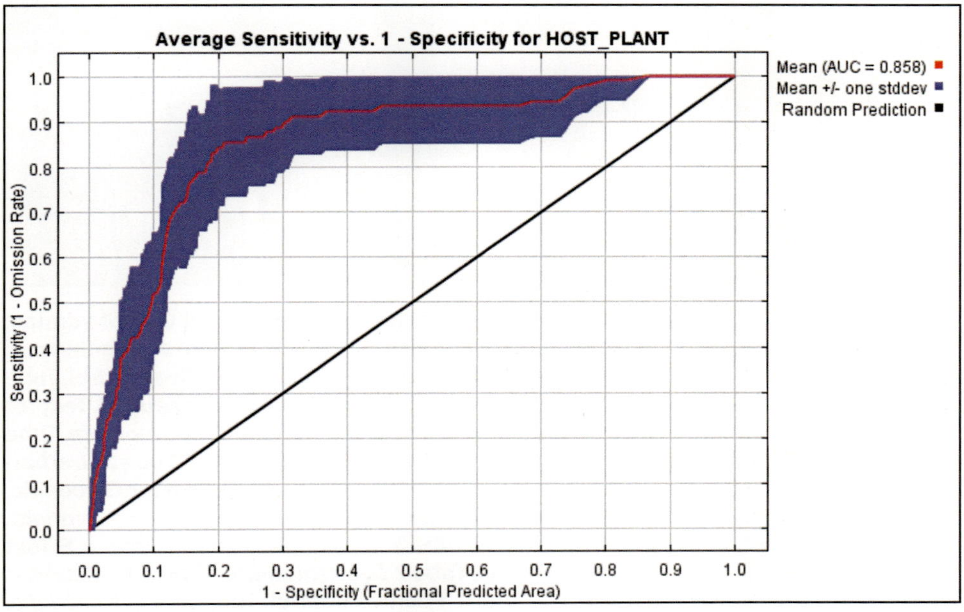

Figure 15.3: Average Sensitivity vs Specificity. p. (173)

Nyctanthes arbor-tristis

Datura stramonium

Nerium oleander

Catharanthus roseus

Figure 21.1: Four Medicinal Plants Used in the Study. p. (229)

Figure 21.2: Qualitative Analysis of Four Plant Extracts for Presence of Carbohydrates, Terpenoids, Steroid, Saponin, Flavonoid, Glycosides, Alkaloids, Phenol and Tannins. A. Test for Carbohydrate, Tube 1 Positive control (Glucose), Tube 2 Negative control, Tube 3*N.arbor-tristis*, Tube 4 *D. stramonium*, Tube 5 *N. oleander*, Tube 6 *C. roseus*. B. Test for Terpenoid, Tube 1 Positive control, Tube 2 *N. arbor-tristis*, Tube 3 *D. stramonium*, Tube 4*N.oleander*, Tube 5 *C. roseus*. C. Test for Steroid, Tube 1 Positive control, Tube 2 *N. arbor-tristis*, Tube 3 *D. stramonium*, Tube 4 *N. oleander*, Tube 5 *C. roseus*. D. Test for Glycoside, Tube 1 *N. arbor-tristis*, Tube 2 *D. stramonium*, Tube 3 *N. oleander*. E. Test for Flavonoid, Tube 1 *N. arbor-tristis*, Tube 2 *D. stramonium*, Tube 3 *N. oleander*, Tube 4 *C. roseu*. F. Test for Phenol and tannin, Tube 1 *N. arbor-tristis*, Tube 2 *D. stramonium*, Tube 3 *N. oleander*, Tube 4 *C. roseus*. G. Test for Alkaloid, Tube 1 Positive control, Tube 2 *N. arbor-tristis*, Tube 3 *D. stramonium*, Tube 4 *N. oleander*, Tube 5 *C. roseus*. H. Test for Saponin, Tube 1 *N. arbor-tristis*, Tube 2 *D. stramonium*, Tube 3 *N. oleander*, Tube 4 *C. roseus*. p. (231)

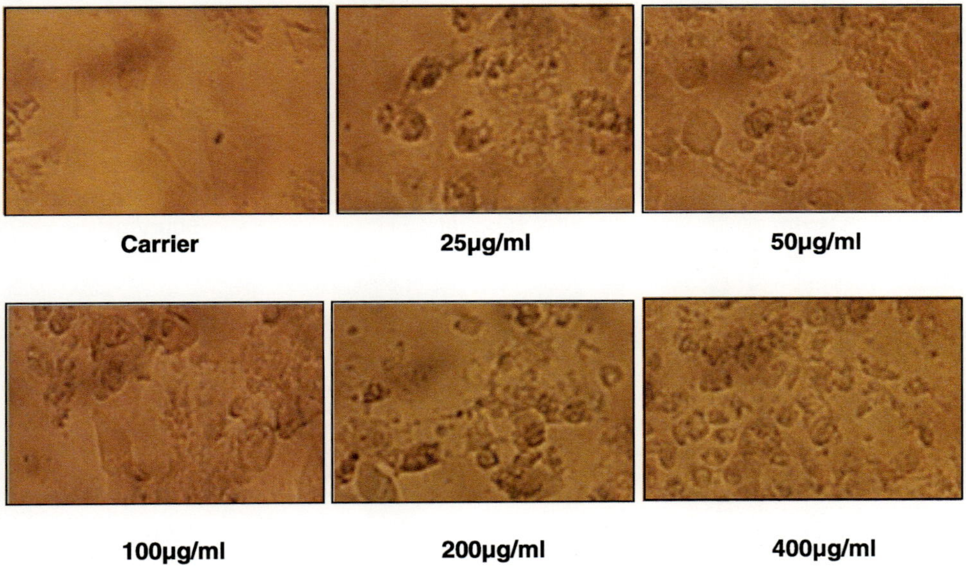

Figure 21.3A: Chloroform Extract of *N. arbotristis* Treated T98 Human Glioma Cells. The cells were treated with carrier, 25µg/ml, 50µg/ml, 100µg/ml, 200µg/ml and 400µg/ ml. The Chloroform extract causes significant cell death of T98G cells. p. (232)

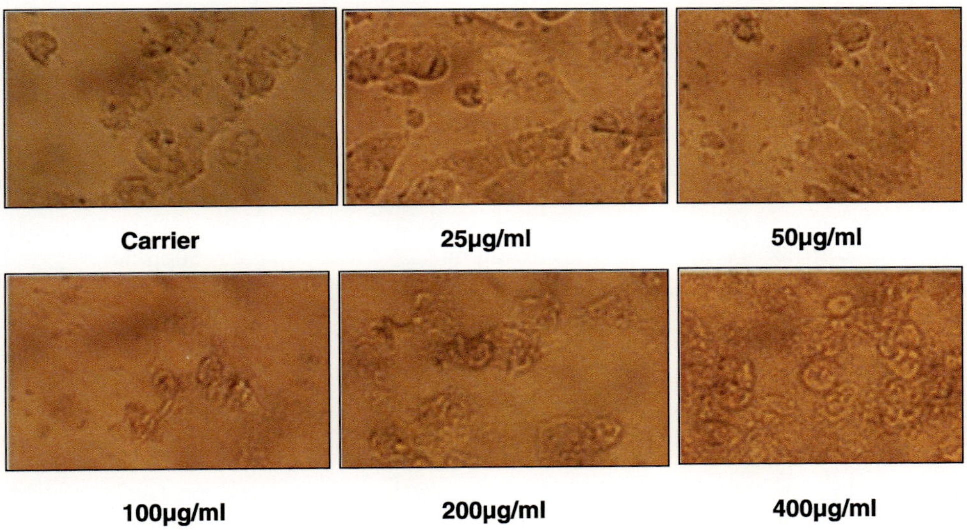

Figure 21.3B: Hexane Extract of *N. arbor-tristis* Treated T98 Human Glioma Cells. The cells were treated with carrier, 25µg/ml, 50µg/ml, 100µg/ml, 200µg/ml and 400µg/ ml and images were taken. The Hexane extract causes significant cell death in T98G cells. p. (232)

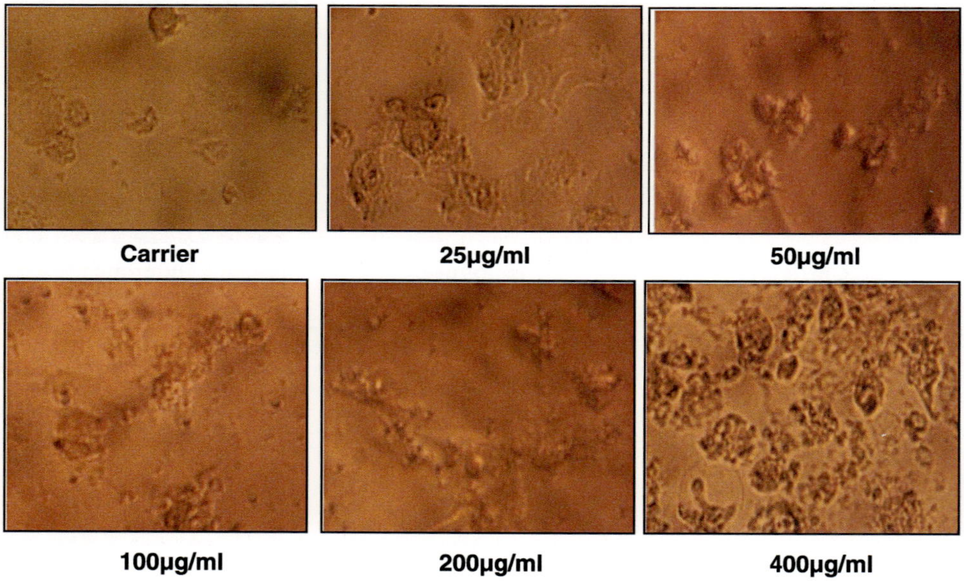

Figure 21.4: Chloroform Extract of *D. stramonium* Treated T98 Human Glioma Cells. The cells were treated with carrier, 25µg/ml, 50µg/ml, 100µg/ml, 200µg/ml and 400µg/ml and images were taken after 48h. The Chloroform extract causes significant apoptosis to T98G cells. p. (233)

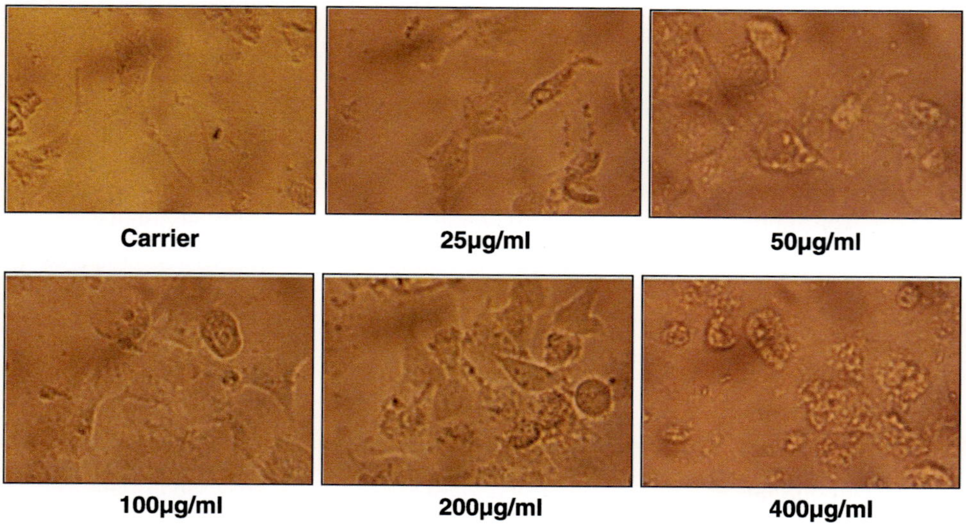

Figure 21.5A: Chloroform Extract of *N. oleander* Treated T98 Guman Glioma Cells. The cells were treated with carrier, 25µg/ml, 50µg/ml, 100µg/ml, 200µg/ml and 400µg/ml and images were taken after 48h. The Chloroform extract causes significant cell death in T98G cells. p. (233)

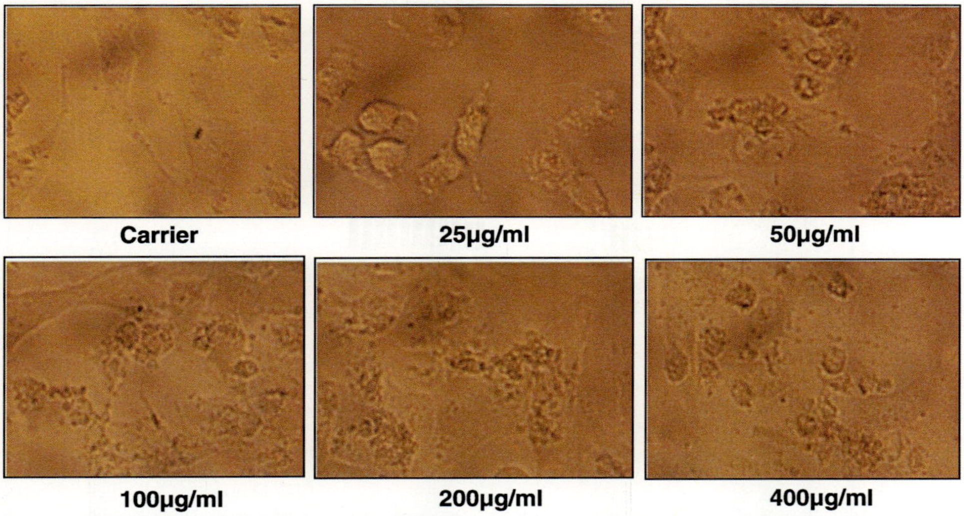

Figure 21.5B: Photo Imaging of Hexane Extract of *Nerium oleander* Treated T98 Human Glioma Cells. The cells were treated with carrier, 25μg/ml, 50μg/ml, 100μg/ml, 200μg/ml and 400μg/ml. The Hexane extract causes significant apoptosis to T98G cells. p. (234)

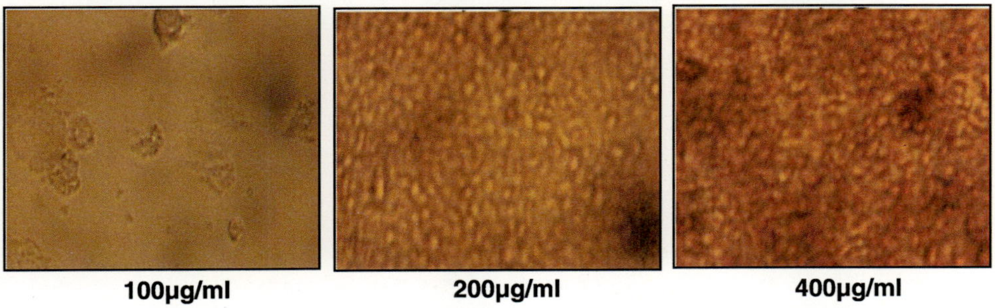

Figure 21.6: Photo Imaging of Chloroform Extract of *C. roseus* Treated T98 Human Glioma Cells. The cells were treated with carrier, 200μg/ml and 400μg/ml. The Chloroform extract causes significant apoptosis to T98G cells. p. (234)

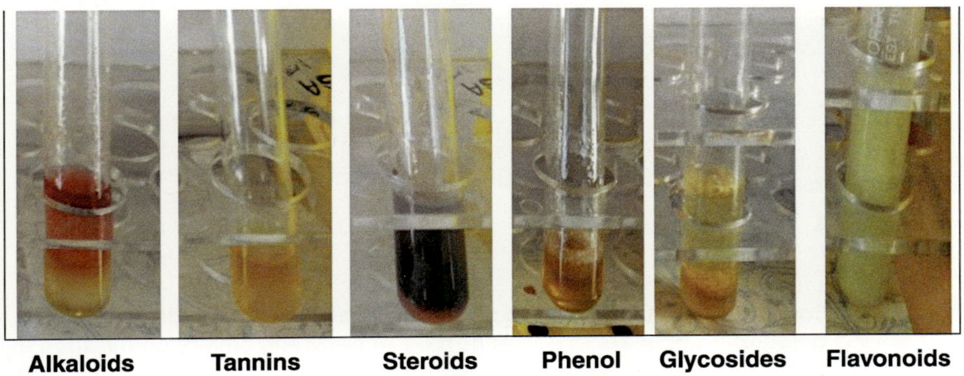

Figure 22.1: Phytoconstituents Presence in Phytochemical Tests of Chloroform Extract. p. (243)

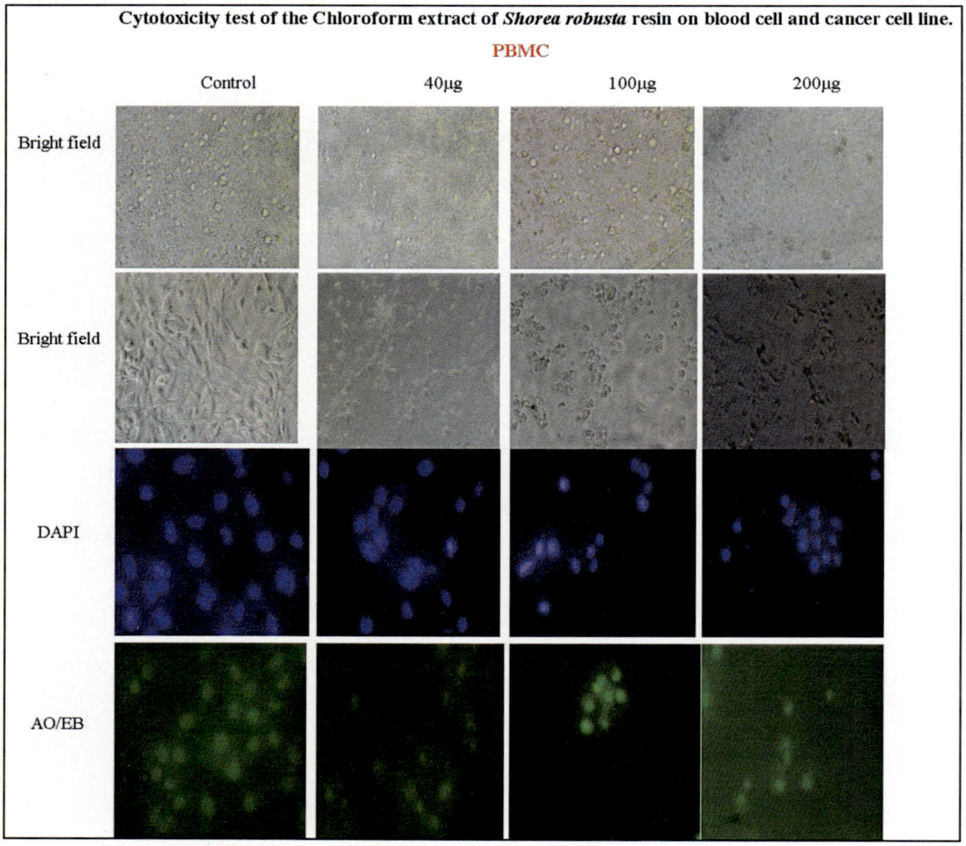

Figure 22.4: Morphological Data of PBMC and Tc1 Cell. p. (245)

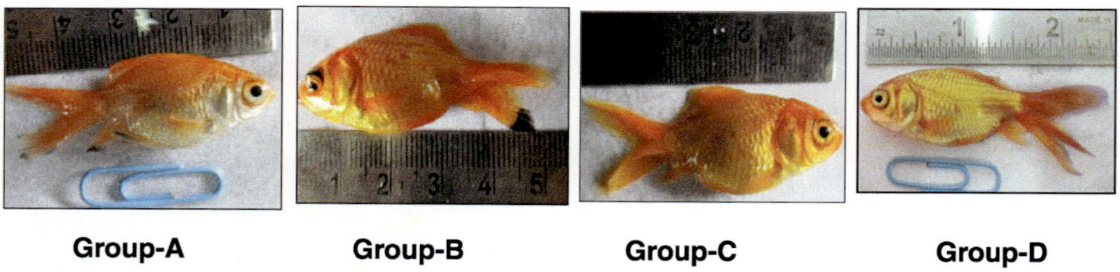

Figure 24.1: Colour Variation of Fishes in different Experimental Groups. In Group-A, light orange colouration appears in dorsal portion of body and silvery colouration in abdomen portion. Group-B fishes showed deep orange colouration throughout the body. Group-C fishes showed deep orange colouration but lighter than the group-B. Group-D possess light orange colouration in whole body.

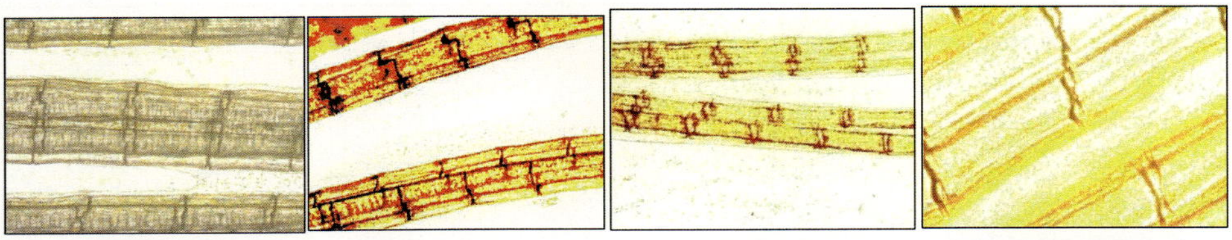

Group-A Group-B Group-C Group-D

Figure 24.2: Magnified View (400X) of Fin of Fishes in different Experimental Groups. Group-A fishes showed weak colour pigments appearance in fin-rays with greyish-yellow colouration. In Group-B, fin-rays developed bright colour pigmentation with reddish-yellow colour combination. Appearance of moderate colour pigmentation with light reddish-yellow colour observed in Group-C fishes. The fin-rays of Group-D fishes showed yellow colouration with stripes of reddish pigmentation. p. (262)

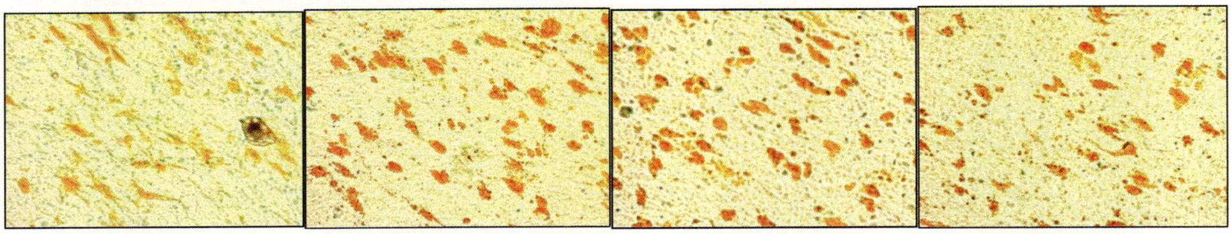

Group-A Group-B Group-C Group-D

Figure 24.3: Magnified View (400X) of Skin of Fishes in different Experimental Groups. Group-A showed lighter orange colouration with lesser pigmentation. Skin of Group-B fishes develop deep and bright reddish pigmentation. Intermediate pigmentation with moderate reddish colouration observed in the skin of Group-C fishes. Skin of Group-D fishes showed with lesser pigmentation with light reddish colouration. p. (263)

Figure 26.1: Some of the Wild Medicinal Mushrooms of Simlipal (A) *Ganoderma applanatum* (B) *Ganoderma tsugae* (C) *Volvorella volvacea* (D) *Astraeus hygrometricus* (E) *Geastrum* sp. (F) *Lentinus sajor-caju* p. (287)

Figure 26.1: Some of the Wild Medicinal Mushrooms of Simlipal (G) *Russula delica* (H) *Russula vesca* (I) *Termitomyces eurrhizus*. p. (287)

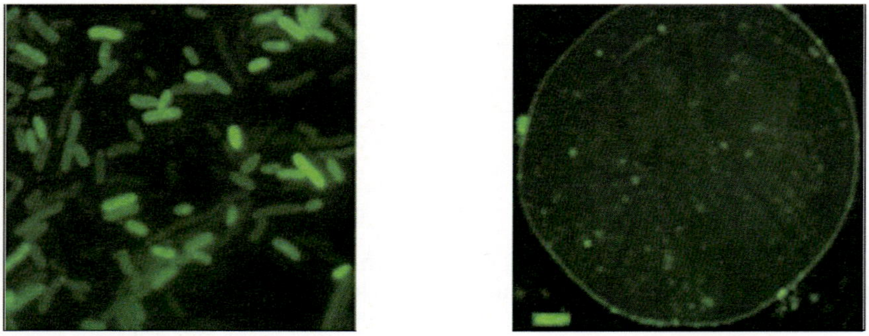

Figure 28.2: Comparison between the Size of Bacteria and Size of GUVs. Bacteria (*e.g. E.coli*) are 1-2μ in length to be visualized in minute detailed (left). GUV prepared in our laboratory stained with 10-N, Nonyl-Acridine Orange (NAO) (inset: a single bacterium for comparision of size). Scale: 1 μ. The images were taken using a Nikkon inverted fluorescence microscope (Japan) using 100 X objective. p. (312)

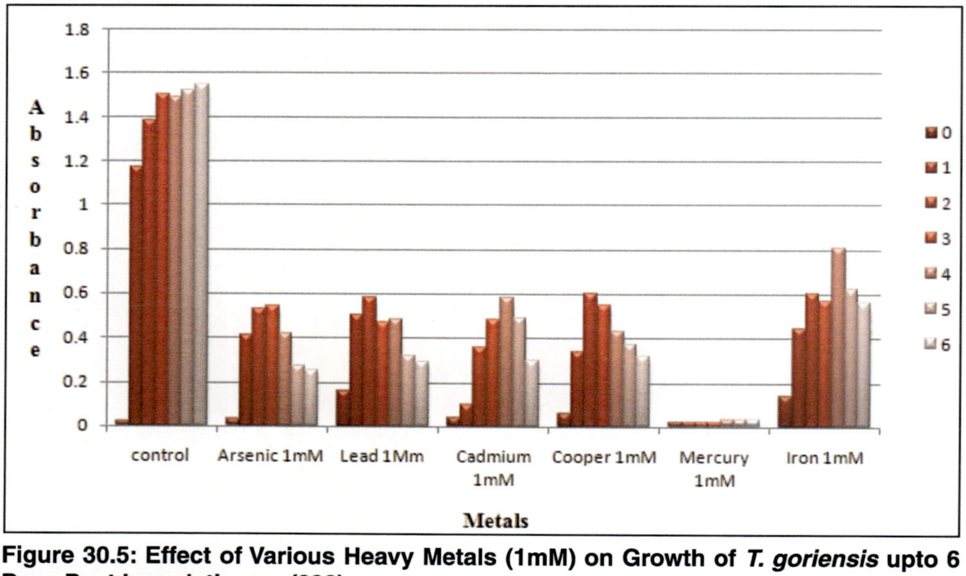

Figure 30.5: Effect of Various Heavy Metals (1mM) on Growth of *T. goriensis* upto 6 Days Post Inoculation. p. (338)